AF715010

UAV-Based Remote Sensing
Volume 2

Special Issue Editors

Felipe Gonzalez Toro
Antonios Tsourdos

MDPI • Basel • Beijing • Wuhan • Barcelona • Belgrade

MDPI

Special Issue Editors
Felipe Gonzalez Toro
Queensland University of Technology
Australia

Antonios Tsourdos
Cranfield University
UK

Editorial Office
MDPI
St. Alban-Anlage 66
Basel, Switzerland

This edition is a reprint of the Special Issue published online in the open access journal *Sensors* (ISSN 1424-8220) from 2016–2017 (available at: http://www.mdpi.com/journal/sensors/special_issues/UAV_remote_sensing).

For citation purposes, cite each article independently as indicated on the article page online and as indicated below:

Lastname, F.M.; Lastname, F.M. Article title. *Journal Name* **Year**, *Article number*, page range.

First Edition 2018

Volume II
ISBN 978-3-03842-855-8 (Pbk)
ISBN 978-3-03842-856-5 (PDF)

Volume I–II
ISBN 978-3-03842-857-2 (Pbk)
ISBN 978-3-03842-858-9 (PDF)

Table of Contents

About the Special Issue Editors

Felipe Gonzalez Toro, Associate Professor at the Science and Engineering Faculty, Queensland University of Technology (Australia), with a passion for innovation in the fields of aerial robotics and automation and remote sensing. He creates and uses aerial robots, drones or UAVs that possess a high level of cognition using efficient on-board computer algorithms and advanced optimization and game theory approaches that assist us to understand and improve our physical and natural world. Dr. Gonzalez leads the UAVs-based remote sensing research at QUT. As of 2017, he has published nearly 120 peer reviewed papers. To date, Dr. Gonzalez has been awarded $10.1M in chief investigator/partner investigator grants. This grant income represents a mixture of sole investigator funding, international, multidisciplinary collaborative grants and funding from industry. He is also a Chartered Professional Engineer, Engineers Australia—National Professional Engineers Register (NPER), a member of the Royal Aeronautical Society (RAeS), The IEEE, American Institute of Aeronautics and Astronautics (AIAA) and holder of a current Australian Private Pilot Licence (CASA PPL).

Antonios Tsourdos obtained a MEng on Electronic, Control and Systems Engineering, from the University of Sheffield (1995), an MSc on Systems Engineering from Cardiff University (1996) and a PhD on Nonlinear Robust Autopilot Design and Analysis from Cranfield University (1999). He joined the Cranfield University in 1999 as lecturer, was appointed Head of the Centre of Autonomous and Cyber-Physical Systems in 2007 and Professor of Autonomous Systems and Control in 2009 and Director of Research—Aerospace, Transport and Manufacturing in 2015. Professor Tsourdos was a member of the Team Stellar, the winning team for the UK MoD Grand Challenge (2008) and the IET Innovation Award (Category Team, 2009). Professor Tsourdos is an editorial board member of: Proceedings of the IMechE Part G Journal of Aerospace Engineering; IEEE Transactions of Aerospace and Electronic Systems; Aerospace Science & Technology; International Journal of Systems Science; Systems Science & Control Engineering; and the International Journal of Aeronautical and Space Sciences. Professor Tsourdos is Chair of the IFAC Technical Committee on Aerospace Control, a member of the IFAC Technical Committee on Networked Systems, Discrete Event and Hybrid Systems, and Intelligent Autonomous Vehicles. Professor Tsourdos is also a member of the AIAA Technical Committee on Guidance, Control and Navigation; AIAA Unmanned Systems Program Committee; IEEE Control System Society Technical Committee on Aerospace Control (TCAC) and IET Robotics & Mechatronics Executive Team.

Preface to "UAV-Based Remote Sensing"

Active technological development has fuelled rapid growth in the number of Unmanned Aerial Vehicle (UAV) platforms being deployed around the globe. Improved sensors and enhanced image processing techniques have consolidated and confirmed UAVs as the technology of choice. Many jurisdictions have regulated stringent restrictions on flying UAVs in numerous scenarios where they could be of great value. Despite the increased regulation, UAVs or drones have rapidly become the tool of choice for both the environmental science and the remote sensing communities. This is due, in no small part, to a lowering cost of entry. The last few years has seen significant technological development in UAV platforms, sensor miniaturisation, and robotic sensing. UAV technology progresses apace. The design of novel UAV systems and the use of UAV platforms integrated with RGB, multispectral, hyperspectral, thermal imaging, gas sensing and/or laser scanning sensors have now been demonstrated in both research and practical applications. Novel UAV platforms, UAV-based sensors, robotic sensing and imaging techniques, the development of processing workflows, as well as the capacity of ultra-high temporal and spatial resolution data, provide both opportunities and challenges that will allow engineers and scientists to address novel and important scientific questions in UAV and sensor design, remote sensing and environmental monitoring. This work features papers on UAV sensor design; improvements in UAV sensor technology; obstacle detection, methods for measuring optical flow; target tracking; gimbal influence on the stability of UAV images; augmented reality tools; segmentation in digital surface models for 3D reconstruction; detecting the location and grasping objects; multi-target localization; vision-based tracking in cooperative multi-UAV systems; noise suppression techniques; rectification for oblique images; two-UAV communication system; fuzzy-based hybrid control algorithms; pedestrian detection and tracking as well as a range of atmospheric, geological, agricultural, ecological, reef, wildlife, building and construction; coastal area coverage; search and rescue (SAR); water plume temperature measurements; aeromagnetic and archaeological survey applications.

Felipe Gonzalez Toro, Antonios Tsourdos

Special Issue Editors

sensors

MDPI

Article

An Analysis of the Influence of Flight Parameters in the Generation of Unmanned Aerial Vehicle (UAV) Orthomosaicks to Survey Archaeological Areas

Francisco-Javier Mesas-Carrascosa *, María Dolores Notario García, Jose Emilio Meroño de Larriva and Alfonso García-Ferrer

Department of Graphic Engineering and Geomatics, University of Cordoba, Campus de Rabanales, Córdoba 14071, Spain; maranotario@gmail.com (M.D.N.G.); jemerono@uco.es (J.E.M.d.L.); agferrer@uco.es (A.G.-F.)

* Correspondence: fjmesas@uco.es; Tel.: +34-957-218-537

Academic Editors: Felipe Gonzalez Toro and Antonios Tsourdos
Received: 28 July 2016; Accepted: 26 October 2016; Published: 1 November 2016

Abstract: This article describes the configuration and technical specifications of a multi-rotor unmanned aerial vehicle (UAV) using a red–green–blue (RGB) sensor for the acquisition of images needed for the production of orthomosaics to be used in archaeological applications. Several flight missions were programmed as follows: flight altitudes at 30, 40, 50, 60, 70 and 80 m above ground level; two forward and side overlap settings (80%–50% and 70%–40%); and the use, or lack thereof, of ground control points. These settings were chosen to analyze their influence on the spatial quality of orthomosaicked images processed by Inpho UASMaster (Trimble, CA, USA). Changes in illumination over the study area, its impact on flight duration, and how it relates to these settings is also considered. The combined effect of these parameters on spatial quality is presented as well, defining a ratio between ground sample distance of UAV images and expected root mean square of a UAV orthomosaick. The results indicate that a balance between all the proposed parameters is useful for optimizing mission planning and image processing, altitude above ground level (AGL) being main parameter because of its influence on root mean square error (RMSE).

Keywords: unmanned aerial vehicle (UAV); positional quality; orthomosaick; archeology

1. Introduction

Traditional archaeological site surveys are a time-consuming effort proportional to the difficulty of access of the site being investigated. Prospective sites are identified based on oral tradition, written records or inspection using images [1]. Once a prospective site is identified, fieldwork starts to detect evidence of human activity. With images, these have usually been registered by sensors on board two traditional platforms, satellite and manned aircraft [2–4]. These platforms present problems with spatial resolution applied to archaeology. One of the limiting factors of satellite images is the difficulty to detect small- or even medium-sized details like sites smaller than a hectare [3]. There are also reasons to be cautious about the effectiveness of satellite imagery in detecting prehistoric sites lacking a long history of occupation [5]. On the other hand, manned aircraft are able to supply images with better spatial resolution but no higher than at a 1:500 scale. Moreover, the economical cost of aerial photogrammetry is usually too high for small surveyed areas [6].

Currently, unmanned aerial vehicles (UAVs) are an alternative for the acquisition of images with a very high spatial resolution for documenting archeological areas [7]. UAVs are classified with different characteristics like range, endurance, mass and architecture. Generally, most common UAV categories used in civil applications are micro and mini UAVs with a mass of less than 5 and 150 kg, respectively.

Other characteristics, like endurance or range, depend mainly on the type of platform architecture, for example multirotor, fixed wing and balloon.

Different types of UAVs have been used successfully to survey archeological areas such as helium balloons [8], blimp [9], kites [10], fixed wing [11] and rotor wing [12]. In mapping, UAV flight parameters are critical in obtaining adequate spatial quality on the derived geomatic products to survey archeological sites. The correlation between flight parameters, spatial quality and photogrammetric processing of images acquired by metric sensors have been well studied in classic photogrammetry [13]. Accuracy assessment of digital elevation models [14] or the influence of Ground Control Points (GCPs) in aerial-triangulation [15], among others, have contributed to defining a standardized processing framework.

In contrast, UAV photogrammetry for research applications is still at an early stage [16]. One consequence of this is that operational frameworks for working with UAV platforms are not defined in some aspects and applications. The operational framework depends on the type of UAV platform, sensors on board and case of use. Ref. [17] analyzes the potential of UAVs for measuring area of land plots for monitoring land policies, Ref. [18] explored the positional quality of orthophotos obtained by a UAV following the requirements of National Mapping Agencies, and Ref. [19] defines specifications to acquire remote images using a six-band multispectral sensor on board a UAV for use in precision agriculture. Regarding cultural heritage, UAVs have rarely been used in scientific research [20]. Parameters like altitude above ground level (AGL), number of GCPs, or the percentage of forward-lap and side-lap determine the spatial quality of orthomosaics.

One of the most important parameters in an UAV flight is altitude AGL. It determines the pixel size on the registered images, flight duration and area covered. Firstly, it is necessary to define the spatial quality requirements for the orthomosaics to achieve the ideal pixel size in the images registered by the sensor. In general, at least four pixels are required to detect the smallest detail in an image [21]. In selecting altitude AGL, sufficiently fine spatial resolution has to be guaranteed and, as the same time, as much surface as possible has to be covered. Very low altitude AGL UAV flights generate very high spatial resolution images but cover a limited area and therefore increase flight duration. As a result, the UAV operation has to be fragmented into different flights due to battery life, causing variations in illumination, the appearance or disappearance of shadows, saturated images, depending on the type of materials present, and so on.

As in traditional photogrammetry, the algorithms used process overlapping images acquired from multiple viewpoints. Mainly, these techniques are based on imaging techniques called structure from motion (SfM) [22]. SfM photogrammetry differs from conventional photogrammetric approaches by calculating internal camera parameters (focal length, principal point and distortion coefficients), camera position and orientation. SfM algorithms need a large number of overlapping images to cover the area of interest [23,24], which impacts flight duration. High percentages of forward and side overlap increase flight duration because it is necessary to capture more images for each individual lap and to increase the number of total laps. However, this improves the spatial quality of geomatic products. All of these parameters affect battery life and thus a balance between spatial quality, forward and side overlap and flight time duration is necessary.

Finally, GCP distribution and its influence on the spatial quality of orthomosaics in traditional platforms is well described [25]. With UAVs, the number and distribution of GCPs are not standardized, being analyzed by [26,27]. The consequence is that the number of GCPs may covers a broad range, from just four GCPs to more than 100 GCPs [27,28]. In addition, Ref. [29] analyzes the accuracy of UAV orthomosaics without GCPs, with the resulting root mean square error (RMSE) being higher than one meter.

To our knowledge, no detailed investigation has been conducted regarding the influence of UAV flight parameters such as altitude AGL, forward and side overlap, and the use or lack of GCPs on the spatial quality of orthomosaics using a red–green–blue (RGB) on board a multi-rotor UAV to be used in surveying archaeological areas. This paper defines the technical specifications for working with a multi-rotor UAV to obtain an accurate spatial orthomosaics to be used to survey archeological areas.

The manuscript is divided in the following sections: in Section 2, the technology, study area, and the materials and methods are described. In Section 3, results are presented, followed by conclusions in Section 4.

2. Materials and Methods

2.1. UAV and Sensor Description

The unmanned aerial vehicle used for mapping was a MD4-1000 multi-rotor (Microdrones GmbH, Siegen, Germany) (Figure 1). This UAV is a vertical takeoff and landing aircraft of an entirely carbon design. The system has a maximum payload of 1.2 kg. It uses 4 × 250 W gearless brushless motors powered by a 22.2 V battery. It reaches a cruising speed of 12.0 m/s and a maximum climb speed of 7.5 m/s. Its maximum wind tolerance is up to 12.0 m/s, registering steady picture up to 6 m/s. Flight duration depends on sensor weight and weather conditions. For this project, the multi-rotor UAV was equipped with a Sony NEX-7 RGB sensor (Sony Corporation, Minato, Tokyo, Japan) with a 16 mm lens. The camera weighs 353 g including the camera body, card and battery and provides a 23.5 × 15.6 mm image (6000 × 4000 pixels). The sensor was field calibrated and the results used in this research are summarized in Table 1. With this sensor, the UAV's flight duration is approximately 30 min. During the flight, the sensor registers vertical images. The image trigger is activated by the UAV's autopilot flight settings. For each shot, the UAV autopilot sends a signal to the sensor to register an image and simultaneously timestamps and records the GPS location and navigation angles (yaw, roll and pitch) on a Secure Digital Card (SD-Card). This information will be used for the initial values in photogrammetric processing.

Figure 1. The MD4-1000 multi-rotor (Microdrones GmbH, Siegen, Germany) taking off over the study site.

Table 1. Results of the field-calibrated Sony NEX-7 (Sony Corporation, Minato, Tokyo, Japan).

Parameters	Value	Parameters	Value
Focal length (mm)	16.6286	Radial Distortion K2	0.0290259
Principal point—X (mm)	12.2712	Radial Distortion K3	−0.0338008
Principal point—Y (mm)	7.76064	Tangential Distortion T1	−0.00162188
Radial Distortion K1	−0.0108059	Tangential Distortion T2	−0.00094999

2.2. Study Site and UAV Flights

The study area was approximately 1.13 ha (101 × 112 m) in size and was conducted in Torreparedones, an old Iberian and Roman town situated between the Guadalquivir river and the Guadajoz river in the province of Córdoba (Southern Spain) (37°45′ N, 4°22′ W) (Figure 2). This settlement has been continuously populated between the 2nd Century Before Christ (BC) and the 16th Century Anno Domini (AD). It reached its peak during the Iberian and Roman period as a municipality and played an important role in the commercial trade routes throughout the southern and eastern territories of the Peninsula as well as in the development of metallurgy. Evidence of public buildings like shrines and a forum have been found in the archaeological remains.

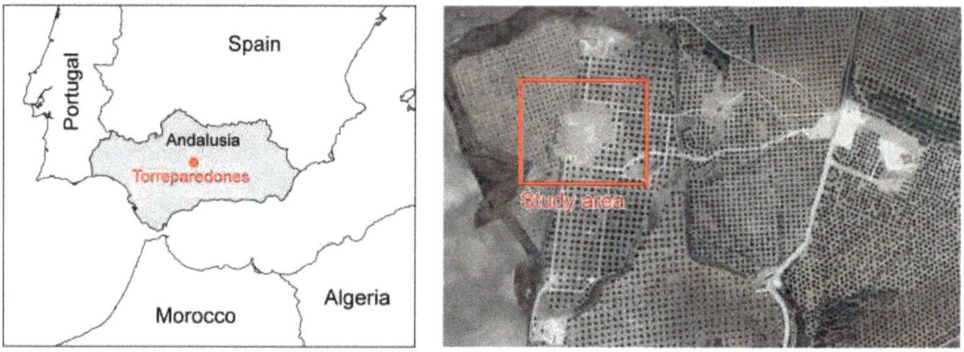

Figure 2. Overview of the study site.

Several flights above the remains were planned following the scheme presented below in Figure 3, which combines different flight altitudes, overlap settings and the use, or lack thereof, of GCPs. Descriptions of flight parameters and their formula are widely available, for example [30].

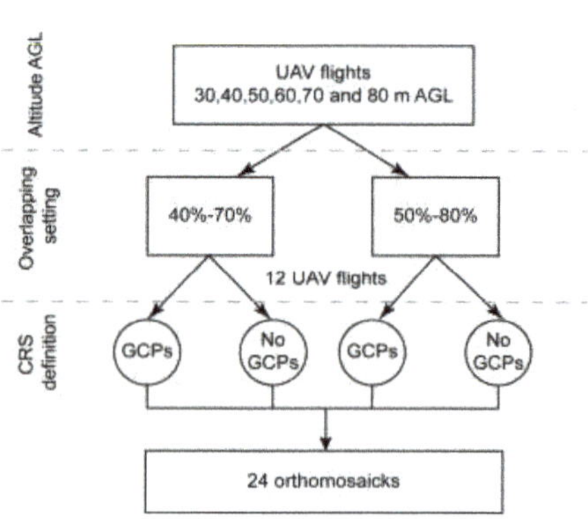

Figure 3. Scheme of unmanned aerial vehicle (UAV) flights and processing.

One of the most important flight parameters is altitude AGL. In this study, a set of flight missions were flown at altitudes of 30, 40, 50, 60, 70 and 80 m AGL. Each altitude AGL is linked to a specific ground sample distance (GSD) value. In this study, GSD ranged from 0.7 cm × pixel^{-1} at 30 m AGL, to 2.0 cm × pixel^{-1} at 80 m AGL. Additionally, two different forward-lap and side-lap settings were used: 80%–50% and 70%–40%. In combining these settings, twelve missions were flown in total. All UAV flights were carried out under the same wind conditions, the wind speed being equal to 2 m/s.

In addition, all UAV flights were planned in such a way that each point in the study area was captured in at least 3 images. Thus, the accuracy of the orthomosaicks was only dependent on altitude AGL and forward and side lap settings.

Afterwards, each UAV flight was processed with and without GCPs. In the former case, GCPs and the georreferentiation information registered by the UAV's autopilot were used in the aerial triangulation phase to accurately place the photogrammetric block into a coordinate reference system (CRS). In the latter case, the aerial triangulation was processed with the information registered by the UAV's autopilot, and nothing more. In using, or not using, GCPs with the information from the 12 flights, a total of 24 orthomosaics were produced to assess spatial quality.

The coordinates of each GCP in the study area were determined by using traditional topography methodologies instead of global navigation satellite system (GNSS) sensors. This decision was made because the precision and accuracy of GCP coordinates have to be greater than the GSD of UAV flights. In this context, a GNSS sensor receiving real-time corrections does not obtain results greater than 2 cm. This value is higher or equal to the GSD of UAV flights, and, therefore, GNSS was rejected. GCPs were chosen in the corners of the study area, one for each corner, and another in the center. Each GCP was set with an artificial target and measured using a total station Leica TC805 (Leica Geosystem AG, Heerbrugg, Switzerland) (Figure 4a) with an angle accuracy equal to 5″ and a distance measurement precision equal to ±(3 mm + 2 ppm).

(a) (b)

Figure 4. Assessing spatial resolution: (**a**) Measuring using the total station; and (**b**) Samples of spatial details of the ground measurements.

2.3. Photogrammetric Processing

The photogrammetric processing is divided into 4 phases: (1) aerial triangulation; (2) Digital Surface Model (DSM) generation; (3) rectification of individual images; and, lastly (4) orthomosaicking.

Aerial triangulation is the basic method for analyzing aerial images in order to calculate the three-dimensional coordinates of object points and the exterior orientation of the images [31]. This process allows the absolute orientation of the entire photogrammetric block to be calculated. To perform the bundle adjustment, algorithms based on "Structure from Motion" (SfM) techniques are used. SfM algorithms operate with the same basic fundamentals of stereoscopic photogrammetry.

However, it differs from traditional photogrammetry in the geometry of the scene, camera positions and orientation. Using UAV platforms, this is resolved using highly redundant information extracted from a set of multiple high percentage overlaps that register the three-dimensional structure of the scene [32]. In a first stage, SfM techniques extract individual features in each image of the photogrammetric block, which are afterward matched to their corresponding feature in the other images of the photogrammetric block. These features are used to determine the relative position of the sensor during the flight, which allows the position and orientation for each individual sensor to be calculated. At this stage, the spatial quality of the results depends on the quality of the geolocation sensor, GNSS sensor and IMU sensor. In general, the attributed geolocational accuracy of images taken on commercial UAVs is medium to low. Therefore, in this study, geolocation was calculated via aerial-triangulation. To improve the spatial quality of the results, a group of GCPs were distributed over the study area. These GCPs were measured on field with a greater spatial accuracy than GSD.

Once aerial-triangulation was calculated, a DSM was generated in three stages: feature extraction, multi-image matching and blunder detection [6]. DSM and external orientation were used to orthorectify each image. Finally, individual orthorectified images were mosaicked to obtain an UAV orthomosaic of the entire study area. Each orthomosaic was produced with a GSD equal to the corresponding GSD of each UAV flight.

The photogrammetric processing was performed using Inpho UASMaster (Trimble, CA, USA) [33].

2.4. *Assessment of Spatial Quality*

Spatial accuracy is the accuracy of the position of a feature related to Earth [34] and can be described in absolute or relative terms. Absolute accuracy is defined as the closeness of reported coordinate values to values accepted as or being true. Relative accuracy is defined as the closeness of the relative spatial positions of features in a dataset to their respective relative spatial positions accepted as or being true.

Before the UAV flights, 150 check points were measured in the study area to assess the absolute and relative spatial accuracies (Figure 4b). The check points' coordinates were obtained using a total station in the same manner as the GCPs (Figure 4a) and were well-defined and well distributed. These coordinates were used as ground reference values to assess the spatial quality of the orthomosaics.

All check point locations were digitized on screen via the produced orthomosaics. These coordinates were obtained using Quantum GIS (QGIS) [35]. Both sets of ground and orthomosaic coordinates were compared to determine the spatial quality.

Absolute positional accuracy was assessed by RMSE, which is used to estimate positional accuracy [36]. Relative positional accuracy was assessed using the methodology developed by the Department of Defense of the United States [37]. Subsequently, all possible check point pair combinations were determined. Afterwards, the absolute and relative errors in the X and Y dimensions of each check point were calculated. These errors were used to calculate both the relative standard deviations on each axis and the relative horizontal standard deviation (RHSD).

3. Results

A total of two single flights missions were flown, one for each forward and side lap setting. Once the UAV completed the initial altitude, it ascended 10 m and flew the same area again. This process repeated until all programmed altitudes were covered. Table 2 summarizes the duration of, and number of images taken from, each UAV flight with each flight having a different altitude AGL and forward and side overlap setting. In Table 2, time duration expresses the duration of the flight for an individual altitude AGL, without time spent taking off and landing. Table 2 demonstrates, as altitude AGL increases, flight duration and the number of images taken as a decrease because each image covers more area, and, therefore, fewer laps and images are needed. This occurs independently of the forward and side overlap settings, although higher percentages increase flight time and number of images taken. The longest UAV flight was 7 min and 35 s at 30 m AGL with 80%–50% forward and side overlap, while the shortest was 33 s at 80 m AGL with 70%–40% forward and side overlap. Figure 5 shows an exponential correlation between altitude AGL, flight duration (Figure 5a) and the number of images taken (Figure 5b). Inversely, as altitude AGL was reduced, flight duration and the number of images taken exponentially increased.

Table 2. Flight durations and the number of images taken at different altitudes Above Ground Level (AGL) and forward and side lap settings.

Altitude AGL (m)	Forward/Side Lap 70%–40%		Forward/Side Lap 80%–50%	
	Time Duration	Num Images	Time Duration	Num Images
30	0:07:12	27	0:07:35	34
40	0:02:40	12	0:06:03	27
50	0:02:12	10	0:03:08	14
60	0:01:52	8	0:02:44	12
70	0:00:46	4	0:00:56	5
80	0:00:33	3	0:00:56	5

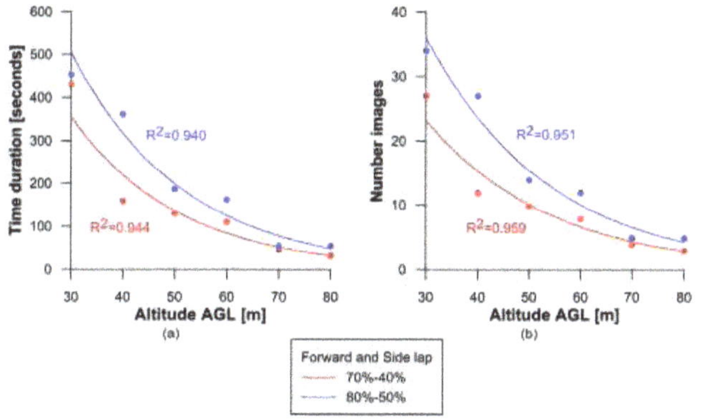

Figure 5. Relationship between altitude above ground level (AGL) and forward and side lap settings on: (**a**) Flight duration and (**b**) Number of images taken.

The shape of the study area in relation to the percentage of forward and side overlap also affects the duration of UAV flights. In this study, there was a significant time difference at 40 m AGL due to the forward and side overlap settings. Fewer laps were needed to cover the area of study at 70%–40% because more distance needed to be covered between laps at 80%–50%. At higher altitudes AGL, for example 70 or 80 m, the differences in flight duration were reduced due to only one lap being needed to capture both forward and side overlap, which reduced the number of images taken.

Another factor that affects flight duration is illumination. Sometimes, elements may appear in the study area which can produce shadows depending on the direction of the flight. Moreover, some materials, like marble, reflect light intensely, resulting in highly saturated images. Consequently, flying within limited timeframes may be necessary to avoid problems caused by illumination. Therefore, reducing flight times while simultaneously maintaining orthomosaic spatial quality is of interest.

Figure 6 compares images taken in the early morning (7:15 a.m.) and again close to midday (11:30 a.m.) at the same altitude AGL. Elements such as walls (Figure 6a,b) or columns (Figure 6c,d) project shadows if the images are taken at midday (Figure 6a,c). On the other hand, if taken in the early morning, images do not contain shadows (Figure 6b,d). Because sun elevation is reduced and objects do not drop shadows, as such images are darker. One option, to avoid shadows, is to fly when the Sun is at the zenith position. In this case, it is necessary to take into account the coordinates of study area and day of the year to know when the Sun is at this position. However, in this case, problems can arise with materials like marble, which can saturate images leading difficulties in visual interpretation. As an example, in Figure 6c, it is more difficult to identify individual elements at the top of the wall compared to Figure 6d.

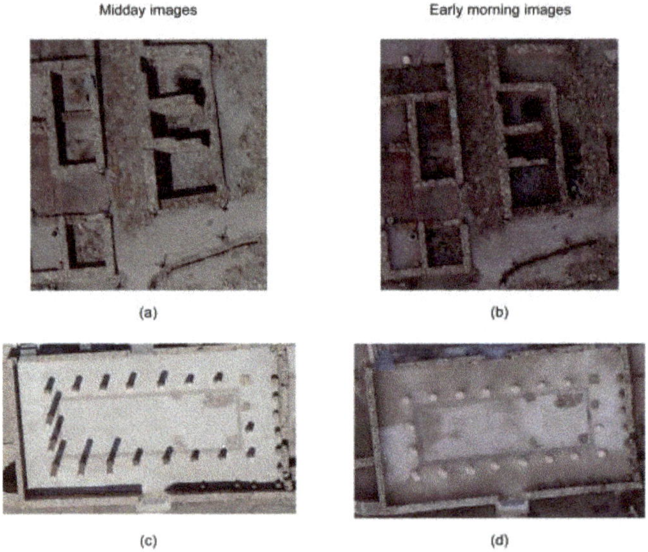

Figure 6. (**a–d**) The effects of illumination on images taken by UAV flights at midday (**a**,**c**) and early morning (**b**,**d**).

Figure 7 shows an example of two sets of images of two different areas taken at different altitudes AGL and demonstrates that, as altitude AGL increases, the area covered by each image increases, reducing flight duration. On the other hand, the quality of spatial resolution and border definition of individual elements improve as altitude AGL decreases. At 30 m, element boundaries are well defined and recognizable in an individual context, while at higher altitude AGL, definition incrementally diffuses, and it becomes more difficult to define individual elements. However, this does not mean that UAV flights at high altitude AGL are not useful in archaeological utilities, and it depends on the need of the user. The geomatic product requirements of an archaeological area needing only a general map, which typically meets or exceeds user expectations, are not the same as those of a specific site prospection. Therefore, product features are said to posses "fitness for use" if they are able to serve their purposes [38].

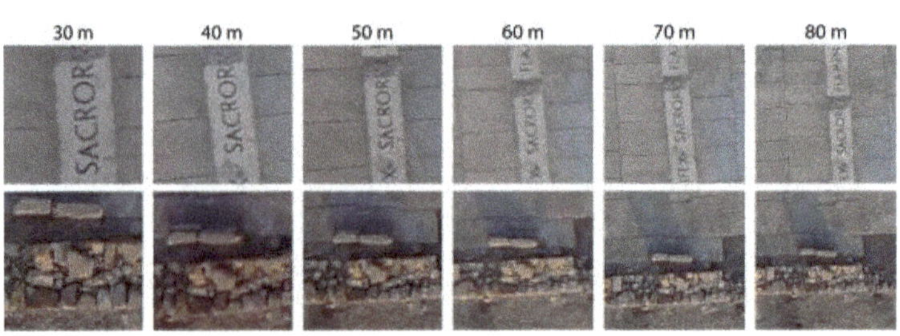

Figure 7. Effect of flight altitudes AGL on image coverage and quality.

From the point of view of image interpretation, altitudes equal to 80 m AGL or higher allow valuable information to be gained for a general analysis of the entire work area and surroundings although details less than 2 cm in size are more difficult to properly identify. Conversely, lower altitude AGL flights allow details to be better studied at the cost of increased flight duration.

3.1. Assesement of Absolute Positional Accuracy

Table 3 summarizes the results of the absolute positional accuracy assessment factoring: (1) altitude AGL; (2) forward and side lap settings; and (3) processing with and without GCPs. Error ranges from 3.8 cm (30 m altitude AGL and 80%–50% overlap settings with GCPs) to 934.2 cm (80 m altitude AGL and 70%–40% overlap settings without GCPs). In Figure 8, the same factors from Table 3 are applied to error box plots. Figure 8a,b show a lower RMSE where GCPs were used than Figure 8c,d where GCPs were not used, these results being independent of altitude AGL and forward and side lap setting.

Table 3. Absolute positional accuracy results factoring altitude AGL, percentage of forward and side overlap and processing with or without GCPs.

Altitude AGL (m)	GSD (cm)	Forward/End Lap (%)	GCP RMSE (cm)	No GCP RMSE (cm)
30	0.7	80%/50%	3.8	507.8
		70%/40%	5.4	178.0
40	1	80%/50%	5.9	138.4
		70%/40%	6.3	73.3
50	1.2	80%/50%	6.2	99.2
		70%/40%	6.5	53.8
60	1.5	80%/50%	6.9	120.7
		70%/40%	6.8	151.9
70	1.7	80%/50%	9.6	179.6
		70%/40%	9.2	229.0
80	2	80%/50%	9.5	179.6
		70%/40%	10.0	934.2

AGL: Above Ground Level, GSD: Ground Sample Distance, GCP: Ground Control Point, RMSE: Root Mean Square Error.

The absolute positional accuracy of orthomosaics produced without GCPs depends on the accuracy of the navigation system of the UAV. Currently, these systems generally have an accuracy of about 1 to 2 m, which is not accurate enough for direct georeferencing. Therefore, GCPs are necessary to properly define the coordinate reference system. Alternatively, integrating an accurate direct georeferencing system onto a UAV platform would allow the elimination of GCPs [39]. Although most commercial UAVs are not equipped with an accurate direct georeferencing system, there are UAVs in the market with this capability, which will likely be a more common solution in the future.

As altitude AGL increased, errors also increased when GCPs were used (Figure 8a,b). This behavior was constant independent of forward and side overlap settings. From 30 m to 40 m AGL, flights showed a positional accuracy of less than 5 cm. From 50 m to 60 m AGL, the RMSE was around 6 cm. RMSE was higher than 9 cm with altitudes 70 m AGL and up. This suggests that, as the altitude AGL increases, GSD of images increases, which is reflected in RMSE. On the other hand, the orthomosaics where GCPs were not used (Figure 8c,d) showed random RMSE behavior due to the lack of geometric constraints of calculating aerial-triangulation which was because the RMSE depended on the accuracy of the UAV's navigational system, suggesting lower altitude AGL flights and the use of GCPs give better absolute positional accuracy.

With the forward and side overlap settings, higher percentages (Figure 8a) resulted in lower RMSE for all UAV flights. The cause of this may be that there was more redundant information to extract tie points. SfM algorithms applied to UAV flights show better results when using a high redundant bundle adjustment based on matching features in multiple overlapping images. As in [40], all individual flights with a forward and side lap equal to 80%–50%, respectively, showed better results than a 70%–40% configuration. These improved results were more evident at lower altitudes AGL.

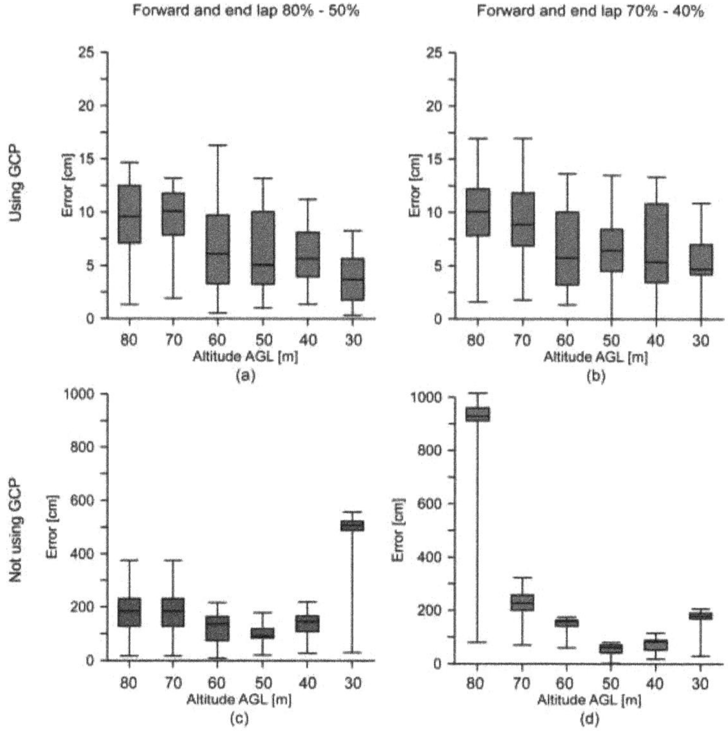

Figure 8. (**a–d**) Root mean square error (RMSE) box plot graph factoring altitude AGL; forward and side lap (**a**,**c**) 80%–50% (**b**,**d**) 70%–40% and; processing (**a**,**b**) with or (**c**,**d**) without ground control points (GCPs).

Figure 9 shows the forward and side overlap settings related to altitude AGL and RMSE in the flights where GCPs were used. Altitude AGL and RMSE show a linear relationship with a correlation coefficient higher than 0.9, independently of forward and side lap settings. The two linear models represented tend to converge. At lower altitudes AGL, the distance between both adjusted lines is greater while tending to converge as altitudes AGL increases. The higher forward and side overlap settings correlate with a lower RMSE having more influence on RMSE at lower altitudes AGL. On the other hand, Figure 9 also shows that altitude AGL has more impact on RMSE than forward and side overlap settings.

Also in Figure 9, GSD is represented by a continuous line, which has a moderate slope compared to the linear models of error, representing its correlation to RMSE. The mean ratio between RMSE and GSD of all UAV flights in this study was 5, suggesting that the expected RMSE of an UAV orthomosaic is five times greater than GSD of images. Ref. [41] obtained a ratio of 3.7 in an experiment on an archaeological site. This difference between results may be due to the fact that they used circular targets as their well-defined check points, while, in this study, in order to approximate an applied assessment, elements of archaeological interest were used as intersection and corner check points that are more diffuse and therefore more difficult to locate and measure.

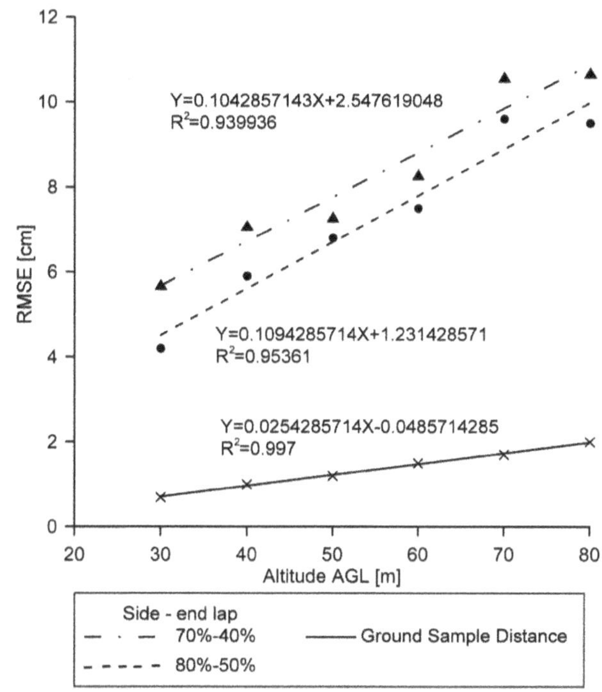

Figure 9. Linear model analyzing forward and side lap settings against altitude AGL and RMSE.

3.2. *Assesment of Relative Positional Accuracy*

Referring to relative positional accuracy (Table 4), RHSD was stable when GCPs were used to define the coordinate reference system in the aerial triangulation phase, ranging from 4.5 cm to 6.5 cm, increasing as altitudes AGL increased and with both forward and side lap settings. RHSD showed lower values with the higher forward and side lap setting.

Table 4. Relative horizontal spatial deviation.

GCPs	Forward–End Lap	Altitude AGL (m)					
		30	40	50	60	70	80
With	70%–40%	5.1	5.5	6.1	6.2	6.2	6.5
	80%–50%	4.5	4.5	4.9	5.0	5.3	5.1
Without	70%–40%	44.0	63.3	56.7	38.1	44.4	88.8
	80%–50%	21.4	33.4	19.7	6.9	42.5	16.6

Without GCPs, RHSD showed random behavior similar to the results obtained in the absolute positional accuracy assessment above. As such, the orthomosaics obtained only using data from the UAV navigation system were rotated, translated and scaled respecting the coordinate reference system. These products were not useful even in a relative coordinate system because any linear or surface measurement is not going to appear correct because the coordinate system was not well defined.

Therefore, designing a UAV flight plan requires defined technical specifications related to illumination, resolution and spatial quality. These parameters have to be considered equally to produce an adequate UAV orthomosaic. Illumination and material effects e.g., marble and image saturation, define the time frames for flying, which is important if the time frames are narrow and flights have to be short. In regards to spatial accuracy, the expected RMSE is five times greater than the GSD registered on flight. AGL is the parameter that mainly influences RMSE. While using higher forward and side overlap settings guarantee greater positional accuracy, flight duration will be increased. Finally, if the navigation system of the UAV is inaccurate, it is necessary to use GCPs, even if the orthomosaics are going to be used in a relative coordinate system.

In this manuscript, spatial resolution of UAV orthomosaicks has been assessed to survey archaelogical areas. Another useful geomatic product in archaeology is model digital surface (MDS). These models can be generated using passive sensors, as we explain and use in this manuscript, or by active sensors like LiDAR. LiDAR sensors can be mounted on manned platforms, airborne and terrestrial [42] or unmanned platforms [43], and have been used successfully in archaelogical prospection [44]. Future works should be performed to compare MDS obtained by LiDAR and aerial imagery and assessing spatial resolution of LiDAR sensor onboard UAV and the influence of flight parameters.

4. Conclusions

This study has shown that UAV systems are useful complements for archaeological mapping, such as GNSS measurements or aerial photogrammetry among others. The main objective of this investigation was to analyze the configuration and technical specifications of a multi-rotor UAV equipped with a RGB sensor to produce accurate orthomosaics to be applied in archaeological applications.

Concerning spatial resolution, flight altitude AGL is an important parameter because of the degree detail achieved in the orthomosaic image to be used to for analysis and study in an archaeological context. Additionally, adequate values of altitude AGL and forward and side overlap settings have to be applied because of their impact on flight duration and positional accuracy of absolute and relative RMSE obtained. Our results have shown a ratio between RMSE and GSD of UAV flights equal to 5. Whenever possible, higher percentages of forward and side overlap are recommended for UAV flights. Other flight planning configurations, including transversal laps, can be carried out in future works to study their impact on flight duration and accuracy of results. Moreover, if the UAV's navigation system is not accurate enough, GCPs can be used instead, recalling that even if working in a relative coordinate reference system, any linear or superficial measurement is not going to be accurate without GCPs. The use of navigation systems based on differential-GPS can be an alternative to GPC measurements to be taken into account, assessing its influence in spatial resolution.

The results herein presented can be used to configure flight missions using a RGB sensor onboard a multi-rotor UAV to maximize the spatial positional accuracy of orthomosaics to be used in archaeological mapping.

Acknowledgments: The authors would like to thank the Campus de Excelencia Internacional en Patrimonio for financial support via project "Determinación de parámetros óptimos en vuelos UAV aplicados a la caracterización métrica de yacimientos arqueológicos" of the "Proyectos de referencia internacional Campus de excelencia internacional en patrimonio, PATRIMONIUN-10".

Author Contributions: F.-J.M.-C. and A.G.-F. conceived and designed the experiments; F.-J.M.-C. and M.D.N.G. performed the experiments; F.-J.M.-C., M.D.N.G., J.E.M.L. and A.G.F. analyzed the data; and F.-J.M.-C. and M.D.N.G. wrote the paper.

Conflicts of Interest: The authors declare no conflict of interest.

References

1. Schindling, J.; Gibbes, C. Lidar as a tool for archaeological research: A case study. *Archaeol. Anthropol. Sci.* **2014**, *6*, 411–423. [CrossRef]
2. Fryer, J.; Mitchell, H.; Chandler, J. *Applications of 3D Measurements from Images*; Whittles Publishing: Caithness, UK, 2007.

3. Altaweel, M. The use of ASTER satellite imagery in archaeological contexts. *Archaeol. Prospect.* **2005**, *12*, 151–166. [CrossRef]
4. Rosa, L.; Nicola, M. On the potential of quickbird data for archaeological prospection. *Int. J. Remote Sens.* **2006**, *27*, 3607–3614.
5. Wilkinson, K.N.; Beck, A.R.; Philip, G. Satellite imagery as a resource in the prospection for archaeological sites in central Syria. *Geoarchaeology* **2006**, *21*, 735–750. [CrossRef]
6. Chiabrando, F.; Nex, F.; Piatti, D.; Rinaudo, F. UAV and RPV systems for photogrammetric surveys in archaelogical areas: Two tests in the Piedmont region (Italy). *J. Archaeol. Sci.* **2011**, *38*, 697–710. [CrossRef]
7. Verhoeven, G.J.J. Providing an archaeological bird's-eye view—An overall picture of ground-based means to execute low-altitude aerial photography (LAAP) in Archaeology. *Archaeol. Prospect.* **2009**, *16*, 233–249. [CrossRef]
8. Mozas-Calvache, A.T.; Pérez-García, J.L.; Cardenal-Escarcena, F.J.; Mata-Castro, E.; Delgado-García, J. Method for photogrammetric surveying of archaeological sites with light aerial platforms. *J. Archaeol. Sci.* **2012**, *39*, 521–530. [CrossRef]
9. Gomez-Lahoz, J.; Gonzalez-Aguilera, D. Recovering traditions in the digital era: The use of blimps for modelling the archaeological cultural heritage. *J. Archaeol. Sci.* **2009**, *36*, 100–109. [CrossRef]
10. Bogacki, M.; Malkowski, W.; Misiewicz, K. Kite Aerial Photography (KAP) as a Tool for Completing GIS Models. Ptolemais (Libya) Case Study. In *Remote Sensing for Archaeology and Cultural Heritage Management*, Proceedings of the 1st International EARSeL Workshop, CNR, Rome, Italy, 30 September–4 October 2008; Lasaponara, R., Masini, N., Eds.; pp. 329–332.
11. Bendea, H.; Chiabrando, F.; Giulio Tonolo, F.; Marenchino, D. Mapping of archaeological areas using a low-cost UAV. The augusta bagiennorum test site. In Proceedings of the XXI International CIPA Symposium, Athens, Greece, 1–6 October 2007.
12. Sauerbier, M.; Eisenbeiss, H. UAVs for the documentation of archaeological excavations. *Int. Arch. Photogramm. Remote Sens. Spat. Inf. Sci.* **2010**, *38*, 526–531.
13. Paparoditis, N.; Souchon, J.-P.; Martinoty, G.; Pierrot-Deseilligny, M. High-end aerial digital cameras and their impact on the automation and quality of the production workflow. *ISPRS J. Photogramm. Remote Sens.* **2006**, *60*, 400–412. [CrossRef]
14. Müller, J.; Gärtner-Roer, I.; Thee, P.; Ginzler, C. Accuracy assessment of airborne photogrammetrically derived high-resolution digital elevation models in a high mountain environment. *ISPRS J. Photogramm. Remote Sens.* **2014**, *98*, 58–69. [CrossRef]
15. Ackermann, F. Operational rules and accuracy models for GPS-aerotriangulation. *Arch. ISPRS* **1992**, *1*, 691–700.
16. Pajares, G. Overview and current status of remote sensing applications based on unmanned aerial vehicles (UAVs). *Photogramm. Eng. Remote Sens.* **2015**, *81*, 281–329. [CrossRef]
17. Mesas-Carrascosa, F.J.; Notario-García, M.D.; de Larriva, J.E.M.; de la Orden, M.S.; Porras, A.G.-F. Validation of measurements of land plot area using UAV imagery. *Int. J. Appl. Earth Observ. Geoinf.* **2014**, *33*, 270–279. [CrossRef]
18. Mesas-Carrascosa, F.J.; Rumbao, I.C.; Berrocal, J.A.B.; Porras, A.G.-F. Positional quality assessment of orthophotos obtained from sensors onboard multi-rotor UAV platforms. *Sensors* **2014**, *14*, 22394–22407. [CrossRef] [PubMed]
19. Mesas-Carrascosa, F.-J.; Torres-Sánchez, J.; Clavero-Rumbao, I.; García-Ferrer, A.; Peña, J.-M.; Borra-Serrano, I.; López-Granados, F. Assessing optimal flight parameters for generating accurate multispectral orthomosaicks by UAV to support site-specific crop management. *Remote Sens.* **2015**, *7*, 12793–12814. [CrossRef]
20. Hendrickx, M.; Gheyle, W.; Bonne, J.; Bourgeois, J.; De Wulf, A.; Goossens, R. The use of stereoscopic images taken from a microdrone for the documentation of heritage—An example from the Tuekta burial mounds in the Russian Altay. *J. Archaeol. Sci.* **2011**, *38*, 2968–2978. [CrossRef]
21. Hengl, T. Finding the right pixel size. *Comput. Geosci.* **2006**, *32*, 1283–1298. [CrossRef]
22. Szeliski, R. Structure from motion. In *Computer Vision: Algorithms and Applications*; Springer: London, UK, 2011; pp. 303–334.

23. Micheletti, N.; Chandler, J.H.; Lane, S.N. Structure from Motion (SFM) Photogrammetry. In *Geomorphological Techniques*; Clarke, L.E., Nield, J.M., Eds.; British Society for Geomorphology: London, UK, 2015; Chapter 2, Section 2.2; pp. 1–12.
24. Fonstad, M.A.; Dietrich, J.T.; Courville, B.C.; Jensen, J.L.; Carbonneau, P.E. Topographic structure from motion: A new development in photogrammetric measurement. *Earth Surf. Process. Landf.* **2013**, *38*, 421–430. [CrossRef]
25. Wang, J.; Ge, Y.; Heuvelink, G.B.M.; Zhou, C.; Brus, D. Effect of the sampling design of ground control points on the geometric correction of remotely sensed imagery. *Int. J. Appl. Earth Observ. Geoinf.* **2012**, *18*, 91–100. [CrossRef]
26. Carvajal, F.; Agüera, F.; Martínez, P.J. Effects of image orientation and GCP distribution on unmanned aerial vehicle photogrammetry projects on a road cut slope. *J. Appl. Remote Sens.* **2016**, in press. [CrossRef]
27. Agüera, F.; Carvajal, F.; Pérez, M.; Orgaz, F. Multi-temporal imaging using an unmanned aerial vehicle for monitoring a sunflower crop. *Biosyst. Eng.* **2015**, *132*, 19–27.
28. Zhang, Y.; Xiong, J.; Hao, L. Photogrammetric processing of low-altitude images acquired by unpiloted aerial vehicles. *Photogramm. Rec.* **2011**, *26*, 190–211. [CrossRef]
29. Küng, O.; Strecha, C.; Beyeler, A.; Zufferey, J.-C.; Floreano, D.; Fua, P.; Gervaix, F. The accuracy of automatic photogrammetric techniques on ultra-light UAV imagery. In Proceedings of the UAV-g 2011-Unmanned Aerial Vehicle in Geomatics, Zürich, Switzerland, 14–16 September 2011.
30. Kraus, K. *Photogrammetry—Geometry from Images and Laser Scans*; Walter de Gruyter: Goettingen, Germany, 2007.
31. Yuan, X.; Fu, J.; Sun, H.; Toth, C. The application of gps precise point positioning technology in aerial triangulation. *ISPRS J. Photogramm. Remote Sens.* **2009**, *64*, 541–550. [CrossRef]
32. Snavely, N.; Seitz, S.M.; Szeliski, R. Photo tourism: Exploring photo collections in 3D. *ACM Trans. Graph.* **2006**, *25*, 835–846. [CrossRef]
33. Inpho Uasmaster. Available online: http://www.trimble.com/geospatial/inpho-uasmaster.Aspx (accessed on 1 October 2016).
34. International Organization for Standardization (ISO). *Geographic Information—Data Quality*; ISO: London, UK, 2013; Volume 19157.
35. QGIS. Available online: http://www.qgis.org/en/site/ (accessed on 27 October 2016).
36. Photogrammetric Engineering & Remote Sensing. ASPRS Positional Accuracy Standards for Digital Geospatial Data. American Society for Photogrammetry and Remote Sensing (ASPRS): Bethesda, MD, USA, 2015; Volume 81, pp. A1–A26.
37. Mapping, Charting and Geodesy Accuracy. Available online: http://earth-info.nga.mil/publications/specs/printed/600001/600001_Accuracy.pdf (accessed on 1 June 2016).
38. Juran, J.M.; De Feo, J.A. *Juran's Quality Handbook: The Complete Guide to Performance Excellence*, 6th ed.; McGraw-Hill Education: New York, NY, USA, 2010.
39. Turner, D.; Lucieer, A.; Wallace, L. Direct georeferencing of ultrahigh-resolution UAV imagery. *IEEE Trans. Geosci. Remote Sens.* **2014**, *52*, 2738–2745. [CrossRef]
40. Westoby, M.J.; Brasington, J.; Glasser, N.F.; Hambrey, M.J.; Reynolds, J.M. 'Structure-from-Motion' photogrammetry: A low-cost, effective tool for geoscience applications. *Geomorphology* **2012**, *179*, 300–314. [CrossRef]
41. Nex, F.; Remondino, F. UAV for 3D mapping applications: A review. *Appl. Geomat.* **2014**, *6*, 1–15. [CrossRef]
42. Williams, K.; Olsen, M.J.; Roe, G.V.; Glennie, C. Synthesis of transportation applications of mobile LiDAR. *Remote Sens.* **2013**, *5*, 4652–4692. [CrossRef]
43. Wallace, L.; Lucieer, A.; Watson, C.; Turner, D. Development of a UAV-LiDAR system with application to forest inventory. *Remote Sens.* **2012**, *4*, 1519–1543. [CrossRef]
44. Carter, W.E.; Shrestha, R.L.; Fernandez-Diaz, J.C. Archaeology from the air. *Am. Sci.* **2016**, *104*, 28–35. [CrossRef]

sensors

MDPI

Article

Development of an Unmanned Aerial Vehicle-Borne Crop-Growth Monitoring System

Jun Ni [1], Lili Yao [1], Jingchao Zhang [2], Weixing Cao [1], Yan Zhu [1,*] and Xiuxiang Tai [1]

[1] National Engineering and Technology Center for Agriculture/Jiangsu Key Laboratory for Information Agriculture/Collaborative Innovation Center for Modern Crop Production/Jiangsu Collaborative Innovation Center for the Technology and Application of Internet of Things, Nanjing Agriculture University, Nanjing 210095, China; nijun@njau.edu.cn (J.N.); 2015101038@njau.edu.cn (L.Y.); caow@njau.edu.cn (W.C.); 11114416@njau.edu.cn (X.T.)

[2] Nanjing Institute of Agricultural Mechanization of National Ministry of Agriculture, Nanjing 210014, China; zhangjc9@163.com

* Correspondence: yanzhu@njau.edu.cn; Tel./Fax: +86-25-8439-6598

Academic Editors: Felipe Gonzalez Toro and Antonios Tsourdos
Received: 10 December 2016; Accepted: 24 February 2017; Published: 3 March 2017

Abstract: In view of the demand for a low-cost, high-throughput method for the continuous acquisition of crop growth information, this study describes a crop-growth monitoring system which uses an unmanned aerial vehicle (UAV) as an operating platform. The system is capable of real-time online acquisition of various major indexes, e.g., the normalized difference vegetation index (NDVI) of the crop canopy, ratio vegetation index (RVI), leaf nitrogen accumulation (LNA), leaf area index (LAI), and leaf dry weight (LDW). By carrying out three-dimensional numerical simulations based on computational fluid dynamics, spatial distributions were obtained for the UAV down-wash flow fields on the surface of the crop canopy. Based on the flow-field characteristics and geometrical dimensions, a UAV-borne crop-growth sensor was designed. Our field experiments show that the monitoring system has good dynamic stability and measurement accuracy over the range of operating altitudes of the sensor. The linear fitting determination coefficients (R^2) for the output RVI value with respect to LNA, LAI, and LDW are 0.63, 0.69, and 0.66, respectively, and the Root-mean-square errors (RMSEs) are 1.42, 1.02 and 3.09, respectively. The equivalent figures for the output NDVI value are 0.60, 0.65, and 0.62 (LNA, LAI, and LDW, respectively) and the RMSEs are 1.44, 1.01 and 3.01, respectively.

Keywords: unmanned aerial vehicle sensor; crop-growth model; computational fluid dynamics; flow field analysis; monitoring system; field experiment

1. Introduction

Real-time, non-destructive, and high-throughput acquisition of crop-growth information is the most important requirement for precision management of crop production. Traditional detection methods which rely on the destructive sampling of plants and indoor physical and chemical analyses, are time-consuming, laborious, and have poor timeliness. In recent years, technologies based on feature recognition using reflection spectra have proven to have several advantages over the traditional methods: non-destructibility, convenient access to information, and good real-time performance. Therefore, this kind of technology has been widely used in research on the mechanisms of monitoring crop growth [1–11].

At present, research institutions around the world have gained access to reflection spectra of crop canopies obtained using various devices (e.g., Cropscan multispectral radiometers [12], ASD FieldSpec 3 hyper-spectrometers [13,14], GreenSeeker sensors [15,16], and CropCircle ACS-470) [17]. The research undertaken has indicated out that there is a good correlation between the reflection spectra of crop canopies and crop nutrients. Hand-held sensors are usually used to statically acquire information about the crop canopy. Although these types of sensors can produce a detailed determination of the spectral characteristics of a crop's biochemical components, they have several disadvantages including small monitoring range, large labor intensities, and a monitoring regime that is discontinuous. Therefore, these methods cannot provide the high-throughput of information needed for real-time decisions to be made in the production and management of crops spread over large areas in the field. To address this problem, research institutes have started to develop crop-growth monitoring equipment based on vehicle platforms.

The German Yara and Japanese Topcon companies have designed ways to determine the nitrogen content of crops (using their proprietary N-Sensor [18,19] and laser modulated light source sensor CropSpec [20], respectively). In addition, the Trimble Navigation Company based in the United States has also produced the GreenSeeker-RT200 sensor to determine the normalized difference vegetation index (NDVI) of crops [21,22]. Such equipment can continuously gather information on crop growth with a high-throughput and high labor efficiency. However, the vehicle platform causes a certain amount of destruction of crops in operation. Also, operation is not flexible and is easily limited by the size and terrain of the farmland.

The use of unmanned aerial vehicles (UAVs) for such operations promises to have several advantages including high efficiency, good flexibility, convenient operation, and strong adaptability to terrain. Thus, UAVs are becoming more extensively applied to the monitoring of crop growth [23]. By utilizing a miniature hyperspectral infrared thermograph on a UAV, Zarco-Tejada et al. [24] obtained hyperspectral image information on a citrus canopy of large area. The water-stress state of the citrus trees was analyzed offline using remote sensing image processing software including the Environment for Visualizing Images (ENVI). By fixing a color camera onto a UAV, Bendig et al. [25] acquired real-color images of a tree canopy and established a three-dimensional (3D) geometrical model of the trees. Moreover, crop vegetation indices and plant heights could be measured with the use of a ground-based hyper-spectrometer.

By using UAVs with spiral and fixed wings equipped with a real-color camera and a color-near infrared camera, respectively, Rasmussen et al. [26] obtained information on a crop canopy under different lighting environments. Image processing software was then used to splice and interpret the information obtained so that crop vegetation indices could be obtained. Moreover, it was verified that consumer-grade color cameras could be used to reliably acquire images to allow vegetation indices to be retrieved. A multispectral camera carried on a UAV was used by Caturegli et al. [27] to obtain multispectral images of lawns. By utilizing ENVI software to process the images, information on the vegetation index of the lawns could be extracted to evaluate their nitrogen nutrition status.

Most of the abovementioned research used a UAV as a platform to carry various types of imaging spectroradiometers to obtain images containing crop information. This information was then corrected offline and spliced using special remote-sensing analysis software in order to interpret the crop-growth information. Due to the complexity of the procedures employed, such an operation needs remote-sensing specialists and is mainly used in scientific research. Furthermore, any possible interpretation of the crop-growth information is delayed and the images cannot be directly used in agricultural production. Besides this, the approach cannot be popularized in agricultural production settings due to the high price of the equipment involved (mainly the various imaging spectrometers).

NDVI and the ratio vegetation index (RVI) are two commonly used indices when inverting crop-growth parameters in the existing UAV-based remote sensing field. Gao et al. [28] carried out experiments using a multi-rotor UAV as the platform from which a crop-growth monitoring system composed of a Canon PowerShot G16 camera and an ADC-Lite multispectral sensor was trialled. In the experiments, remotely sensed images of soybean in its podding and seed-filling stages were obtained. On this basis, by using vegetation indices, including NDVI and RVI, and combining them with LAI data synchronously measured in the field, they constructed univariate and multivariate LAI inversion models using empirical methods. By using an improved camera with an infrared filter borne on an UAV, Ghazal et al. [29] acquired NDVI videos which were then processed to obtain the area of crop growing spots, and relevant agronomic parameters were inverted at the same time. Tian et al. [30] acquired remote sensing images of winter wheat using an UAV-borne ADC air vegetation canopy camera. Based on the spectral characteristics of the images and the changing threshold of NDVI, they proposed a quick classification and extraction method for crops. The results show that, using the method to extract classification information of crops of different types from high-resolution images collected by UAVs presents high accuracy and universality. All this suggests that, while being used for inverting crop agronomic parameters and classifying crop characteristics, NDVI and RVI show high accuracy and potential application value.

In this study, we present a new UAV-borne crop-growth monitoring system based on research achievements of the Nanjing Agricultural University in China relating to crop-growth sensors [31–34]. The work is aimed at meeting the demands for a high-throughput, continuous, and online real-time method of acquiring crop-growth information. Using a four-rotor UAV (DJI Phantom, SZ DJI Technology Co., Ltd. Shenzhen, China) as the operating platform, we independently design a UAV-borne crop-growth sensor and a matching ground-based data processor to complement the platform. We subsequently used the system to determine, in real-time and online, the major growth indices of a crop canopy including the NDVI, ratio vegetation index (RVI), leaf nitrogen accumulation (LNA), leaf area index (LAI), and leaf dry weight (LDW). This research thus provides a new technological means of acquiring high-throughput growth information for crops covering large areas.

2. Design of the UAV-Borne Crop-Growth Monitoring System

2.1. Overall Design of the System

The UAV-borne crop-growth monitoring system consists of a UAV platform, a UAV-borne crop-growth sensor, and a ground-based data processor. The UAV-borne crop-growth sensor is fixed to the UAV platform and used to obtain reflection spectra of a crop canopy in real time. The data collected are wirelessly transmitted to the data processor on the ground. The ground-based data processor receives spectral information on the crop canopy which is input into a crop-growth monitoring model. Derived information, including the NDVI, RVI, LNA, LAI, and LDW of the crop canopy, is presented online. The system structure is shown in Figure 1.

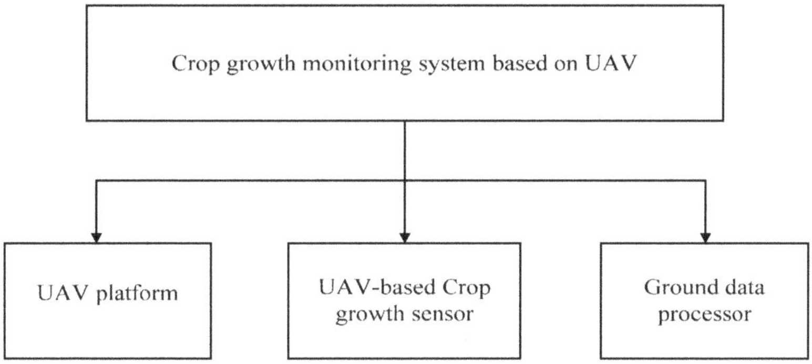

Figure 1. The structure of the UAV-borne crop-growth monitoring system.

2.2. Optimization of the UAV Platform

At present, UAVs intended for agricultural use are mainly of the fixed-wing and multi-rotor type [35]. The former drives the aircraft forwards according to the thrust produced by spiral wings or turbine generators. The force of elevation required is generated by the relative movement of the wing and the air. Before flying, the rotors of a UAV need to have a certain initial speed. Common takeoff modes include catapult takeoff and running takeoff. The latter relies on the elevating forces generated by the rotation of multiple spiral wings to balance the weight of the aircraft. This mode does not need the UAV to have an initial speed for takeoff, and is able to realize vertical takeoff.

Although fixed-wing UAVs are fast and have long cruise durations, the flight speed is hard to adjust according to demand. In addition, they cannot hover and are used merely to carry loads of low weight. In contrast, the flight speed and height of a spiral-wing UAV are adjustable and, in addition, its takeoff mode is simple and places no limits on the takeoff and landing sites [36]. Therefore, spiral-wing UAVs are more suitable for obtaining reflectance spectra of crop canopies in the field. In this study, we used a DJI phantom UAV which is an ideal low-cost platform for use in crop-growth monitoring (Figure 1). The aircraft's mass, maximum load, flight speed, vertical hovering precision, and horizontal hovering precision are 0.92 kg, 1.2 kg, 10 m/s, 0.8 m, and 2.5 m, respectively. Furthermore, its maximum angular spin velocity, maximum tilt angle, rotor radius, motor speed, and battery life are 1.11 rad/s, 35°, 10.30 cm, 16 r/s, and 18 min, respectively. The DJI phantom UAV is illustrated in Figure 2.

Figure 2. The DJI phantom UAV platform.

2.3. UAV-Borne Crop-Growth Sensor

The UAV-borne crop-growth sensor consists of a multispectral crop-growth sensor, a sensor support, and a sensor signal-processing circuit. The multispectral crop-growth sensor works on the same measurement principles that ground-based object spectrometers work on. We suppose that the reflection from the crop canopy shows Lambertian reflection characteristics to obtain the bi-directional reflectance of the canopy spectra. Measurement was conducted on sunny days in the absence of heavy cloud cover and strong winds so that the crop canopy remains relatively static. The surface of the crop canopy is close to being a Lambertian reflector and the sensor is placed 1.0–1.2 m directly above the crop canopy in order to capture reflection spectra from it. The sensor support is used for installation of the crop-growth sensor and fixing to the UAV. During low-altitude flight, the rotors of the spiral-wing UAV produce strong airflow fields below the body of the UAV. These may disturb the crop canopy and damage the Lambertian reflectance characteristics. The sensor support ensures that the detection fields of the multispectral sensor relate to the canopy in the absence of airflow disturbance.

2.3.1. Multispectral Crop-Growth Sensor

The multispectral sensor consists of two kinds of lens (for detection at 720 and 810 nm), which are used to measure the spectral reflectance of the crop canopy. The sensor system utilizes sunlight as the light source which is split using an optical filter. Structurally, there is a solar sensor and a two-band sensor. The former is employed to collect radiation information from the sunlight at 720 nm and 810 nm and to conduct cosine correction. The latter is utilized to collect information on the radiation reflected from the crop canopy at the same wavelengths.

The key to designing a multispectral sensor is to determine the correct aperture parameters required for the detection lens. These need to guarantee the sensor system has a high resolution but should also ensure that the signal from the sensor is sufficiently strong. The aperture parameters used in our design are 12.8 mm, 26 mm, and 27°, for the aperture diameter, hole depth, and field of view of the detection lens. The performance parameters correspond to a spectral filter bandwidth of 10 nm and transmittance of 65%–70%. Furthermore, the sensitivity and spectral response of the photoelectric detector selected are 0.55 A/W and 0.011 A/(W/cm^2), respectively. With this combination of parameters, each detection lens comprises a spectral filter and a photoelectric detector with a simple light path. This guarantees good reliability of the transmitted signal and makes integration and transplantation convenient. The design thus tackles some of the disadvantages of previous crop-growth sensors (complex light paths and heavy use of optical devices). The sensor is packaged within a cylindrical aluminum case, which is highly appropriate for field application. The measurement principles underlying operation of the multispectral sensor are shown in Figure 3.

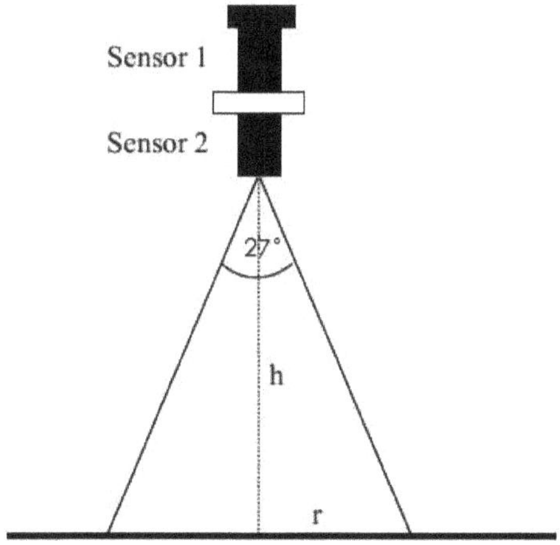

Figure 3. Measurement principles of the multispectral crop-growth sensor. Note: Sensor 1 and sensor 2 represent the solar sensor and two-band sensor, respectively.

As already mentioned, the maximum load of the UAV is 1.2 kg. In order to ensure that the UAV equipped with the multispectral crop-growth sensor is capable of stable flight, the sensor design should be as lightweight as possible. On the premise of retaining the optical structure of the sensor, synthetic fiber (nylon) was adopted in place of the aluminum package holding the sensor to greatly reduce the weight of the sensor (the weight of the improved sensor dropped from 142.7 to 11.34 g, which clearly meets the maximum payload requirements of the DJI Phantom UAV). The structure of the improved sensor is illustrated in Figure 4.

Figure 4. The lightweight structure used for the multispectral crop growth sensor.

2.3.2. Design of the Sensor Support

The sensor support is used to install the two-band sensor of the multispectral crop-growth sensor to measure the Lambertian reflection intensity of the crop canopy. To ensure measurement accuracy and sensitivity, the working height was set to 1.0–1.2 m above the crop canopy. The canopy also needs to be relatively static. At such low altitudes, the strong down-wash flow fields produced by the spiral wings of the UAV may be expected to affect the sensing process. In particular, the leaves of the canopy can be expected to be displaced towards a common direction under the effects of the airflow field, thus showing Fresnel reflection characteristics and damaging the original Lambertian reflection characteristics of the canopy structure. As a result, the multispectral crop-growth sensor would be unable to correctly obtain the bidirectional reflectance of the canopy spectra. Therefore, the spatial distribution of the down-wash flow field generated by the UAV on the surface of the crop canopy needs to be analyzed. The aim is to determine the optimum length of the sensor support and position of installation of the two-band sensor on the support to ensure that the working field of view of the two-band sensor with respect to the crop canopy is not disturbed by the airflow to a significant extent.

In recent years, with the rapid development of computer technologies and fluid turbulence models, computational fluid dynamics (CFD) has gradually become a powerful tool for studying the distribution of airflow fields associated with UAVs [37,38]. Here, we use a 3D CFD method of numerical simulation to analyze the distribution of the down-wash flow fields produced by the UAV's rotors on the surface of the crop canopy. The results are used to propose an optimization scheme for the design of the sensor support.

Model Establishment

(1) Physical model of the DJI Phantom UAV

The situation is complicated by the irregular nature of the surfaces of the UAV's rotors, making it difficult to measure linear data. A 3D scanner was therefore used to scan the rotors in order to obtain a uniform point diagram of a rotor blade. By utilizing reverse-engineering software (Imageware, Siemens, Berlin, Germany), the uniform point diagram was then substantialized and the boundaries trimmed to convert it into a blade entity model. Finally, the whole UAV body was modeled (based on measured data) using appropriate modeling software (Creo, Parametric Technology Corporation, MA, USA), and combined with the scanned blade entity, thus obtaining a 3D entity model of the UAV (Figure 5).

(2) The aerodynamic model

When the UAV hovers, the down-wash airflow exhibits 3D turbulence. The airflow can be described using a series of mass, momentum, and energy conservation equations. We used the standard k–ε model to solve this problem, resulting in a control equation which can be expressed in the following common form [39,40]:

$$\frac{\partial(\rho\varphi)}{\partial t}+\frac{\partial(\rho u\varphi)}{\partial x}+\frac{\partial(\rho v\varphi)}{\partial y}+\frac{\partial(\rho w\varphi)}{\partial z}=\frac{\partial}{\partial x}\left(\Gamma\frac{\partial\varphi}{\partial x}\right)+\frac{\partial}{\partial y}\left(\Gamma\frac{\partial\varphi}{\partial y}\right)+\frac{\partial}{\partial z}\left(\Gamma\frac{\partial\varphi}{\partial z}\right)+S \quad (1)$$

where φ, Γ, and S represent the generalized variable, diffusion coefficient, and source term, respectively, and u, v, and w indicate the speeds in the x, y, and z directions (m/s), respectively. Furthermore, ρ and t stand for the density (kg/m^3) and time (s), respectively.

(**a**)

(**b**)

Figure 5. 3D models used for the DJI phantom UAV. (**a**) Rotor blade; (**b**) UAV.

Numerical Simulations

(1) Grid divisions

The flow fields are stable when the UAV is hovering and the whole of the circumferential flow field tends to be consistent. Therefore, the flow fields of the hovering UAV rotor were simulated as a cylindrical flow field. The calculation domain of this flow field is divided into two parts: an inner, rotating flow field due to the four rotors, and an outer, static flow field which includes the UAV body and airflow fields. The diameter and height of the static field are 1.2 and 1.85 m, respectively, while those of the rotating field are 27.5 cm and 1.8 cm. The UAV rotors are taken to be 1.3 m from the ground.

The interfaces between the rotating regions in the rotating inner field are axially averaged adopting multiple reference frames, so as to make the values of the flow fields in the circumferential position be identically the same at the same elevation. The interfaces of the rotating regions are rotating walls and the rotating flow field rotates at the rated speed of the UAV rotors in the direction determined by the right-hand rule.

Due to the complex 3D shape of the UAV structure, it is difficult to divide the whole model into structured grids. Thus, we assume that the structure can be reasonably divided into unstructured grids and use the integrated computer engineering and manufacturing code for computational fluid dynamics (ICEM-CFD) to divide the body-fitted grids of the model. Furthermore, the high rotation speeds of the rotor blades leads to a large airflow speed gradient in the inner flow field. However, the outer flow field is only slightly affected. Based on this, dense and sparse grids were generated for the rotor flow field and outer flow field, respectively. In order to improve the accuracy and completeness of data transmission at the interfaces, the grid numbers in the interface regions should be as close as possible. The grid division used is demonstrated in Figure 6.

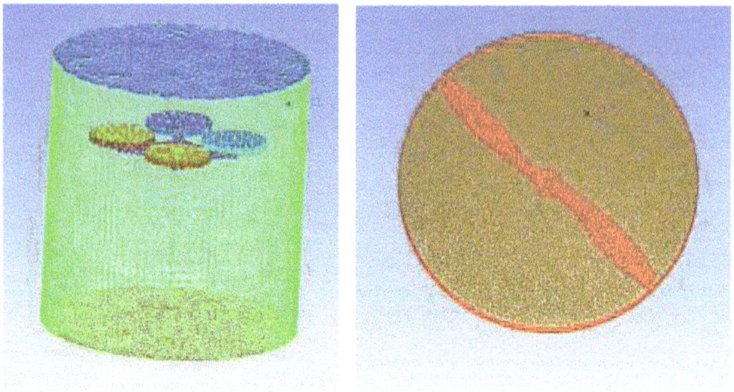

(a) (b)

Figure 6. The grid divisions of the flow fields. (**a**) Grid division of the outer flow field; (**b**) Grid division of the inner flow field.

(2) Boundary conditions

As the UAV is to hover in a flow field corresponding to open space, the outer boundary of the cylinder is set as open. The pressure boundary condition is set to one atmosphere and air at 25° is adopted as the fluid. The rotors are rotating bodies and rotate at a given speed (960 r/min). The interface between the two fields is set as the interface and the reference pressure is taken to be one atmosphere. The roughness of the gas boundary surface is assumed to be zero (i.e., wall without slip). A second-order upwind scheme is utilized for the momentum, turbulent energy, and dissipation equations. To improve the accuracy of the calculations, the residual error is set to have an order of magnitude of 10^{-4}.

Calculation Results and Analysis

Using the parameters set above, CFX software was used for the numerical calculation and the post-processing modules of the CFX package were utilized for display purposes. Figure 7 shows the distribution of the velocity vectors on the axial section of the down-wash flow fields when the UAV rotors are 1.3 m above the canopy.

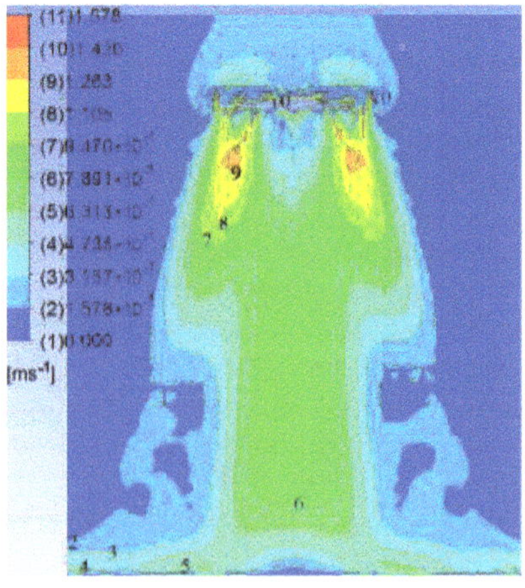

Figure 7. Velocity vector distribution for the down-wash flow fields on the *Z*-*Y* section.

Figure 8 illustrates the airflow velocities in horizontal planes 0.4, 0.6, 0.8, 1.0, and 1.2 m below the rotors.

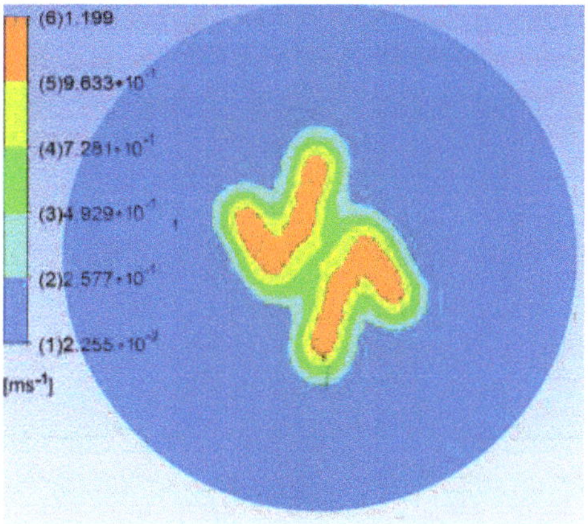

(a)

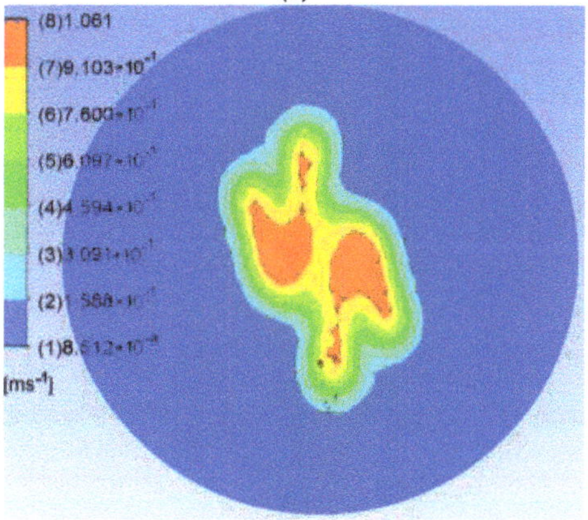

(b)

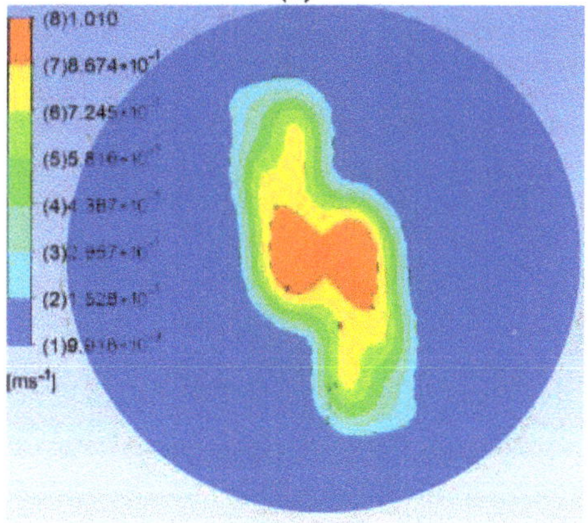

(c)

Figure 8. *Cont.*

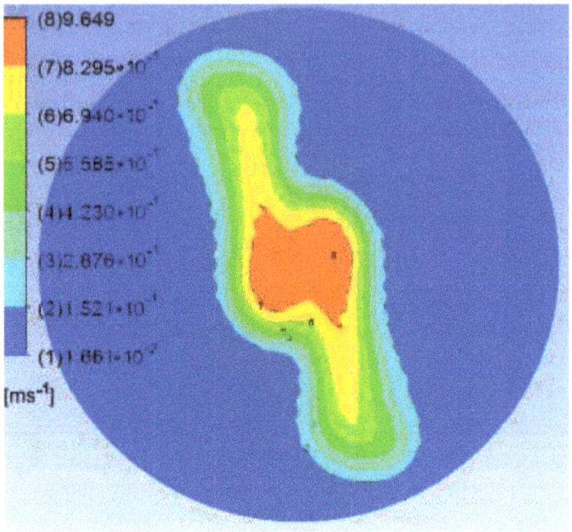

(**d**)

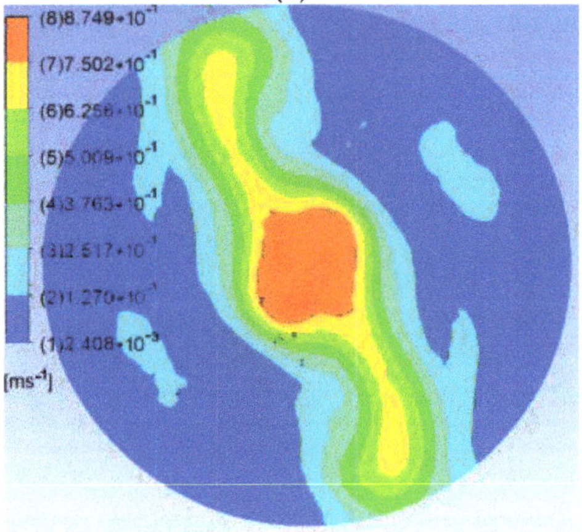

(**e**)

Figure 8. Air velocities in given horizontal planes below the rotors. (**a**) 0.4 m; (**b**) 0.6 m; (**c**) 0.8 m; (**d**) 1.0 m; (**e**) 1.2 m.

It can be seen from Figure 7 that the air above the rotors shrinks and sinks under the rotating action of the high-speed rotors. On the one hand, the air flow is thrown outwards by the high-speed rotors. On the other hand, the flow is squeezed by the rotors and forms high-speed flowing regions adjacent to the rotors. The airflow here has high velocity with many axial components. Moreover, the airflow in the down-wash flow fields is concentrated under the rotors and the airflow velocity below the UAV abdomen is significantly slower. After reaching the crop canopy, the airflow spreads around and its velocity falls.

As shown in Figure 8, the flow fields in horizontal planes with different elevations below the rotors show basically consistent forms and movement behavior. Because the high-speed rotors affect the down-wash flow fields, a trail is formed in the circumferential direction whose speed gradually decreases. The velocity fields are symmetrical around the central axis and the greater the distance to the central axis, the smaller are the values and gradients of the velocity. The airflow velocities in horizontal surfaces below the rotors gradually drop with increasing vertical distance from the rotors. The maximum speed is 1.20 m/s in the plane 0.4 m from the rotors. This drops to 0.87 m/s in that

1.2 m from the rotors. Furthermore, the area of action of the airflow gradually increases and the maximum width of the flow field is 0.60 m in the horizontal plane 1.2 m from the rotors. Considering the constraints placed on the balance and stability of the UAV, the sensor support should be installed passing through the geometrical center of the UAV abdomen and the solar sensor and two-band sensor should be installed at the two ends of the support. As the field angle of the multispectral crop-growth sensor is 27°, when the operating altitude of the sensor is 1.2 m, the field radius is 0.29 m. To avoid the down-wash flow fields disturbing the crop canopy (considering the maximum width of the down-wash flow fields, the size of the UAV body, and field radius of the sensor), the designed length of the sensor support is set to 1.5 m, as shown in Figure 9.

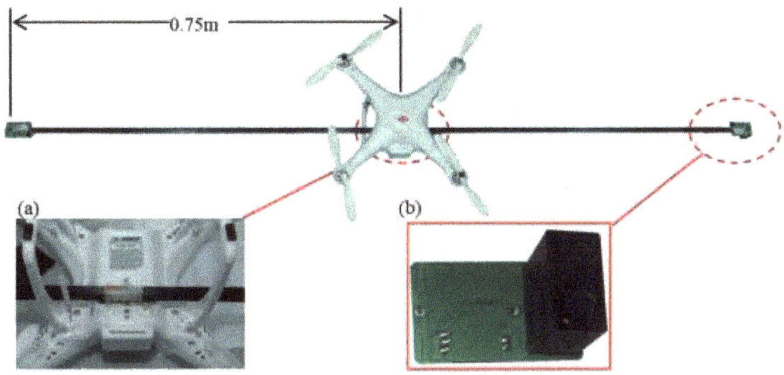

Figure 9. The UAV-borne crop-growth sensor. (**a**) Installation of sensor support; (**b**) Two-band sensor.

2.3.3. Sensor Signal Processing Circuit

The signal processing circuit carries out photoelectric conversion, amplification, and filtering of the optical information output by the solar sensor and two-band sensor. Then, the characteristic spectral information must be extracted and wirelessly transmitted to the ground-based controller. The circuit therefore includes a photoelectric conversion circuit, an amplifier circuit, a filter and pulse-shaping circuit, and a wireless communication circuit. The radiation of the crop canopy are collected by the two-band sensor and transformed from photonic energy to electrical energy using a photodiode. However, after photoelectric conversion, the electrical signal is very weak. To ensure that the system has high stability and is not likely to be self-excited when the conditioning circuit has a high gain, we designed a T-type amplifier circuit with integral resistance to amplify and filter the electrical signals in this study. The circuit's principles are displayed in Figure 10.

2.4. Ground-Based Data Processor

The ground-based data processor is mainly used to collect and process the signals output by the solar sensor and two-band sensor and to control the two sensors by configuring the wireless communication modules. The processor also calculates certain vegetation indices (RVI and NDVI) and obtains the major growth indices (including LNA, LAI, and LDW) by coupling the crop-growth parameters with the spectral monitoring model. Furthermore, by controlling press keys, the results are displayed on a liquid-crystal display (LCD).

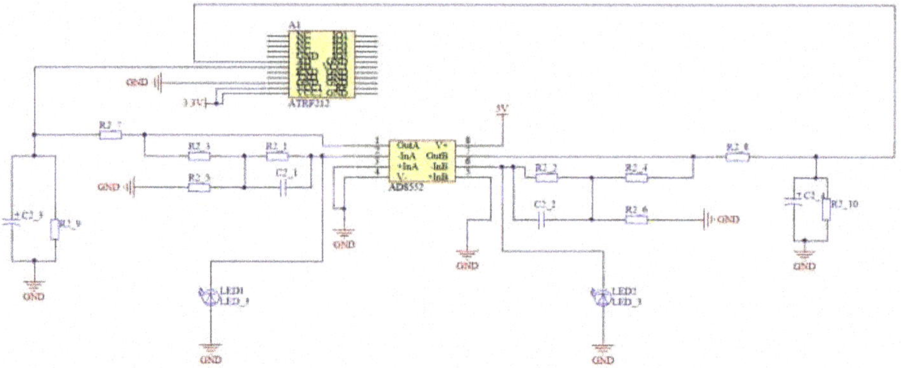

Figure 10. Principles used in the sensor signal processing circuit.

2.4.1. Hardware System

The hardware mainly consists of a controller module, a signal collection module for the solar sensor, a wireless communication module, a key detection module, an LCD display module, and a system power module. An Atmega328P-AU single-chip microcomputer (Atmel, San Jose, CA, USA) was used as the processing core. The controller module receives and processes the data from the solar sensor through a driving analog I/O port. In addition, it drives the digital I/O port to control the key detection module and the LCD display module. Furthermore, by driving the TTL serial ports, the XCBeep wireless communication module can be controlled to receive and send the data collected by the two-band sensor. The communication states and fault test results can be displayed using a light-emitting diode (LED) and configured by the I/O data port of the single-chip microcomputer. The overall connection structure of the hardware system is illustrated in Figure 11.

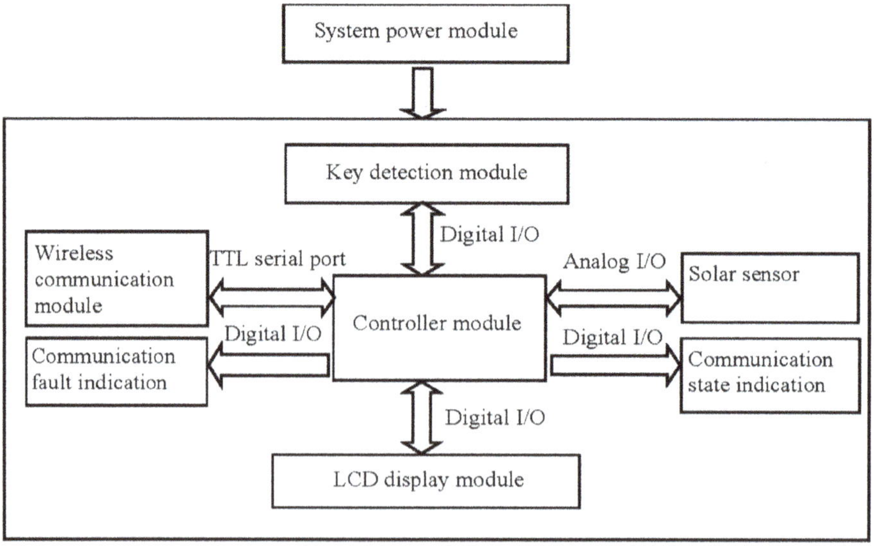

Figure 11. The overall connection structure of the hardware system.

2.4.2. Software System

The software system is comprised of three modules: a program initialization module, a resource control module for the I/O port, and an application program. The program initialization module was used for power-on self-testing and initialization of the Atmega328P-AU single-chip microcomputer and initialization of the XCBeep wireless communication module (Xiangce Intelligent Technology Co., Ltd., Nanjing, China) and the LCD. The resource control module of the I/O port was utilized for key detection and function switching, LED indication control, and the LCD display. The application program module was employed to collect radiation from the solar and the crop canopy, preprocess the spectral information, calculate the reflectance of the crop canopy, and couple the vegetation indices and crop-growth model. The software system as a whole adopts a modular design, which is convenient for debugging, transplanting, and future upgrades.

The ground-based controller has function keys in three modes: measurement, calculation, and reset. In the measurement mode, the Atmega328P-AU single-chip microcomputer receives information from the solar sensor through the analog I/O port. The control command "Receiving" is sent to the XCBeep wireless communication module through the serial port to communicate with the two-band sensor. Received spectral information is preprocessed and displayed in real-time. The information obtained by the two-band sensor is transmitted to the ground-based controller (whereupon the LED indicating the communication state flickers at a frequency of 1 kHz). After data transmission, the LED remains lit. When data packets are dropped and lost in transmission, the LED indicating communication faults flickers at a frequency of 1 kHz. At this time, the measurement key needs to be pressed to recollect data from the two-band sensor.

In the calculation mode, the Atmega328P-AU single-chip microcomputer calculates the major growth indices including the canopy's spectral reflection, the vegetation's RVI and NDVI values, which are then coupled with the crop-growth monitoring model to calculate the LNA and LAI of the crops. These indices are displayed on the LCD. In the reset mode, the resources of the controller and the external I/O ports are restored to their initial states.

3. Tests and Analysis of Results

3.1. Test Design

Systematic field tests were conducted in experimental wheat fields in Sihong County, Suqian City, Jiangsu Province, China from March to May 2016. The test varieties Ningmai 13 and Huaimai 20 were fertilized using five levels of nitrogen application, namely, N_0 (0 kg/hm^2), N_1 (90 kg/hm^2), N_2 (180 kg/hm^2), N_3 (270 kg/hm^2), and N_4 (360 kg/hm^2), each of which was repeated three times. Each separate plot covered an area of 42 m^2 (6 m $\times$ 7 m plots). Moreover, 135 kg/hm^2 of potash fertilizer (K_2O) was applied so the ratio of nitrogenous to potash fertilizers was 5:5. The basic fertilizers were applied before seeding, while the topdressing fertilizers were fertilized at the jointing stage. In addition, 105 kg/hm^2 of P_2O_5 base fertilizer was used for one time during soil preparation. The other cultivation and management measures undertaken were the same as those commonly employed in high-yield fields.

3.2. Test Methods

3.2.1. UAV-Borne Crop-Growth Sensor Measurements at Different Elevations

Before flying the UAV, static tests were carried out. This was done by holding the UAV-borne crop-growth sensor at different elevations to verify the detection performance of the sensor after making the improvement in system weight. The static tests made using the sensor in hand-held mode remove the potential effects of several factors including the shake of the UAV body and rotor wind fields. Tests were carried out at the tillering and jointing stages from 10:00 to 14:00 on sunny days. For each test, the hand-held UAV-borne crop-growth sensor and a commercial ASD spectrometer were simultaneously employed to determine the reflection spectra of the canopy of the wheat. For these measurements, the vertical distance between the two-band sensor and the wheat canopy was either 0.4 or 1.0 m. Three locations in each separate region were measured four times and the average values computed. Then, the NDVI and RVI values determined using the ASD spectrometer at 720 and 810 nm and those output from the UAV-borne crop-growth sensor were recorded.

Afterwards, the crop-growth sensor was fixed onto the UAV for the next set of measurements. The initial flight tests made using the UAV/sensor combination were intended to verify the immunity of the designed UAV-borne sensor to the effects of vibration encountered during the flight and disturbance created by the down-wash wind fields from rotors. At the jointing, booting, and heading stages, the reflection spectra of the canopy were dynamically tested at different elevations using the UAV-borne crop-growth sensor from 10:00 to 14:00 on sunny days in the absence of wind. In these tests, the flying height of the UAV was adjusted so that the vertical distance from the two-band sensor to the wheat canopy was 0.4, 0.7, 1.0, and 1.2 m. At each height, the UAV was made to hover by keeping its rotors rotating at the rated speed. In this way, the NDVI and RVI values as output by the UAV-borne crop-growth sensor were recorded at the different elevations used. The field tests are shown in Figure 12.

Figure 12. Field tests based on UAV-borne crop-growth monitoring system.

3.2.2. Performance Tests

Performance tests were conducted between 10:00 and 14:00 on sunny (windless) days at the tillering, jointing, booting, and heading stages of wheat growth. During the tests, the UAV-borne crop-growth monitoring system was used to measure reflection spectra of the wheat canopies. As a further check, the ASD spectrometer was simultaneously used to detect the reflection spectra of the canopies. The flight height of the UAV was adjusted so that the vertical distance between the two-band sensor and wheat canopy being measured was 1.0 m. Three points in each separate region were measured with four repetitions to permit more representative average values to be calculated. The NDVI and RVI values measured using the ASD spectrometer at 720 and 810 nm and the values output by the UAV-borne crop-growth sensor were recorded. At the same time that the spectral measurements were made, 20 single stems were selected from each region and separated according to their organs in the laboratory. A leaf area meter (model: LAI3000C) was used to measure the leaf area and thereby the LAI of the whole field region could be calculated. Afterwards, the samples were heated to 105 °C for 30 min (as green-killing treatment) and then dried to constant weight at 80 °C. Thus, the LDW could be determined. After smashing the samples, the Kjeldahl nitrogen method was used to determine their LNA values.

3.3. Data Analysis

The data from the tests were statistically analyzed by using appropriate software (Excel 2010). The correlation of the model was evaluated by calculating the root mean square errors (RMSEs) and determination coefficients. The stability of the method was assessed through the variances and variation coefficients derived. The necessary formulas required for the calculations are:

$$u = \frac{\sum_{i=1}^{n} x_i}{n} \tag{2}$$

$$s = \sqrt{\frac{\sum_{i=1}^{n} (x_i - u)^2}{n}} \tag{3}$$

$$CV = \frac{s}{u} \tag{4}$$

$$RMSE = \sqrt{\frac{\sum_{i=1}^{n} {d_i}^2}{n}} \tag{5}$$

In Equations (2)–(5), x_i, μ, s, CV, d_i, and $RMSE$ represent the ith measured value, mean value, standard deviation, deviation coefficient, difference between the ith measured and ith true value, and the root mean square error, respectively.

3.4. Results and Discussion

3.4.1. Elevation Test Results

Figure 13 shows the NDVI values measured when the hand-held crop-growth sensor is 0.4 and 1.0 m from canopy for wheat at its tillering and jointing stages. It can be seen that the variation exhibited by the NDVI curves is consistent at both of the elevations used. By calculating the deviation coefficients, the stability variance of the NDVI values, as measured by the sensor at the two different elevations, is found to be 0.03. The maximum deviation coefficient is 3.78%, so the difference is small. This is a good indication of the high stability of the weight-improved sensor over its intended range of operating altitudes.

The RVI and NDVI values measured by the ASD spectrometer and the UAV-borne crop-growth sensor at an elevation of 1 m (relative to the canopy) were fitted to a one-variable linear function using a least-squares fitting procedure (Figure 14).

The figure shows that a good linear relationship exists between the RVI and NDVI values output by the UAV-borne crop-growth sensor and the ASD spectrometer. The coefficients are determined to be 0.82 and 0.77 and the RMSEs are 0.17 and 0.05, respectively. Thus, the measurements made using the lightweight sensor can be seen to have a high level of precision. Figure 15 displays the NDVI values measured when the UAV hovered at 0.4, 0.7, 1.0, and 1.2 m over the canopy (by adjusting to the rated rotation) at the jointing, booting, and heading stages of the wheat. As can be seen from the figure, the changes observed in the NDVI values are consistent at the different measurement elevations employed. The deviation coefficients were then calculated. The variance of stability for the NDVI values measured using the sensor at different elevations is 0.0034 and the maximum deviation coefficient is 5.30%, so the differences are small. This suggests that the sensor installation position is reasonable, and that the effects on the sensor of the UAV vibrations and down-wash flow fields are small. Furthermore, the system has good dynamic stability over the range of operating altitudes of the sensor.

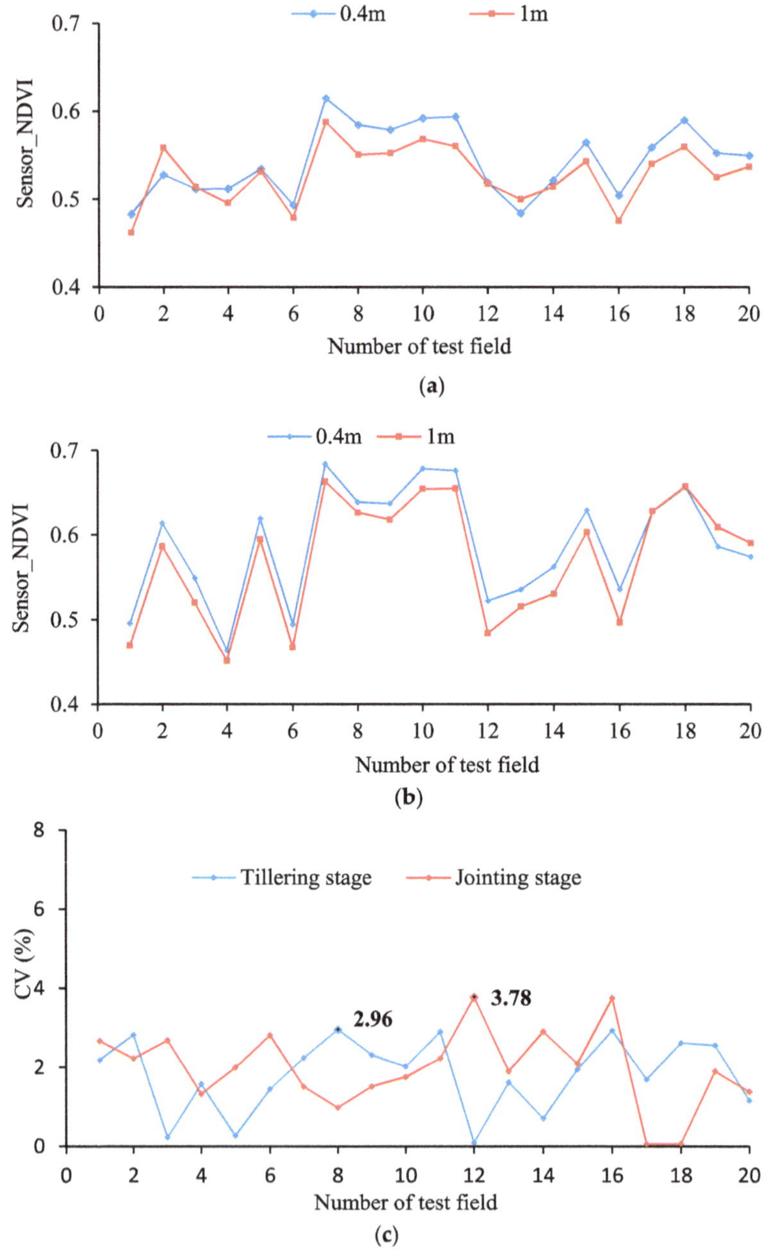

Figure 13. NDVI values measured using the hand-held sensor at different elevations. (**a**) Tillering stage; (**b**) Jointing stage; (**c**) Deviation coefficients of the NDVI values measured.

The RVI and NDVI values measured 1 m from the canopy by the ASD spectrometer and UAV-borne crop-growth sensor were also fitted to a one-variable linear polynomial using least-squares regression (Figure 16). The figure shows there is a good linear relationship between the RVI and NDVI values output by the UAV-borne crop-growth sensor and those from the ASD spectrometer. The coefficients are determined to be 0.74 and 0.75 and the RMSEs are 0.18 and 0.04, respectively. This shows that the sensor support designed according to the numerical simulation of the down-wash flow fields can effectively avoid disturbance from the wind fields. In addition, the UAV-borne crop-growth sensor can be used to make dynamic measurements with high precision.

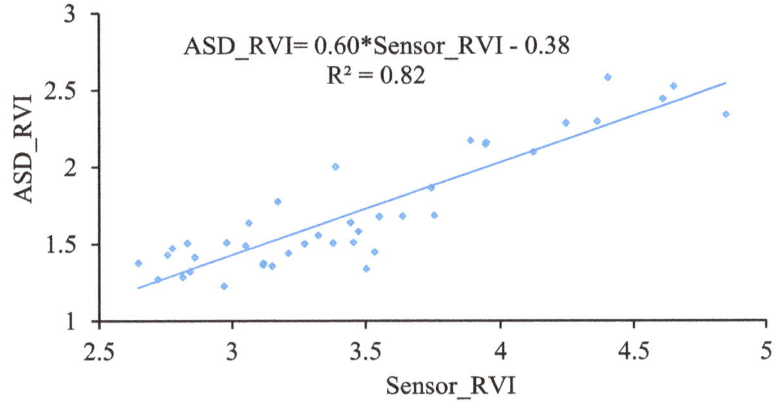

(a)

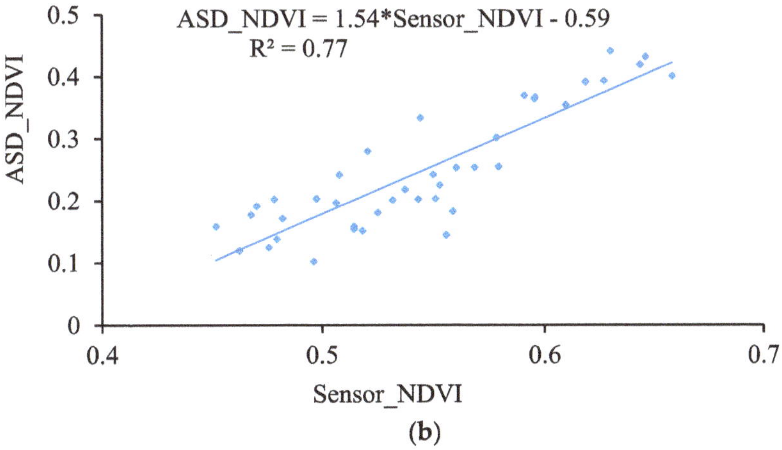

(b)

Figure 14. Fitting curves for the hand-held sensor and ASD data. (**a**) NDVI; (**b**) RVI.

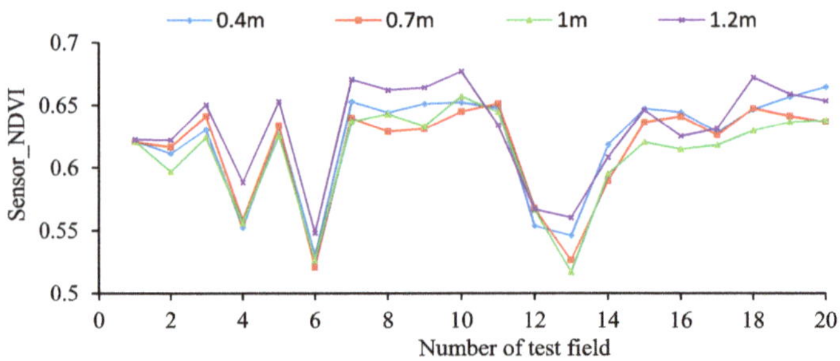

(**a**)

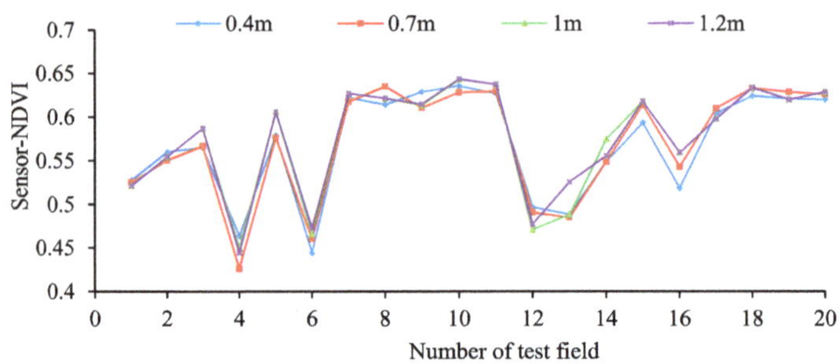

(**b**)

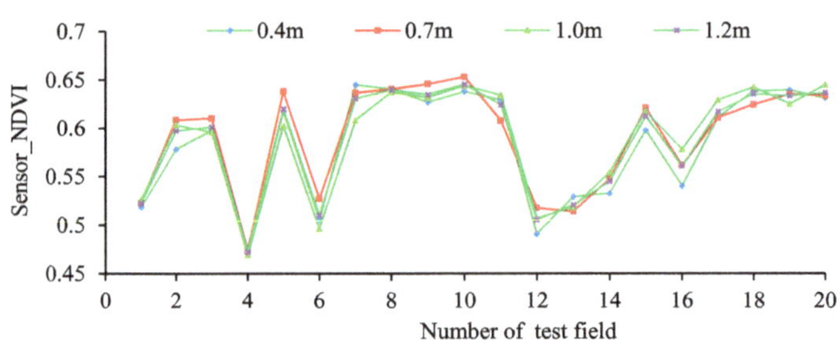

(**c**)

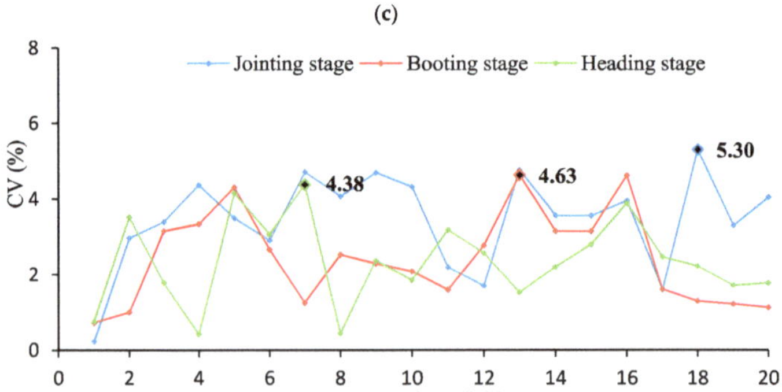

(**d**)

Figure 15. NDVI values measured using the sensor fixed onto the UAV for different elevations. (**a**) Jointing stage; (**b**) Booting stage; (**c**) Heading stage; (**d**) Deviation coefficients of the NDVI values measured.

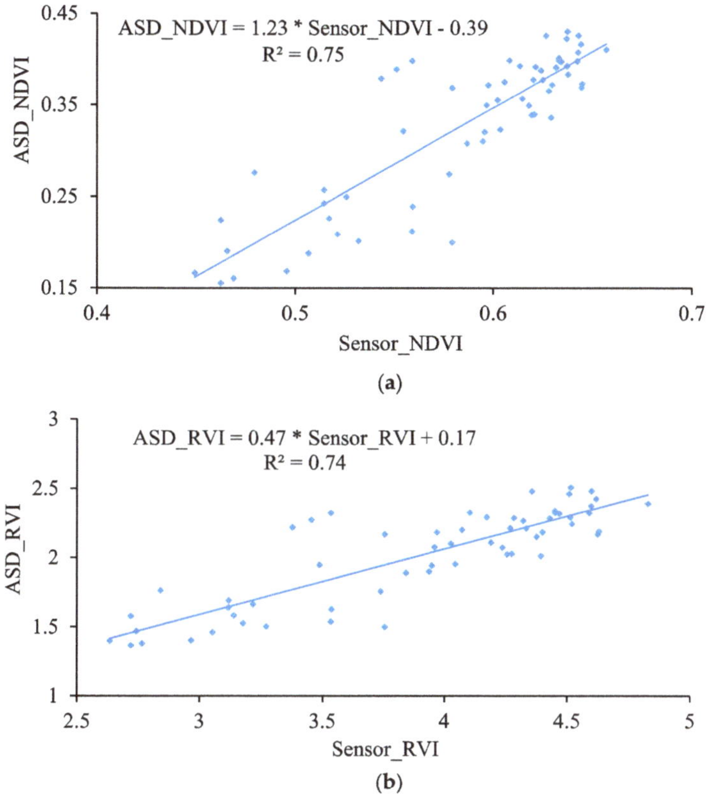

Figure 16. Fitting curves for the UAV-borne sensor and ASD data. (**a**) NDVI; (**b**) RVI.

3.4.2. Performance Test Results for the UAV-Borne Monitoring System

As shown in Figure 17, the UAV-borne crop-growth monitoring system can accurately reflect the changes in the wheat growth indices (the measured RVI and NDVI values show good linear relationships with LNA, LAI, and LDW). The determination coefficients R^2 of the RVI values with respect to LNA, LAI, and LDW are 0.63, 0.69, and 0.66, and the RMSEs are 1.42, 1.02, and 3.09 respectively. Similarly, the determination coefficients R^2 of the NDVI values with respect to LNA, LAI, and LDW are 0.60, 0.65, and 0.62 and the RMSEs are 1.44, 1.01 and 3.01, respectively. The fitting equations thus established were subsequently stored in the control chips of the ground-based data processor. The system thus calibrated is capable of quickly, non-destructively, and online quantitatively analyzing the growth information subsequently collected on the wheat.

The research developed a UAV-borne crop-growth monitoring system for on-line, and real-time, acquisition of continuous, high-throughput, information about crop growth. Much attention was paid to the design of the matching UAV-borne crop-growth sensor and the crop-growth monitoring system for UAVs. For the former, the key point in the design is to ensure that the working field of view of the downward-looking optical sensor is crop canopies without airflow disturbance. Through CFD simulations, spatial distributions were obtained for the UAV down-wash flow fields on the surface of the crop canopy. Influenced by the high-speed rotation of the rotors, the down-wash flow fields form a trail in the circumferential direction whose speed gradually decreases. When the UAV is 1.2 m from the crop canopy, the maximum velocity is 0.87 m/s on the surface of the crop canopy and the maximum width of the flow field is 0.60 m. Owing to the field angle of the multispectral crop-growth sensor being 27°, the field radius is about 0.29 m when the UAV hovers at 1.2 m above the canopy. In addition, considering the maximum width of the down-wash flow fields, the size of the UAV body, and field radius of the sensor, the sensor support was designed to be 1.5 m long, with which the multispectral crop-growth sensor was integrated with the UAV. It overcomes shortcomings of hand-held multispectral crop-growth sensors such as their small monitoring region, labour-intensity, and discontinuous monitoring: it also improves the test efficiency. As for the UAV-borne crop-growth monitoring system, it needs to be designed to be capable of timeous processing and on-line interpretation of the acquired data. To this end, wireless communication technology is used to transmit information obtained by the UAV-borne crop-growth sensor to the ground-based data processor in real-time. In addition, with the application of a single-chip microcomputer, the information obtained by the sensor and the crop-growth monitoring model is integrated, which overcomes the hysteresis induced by off-line interpretation of existing UAV-borne remote sensing data.

Meanwhile, this new UAV-borne crop-growth monitoring sensor can be operated at two wavelengths, which remains insufficient for the types of vegetation indices required. Therefore the authors plan to develop crop-growth monitoring sensors capable of working at a greater number of wavelengths in future studies, so as to establish more vegetation indices able to predict crop-growth indices and therefore improve prediction accuracy and stability.

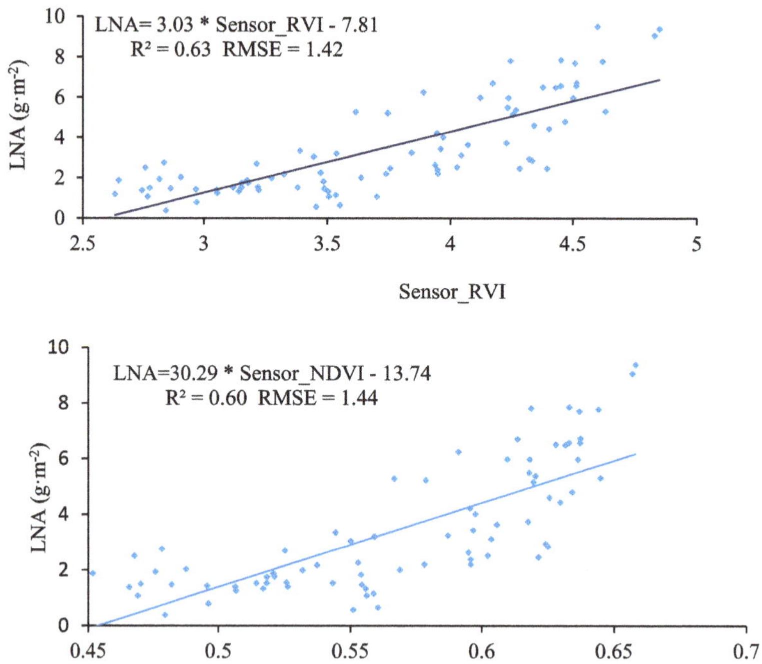

(a)

Figure 17. *Cont.*

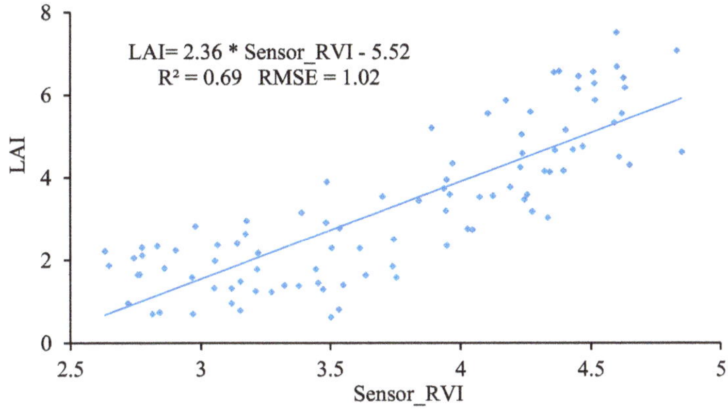

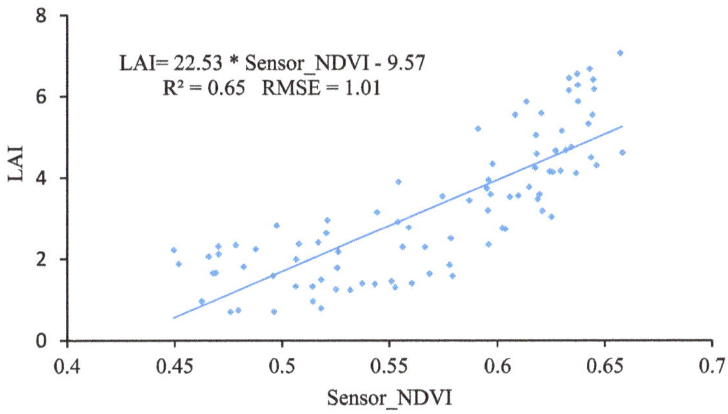

(**b**)

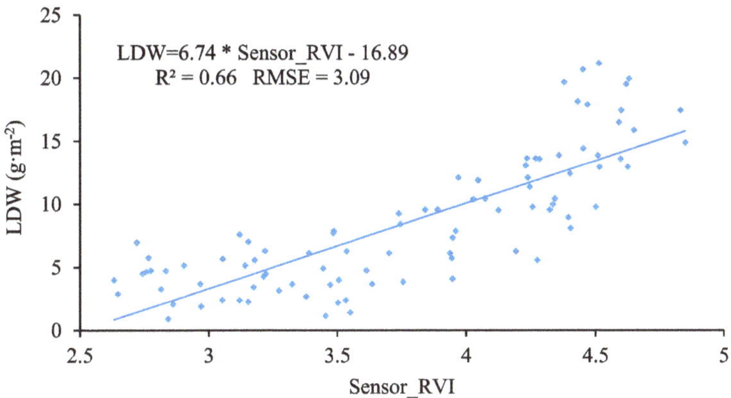

Figure 17. *Cont.*

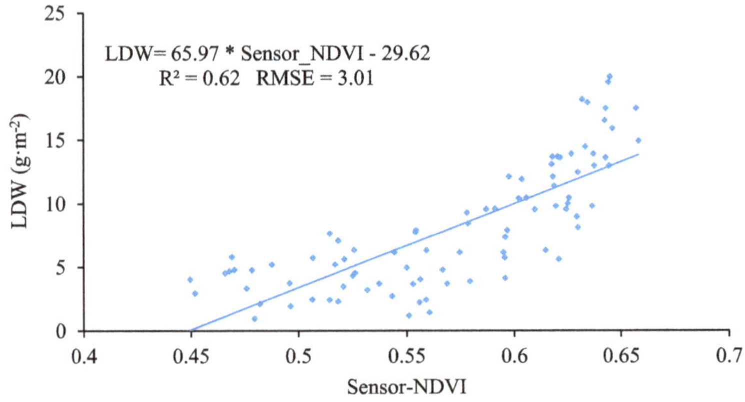

(c)

Figure 17. The spectral model for the UAV-borne crop-growth monitoring system. (**a**) LNA-RVI/NDVI fitting curve; (**b**) LAI–RVI/NDVI fitting curve; (**c**) LDW–RVI/NDVI fitting curve.

4. Conclusions

(1) The DJI Phantom quad-rotor UAV can be used as an operating platform to create a matched crop-growth monitoring system. This complete system combines the UAV platform, a UAV-borne crop-growth sensor, and a ground-based data processor. The system can continuously and conveniently obtain the NDVI and RVI values of the crop canopy online (as well as growth indices including LNA, LAI, and LDW) with high throughput and is not limited by the terrain.

(2) Numerical CFD simulations were conducted to investigate the spatial distribution of the down-wash flow fields from the DJI phantom quad-rotor UAV at the surface of the crop canopy. The results show that the airflow is mainly distributed underneath the rotors, and the speed of the airflow below the UAV body is obviously slower. On reaching the crop canopy, the airflow spreads around and its velocity falls. The airflow velocity in horizontal planes below the rotors gradually decreases as the vertical distance from the rotors increases. The maximum velocity is 1.2 m/s at 0.4 m from the rotors and 0.87 m/s at 1.2 m from the rotors. With increasing vertical distance from the rotors, the airflow area gradually increases. The maximum width of the flow field is 0.60 m in the plane 1.2 m from the rotors. On this basis, the length of the sensor support was chosen to be 1.5 m. The solar sensor and two-band sensors were fixed onto the two ends of the support, and this is installed on the UAV so that it passes through the geometrical center of the UAV's abdomen. This arrangement can effectively avoid the down-wash flow fields below the UAV significantly affecting measurement of the reflection spectra of the crop canopy.

(3) The improved, lightweight UAV-borne crop-growth sensor showed good stability and measurement precision over the range of operating altitudes required of the sensor. When measuring at elevations 0.4 and 1.0 m from the wheat canopy, the stability variance of the NDVI values output by the sensor was determined to be 0.03 and the maximum deviation coefficient was 3.78%. The RVI and NDVI values output by the sensor vary linearly with those obtained by an ASD spectrometer (determination coefficients of 0.82 and 0.77 and RMSEs of 0.17 and 0.05, respectively). Tests of the UAV-borne sensor and UAV show that the designed size and installation position of the sensor support are reasonable and that the effects of in-flight vibration and down-wash are small. Over the operating range of altitudes of the sensor, the monitoring system demonstrated high dynamic stability and measurement precision. When the UAV hovered at 0.4–1.2 m above the canopy (at its rated rotor speed), the stability variance of the NDVI values output by the sensor was determined to be 0.0034 and the maximum deviation coefficient was 5.30%. In addition, the RVI and NDVI values output by

the sensor are linearly related to those obtained by the ASD spectrometer (determination coefficients of 0.74 and 0.75, and RMSEs of 0.18 and 0.04, respectively).

(4) The UAV-borne crop-growth sensor performed well when it came to monitoring the growth indices of wheat. The determination coefficients (R^2) of the linear fits between the output RVI values and LNA, LAI, and LDW values were 0.63, 0.69, and 0.66, respectively, and the RMSEs were 1.42, 1.02 and 3.09, respectively. The equivalent figures for the output NDVI values are 0.60, 0.65, and 0.62 (for LNA, LAI, and LDW, respectively), and the RMSEs are 1.44, 1.01 and 3.01, respectively.

Acknowledgments: The authors thank all those who helped in the course of this research. This project was also supported by the National Natural Science Foundation of China (Grant No. 31371534), Primary Research & Development Plan of Jiangsu Province of China (Grant No. BE2016378), National Key Research and Development Program of China (Grant No. 2016YFD0300606), Jiangsu Agricultural Science and Technology Independent Innovation Fund Project (Grant No. CX(16)1006), and The Priority Academic Program Development of Jiangsu Higher Education Institutions (PAPD).

Author Contributions: Jun Ni, Yan Zhu and Weixing Cao designed research, Jun Ni and Lili Yao performed the research, Jun Ni, Jingchao Zhang and Xiuxiang Tai analyzed the data, Jun Ni and Yan Zhu wrote the paper. All of the authors have read and approved the final manuscript submitted to the Editor.

Conflicts of Interest: The authors declare no conflict of interest.

References

1. Lee, Y.; Yang, C.; Chang, K.; Shen, Y. A Simple Spectral Index Using Reflectance of 735 nm to Assess Nitrogen Status of Rice Canopy. *Agron. J.* **2008**, *100*, 205–212. [CrossRef]
2. Zhu, Y.; Yao, X.; Tian, Y.; Liu, X.; Cao, W. Analysis of Common Canopy Vegetation Indices for Indicating Leaf Nitrogen Accumulations in Wheat and Rice. *Int. J. Appl. Earth Obs. Geoinform.* **2008**, *10*, 1–10. [CrossRef]
3. Guo, J.; Zhao, C.-J.; Wang, X.; Chen, L. Research Advancement and Status on Crop Nitrogen Nutrition Diagnosis. *Soil Fertil. Sci. China* **2008**, *4*, 10–14.
4. Gnyp, M.L.; Miao, Y.; Yuan, F.; Ustin, S.L.; Yu, K.; Yao, Y.; Huang, S.; Bareth, G. Hyperspectral Canopy Sensing of Paddy Rice aboveground Biomass at Different Growth Stages. *Field Crops Res.* **2014**, *155*, 42–55. [CrossRef]
5. Li, F.; Mistele, B.; Hu, Y.; Chen, X.; Schmidhalter, U. Optimising Three-Band Spectral Indices to Assess Aerial N Concentration, N Uptake and Aboveground Biomass of Winter Wheat Remotely in China and Germany. *ISPRS J. Photogramm. Remote Sens.* **2014**, *92*, 112–123. [CrossRef]
6. Li, F.; Mistele, B.; Hu, Y.; Chen, X.; Schmidhalter, U. Reflectance Estimation of Canopy Nitrogen Content in Winter Wheat Using Optimised Hyperspectral Spectral Indices and Partial Least Squares Regression. *Eur. J. Agron.* **2014**, *52*, 198–209. [CrossRef]
7. Holland, K.H.; Schepers, J.S. Use of a Virtual-Reference Concept to Interpret Active Crop Canopy Sensor Data. *Precis. Agric.* **2013**, *14*, 71–85. [CrossRef]
8. Loubet, B.; Laville, P.; Lehuger, S.; Cellier, P. Carbon, Nitrogen and Greenhouse Gases Budgets over a Four Years Crop Rotation in Northern France. *Plant Soil* **2011**, *343*, 109–137. [CrossRef]
9. Erdle, K.; Mistele, B.; Schmidhalter, U. Comparison of Active and Passive Spectral Sensors in Discriminating Biomass Parameters and Nitrogen Status in Wheat Cultivars. *Field Crops Res.* **2011**, *124*, 74–84. [CrossRef]
10. Tow, P.; Cooper, I.; Partridge, I.; Birch, C. Using Conservation Agriculture and Precision Agriculture to Improve a Farming System. In *Rainfed Farming Systems*; Springer: Rotterdam, The Netherlands, 2011.
11. Lang, M.; Kuusk, A.M.; Ttus, M.; Miina, R.; Nilson, T. Canopy Gap Fraction Estimation from Digital Hemispherical Images Using Sky Radiance Models and a Linear Conversion Method. *Agric. For. Meteorol.* **2010**, *150*, 20–29. [CrossRef]
12. Zhu, Y.; Li, Y.; Qin, X.; Tian, Y.; Cao, W. Quantitative Relationship between Leaf Nitrogen Concentration and Canopy Reflectance Spectra in Rice and Wheat. *Acta Ecol. Sin.* **2006**, *26*, 3463–3469.
13. Zheng, H.; Cheng, T.; Yao, X.; Deng, X.; Tian, Y.; Cao, W.; Zhu, Y. Detection of rice phenology through time series analysis of ground-based spectral index data. *Field Crops Res.* **2016**, *198*, 131–139. [CrossRef]
14. Croft, H.; Chen, J.M.; Zhang, Y. The applicability of empirical vegetation indices for determining leaf chlorophyll content over different leaf and canopy structures. *Ecol. Complex.* **2013**, *17*, 119–130. [CrossRef]

15. Ali, A.M.; Thind, H.S.; Sharma, S.; Varinderpal-Singh. Prediction of Dry Direct-Seeded Rice Yields Using Chlorophyll Meter, Leaf Colour Chart and Greenseeker Optical Sensor in Northwestern India. *Field Crops Res.* **2014**, *16*, 11–15. [CrossRef]
16. Walsh, O.S.; Klatt, A.R.; Solie, J.B.; Godsey, C.B.; Raun, W.R. Use of Soil Moisture Data for Refined Greenseeker Sensor Based Nitrogen Recommendations in Winter Wheat. *Precis. Agric.* **2013**, *14*, 343–356. [CrossRef]
17. Lamb, D.W.; Trotter, M.G.; Schneider, D.A. Ultra Low-Level Airborne (ULLA) Sensing of Crop Canopy Reflectance: A Case Study Using a Cropcircle™ Sensor. *Comput. Electron. Agric.* **2009**, *69*, 86–91. [CrossRef]
18. Mayfield, A.H.; Trengove, S.P. Grain yield and protein responses in wheat using the n-sensor for variable rate n application. *Crop Pasture Sci.* **2009**, *60*, 818–823. [CrossRef]
19. Link, A.; Panitzki, M.; Reusch, S.; Robert, P.C. In Hydro n-sensor: Tractor-mounted remote sensing for variable nitrogen fertilization. In Proceedings of the International Conference on Precision Agriculture and Other Precision Resources Management, Minneapolis, MN, USA, 14–17 July 2003.
20. Reusch, S.; Jasper, J.; Link, A. Estimating Crop Biomass And Nitrogen Uptake Using Cropspectm, a Newly Developed Active Crop-Canopy Reflectance Sensor. In Proceedings of the International Conference on Precision Agriculture, Denver, CO, USA, 18–21 July 2010.
21. Thomason, W.E.; Phillips, S.B.; Davis, P.H.; Warren, J.G.; Alley, M.M.; Reiter, M.S. Variable nitrogen rate determination from plant spectral reflectance in soft red winter wheat. *Precis. Agric.* **2011**, *12*, 666–681. [CrossRef]
22. Yang, W.; Wang, X.; Ma, W.; Li, M.Z. Variable-rate fertilizing for winter wheat based on canopy spectral reflectance. *J. Jilin Univ.* **2007**, *37*, 1455–1459.
23. Freeman, P.K.; Freeland, R.S. Agricultural UAVs in the U.S.: Potential, policy, and hype. *Remote Sens. Appl.: Soc. Environ.* **2015**, *2*, 35–43. [CrossRef]
24. Zarco-Tejada, P.J.; González-Dugo, V.; Berni, J.A.J. Fluorescence, temperature and narrow-band indices acquired from a uav platform for water stress detection using a micro-hyperspectral imager and a thermal camera. *Remote Sens. Environ.* **2012**, *117*, 322–337. [CrossRef]
25. Bendig, J.; Yu, K.; Aasen, H.; Bolten, A.; Bennertz, S.; Broscheit, J.; Gnyp, M.L.; Bareth, G. Combining UAV-based plant height from crop surface models, visible, and near infrared vegetation indices for biomass monitoring in barley. *Int. J. Appl. Earth Obs. Geoinform.* **2015**, *39*, 79–87. [CrossRef]
26. Rasmussen, J.; Ntakos, G.; Nielsen, J.; Svensgaard, J.; Poulsen, R.N.; Christensen, S. Are vegetation indices derived from consumer-grade cameras mounted on uavs sufficiently reliable for assessing experimental plots? *Eur. J. Agron.* **2016**, *74*, 75–92. [CrossRef]
27. Caturegli, L.; Corniglia, M.; Gaetani, M.; Grossi, N.; Magni, S.; Migliazzi, M.; Angelini, L.; Mazzoncini, M.; Silvestri, N.; Fontanelli, M. Unmanned aerial vehicle to estimate nitrogen status of turfgrasses. *PLoS ONE* **2016**, *11*, e0158268. [CrossRef] [PubMed]
28. Lin, G. Soybean leaf area index retrieval with UAV (unmanned aerial vehicle) remote sensing imagery. *Chin. J. Eco-Agric.* **2015**, *7*, 868–876.
29. Ghazal, M.; Khalil, Y.A.; Hajjdiab, H. UAV-based remote sensing for vegetation cover estimation using NDVI imagery and level sets method. *IEEE Int. Symp. Signal Process. Inf. Technol.* **2015**, *12*, 332–337.
30. Tian, Z.; Fu, Y.; Liu, S.; Liu, F. Rapid crops classification based on UAV low-altitude remote sensing. *Trans. Chin. Soc. Agric. Eng.* **2013**, *7*, 109–116.
31. Ni, J.; Xia, Y.; Tian, Y.; Cao, W.; Yan, Z. Design and experiments of portable apparatus for plant growth monitoring and diagnosis. *Nongye Gongcheng Xuebao/Trans. Chin. Soc. Agric. Eng.* **2013**, *29*, 150–156.
32. Ni, J.; Dong, J.; Zhang, J.; Cao, W.; Yan, Z. The spectral calibration method for a crop nitrogen sensor. *Sens. Rev.* **2016**, *36*, 48–56. [CrossRef]
33. Ni, J.; Jiang, Q.; Xu, Z.; Cao, W.; Zhu, Y. The Optical System Calibration of the Crop Nitrogen Sensor. *Int. J. Control Autom.* **2015**, *8*, 263–274. [CrossRef]
34. Ni, J.; Wang, T.; Yao, X.; Cao, W.; Zhu, Y. Design and experiments of multi-spectral sensor for rice and wheat growth information. *Trans. Chin. Soc. Agric. Mach.* **2013**, *44*, 207–212.
35. Zou, X.; He, Q.; He, J. Current development and related technologies of UAV. *Aerodyn. Missile J.* **2006**, *10*, 9–14.
36. Jin, W.; Ge, H.; Du, H.; Xu, X. A review on unmanned aerial vehicle remote sensing and its application. *Remote Sens. Inf.* **2009**, *1*, 88–89.

37. Panagiotou, P.; Tsavlidis, I.; Yakinthos, K. Conceptual design of a hybrid solar male UAV. *Aerosp. Sci. Technol.* **2016**, *53*, 207–219. [CrossRef]
38. Panagiotou, P.; Kaparos, P.; Yakinthos, K. Winglet design and optimization for a male UAV using CFD. *Aerosp. Sci. Technol.* **2014**, *39*, 190–205. [CrossRef]
39. Hubvert, P.; Siegfried, W. Navier-Stokes analysis of helicopter rotor aerodynamics in hover and forward flight. *J. Air Craft* **2002**, *39*, 813–821.
40. Wang, F. *Computational Fluid Dynamics Analysis-Principle and Application of CFD Software*; Tsinghua University Press: Beijing, China, 2004; p. 125.

sensors

MDPI

Article

Unmanned Aerial Vehicle Systems for Remote Estimation of Flooded Areas Based on Complex Image Processing

Dan Popescu *, Loretta Ichim and Florin Stoican

Department of Control Engineering and Industrial Informatics, University Politehnica of Bucharest, Bucharest 060042, Romania; loretta.ichim@upb.ro (L.I.); florin.stoican@upb.ro (F.S.)

* Correspondence: dan.popescu@upb.ro or dan_popescu_2002@yahoo.com; Tel.: +40-76-621-8363

Academic Editors: Felipe Gonzalez Toro and Antonios Tsourdos

Received: 28 December 2016; Accepted: 20 February 2017; Published: 23 February 2017

Abstract: Floods are natural disasters which cause the most economic damage at the global level. Therefore, flood monitoring and damage estimation are very important for the population, authorities and insurance companies. The paper proposes an original solution, based on a hybrid network and complex image processing, to this problem. As first novelty, a multilevel system, with two components, terrestrial and aerial, was proposed and designed by the authors as support for image acquisition from a delimited region. The terrestrial component contains a Ground Control Station, as a coordinator at distance, which communicates via the internet with more Ground Data Terminals, as a fixed nodes network for data acquisition and communication. The aerial component contains mobile nodes—fixed wing type UAVs. In order to evaluate flood damage, two tasks must be accomplished by the network: area coverage and image processing. The second novelty of the paper consists of texture analysis in a deep neural network, taking into account new criteria for feature selection and patch classification. Color and spatial information extracted from chromatic co-occurrence matrix and mass fractal dimension were used as well. Finally, the experimental results in a real mission demonstrate the validity of the proposed methodologies and the performances of the algorithms.

Keywords: unmanned aerial vehicle; path planning; flood detection; feature selection; image processing; image segmentation; texture analysis

1. Introduction

In the repertory of natural disasters, floods often cause the greatest material damage [1]. For example, floods in 2013 constituted 31% of the economic losses resulting from natural disasters [2]. Therefore, the forecasting, prevention, detection, monitoring, and flood damage assessment are of paramount importance for public authorities and people. Because early warning is essential for limiting the loss of life and property damage, many works examine real time flood detection systems [1,3,4]. For example, the integration of several existing technologies was used in Taiwan to develop a coastal flooding warning system [3].

The problem that we are solving in this paper is the evaluation of small flooded areas in rural zones with the aid of a cheap solution based on processing of images taken by the unmanned aerial system designed by the authors. The result is necessary to evaluate the post-flood damage by the local authorities (to determine relief funds) and assurance companies (to determine payments). Even small flooded areas are investigated and classified. For this purpose two sub-problems must be solved. First is the optimal coverage of the area to be monitored, from the point of view of energy consumption (a limiting factor, especially if the UAV is electrically powered through a battery) and trajectory length. The second sub-problem is the detection and estimation in terms of flooded areas of the covered regions.

For the purpose of flood detection and monitoring, different sensors have been used: optical sensors, multi-spectral sensors and synthetic aperture radars (SARs). Satellite remote sensing imagery offers less spatial and temporal resolution than aircraft and UAVs, but a larger field of view. It was successfully used on large-scale geographic analysis on the flood overflow area. For example, images from Landsat Thematic Mapper/Enhanced Thematic Mapper Plus (TM/ETM+) sensors were used to monitor the floods near Lena River (Siberia) with a spatial resolution of 30 m and a temporal resolution of 2.6 days [5]. On the other hand satellite images in the visible and near infrared spectrum are highly dependent on cloud conditions whereas radar transmitters and receivers can be used independently of weather conditions [6]. Based on the surface water extent, measured with a microwave remote sensor (Radiometer for Earth Observation System AMSR-E and AMSR2), the authors in [7] implemented a method for detecting floods at large scale. In [8] the authors propose to combine aerial thermal data with high resolution RGB images in order to quickly and accurately identify the sub fluvial springs of a stream. Both cameras, thermal and action, are installed on a rotor platform which is able to support more weight, but has a smaller surveillance area.

Combining information from space, aerial and ground [6], an integrated system using different technologies (remote sensing, the Global Navigation Satellite System (GNSS), data transmission, and image processing) was implemented in China for monitoring and evaluating flood disasters. Because of the high cost, aircrafts use SAR only for serious and urgent flood cases. The spatial resolution is of 3 m at 9 km swath and the monitoring is in real time, independent of weather. Among the methods for flood detection, the interpretation of optical remote images is widely used and also gives the best results concerning price and accuracy.

In order to detect the flood by image analysis, three solutions usually appear in the literature: (a) use of images from satellites [9–11]; (b) use of images from fixed cameras on the ground [4,12,13]; and (c) use of images from aircrafts or UAVs [14].

The Advanced Land Observing Satellite Phased Array type L-band Synthetic Aperture Radar (ALOS PALSAR) satellite [11] provides multi-temporal data which maps large zones of flood via a classification into flood and non-flood areas. Based on this classification and on images taken at pre-flood and post-flood time instants, information about the flooding hazard is provided. In general the satellite applications for flood detection, like those presented in [10] from Spot-5 imagery, or in [15] by WorldView-2 satellite imagery, are based on high spatial resolution images and have the disadvantage of being high-cost solutions, hence less approachable for public use. In addition, these solutions have the disadvantage of being sensitive to weather patterns (clouds or other obscuring weather features will render them useless).

An alternative approach for monitoring flood disasters is the system of fixed cameras proposed in [12] which is based on the dynamic detection of floods via intrusions of objects in the video frames. These objects are separated by segmentation from the rest in the image.

To monitor and evaluate the flood disasters, concatenated images, created by photomosaic generation, can be useful. Thus, the gaps or duplications of flooded regions, in different analyzed images, are avoided. In this case, the UAV solution is a cheaper and more flexible one which can ensure superior image resolution even under adverse weather conditions. In this direction, the authors in [16] developed a solution for detection and evaluation of the dynamic evolution of the flood based on a collaborative team of UAVs. More recently a multicopter-based photogrammetry procedure was used to evaluate the effect of an earthquake on complex architectural landscapes [17]. Also, Feng et al. [18] used a UAV for urban flood monitoring and showed that such platforms can provide accurate flood maps. In their proposed method, the authors show how the acquired images are ortho-rectified and combined into a single image. Subsequently, the flood detection is realized through feature extraction from gray co-occurrence matrix and forest tree classifier methods. Boccardo et al. [19] compare the main advantages/disadvantages of fixed-wing UAV versus rotor platforms for area surveillance.

So, rotor platforms can be used only for very small areas or isolated buildings, while for small and medium areas fixed-wing UAVs are recommended. For large areas, UAV teams with pre-positioned stand-by can successfully perform the aerial surveillance of the disaster affected areas. Systems using UAVs are able to operate at lower altitude and to acquire multi-view, repetitive images with high resolutions [20]. These systems (fixed-wing type) are used to provide large image blocks to perform an image-based registration on multi-temporal datasets.

Control of a team of collaborative agents (UAVs in our case) is challenging, especially so under external perturbations, loss or delay of communication, etc. Therefore, the usual approach is to have a hierarchical control structure: the lower-level controller (the "autopilot" implemented on-board tracks a given reference) and the higher-level controller ("mission flight management") provides a reference trajectory [21].

Any mission has specific design requirements for the trajectory generation procedure [22,23]. Foremost in observation missions (surveillance, photogrammetry, target tracking, etc.) is to maintain a constant velocity or to allow variation within a narrow band (such that the photos taken cover uniformly the area under observation [24]). Whenever a team of UAVs is considered, additional issues appear (e.g., collision and avoidance constraints [25]). Not in the least, the trajectory has to be feasible (in the sense that it respects the UAV dynamics) Additional limitations on trajectory length, obstacle and collision avoidance are also encountered.

A promising framework is the flat representation of dynamical systems [26]. This approach expresses the states and inputs independently though a so-called flat output (which "hides" the underlying link between the states and inputs). Relatively recent work, has concentrated on providing flat characterizations which handle well numerical issues and have a manageable complexity [27,28]. In this sense, B-spline functions are an interesting choice: their geometrical properties lead to a good flat output parametrization and allow construct optimization problems which integrate costs and constraints in order to obtain the desired results [29]. Assuming that all low-level control loops are already designed such that a predefined trajectory is followed accurately and the payload is stabilized, we can reduce the path generation problem to an optimization problem where various constraints, parameters and costs are taken into account. To conclude, a flat representation which accounts for the low-level dynamics of the autopilot (approximated by first and second-order dynamics) and uses the properties of B-spline basis functions will provide a comprehensive and flexible framework [30]. In particular, it is possible to penalize trajectory length and energy costs, guarantee obstacle avoidance and pass through or within a pre-specified distance from a priori given way-points.

In order to detect and segment the flooded regions from the acquired images, texture and color analysis can be employed. Texture analysis techniques are a subject extensively studied in literature [31–34], but all suggested solutions are tailored to the specific context of the application considered. Moreover, there is a substantial interest in studying methods using the grey level co-occurrence matrix for texture characterization, but extremely little work is done when multi-spectral (color) co-occurrence is concerned [33,34]. All the above image acquisition strategies impose strict constraints on the photographs' capture during the UAV mission, i.e., photographs have to be captured: at a constant height (low/medium/high—the classification is relative, depending on context and application); such that there is a predefined overlap between neighboring photographs and there are no gaps in the area of interest (such that a photo-mosaiced image covering the entire area is computed). While there are many specialized software applications which can merge photographs with partial overlap to generate a continuous mapping and detect features of interest, there are still open issues in the generation of the flight path to be followed by a UAV [22]. This apparently simple problem has a number of intricacies: turn maneuvers of the UAV should not "cut" into the shape of the area under observation, maximal distance between consecutive path lines has to be respected and, not in the least, the UAV operational costs (energy, time of travel) should be minimized [23,24].

In this paper, we implemented new solutions concerning the system concept, path generation and image segmentation. As first novelty, we propose a multilevel system, with two components, terrestrial and aerial, as support for image acquisition and transmission from a specified region. The terrestrial component comprises a Ground Control Station (GCS, as coordinator or master node at distance), more Ground Data Terminals (GDTs, as a fixed nodes network for data acquisition and transmission), and a launcher. The aerial component contains mobile nodes (UAV—fixed wing type). The communication is established via internet (GDTs—GCS) or direct radio (in rest). This hybrid network has the advantage of extending operational area. The fixed-wing type UAV for image acquisition was developed by the authors in the Multisensory Robotic System for Aerial Monitoring of Critical Infrastructures (MUROS) project [35] funded by the Romanian National Research Program Space Technology and Advanced Research (STAR) from the Romanian Space Agency (ROSA) [36]. The proposed system is completely autonomous, apart from the take-off stage where a human operator is needed, and can be monitored and controlled at distance from the operational field. The area to be monitored is covered with the aid of a trajectory designed by a suitable optimization problem while the acquired images are analyzed in order to detect and assess the extent of floods. The second novelty refers to previous trajectory design implementations. So, the main contributions are: (a) the full dynamics (GCS + autopilot levels)—described in the flat representation and (b) the area under surveillance—partitioned between UAVs, such that the workload is balanced and collision with another UAV is impossible. The third novelty is a new solution for detection and quantitative evaluation of flooded small areas, based on the gliding box algorithm and local image processing. The advantages of this solution are the following: lower cost compared to a manned aircraft or a satellite solution, better resolution than a satellite solution, and the possibility of operating on cloudy conditions. The proposed method simultaneously uses pixel distribution and color information taking into account the chromatic co-occurrence matrix and mass fractal dimension on color components. The features used are not fixed as in [18] but rather they are being adapted to each application and environment condition. Results of the feature selection (especially associated with color channels) eliminate the temporal (colors of vegetation) and geographical influences (soil and vegetation colors, buildings and infrastructures).

The rest of the paper is organized as follows: in Section 2, first, the model of the UAV system based on hybrid wireless network is presented and second, the methodology and algorithms for image processing with the aim of flooded area detection, segmentation and estimation are described and implemented. The results obtained from images acquired with a fixed-wing UAV, designed by a team including the authors, are reported and analyzed in Section 3. For image acquisition, a path generated by the method introduced in Section 2 is used. Finally, the conclusions and discussions are reported in Section 4 and, respectively, Section 5.

2. Instruments and Methods

It is difficult and expensive to obtain precise data of the flood size within a certain small area from aerial photographs. As it was stated in Section 1, in this paper we propose a cheap and accurate solution to estimate the size of the dispersed small flooded areas. The solution is based on image segmentation obtained by a hybrid aerial—ground network integrated in internet. Three important sides are investigated: (a) the configuration of the network (which was partially implemented in the MUROS project [35] and will be finalized in SIMUL project [36]); (b) the trajectory control; and (c) the image processing for flooded area detection and estimation. The entire system is monitored and controlled remotely by GCS, via the Internet.

2.1. UAV System

The proposed system is configured as a hybrid network both with fixed nodes (terrestrial part) and mobile nodes (aerial part). The terrestrial part consists of the following components, which are considered at fixed locations during the mission (Figure 1): Ground Data Terminals (GDTs), Launchers (Ls), Ground Control Station (GCS) and Image Processing Unit (IPU). The aerial part contains mobile nodes (UAVs, fixed wing type) which fly over a specified flooded zone. GCS is located at distance from the operational field and the communication is made via GSM + Internet. Four wireless communication channels were used: GCS-GDT (GSM + internet), UAV-GDT (radio) and L-GDT (radio), and UAV-UAV (radio). GDT-GCS connection uses a modem GPRS/LTE as router via Ethernet interface. It is a Virtual Private Network. The block diagram of the system consists of several modules, wired to a common control bus. Each module contains a Central Processor Unit (CPU), a Power Supply Unit (PSU), and a Controller Area Network (CAN) adaptor. The wireless module is characterized by the following: (a) radio modem; (b) frequency: telemetry—[3.3 GHz–3.5 GHz], video—2.4 GHz; (c) Data rate: telemetry—230 kbps, video—analog PAL; (d) range: telemetry—20 km, video—15 km. The significance of the module abbreviations in Figure 1 and their functions are presented in Table 1. Figures 2 and 3 present the principal components of UAV system, used for flood detection: UAV MUROS, GCS, GDT with ID box, Payload with camera, and Launcher.

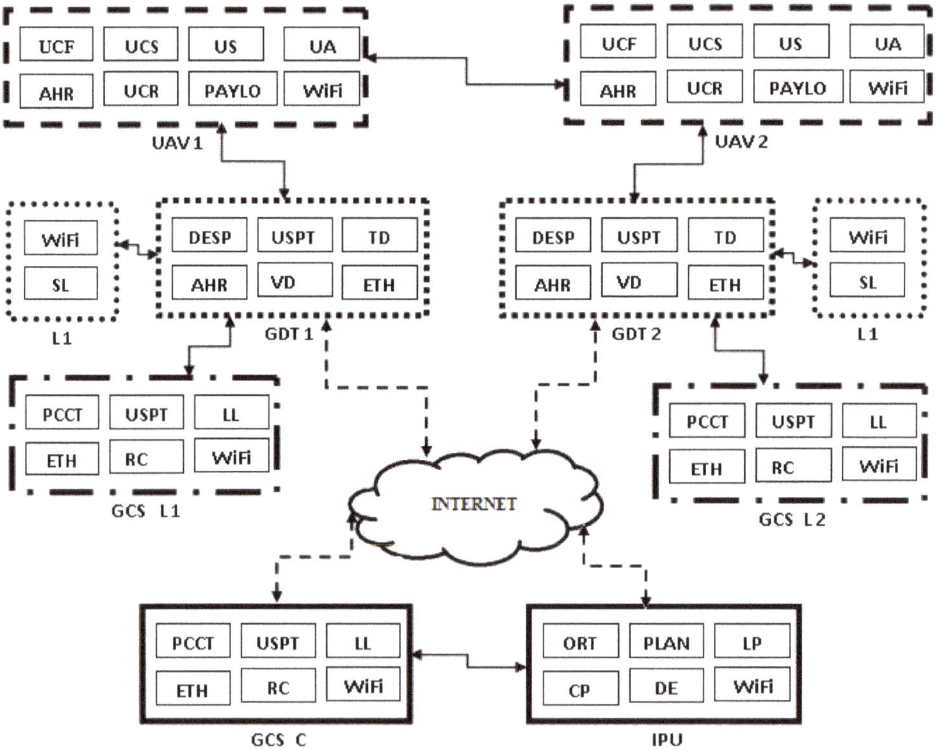

Figure 1. Block diagram of the system.

Figure 2. UAV MUROS on launcher.

Table 1. MUROS UAV—Abbreviations and functionality.

Abbreviation/Module Name	Function
FMCU Flight and Mission Control Unit	-Coordinates the flight mission; -Provides the platform's stability and quick response in case of disturbances that may deflect the drone from its pre-defined route or its removal from the flight envelope; -Allows for manual piloting by an operator on the ground; -Implements the automated low-level control loops which assure path tracking.
AHRS Attitude and Heading Reference System	-Provides information for an autonomous flight; -Contains the sensor subsystem composed of static and dynamic pressure sensors for speed measurement (ADXL352), accelerometer (ASDXRRX005PD2A5), magnetometer (HMC5983), altimeter (MPL3115A2) and gyroscope (ADXRS450); -Data provided by AHRS are used by FMCU.
SU Safety Unit	-Assures the permanent monitoring of the signals sent by other units and interprets the error signals received; -Taking into account the fault-tree and the reported error, the SU may decide the future functioning of the UAV. Thus, it can decide to continue the mission, to return to the launching point or the designated retrieval point, or, as a last step, to deploy the parachute.
PU Power Unit	-Assures the electrical power to the other components of the UAV, especially to the propulsion motor; -Contains power sources and a storage balance sensor used to equilibrate the energy consumed.
VD Video Datalink	-Sends video data from the camera (PS) to the ground (via the GDT, to the GCS). It contains a modem RF (TXR) and a power amplifier RFA.
TD Telemetric Datalink	-Assures a duplex communication for both transmission and reception of telemetry data. It has a structure similar to the VD.
Payload Working load (payload)	-Has a dedicated CPU for device retracted; -Provides high resolution imagery or video HD; -Based on a gyro-stabilized mechanism.
GDT Ground Data Terminal	-Antenna based tracking system; -The operational range is extended by using multiple ground data terminals; -Radio and Internet connections.
GCS C Ground Control Station Coordinator	-Is the main component of the system; -Has a friendly user interface for operational purposes; -Internet connection with GDTs and IPU.
GCS L Local Ground Control Station	-Optional -Transfer the control to operational field for each UAV.
CSU Control for Servomotor Unit	-Ensures the control of the electric actuators; -Provides a feedback on their state.
CRU Control Radio Unit	-Ensures the radio data transmission to and from GDT: telemetry, video/images and control.
DESP Data Exchange & Signal Processing	-Data exchange between GCS and UAV via GDT; -Encoding/ decoding of video data; -Interface with Ethernet IP (ETH).
SPTU Servo Pan Tilt Unit	-Transmission of control to the payload servomotors.
PFCT PC for Flight Control and Telemetry	-Is the main module of GCS and is based on a CPU.
ETH Switch Ethernet	-Ensures the data transmission at distance.

Table 1. *Cont.*

Abbreviation/Module Name	Function
RC Radio Control	-Ensures the control transmission to the GDT.
LL Launcher Link	-Ensures the interface of GCS with the launcher; -Transmits the launch command.
SL Safety Launcher Module	-Assures the start of UAV propulsion, if the speed launch is correct.
IPU Imge Processing Unit	-Processes the images for flood detection -Estimate the size of flooded areas.
ORT Ortho-rectified module	-Creates the ortho-rectified images.
PLAN Ortho-photoplan module	-Creates the ortho-photoplan.
LP Learning module	-Establishes the patches for feature selection; -Establishes the class representatives and features for patch signatures.
CP Clssification module	-Divides the image in patches; -Classifies the patches as flood and non flood.
DE Flood detection and estimation module	-Creates the segmented images -Estimates the flooded area (in percent).
WiFi Module for WiFi communication	-Assures WiFi communication.

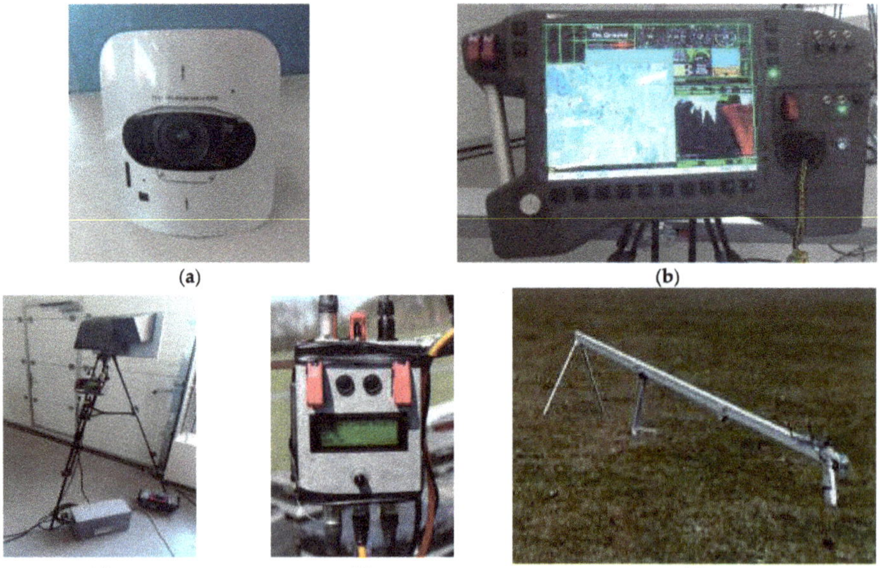

(a) (b) (c) (d) (e)

Figure 3. System components: (**a**) Payload photo; (**b**) GCS; (**c**) GDT; (**d**) ID box; (**e**) Launcher.

2.2. Trajectory Control

For image acquisition, the UAVs must follow specific trajectories such as simultaneously cover the monitored area (Figure 4). As stated earlier, we propose to use flat output characterizations to describe the dynamics of the UAVs and further use B-spline parameterizations of the flat output in order to enforce various constraints and penalize some desired cost in the resulting constrained optimization problem.

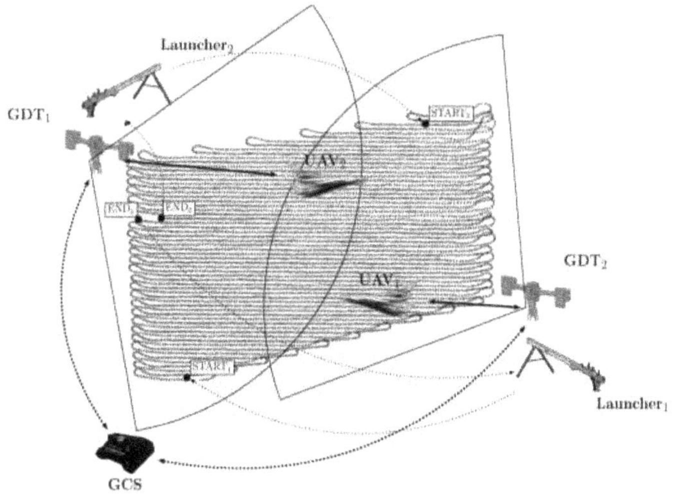

Figure 4. Model for trajectory generation in two-UAV applications.

Let us consider the nonlinear dynamics in standard notations [37]:

$$x'(t) = f(x(t), u(t)) \tag{1}$$

where $x(t) \in R^n$ is the state vector and $u(t) \in R^m$ is the input vector. The system (1) is called differentially flat if there exists $z(t) \in R^m$ such that the states and inputs can be expressed in terms of $z(t)$ and its higher-order derivatives:

$$\begin{aligned} x(t) &= \Theta(z(t), z'(t), ..., z^{(q)}(t)) \\ u(t) &= \Phi(z(t), z'(t), ..., z^{(q+1)}(t)) \end{aligned} \tag{2}$$

where $z(t) = \Upsilon(x(t), u'(t), ..., u^{(q)}(t))$.

Further, let us consider the simplified UAV dynamics with north, east, down directions (p_n, p_e and h) and yaw angle ψ as states:

$$\begin{aligned} p'_n &= V_a \cos\psi \cos\gamma, \quad p'_e = V_a \sin\psi \cos\gamma \\ h' &= V_a \sin\gamma, \quad \psi' = \tfrac{g}{V_a} \tan\varphi \end{aligned} \tag{3}$$

The autopilot is assumed to control directly the fligth-path angle γ, airspeed V_a and roll angle φ through input elements γ^c, $V_a{}^c$ and φ^c, respectively:

$$\gamma' = b_\gamma(\gamma^c - \gamma), \quad V_a' = b_{V_a}(V_a{}^c - V_a), \quad \varphi' = b_\varphi(\varphi^c - \varphi) \tag{4}$$

with parameters b_γ, b_{V_a} and b_φ accordingly chosen. Note that the closed-loop dynamics of the autopilot are simplified to first-order dynamics (a reasonable assumption in many circumstances).

Using the flat output $z = [z_1\, z_2\, z_3]^T = [p_n\, p_e\, h]^T$ we may express the dynamics in their flat representation as follows:

$$\begin{aligned} \psi &= \arctan\frac{z'_2}{z'_1}, & V_a &= \sqrt{z'^2_1 + z'^2_2 + z'^2_3} \\ \gamma &= \arctan\frac{z'_3}{\sqrt{z'^2_1+z'^2_2}}, & \varphi &= \arctan\left(\frac{1}{g}\cdot\frac{z''_2 z'_1 - z''_1 z'_2}{\sqrt{z'^2_1+z'^2_2+z'^2_3}}\right) \end{aligned} \tag{5}$$

together with the auxiliary elements

$$\begin{aligned} V_a^c &= \sqrt{z'^2_1 + z'^2_2 + z'^2_3} + \frac{1}{b_{Va}}\cdot\frac{z'_1 z''_1 + z'_2 z''_2 + z'_3 z''_3}{\sqrt{z'^2_1+z'^2_2+z'^2_3}}, \\ \gamma^c &= \arctan\frac{z'_3}{\sqrt{z'^2_1+z'^2_2}} + \frac{1}{b_\gamma}\cdot\frac{z''_3\left(z'^2_1+z'^2_2\right) - z'_3\left(z'_1 z''_1 + z'_2 z''_2\right)}{\left(z'^2_1+z'^2_2+z'^2_3\right)\sqrt{z'^2_1+z'^2_2}}, \\ \varphi^c &= \arctan\left(\frac{1}{g}\cdot\frac{z''_2 z'_1 - z''_1 z'_2}{\sqrt{z'^2_1+z'^2_2+z'^2_3}}\right) + \frac{1}{b_\varphi}\cdot\frac{1}{1+\left(\frac{1}{g}\cdot\frac{z''_2 z'_1 - z''_1 z'_2}{\sqrt{z'^2_1+z'^2_2+z'^2_3}}\right)^2}\cdot\left(\frac{1}{g}\cdot\frac{z''_2 z'_1 - z''_1 z'_2}{\sqrt{z'^2_1+z'^2_2+z'^2_3}}\right)' \end{aligned} \tag{6}$$

The major difficulty lies in the fact the constraints and costs are expressed as functions of state and input which do not necessarely translate well in formulations involving the flat output z. The usual solution is to parametrize the flat output after some basis functions ($B_{d,i}(t)$):

$$z = \sum_i P_i B_{d,i}(t) = P B_d(t), \tag{7}$$

and to find the parameters P_i which are, in some sense, feasible and optimal. Here, the parameter d denotes the degree of the B-spline functions. That is, each B-spline function can be seen as a piecewise combination of polynomial terms of degree d. Due to the particularities of the construction, a B-spline function of order d is continuous at least up to its (d-1) derivative. B-splines, due to their properties [30], permit to express the constrained optimization problem in terms of their control points P_i (grouped here in column form in matrix P):

$$\begin{aligned} & P^* = \operatorname{argmin} \int_{t_0}^{t_N} ||\Xi(B_d(t), P)||dt \\ & \text{subject to } \Psi_1(B_d(t), P) = 0,\ \Psi_2(B_d(t), P) < 0 \end{aligned} \tag{8}$$

where mappings $\Xi(B_d(t), P)$, $\Psi_1(B_d(t), P)$, $\Psi_2(B_d(t), P)$ are short-hand notations which denote the cost, equality and inequality constraints, respectively. The cost can impose any penalization we deem necessary (length of the trajectory, input variation or magnitude, etc.) and constraints cover way-point validation, input magnitude constraints, etc. In general, a problem like Equation (8) is nonlinear and hence difficult to solve (particular solutions exploit the geometrical properties of the B-spline functions and/or heuristic methods).

Considering multiple UAVs further increases the difficulty of the problem. In particular, we need to decide how the way-points are partitioned between the UAVs. One, rather cumbersome, solution is to attach to each way-point a binary variable and force that at least one of the UAVs passes through it. In practice, this can be relaxed, without any loss of generality to a condition which assumes that each UAV covers a contiguous part of the surveilled region. Moreover, it makes sense to partition the regions into areas of equal length parallel with the direction of travel. Then, each UAV has to cover its own independent region with additional collision avoidance constraints which may become active around the edges (since the UAV make turns which get out from under their surveillance area). To cover this possibility we may consider collision avoidance at the autopilot level (proximity sensors) or, more robustly, at the GCS level by either introducing additional constraints in the trajectory design procedure or, preferably, by changing the start and end points for each of the agent (such that neighboring points are reached at different moments in time).

2.3. Image-Based Flood Detection System

In order to fulfill the mission of detection, segmentation and estimation of the flooded areas, successive images are taken with constant rate on the pre-determined trajectory, like in the above section. For flooded area estimation, a patch-based segmentation was used. So, each image is partitioned in small boxes (e.g., in our application, patches of dimension 50 × 50 pixels), using a partitioning algorithm of images [38]. Note that the patch dimension is chosen depending on the image resolution and the texture of the segmented RoI (in our case, the flood). From a cluster of such patches (boxes), manually selected, a group of efficient features for flood detection is established based on a performance indicator. The features are used to create two classes: flood class (*F*) and non-flood class (*NF*). The propose method for image processing and interpretation has two phases: the learning phase and the mission phase. Both the learning images and test images were captured by the same camera device. Because the characteristics of the flood images can differ for each application, the learning phase is necessary to establish the class representatives and the signature patch structure. In the mission phase, a trajectory covering the investigated area is established. The acquired images are concatenated and processed to create an orthophotoplan without overlapping and without creating gaps. To this end, an overlap of 60% between two adjacent images is necessary to create an orthophotoplan. Then, they are indexed with an ID number in chronological order and are partitioned in the same way as in the learning phase. Based on the features selected in the learning phase, a similarity criterion is used to assign each patch to the class *F* or *NF*. Finally (estimation step), on one hand each patch of *F* is marked with white and is returned to the initial image, and, on the other hand a binary matrix of patches (BMP) is created with logical 1, if the correspondent patch belongs to *F*, and 0, in rest. By counting the "1"s from BMP, taking into account the total number of patches, the relative flooded area is evaluated.

The image characteristics may change as a function of distance from the ground and camera inclination with respect to the vertical axis. To avoid such issues the UAV has to respect a few additional constraints: (a) the altitude remains constant (even through some ground areas, may have different heights, we take as reference the water level, which remains constant). Floods are approximately at the same distance from the UAV, hence, if the flight plan is accurately followed, the resolution remains approximately the same for a given reference altitude; (b) the payload camera has to be oriented such that the lens are perpendicular to the surface of the Earth.

For each UAV there is a channel in GCS for image acquisition and, at the end of the mission, the images from all the UAVs are stored and processed in IPU. The methodology for flood evaluation based on patch analysis consists in the following steps:

1. In IPU, ortho-rectified images are created and then they are combined into a single image without overlapping and without gaps (orthophotoplan).
2. From the orthophotoplan, adjacent cropped images of dimension 6000 × 4000 pixels are investigated for flood evaluation.
3. non-overlapping box decomposition of the tested image is made. So, a grid of boxes is created and its dimension will represent the resolution of flood segmentation. Thus, if the image dimension is $R \times M = 2^r \times 2^m$ and the box dimension is $2^u \times 2^v$, then the resolution of segmentation (BMP dimension) is $2^{r-u} \times 2^{m-v}$.
4. The flood segmentation is made by patch classification in two regions of interest (flood—*F* and non flood—*NF*) taking into account the patch signatures and class representatives, which contain information about color and texture. As we mentioned earlier, the process has two phases: the learning phase (for feature selection and parameter adjustment) and the mission phase (for flood

detection, segmentation and evaluation). Flood evaluation is made for each cropped image and, finally, the sum of partial results is calculated.

2.3.1. Learning Phase

Generally, the aerial images taken from UAVs have textural aspects. Moreover, the remote images for water (and particularly for flood) are characterized by high contrast in texture behavior between the flooded zones and the remaining soil. Therefore, the texture information and, in particular, texture features can be used for flood detection. The selection of effective features must group the patches with flood and differentiating them from the non-flood ones (it must increase the between-class separability and decrease the within-class variance). To this end, a set of significant texture features were analyzed in the learning phase, in order to select the most efficient ones for the classification process. The tested features are of different types: mean intensity (*Im*), contrast (*Con*), energy (*En*), entropy (*Ent*), homogeneity (*Hom*), correlation (*Cor*), variance (*Var*), mass fractal dimension (*Dm*), lacunarity (*L*) and histogram of Local Binary Pattern (*LBP*). They take into account the chromatic information as well (on R, G, B, H, S and V color components). The general formulas for the most used features in texture classification are given in Table 2, where: R is the number of rows of the image representation (matrix I), M is the columns of I and K represents the levels on color channels. C_d is the normalized co-occurrence matrix [38] calculated as an average of the co-occurrence matrices $C_{d,k}$ taken on eight directions, $k = 1, 2, \ldots, 8$ (for $\theta = 0°, 45°, 90°, 135°, 180°, 225°, 270°$ and $315°$, respectively) at distance d (in pixels). The notations: $H_0, H_1, \ldots, H_{s-1}$ represents the components of *LBP* histogram [39]. *Dm* (15) is calculated, based on Differential Box-Counting (DBC), for monochrome images, in [40]. A grid of boxes is created with the image divided in boxes with the factor r. For a box in position (u, v), the difference $n_r(u, v)$ between the maximum value $p(u, v)$ and minimum value $q(u, v)$ of the intensity are considered. Then, the sum of all the differences (17) is used to evaluate *Dm*. Similarly, the lacunarity $L(r)$ is calculated as in [38].

Table 2. Analyzed features.

Energy	$En_d = \sum_{i=1}^{K}\sum_{j=1}^{K} C_d(i,j)^2$	Contrast	$Con_d = \sum_{i=1}^{K}\sum_{j=1}^{K} (i-j)^2 C_d(i,j)$
Entropy	$Ent_d = -\sum_{i=1}^{K}\sum_{j-1}^{K} C_d(i,j) \cdot \log_2[C_d(i,j)]$	Correlation	$\sum_{i=1}^{K}\sum_{j=1}^{K} \frac{i \cdot j \cdot C_d(i,j) - \mu_x \mu_y}{\sigma_x \sigma_y}$
Homogeneity	$Hom_d = \sum_{i=1}^{K}\sum_{j=1}^{K} \frac{C_d(i,j)}{1+\lvert i-j \rvert}$	Mean intensity	$\mathrm{Im} = \frac{1}{M \times R}\sum_{i=1}^{R}\sum_{j=1}^{M} I(i,j)$
Variance	$\sum_{i=1}^{K}\sum_{j=1}^{K} (i-\mu)^2 \cdot C_d(i,j)$	LBP Histogram	$H = [H_0, H_1, \ldots, H_{n-1}]$
Mass fractal dimension	$Dm = \frac{\log(\sum_u \sum_v n_r(u,v))}{\log r}$	Lacunarity	$L(r) = \frac{\sum_n n^2 \cdot P(n,r)}{[\sum_n n \cdot P(n,r)]^2}, \quad n = \sum_u \sum_v n_r(u,v)$

To evaluate the characteristics derived from co-occurrence matrix, besides the classical gray level co-occurrence matrix [33], applied on each color channel, we used the mean color co-occurrence matrix (CCM), between pairs of two spectral components of an input image [41]. So, in H, S, V decomposition, the image I is seen as a three-dimensional array with R rows, M columns and 3 layers (spectral bands). Each array element can take L positive integer (discrete values representing the color component's intensity of each pixel). The image I can be mathematically defined as: $I \in N^{R \times M \times 3}$).

Let H and S two components of a color space H, S, V. So, the mean CCM is considered as a square matrix, having $L \times L$ elements in N. It has two parameters: the distance d (the co-occurrence is the same as in GLCM case), and the component-pair (H, S) between which it is calculated. Each element of the mean color co-occurrence matrix $CMM_d^{HS}(i,j)$ represents how many times a pixel of the H component, having an intensity level of i, is located near a pixel with intensity j in the spectral component S, at a d distance. Then, the elements of the mean CCM are [37]:

$$CCM_d^{HS}(i,j) = \frac{1}{8}\sum_{x=0}^{n-1}\sum_{y=0}^{m-1}\left\{ \begin{array}{cc} 1, \text{ if } H(x,y) = i & \text{and } S(x+d,y+d) = j \\ 0, & \text{otherwise} \end{array} \right\} \tag{9}$$

Obviously, the next symmetry can be easily demonstrated:

$$CCM_d^{HS} = \left[CCM_{d,k+4}^{SH}\right]^T, k = 1,2,3,4 \tag{10}$$

A simple example of calculating the mean CCM is given in Figure 5, where we consider two image components $H, S \in N^{3\times 4}$, having 4 levels of pixel intensity, and the mean CCM computed between these two components, along a distance d = 1:

$$H = \begin{bmatrix} 1 & 1 & 1 & 2 \\ 0 & 3 & 3 & 1 \\ 2 & 1 & 0 & 0 \end{bmatrix} \quad and \quad S = \begin{bmatrix} 3 & 3 & 2 & 2 \\ 0 & 0 & 3 & 2 \\ 1 & 0 & 3 & 0 \end{bmatrix} \quad \Rightarrow C_{11} = CMM_1^{H,S} = \frac{1}{8}\begin{bmatrix} 5 & 1 & 2 & 5 \\ 8 & 1 & 5 & 6 \\ 3 & 0 & 2 & 1 \\ 5 & 1 & 4 & 6 \end{bmatrix}$$

Figure 5. Example of calculating mean CMM.

The algorithm for calculating CMM is presented [41] and the pseudocode in Appendix A. In order to establish the features to be selected, a cluster of 20 patches containing only flood (*PF*) are considered to form the representatives of the class "flood" (*F*) and 20 patches containing non flood elements (*PNF*), e.g., buildings and vegetation, are considered for the class "non flood" (*NF*). Each candidate feature T_i to flood signature is investigated according to the following algorithm:

i. T_i is calculated for all the learning patches (*PF*) and the confidence interval $[m_i - 3\sigma_i, m_i + 3\sigma_i] = \Im_i$ is determined, where m_i and σ_i represents, respectively, the mean and the standard deviation of T_i.
ii. Similarly, T_i is calculated for all the learning patches from *PNF* and the resulting set of values is noted as NF_i.
iii. A confidence indicator for feature T_i, CI_i is created:

$$CI_i = \left\{ \begin{array}{l} 1, \quad if \quad \Im_i \cap NF_i = \varphi \\ 1 - \frac{\eta(\lambda_i)}{\eta(PNF)}, \quad if \quad \Im_i \cap NF_i = \lambda_i \end{array} \right. \tag{11}$$

where, $\eta(A)$ is the cardinal number of the set A.

iv. The features T_i with greatest CI_i are selected in decreasing order, until the fixed number of features imposed for flood signature is reached. For example, in Section 3, a signature T, with 6 elements is considered (12):

$$T = [T_1, T_2, T_3, T_4, T_5, T_6] \tag{12}$$

v. As a consequence of the signature T, a set $\Im$ of confidence intervals is created (13). $\Im$ will be the representative of the class F:

$$\Im = [\Im_1, \Im_2, \Im_3, \Im_4, \Im_5, \Im_6] \tag{13}$$

where:

$$\Im_i = [m_i - 3\sigma_i, m_i - 3\sigma_i] \tag{14}$$

vi. For each selected feature T_i a weight w_i is calculated as follows. Another set of 100 patches (50—flood and 50—non flood) is considered and the confusion matrix for the feature T_i is calculated based on a preliminary classification criterion: the patch $B \in F$ if $T_i \in \Im_i$.

The weight w_i is established as in Equation (15):

$$w_i = \frac{F,F + NF,NF}{F,F + F,NF + NF,NF + NF,F} \quad (15)$$

where *F,F* represents the number of patches manually selected as belonging to class *F* and classified to class *F* after feature T_i. Similarly, *F,NF* represents the number of patches manually selected as belonging to class lF and classified to class *NF* after feature T_i.

Observations:

a. Obviously, $CI_i = 1$ represents an ideal situation and are not encountered.
b. If $\lambda_i = \lambda_j$, then T_i and T_j are redundant and one can be eliminated.

2.3.2. Mission Phase

In the mission phase, the images from orthophotoplan are decomposed in patches with dimension of 50 × 50 pixels. Each patch (box) is indicated by a pair (row number, column number) in the squared grid of the image with an ID number. The mission phase has three steps: patch classification, image segmentation and flood estimation.

For classification of a box (*B*) of as flooded, a weighted vote *D* is considered (16), where *D*(*B*) is the sum of partial weighted vote for each selected feature (17):

$$D(B) = \sum_{i=1}^{s} D_i(B) \quad (16)$$

where:

$$D_i(B) = \begin{cases} w_i & if \quad T_i \in \Im_i \\ 0 & else \end{cases} \quad (17)$$

The patch *B* is considered as flood (18) if the weighted vote is greater than 0.8 from the sum of all weights (the maximum of *D*):

$$B \in F \quad if \quad D(B) \geq 0.8 \cdot (\sum_{i=1}^{s} w_i) \quad (18)$$

where 0.8 is an experimentally chosen threshold.

Inside of the analyzed image, a segmentation process is done with the aid of the detected flood patches. For visualization purposes, the flood boxes are marked with white. With the patches from an image, an associate matrix *BMP* is obtained. Each patch corresponds to an element in *BMP*; so, for an image dimension of 4000 × 6000 pixels and a patch of 50 × 50 pixels, then the *BMP* matrix dimension is dim *BMP* = 80 × 120. If the number of marked boxes is *n*, then the percentage of flood zone in the analyzed image is *PF* (22):

$$PF = \frac{n}{\text{dim}BMP} \times 100 \quad [\%] \quad (19)$$

2.3.3. Algorithm for Flood Detection

The proposed algorithm has two phases: the Learning Phase—Algorithm 1 and the Mission Phase—Algorithm 2.

Algorithm 1: Learning Phase

Inputs: Learning patches (40 patches for feature selection—set 1 and 100 patches for weight establishing—set 2), set of feature to be investigated;
Outputs: Selected features T_i, the weights for selected features w_i, and the intervals $\Im_i$, $i = 1, \ldots, 6$.

For each patch of the first set:
1. Image decomposition on color channels (*R*, *G*, *B*, *H*, *S*, *V*) of patches;
2. Reject noise with median local filter;
3. Calculate the features: *Im*, *Con*, *En*, *Hom*, *Ent*, *Var*, *Dm* and *L* on color channels;
4. Until end of set 1;
5. Calculate the Confidence Indicator CI_i for each feature based on Equation (11);
6. Feature selection: T_i, $i = 1, \ldots, 6$;
7. Determine the intervals for flood class representative $\Im_i$, Equations (13) and (14)
   For each *Ti*:
8. Calculate the confusion matrices CM_i from the set 2;
9. Calculate the weights w_i, $i = 1, \ldots, 6$; Equations (15) and (17)
10. Return $\{T_i, w_i\}$.

Algorithm 2: Classification Phase

Inputs: Images to be analyzed, Selected features T_i, the weights for selected features w_i, and the intervals $\Im_i$, $i = 1, \ldots, 6$;
Outputs: Segmented images and percent of flooded areas

For each image *I*:
1 Image decomposition in small non-overlapping patches (50 × 50 pixels);
   For each patch *B*
   2. Calculate the selected features Im_R, Con_{HH}, En_{HS}, Hom_{HH}, Dm_G and L_R;
   3. Calculate $D_i(B)$;
   4. Patch classification based on voting scheme (18);
   5. Until end of patches from image I_i;
   6. Create the matrix of patches for each feature;
   7. Noise rejection based on local median filter in matrices of patches;
8. Create the final matrix of patches based on voting scheme;
9. Create segmented image;
10. Calculate the percent of flooded area from image with Equation (19);
11. Until end of images to be analyzed;
12. Return the segmented images and percent of flooded area.

Algorithm 1 is executed only once, at the beginning of the mission, while Algorithm 2 runs continuously throughout the mission. Both are implemented in deep neural networks (DNN). The DNN for Algorithm 2 is presented in Figure 6 and contains, besides the input and output layers, other three layers.

Layer 1 is dedicated to simultaneously calculate the features of patches and create the corresponding binary matrices of patches. Layer 2 is dedicated to local filtering of matrices from Layer 1, in order to eliminate the noise from BMP. Layer 3 creates the final BMP by voting scheme. Finally, the Output layer provides the segmented image and the relative flood size.

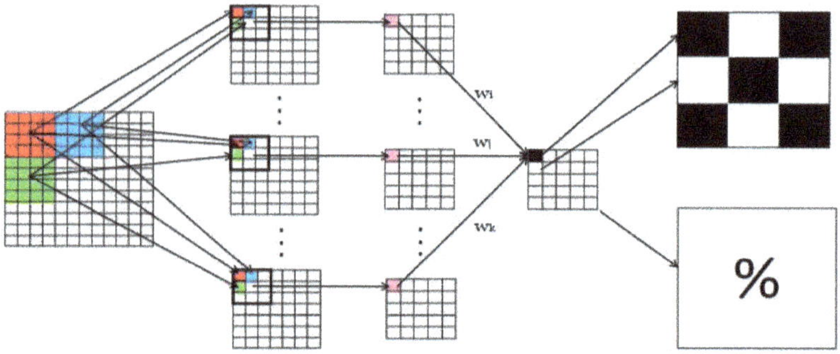

Figure 6. The neural network for the mission phase.

3. Experimental Results

For experimental results we used a UAV, designed, as coordinator, by University POLITEHNICA of Bucharest, MUROS project [35]. The main characteristics and technical specifications of UAV MUROS, as mobile node for image acquisition, are presented in Table 3.

Table 3. MUROS UAV—Characteristics and technical specifications.

Characteristics	Technical Specifications
Propulsion	Electric
Weight	15 kg
Wingspan	4 m
Endurance	120 min
Operating range	15 km in classical regime and 30 km in autopilot regime
Navigation support	GIS
Navigation	manual/automatic
Communication	antenna tracking system
Payload	retractable and gyro-stabilized
Mission	Planning software
Recovery system	Parachute
Maximum speed	120 km/h
Cruise speed	70 km/h
Maximum altitude	3000 m
Maximum camera weight	1 kg
Camera type	Sony Nex7, objective 50 mm, 24.3 megapixels, 10 fps
Parameters for flood detection	Flight speed of 70 km/h and flight level 300 m
Typical applications	Monitoring of critical infrastructures, reconnaissance missions over the areas affected by calamities (floods, earthquakes, fires, accidents, etc.), camera tracking, photography and cartography

To evaluate the algorithms presented in Section 2, an image dataset of a flooded area was gathered with MUROS. The photographs have been captured along a path generated as in Section 2, with distances between lines $d = 75$ m and height of flight de = 100 m (wind strength was considered to be negligible). The portion from the orthophotoplan of an application near Bucharest, during a flood, is presented in Figure 7. The images analyzed with the algorithm described in Section 2 are marked with the specified IDs.

In the learning phase, for patch signature determination T, the first set of 40 patches of dimension 50×50 pixels (20 patches for flood and 20 for non-flood), manually selected, was used (Figure 8). From this set, a cluster of 20 patches containing only flood (*PF*) are considered to form the prototypes for the class "flood" (*F*) and 20 patches, containing non flood elements (*PNF*), e.g., buildings and vegetation, are considered for the class "non-flood" (*NF*).

The results obtained in the learning phase (Table 4) show that the selected features (with *CI* criterion) are: *ImR*, *ConHH*, *HomHH*, *EnHS*, *DmG* and *LR*, where R, G, H, and S are the components of the color spaces. Thus, features on different types (first order statistics, second order statistics and fractal), on different channel color are selected. If *CI* falls below 0.80, then the accuracy can also decrease. It must be mentioned that the list of selected features can be changed in the learning phase, upon the requirements of the application. The fractal dimension was calculated by means of FracLac [42] plug-in of ImageJ and the features extracted from co-occurrence matrix were computed using MATLAB software. In Table 4, the values marked with * are those that are not within the corresponding confidence intervals.

Next step is the calculation of the confusion matrices for the selected features (Table 5). To this end, we used the second set (100 patches) for the learning phase, which contains 50 patches marked as flood (actually) and 50 patches marked as non-flood. From the confusion matrices we calculate the weights w_i which will be used further for patch classification.

Figure 7. Image created from acquired images (with yellow ID) without overlapping or gaps. The image was generated with Agisoft Photoscan Professional Edition (www.agisoft.com).

So, the signature of the patch is:

$$T = [T_1, T_2, T_3, T_4, T_5, T_6] = [\text{Im}R, ConHH, HomHH, EnHS, DmG, LR]$$

and the associate weights are:

$$[w_1, w_2, w_3, w_4, w_5, w_6] = [0.95,\ 1,\ 1,\ 1,\ 0.90,\ 0.95]$$

The representative of the class F is:

$$[\Im_1, \Im_2, \Im_3, \Im_4, \Im_5, \Im_6] = [[0.418; 0.535], [0.994; 1.002], [-0.004; 0.011], [0.896; 0.951], [2.605; 2.709], [0.344; 0.502]]$$

In order to analyze the performances of the algorithm for flood detection, a set of 50 images with flood was investigated (see orthophotoplan from Figure 7). Random patches of flood and non flood types (Figure 9) are classified based on the voting scheme and the results are presented in Table 6. Here, $D(B)$ is calculated as in (16) and compared with maximum value of $0.8 \cdot (\sum_{i=1}^{s} w_i) = 5.59$ as in (17). For example, patches B6_F and B10_F with flood are wrongly classified as non flood. For the mission phase, an example of 6 images used for the algorithm test is presented in Figure 10 and the result of the segmentation, in Figure 11. Figure 12 overlaps the RGB images with masks generated by segmented images.

The random errors of the classification process are characterized by sensitivity, specificity, and accuracy [10,43] which are calculated in Table 7, where: *TP* is the number of true positive cases, *TN* is the number of true negative cases, *FP* is the number of false positive cases, and *FN* is the number of false negative cases. In [12] an accuracy of 87% is obtained using RGB information and six texture features (fixed) extracted from gray level co-occurrence matrix. Our method uses selected features (selected by a performance criterion at the beginning of the segmentation operation) on color channels (chromatic co-occurrence matrix and fractal type) and the accuracy was of 98.1%.

Figure 8. Patches for establish the flood signature (Pi_F as patch with flood and Pj_NF as non flood patch).

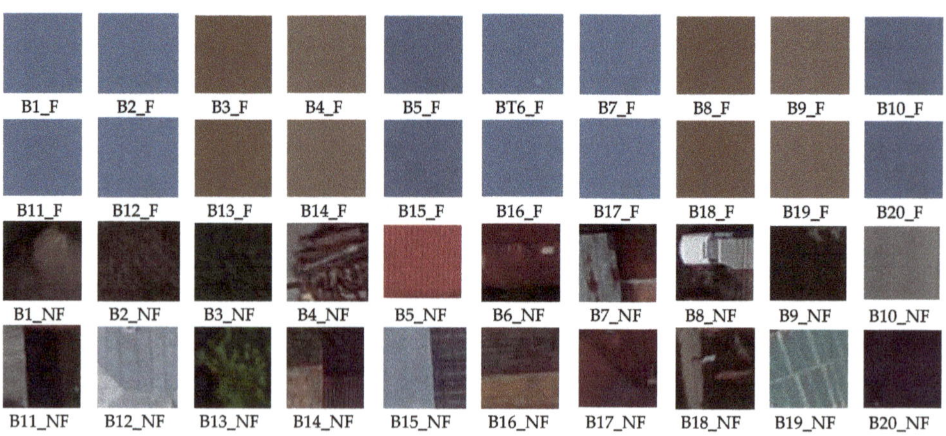

Figure 9. Patches for establish the weight signature (Bi_F as patch with flood and Bj_NF as non flood patch).

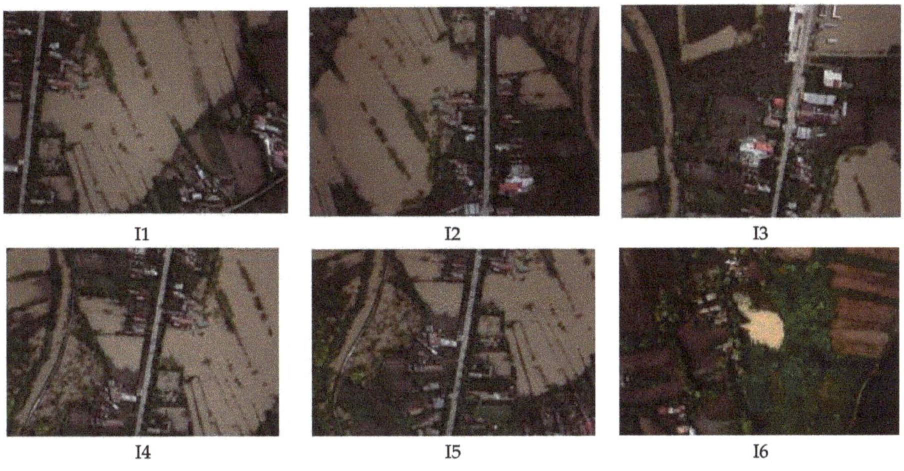

Figure 10. Images acquired by UAV MUROS to be evaluate for flood detection.

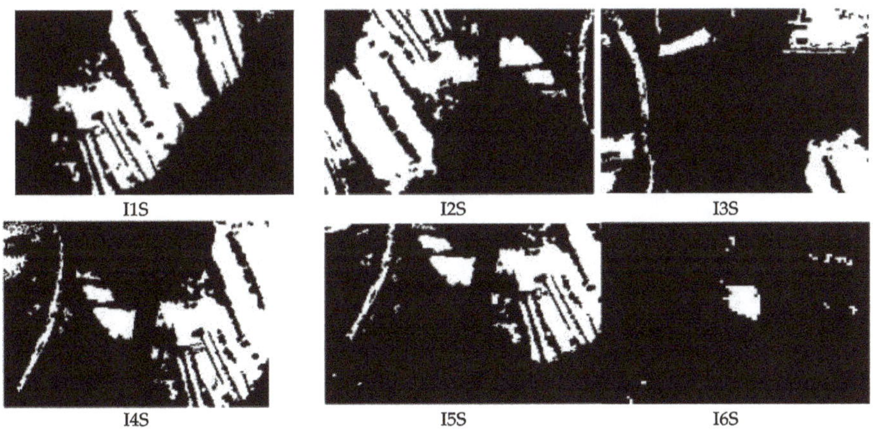

Figure 11. Images segmented for flood evaluation. White—flooded areas; black—non flooded areas.

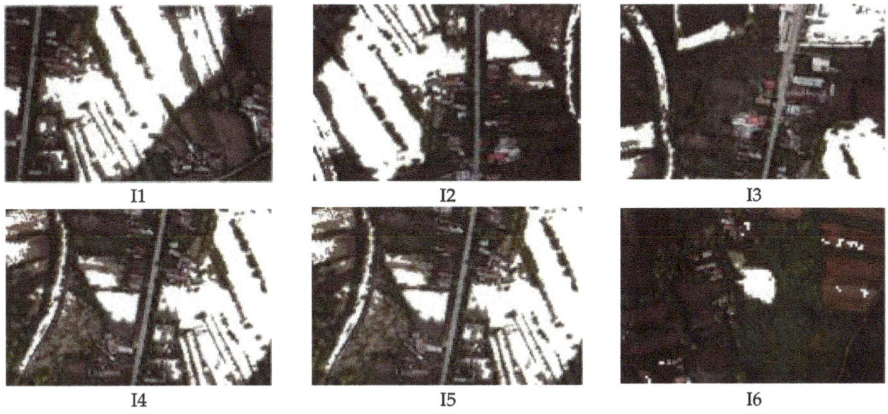

Figure 12. The overlap of RGB images with the segmented images.

Table 4. The selected features, their confidence indicators and the representatives for the class *F*.

Patch	*ImR*	*HomHH*	*ConHH*	*EnHS*	*DmG*	*LR*
P1_F	0.460	0.999	0.001	0.916	2.667	0.445
P2_F	0.472	0.997	0.003	0.921	2.690	0.432
P3_F	0.484	0.998	0.001	0.911	2.665	0.387
P4_F	0.504	0.998	0.007	0.932	2.641	0.455
P5_F	0.478	0.999	0.007	0.926	2.668	0.485
P6_F	0.488	0.996	0.001	0.919	2.643	0.415
P7_F	0.475	0.996	0.008	0.915	2.639	0.395
P8_F	0.485	0.997	0.002	0.912	2.635	0.401
P9_F	0.506	0.999	0.001	0.928	2.664	0.413
P10_F	0.443	0.998	0.001	0.926	2.671	0.398
P11_F	0.433	0.995	0.001	0.934	2.648	0.446
P12_F	0.486	0.997	0.002	0.924	2.685	0.432
P13_F	0.479	0.999	0.003	0.909	2.645	0.457
P14_F	0.502	0.996	0.003	0.914	2.654	0.395
P15_F	0.477	0.996	0.001	0.921	2.675	0.438
P16_F	0.491	0.997	0.002	0.929	2.632	0.442
P17_F	0.465	0.999	0.007	0.941	2.643	0.413
P18_F	0.451	0.998	0.005	0.937	2.642	0.428
P19_F	0.462	1.000	0.006	0.938	2.685	0.391
P20_F	0.498	0.999	0.004	0.917	2.650	0.394
m_i	0.476	0.997	0.003	0.923	2.635	0.423
$\Im_i$	**[0.418; 0.535]**	**[0.994; 1.002]**	**[−0.004; 0.011]**	**[0.896; 0.951]**	**[2.605; 2.709]**	**[0.344; 0.502]**
P1_NF	0.161	0.195	0.392	0.415	2.601	0.177
P2_NF	0.302	0.176	0.591	0.580	2.581	0.182
P3_NF	0.226	0.187	0.560	0.602	2.592	0.164
P4_NF	0.201	0.588	0.621	0.604	2.557	0.161
P5_NF	0.241	0.576	0.399	0.424	2.569	0.345 *
P6_NF	0.151	0.192	0.581	0.522	2.590	0.194
P7_NF	0.160	0.184	0.395	0.589	2.583	0.176
P8_NF	0.215	0.177	0.581	0.449	2.596	0.167
P9_NF	0.210	0.583	0.632	0.608	2.562	0.155
P10_NF	0.151	0.593	0.481	0.625	2.568	0.174
P11_NF	0.356	0.192	0.492	0.519	2.656 *	0.255
P12_NF	0.152	0.201	0.353	0.450	2.592	0.162
P13_NF	0.169	0.171	0.372	0.561	2.590	0.175
P14_NF	0.211	0.581	0.367	0.382	2.577	0.145
P15_NF	0.205	0.544	0.624	0.613	2.573	0.198
P16_NF	0.174	0.193	0.368	0.402	2.590	0.207
P17_NF	0.195	0.576	0.634	0.634	2.562	0.184
P18_NF	0.382	0.476	0.587	0.596	2.606 *	0.195
P19_NF	0.421 *	0.425	0.456	0.545	2.584	0.198
P20_NF	0.203	0.543	0.429	0.512	2.597	0.178
$\eta(\lambda_i)$	1	0	0	0	2	1
$\eta(PNF)$	20	20	20	20	20	20
CI	0.95	1	1	1	0.90	0.95

*: The values are not within the corresponding confidence intervals.

Table 5. The confusion matrices and the resulting weights for the selected features.

$ImR = T_1$	$HomHH = T_2$	$ConHH = T_3$	$EnHS = T_4$	$DmG = T_5$	$LR = T_6$
45 5 4 46	46 4 3 47	50 0 4 46	49 1 2 48	46 4 2 48	47 3 3 47
$w_1 = 0.91$	$w_2 = 0.93$	$w_3 = 0.96$	$w_4 = 0.97$	$w_5 = 0.88$	$w_6 = 0.94$

Table 6. Some experimental results concerning the patch classification based on voting scheme. Gray rows mean wrong classification.

Patch (Actual)	*ImR/D*(B_1)	*HomHH/D*(B_2)	*ConHH/D*(B_3)	*EnHS/D*(B_4)	*DmG/D*(B_5)	*LR/D*(B_6)	*D*(*B*)/*F*,*NF* $T = 0.8 \times 5.59$
B1_F	0.494/0.91	0.996/0.93	0.001/0.96	0.942/0.97	2.661/0.88	0.372/0.94	5.59/F
B2_F	0.506/0.91	0.998/0.93	0.003/0.96	0.9340.97	2.637/0.88	0.421/0.94	5.59/F
B3_F	0.457/0.91	0.999/0.93	0.006/0.96	0.9610.97	2.643/0.88	0.446/0.94	5.59/F
B4_F	0.464/0.91	0.999/0.93	0.005/0.96	0.9160.97	2.701/0.88	0.497/0.94	5.59/F
B5_F	0.515/0.91	0.997/0.93	0.004/0.96	0.9520.97	2.621/0.88	0.480/0.94	5.59/F
B6_F	0.398/0	0.995/0.93	0.021/0	0.899/0.97	2.587/0	0.346/0.94	2.84/NF
B7_F	0.437/0.91	0.998/0.93	0.003/0.96	0.9190.97	2.678/0.88	0.405/0.94	5.59/F
B8_F	0.493/0.91	0.997/0.93	0.004/0.96	0.9310.97	2.671/0.88	0.417/0.94	5.59/F
B9_F	0.476/0.91	0.995/0.93	0.003/0.96	0.9150.97	2.682/0.88	0.482/0.94	5.59/F
B10_F	0.350/0	0.992/0	0.013/0	0.850/0	2.623/0.88	0.321/0	0.88/NF
B1_NF	0.172/0	0.204/0	0.387/0	0.423/0	2.599/0	0.167/0	0/NF
B2_NF	0.137/0	0.189/0	0.582/0	0.502/0	2.579/0	0.202/0	0/NF
B3_NF	0.224/0	0.526/0	0.353/0	0.412/0	2.564/0	0.327/0	0/NF
B4_NF	0.198/0	0.537/0	0.624/0	0.623/0	2.538/0	0.211/0	0/NF
B5_NF	0.249/0	0.592/0	0.617/0	0.589/0	2.521/0	0.149/0	0/NF
B6_NF	0.335/0	0.213/0	0.457/0	0.501/0	2.599/0	0.268/0	0/NF
B7_NF	0.186/0	0.555/0	0.602/0	0.654/0	2.556/0	0.172/0	0/NF
B8_NF	0.139/0	0.185/0	0.366/0	0.573/0	2.572/0	0.161/0	0/NF
B9_NF	0.231/0	0.593/0	0.401/0	0.438/0	2.569/0	0.339/0	0/NF
B10_NF	0.391/0	0.821/0	0.009/0.96	0.722/0	2.651/0.88	0.311/0	1.84/NF

Table 7. Statistic for flooded area in images: 1000 pathces (500—flood, 500—non flood).

TP	*TN*	*FP*	*FN*	Sensitivity	Specificity	Accuracy
486	495	5	14	97.2%	99%	98.1%

4. Discussion

Because we considered only complete flooded boxes, the approximation will be underestimated. Similarly, if mixed boxes are considered, an over approximation will be obtained. Table 8 presents the number of patches considered as *F* and the corresponding percent of flooded area for each images. The main cause was the patches from the contour of flooded which appear as mixed ones. Further studies will consider the decomposition of these patches in boxes increasingly small.

Table 8. Percent of flooded area.

Images	IS1	IS2	IS3	IS4	IS5	IS6
Percent	32.88	32.79	16.85	28.07	21.57	2.44
No. patches	3156	3148	1617	2695	2071	234

On the other hand, by properly choosing textural features on color channels and patch dimension, the proposed algorithm can be extended to more classes like: road, vegetation, buildings, etc. Combining thermal camera with video, the system is able to detect possible persons in difficulty and to monitor the rescue operation. In this case, a flexible and dynamic strategy for trajectory design is necessary. Also the collaboration between the mobile nodes (UAVs) will improve the mission efficiency. The algorithms used for trajectory design minimize total path length while in the same time passing through (or within predefined distance) of a priori given way-points. In further work we plan to: (i) reconfigure trajectories on the fly such that the flooded areas are covered efficiently; and (ii) partition the workload of the UAVs such that total time/effort is minimized (for now we simply divide the area of observation into disjoint regions, one per each UAV).

5. Conclusions

The paper presented a comprehensive system and methodology for the detection and segmentation of flooded areas in a pre-determined zone. The contributions are focused on two important objectives: the planning of an optimal trajectory to cover the area under investigation and the image processing required to detect and to evaluate the flood spread. For the first the novelty lies in the analysis and computation of an optimal path covering the area of interest and for the second, the novelty lies in combining the information for different color channels with information about spatial pixel distribution obtained from chromatic co-occurrence matrix and mass fractal dimension. First, the paper studied a typical photogrammetry problem through the prism of control and optimization theory. That is, for a given polyhedral region which has to be covered by parallel lines (along which photographs are taken) we have given both an estimation of the required number of photographs and provided a minimum-length path covering the area. For the latter case we formulated a constrained optimization problem where various constraints and parameters were considered in order to obtain a minimum-length path. We took into account the maximum distance between consecutive lines and turn conditions (such that the UAV is guaranteed to follow the interior lines). We have also discussed the path generation problem in the presence of wind and for regions with non-convex shapes. Second, a methodology for the detection, segmentation and evaluation of flooded areas from the acquired images was presented. A color co-occurrence matrix was introduced and some efficient features. Furthermore, we illustrated that fractal type features on color component improve the local classification process on flooded and non-flooded boxes. The algorithm was tested on a large number of sub-images and the results showed good performances. We conclude that, by including the color information to texture analysis, by selection of feature based on maximum criterion and by using the fractal techniques, the accuracy of the detection of flooded boxes was increased up to 99.12%.

Acknowledgments: The work has been funded by National Research Program STAR, project 71/2013: Multisensory robotic system for aerial monitoring of critical infrastructure systems—MUROS and by National Research Program BRIDGE GRANT, project PN-III-P2-2.1-BG-2016-0318: Multi-drone system for flooded evaluation—SIMUL. The publication of this paper was supported by the Data4Water H2020, TWINN 2015 Project.

Author Contributions: Dan Popescu conceived the system. Dan Popescu, Loretta Ichim and Florin Stoican contributed to system implementation and wrote the paper. Dan Popescu and Loretta Ichim designed and perform the experiments for image processing. Florin Stoican elaborates the trajectory control.

Conflicts of Interest: The authors declare no conflict of interest. The founding sponsors had no role in the design of the study; in the collection, analyses, or interpretation of data; in the writing of the manuscript, and in the decision to publish the results.

Appendix A. The Pseudo Code (Matlab) for the Proposed Algorithm for Computing CCM between Spectral Components A and B

```
BEGIN Compute(A, B, xdimension, ydimension,~n)
A = normalize(A,n);
B = normalize(B,n);
Initialize result as a n × n array of 0;
bExtended = extendMatrix(B, xdimension, ydimension);
offset = calculateOffset([xdimension, ydimension]);
for i = 0 to n
      for j = 0 to n
            positions = searchAppearences(A, i);
            if positions ! = null
                 for k = 0 to positions[0].length
                 bRow =positions[0][k] +offset[1] + xdimension;
                 bCol = positions[1][k] + offset[2] + ydimension;
                 if bExtended[bRow][bCol] == j
```

```
                              result[i][j] ++;
return result;
END
BEGIN calculateOffset(offsetIn)
for i = 1 to offsetIn.length
      if offsetIn(i) >= 0
            offsetOut(i) = 0;
      else offsetOut(i) = abs(offsetIn(i));
END
BEGIN searchAppearences(A, x)
count = countAppearences(A, x);
initialize positions as a 2 x count array;
if count == 0
     return null;
int k = 0;
for i = 0 to rows
for j = 0 to cols
      if A[i][j] == x
            positions[0][k] = i;
      positions[1][k] = j;
      k++;
return positions;
END
BEGIN countAppearences(A, int x)
count = 0;
for i = 0 to rows
      for j = 0 to cols
            if A[i][j] == x
                  count++;
return count;
END
BEGIN extendMatrix(A, noRows, noCols)
initialize result as an array (A.length + abs(noRows))x(A[0].length + abs(noCols));
offset = calculateOffset([rowsNo, colsNo]);
for i = 0+offset[1] to A.length+offset[1]
      for j = 0+offset[2] to A[0].length+offset[2]
            result[i][j] = A[i][j];
for i = A.length to A.length + noRows
      for k = 0 to A[0].length + noCols
            result[i][k] = −1;
for i = A[0].length to A[0].length + noCols
      for k = 0 to A.length + noRows
            result[k][i] = −1;
return result;
END
```

References

1. Doong, D.J.; Chuang, L.H.; Wu, L.C.; Fan, Y.M.; Kao, C.; Wang, J.H. Development of an operational coastal flooding early warning system. *Nat. Hazards Earth Syst. Sci.* **2012**, *12*, 379–390. [CrossRef]

2. Benfield, A. *Annual Global Climate and Catastrophe Report*; Impact Forecasting: Chicago, IL, USA, 2013.
3. Basha, E.; Rus, D. Design of early warning flood detection systems for developing countries. In Proceedings of the International Conference on Information and Communication Technologies and Development (ICTD 2007), Bangalore, India, 15–16 December 2007; pp. 1–10.
4. Koschitzki, R.; Schwalbe, E.; Maas, H. An autonomous image based approach for detecting glacial lake outburst floods. *ISPRS Int. Arch. Photogramm. Remote Sens. Spat. Inf. Sci.* **2014**, *XL-5*, 337–342. [CrossRef]
5. Sakai, T.; Hatta, S.; Okumura, M.; Hiyama, T.; Yamaguchi, Y.; Inoue, G. Use of Landsat TM/ETM+ to monitor the spatial and temporal extent of spring breakup floods in the Lena River, Siberia. *Int. J. Remote Sens.* **2015**, *36*, 719–733. [CrossRef]
6. Zhang, J.; Zhou, C.; Xu, K.; Watanabe, M. Flood disaster monitoring and evaluation in China. *Glob. Environ. Chang. Part B Environ. Hazards* **2002**, *4*, 33–43. [CrossRef]
7. Revilla-Romero, B.; Thielen, J.; Salamon, P.; De Groeve, T.; Brakenridge, G. Evaluation of the satellite-based Global Flood Detection System for measuring river discharge: Influence of local factors. *Hydrol. Earth Syst. Sci.* **2014**, *18*, 4467–4484. [CrossRef]
8. Aicardi, I.; Chiabrando, F.; Lingua, A.M.; Noardo, F.; Piras, M.; Vigna, B. A methodology for acquisition and processing of thermal data acquired by UAVs: A test about subfluvial springs' investigations. *Geomat. Nat. Hazards Risk* **2016**. [CrossRef]
9. Pandey, R.K.; Crétaux, J.F.; Bergé-Nguyen, M.; Tiwari, V.M.; Drolon, V.; Papa, F.; Calmant, S. Water level estimation by remote sensing for the 2008 flooding of the Kosi river. *Int. J. Remote Sens.* **2014**, *35*, 424–440. [CrossRef]
10. Khurshid, H.; Khan, M.F. Segmentation and Classification Using Logistic Regression in Remote Sensing Imagery. *IEEE J. Sel. Top. Appl. Earth Obs. Remote Sens.* **2015**, *8*, 224–232. [CrossRef]
11. Yulianto, F.; Sofan, P.; Zubaidah, A.; Sukowati, K.A.D.; Pasaribu, J.M.; Khomarudin, M.R. Detecting areas affected by flood using multi-temporal ALOS PALSAR remotely sensed data in Karawang, West Java, Indonesia. *Nat. Hazards* **2015**, *77*, 959–985. [CrossRef]
12. Lo, S.W.; Wu, J.H.; Lin, F.P.; Hsu, C.H. Cyber surveillance for flood disasters. *Sensors* **2015**, *15*, 2369–2387. [CrossRef] [PubMed]
13. Pentari, A.; Moirogiorgou, K.; Livanos, G.; Iliopoulou, D.; Zervakis, M. Feature analysis on river flow video data for floating tracers detection. In Proceedings of the IEEE International Conference on Imaging Systems and Techniques (IST), Santorini Island, Greece, 14–17 October 2014; pp. 287–292.
14. Lee, J.N.; Kwak, K.C. A trends analysis of image processing in unmanned aerial vehicle. *Int. J. Comput. Inf. Sci. Eng.* **2014**, *8*, 261–264.
15. Scarsi, A.; Emery, W.J.; Moser, G.; Pacifici, F.; Serpico, S.B. An automated flood detection framework for very high spatial resolution imagery. In Proceedings of the IEEE International Geoscience and Remote Sensing Symposium (IGARSS), Quebec City, QC, Canada, 13–18 July 2014; pp. 4954–4957.
16. Abdelkader, M.; Shaqura, M.; Claudel, C.G.; Gueaieb, W. A UAV based system for real time flash flood monitoring in desert environments using Lagrangian microsensors. In Proceedings of the International Conference on Unmanned Aircraft Systems (ICUAS), Atlanta, GA, USA, 28–31 May 2013; pp. 25–34.
17. Achille, C.; Adami, A.; Chiarini, S.; Cremonesi, S.; Fassi, F.; Fregonese, L.; Taffurelli, L. UAV-Based photogrammetry and integrated technologies for architectural applications-methodological strategies for the after-quake survey of vertical structures in Mantua (Italy). *Sensors* **2015**, *15*, 15520–15539. [CrossRef] [PubMed]
18. Feng, Q.; Liu, J.; Gong, J. Urban flood mapping based on unmanned aerial vehicle remote sensing and random forest classifier—A case of Yuyao, China. *Water* **2015**, *7*, 1437–1455. [CrossRef]
19. Boccardo, P.; Chiabrando, F.; Dutto, F.; Tonolo, F.G.; Lingua, A.M. UAV deployment exercise for mapping purposes: Evaluation of emergency response applications. *Sensors* **2015**, *15*, 15717–15737. [CrossRef] [PubMed]
20. Aicardi, F.; Nex, F.C.; Gerke, M.; Lingua, A.M. An image-based approach for the co-registration of multi-temporal UAV image datasets. *Remote Sens.* **2016**, *8*, 779. [CrossRef]
21. Beard, R.W.; McLain, T.W. *Small Unmanned Aircraft: Theory and Practice*; Princeton University Press: Princeton, NJ, USA, 2012.
22. Siebert, S.; Teizer, J. Mobile 3D mapping for surveying earthwork projects using an Unmanned Aerial Vehicle (UAV) system. *Autom. Constr.* **2014**, *41*, 1–14. [CrossRef]

23. Obermeyer, K.J. Path planning for a UAV performing reconnaissance of static ground targets in terrain. In Proceedings of the AIAA Guidance, Navigation, and Control Conference, Chicago, IL, USA, 10–13 August 2009; pp. 10–13.
24. Eisenbeiss, H.; Sauerbier, M. Investigation of UAV systems and flight modes for photogrammetric applications. *Photogramm. Rec.* **2011**, *26*, 400–421. [CrossRef]
25. Zhang, B.; Liu, W.; Mao, Z.; Liu, J.; Shen, L. Cooperative and Geometric Learning Algorithm (CGLA) for path planning of UAVs with limited information. *Automatica* **2014**, *50*, 809–820. [CrossRef]
26. Levine, J. *Analysis and Control of Nonlinear Systems: A Flatness-Based Approach*; Springer Science & Business Media: Berlin, Germany, 2009.
27. Suryawan, F. Constrained Trajectory Generation and Fault Tolerant Control Based on Differential Flatness and B-Splines. Ph.D. Thesis, The University of Newcastle, Callaghan, Australia, August 2011.
28. Stoican, F.; Prodan, I.; Popescu, D. Flat trajectory generation for way-points relaxations and obstacle avoidance. In Proceedings of the 23th Mediterranean Conference on Control and Automation (MED), Torremolinos, Spain, 16–19 June 2015; pp. 695–700.
29. Gordon, W.J.; Riesenfeld, R.F. B-spline curves and surfaces. *Comput. Aided Geom. Des.* **1974**, *167*, 95.
30. Stoican, F.; Popescu, D. Trajectory generation with way-point constraints for UAV systems. In *Advances in Robot Design and Intelligent Control*; Borangiu, T., Ed.; Springer: Basel, Switzerland, 2016; pp. 379–386.
31. Haralick, R.M. Statistical and structural approaches to texture. *Proc. IEEE* **1979**, *67*, 786–804. [CrossRef]
32. Haralick, R.M.; Shanmugam, K.; Dinstein, I.H. Textural features for image classification. *IEEE Trans. Syst. Man Cybern.* **1973**, *3*, 610–621. [CrossRef]
33. Khelifi, R.; Adel, M.; Bourennane, S. Multispectral texture characterization: Application to computer aided diagnosis on prostatic tissue images. *EURASIP J. Adv. Signal Process.* **2012**, *2012*, 118. [CrossRef]
34. Losson, O.; Porebski, A.; Vandenbroucke, N.; Macaire, L. Color texture analysis using CFA chromatic co-occurrence matrices. *Comput. Vis. Image Underst.* **2013**, *117*, 747–763. [CrossRef]
35. MUROS—Teamnet International. Available online: http://www.teamnet.ro/grupul-teamnet/cercetare-si-dezvoltare/muros/ (accessed on 15 December 2016).
36. Programul CDI Tehnologie Spatiala si Cercetare Avansata STAR. Available online: https://star.rosa.ro (accessed on 15 December 2016).
37. Fliess, M.; Lévine, J.; Martin, P.; Rouchon, P. *On Differentially Flat Nonlinear Systems, Nonlinear Control Systems Design*; Pergamon Press: Oxford, UK, 1992.
38. Popescu, D.; Ichim, L.; Dobrescu, R. Sliding box method for automated detection of the optic disc and macula in retinal images. *Lect. Notes Comput. Sci.* **2015**, *9043*, 250–261.
39. Ojala, T.; Pietikainen, M. Unsupervised texture segmentation using feature distributions. *Pattern Recognit.* **1999**, *32*, 477–486. [CrossRef]
40. Sarkar, N.; Chaudhuri, B. An efficient differential box-counting approach to compute fractal dimension of image. *IEEE Trans. Syst. Man Cybern.* **1994**, *24*, 115–120. [CrossRef]
41. Popescu, D.; Ichim, L.; Gornea, D.; Stoican, F. Complex image processing using correlated color information. *Lect. Notes Comput. Sci.* **2016**, *10016*, 723–734.
42. Karperien, A. *FracLac for ImageJ*; Charles Sturt University: Sydney, Australia, 2013.
43. Fawcett, T. An introduction to ROC analysis. *Pattern Recognit. Lett.* **2006**, *27*, 861–874. [CrossRef]

sensors

MDPI

Article

Detection, Location and Grasping Objects Using a Stereo Sensor on UAV in Outdoor Environments

Pablo Ramon Soria *, Begoña C. Arrue and Anibal Ollero

Robotics, Vision and Control Group, University of Seville, Camino de los Descubrimientos, s/n, Seville 41092, Spain; barrue@us.es (B.C.A.); aollero@us.es (A.O.)
* Correspondence: prs@us.es; Tel.: +34-697-513-380

Academic Editors: Felipe Gonzalez Toro and Antonios Tsourdos
Received: 9 November 2016; Accepted: 3 January 2017; Published: 7 January 2017

Abstract: The article presents a vision system for the autonomous grasping of objects with Unmanned Aerial Vehicles (UAVs) in real time. Giving UAVs the capability to manipulate objects vastly extends their applications, as they are capable of accessing places that are difficult to reach or even unreachable for human beings. This work is focused on the grasping of known objects based on feature models. The system runs in an on-board computer on a UAV equipped with a stereo camera and a robotic arm. The algorithm learns a feature-based model in an offline stage, then it is used online for detection of the targeted object and estimation of its position. This feature-based model was proved to be robust to both occlusions and the presence of outliers. The use of stereo cameras improves the learning stage, providing 3D information and helping to filter features in the online stage. An experimental system was derived using a rotary-wing UAV and a small manipulator for final proof of concept. The robotic arm is designed with three degrees of freedom and is lightweight due to payload limitations of the UAV. The system has been validated with different objects, both indoors and outdoors.

Keywords: UAV; grasping; outdoors

1. Introduction

The use of unmanned aerial vehicles (UAVs) is becoming increasingly popular; not only for military purposes, but also in many other fields, from wildlife and atmospheric research to disaster relief and sports photography [1]. Applications in agriculture and industrial environments are currently being exploited, for example, in inspection and maintenance. They can be used as a fast and safe response in a critical situation against disasters, plagues, or working in dangerous or inaccessible places.

Some important incidents—such as the nuclear disaster in Fukushima in 2011—also clearly illustrated the problem that UAVs are today nearly exclusively used as flying sensors. UAVs are equipped with different types of sensors providing situational awareness, but are so far not equipped with any type of actuators, unlike many ground-based mobile robots. Unfortunately, flying manipulation comes with many unsolved problems. A suitable manipulator that can be attached to a small UAV must also be small and lightweight.

The work presented in this paper is focused on the development of an on-board perception system for autonomous object manipulation using UAVs. The objective is to provide the aerial robot with capabilities to perform inspection and maintenance tasks (which imply manipulation) that can be dangerous for human operators or in places that are difficult to reach. Sending UAVs will minimize the risks and increase the response speed and automation of operations.

Currently, perceiving the environment remains a challenging task. The robot needs to recognize the targeted objects and minimize false positives. Particularly, in aerial vehicles, a wrong detection can result in disastrous consequences for the platform. The use of different machine learning algorithms for object detection has rapidly expanded. Algorithms such as Support Vector Machines (SVM) and Neural Networks (NN) have been used successfully for this task.

For grasping objects, it is also necessary to analyze the 3D information of the object. Estimating its pose accurately is needed so that a manipulator can grasp it. The use of a monocular camera requires the recovery of the 3D information. Utilizing depth sensors, stereo cameras, or lasers can simplify the task, as they directly provide this information.

Once the object is found and located relative to the robot, it is necessary to analyze how to perform the grasp. Usually, the object is defined as a mesh or a primitive shapes (spheres, cylinders, etc). This information can be provided by depth sensors, stereo cameras, or lasers, producing point clouds and processing them. Grasping information is generated using the full 3D shape with approaches like force-closure and/or quality metrics [2]. However, the object might be occluded in real applications, in addition to the fact that these kinds of methods can be time consuming and may be impractical if the object is fully known.

Our work focuses on the task of object detection and pose estimation for real-time tasks on UAVs for grasping objects. We only consider objects that contain textured surfaces and which are graspable by the UAV (for example, the algorithm will not be able to learn a plain white folder, and the UAV the manipulator on the UAV cannot grasp objects larger than the gripper) The algorithm is based on the work presented in [3,4], but varies in the use of stereo cameras to improve the learning process (as they provide 3D information) and introduces improved feature filtering. Furthermore, as proof of concept, the UAV has been equipped with a three degrees of freedom (DOF) arm with a gripper as end-effector.

The remainder of this article is structured as follows. In Section 2 we describe the related work. Section 3 is divided in three subsections: feature extraction and filtering, creation of object model, and how to find the object in new scenes. Section 4 shows the validation tests we performed, and the final section presents the conclusions and future work.

2. Related Work

As mentioned before, for the development of an on-board perception system for autonomous object manipulation using UAVs, the robot needs to detect the targeted tools to be grasped for the task. This process can be split into three parts. The first part involves the detection and recognition of the object, the second is the estimation of its position (and orientation), and the third is the grasp analysis. Object detection is a wide area of research: there exist several algorithms depending on the prior knowledge, the environment, and also on the sensor used. Some authors use machine learning algorithms as neural networks [5], or classifiers such as Support Vector Machine (SVM) [6] using a bag-of-words model. However, for grasping objects, it is also necessary to estimate the position of the object and grasping points. Authors in [7–9] used depth sensors such as Kinect to model the object and segment it using the depth information. Depth information is very useful for the estimation of object position, but the sensors usually have limitations in outdoor environments, as they use IR-structural light. Other authors use monocular cameras aided with geometrical models of the objects, as in [10]. Authors in [3,4] showed how to describe objects using image features to recognize and estimate its position accurately while being robust to occlusions. Finally, it is necessary to analyze how to grasp the object. Authors in [2] show a complete survey about quality metrics for comparing different grasps using the available geometrical information of the object to generate virtual grasps. This geometrical information can be provided by depth sensors [11], cameras [12], or haptic sensors [13]. In [14], authors used SVM to estimate the quality of the grasp.

References [15–18] studied the grasping of objects using UAVs, paying particular attention to the mechanical and control aspects of the grasp. In general, for grasping with aerial robots, the authors assume that the position of the object is known, or they use simple markers such as color or tags [19–21]. In [22], authors studied the dynamics, control, and visual servoing of an aerial vehicle for grasping inspired by animal movements. Nevertheless, they assume that the object is a cylinder with known radius. Conversely, in this article, the proposed algorithm allows the robot to robustly detect more complex objects. In this work, it is assumed that the UAV has an internal control that is sufficient for movement and manipulation. The arm is lightweight, and the movements are slow enough that the control system is able to stabilize the robot.

In a previous work, the authors [23] developed a system with a low-cost stereo camera for object detection and location. In this system, a local map is created to localize a list of object candidates and the relative position of the UAV on the map. Once the candidates are located, their volume is projected into the images, and a machine learning algorithm is run to recognize the object category. This was called a bottom-up approach which firstly detects that there is something in the environment, and then categorizes the object.

In the present paper, a top-down approach is used. The kind of objects that are to be grasped were known, and it was possible to generate a model for each one. In an offline stage, the algorithm learned the model using visual features. Afterwards, in the online stage, the model was used to seek and locate the object (see Figure 1). A small arm was 3D printed to test the grasping results, and everything ran on an on-board Intel NUC computer [24]. Finally, for image acquisition, the UAV was equipped with a commercial stereo camera called ZED [25]. Through its SDK, this camera can be used to compute depth maps using CUDA (NVIDIA parallel computing platform and API for Graphic Processor Units), but this software has range limitations and cannot be used for distances shorter than one meter. Nonetheless, the built-in cameras have high quality, the auto-focus and auto-exposure are remarkable, their construction is stiff, and they provide a versatile combination between resolution and frames per second (FPS). Moreover, the baseline of the camera (120 mm) is good for a reconstruction on relatively short distances (within 20–100 cm). For these reasons, this camera was chosen for this project. The object detection and pose estimation algorithm was tested on-board, and it was shown to be robust to vibrations, occlusions, and illumination changes. However, for security reasons, the robot was hanging on a structure while testing the grasping algorithm.

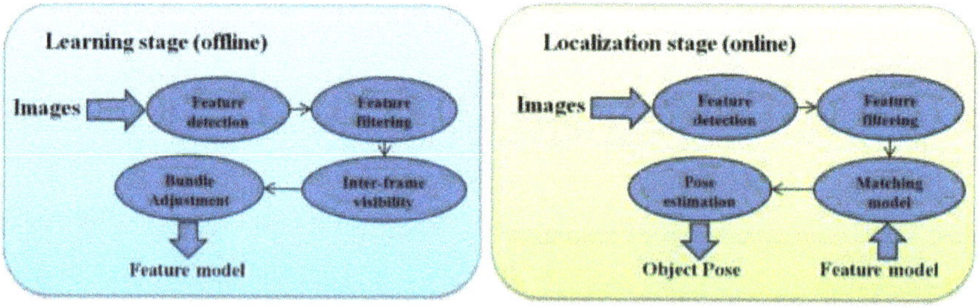

Figure 1. Pipeline of both stages of the algorithm.

3. Object Detection and Pose Estimation

This algorithm is divided into two stages: the learning stage (or offline stage), in which the model of the object is created, and the localization stage (or online stage), in which the object is detected and its position is estimated. Figure 1 summarizes the pipeline of the algorithm in both the modeling and localization stages.

3.1. Feature Extraction and Filtering Using Stereo Cameras

Image features have been widely used and studied. SIFT (Scale-Invariant Feature Transform) [26] is a well-known detector and descriptor, and it has been proven to be robust to scales, rotations, and translations on images. However, it has a high computational time. Authors in [27,28] proposed some optimization to speed it up, but it is sometimes still not fast enough, or some losses of information are caused by their approximation. Vision algorithms for UAVs have a strong speed requirement, as they need a faster response for the control loop than ground robots. Several new feature detectors and descriptors have been developed, such as SURF (speeded up robust features) [29], ORB (Oriented FAST and Rotated BRIEF) [30], DAISY (an efficient dense descriptor applied for wide baseline stereo) [31], BRIEF (Binary Robust Independent Elementary Features) [32], FAST (features from accelerated segment test) [33], and more. These methods have been designed according to different objectives, such as being faster or more robust.

The performance of the image feature detection and matching for the model creation and object position estimation depends on the combination of the detector and descriptor chosen. Nevertheless, for both the object modeling and detection algorithms, the features are completely exchangeable. In Section 4, we show the processing time for the different combinations of detectors and descriptors.

As mentioned before, a ZED stereo camera was used for image acquisition. This particular camera has a wide-angle lens, so a rectification of the images is needed to properly detect and match the features. Initially, features are detected on each pair of images, then these features are matched with either a FLANN (Fast Library for Approximate Nearest Neighbors) [34]-based matcher or just a force brute matcher. The first advantage of using stereo images is to improve the filtering of matches, making it more robust to outliers. Despite this, it needs more process time, as features are computed in both images. The resulting inliers are assumed to be key features of the object. This makes our algorithm more robust to outliers, as they are rejected at the beginning of the process, so only the features that are more invariant are used. Figure 2 shows examples of feature filtering using stereo, as described in this paragraph.

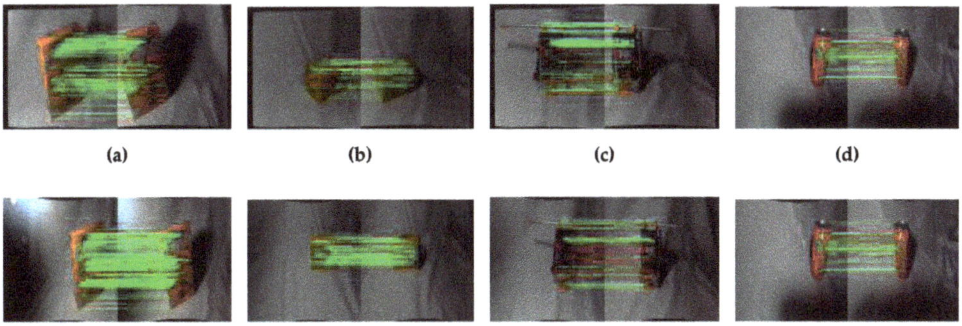

(a) (b) (c) (d)

(e) (f) (g) (h)

Figure 2. Filtering bad features using known stereo geometry. (**a**) Whoopies box 640 × 480; (**b**) Gena box 640 × 480; (**c**) Drilling tool 640 × 480; (**d**) Coke 640 × 480; (**e**) Whoopies box 1280 × 720; (**f**) Gena box 1280 × 720; (**g**) Drilling tool 1280 × 720; (**h**) Coke 1280 × 720.

3.2. Object Modeling

The learning stage (or offline stage) generates a model of an object from a set of images. This approach differs from [3,35] in the use of a stereo camera, automating the learning process (providing 3D information of the real-world scale) and improving the filtering of outliers. The process is summarized in the following points:

- Image rectification from camera calibration.
- Detection of features on both images and matching them. Use of stereo geometry and RANSAC (Random sample consensus) to filter outliers.
- Matching of sequentially filtered features.
- Performance of bundle adjustment to create a 3D model of the object and store the corresponding descriptors.
- Scale model of the object to its real size using stereo information.

The first stages of the object modeling are feature detection and filtering using the stereo system and its calibration following the procedure described in previous section. Once all the images are processed, and the corresponding set of cleaned features is obtained, a bundle adjustment (BA) of both features and camera positions is performed [36] in order to reconstruct the correct object shape. Further, the camera positions—where the images were taken—are obtained (however, this information is not used for the proposed method). We used the library cvSBA as implementation for the Sparse Bundle Adjustment (SBA). It is a wrapper of the SBA library [37] for use with OpenCV.

In order to perform the BA, it is necessary to correlate the points within all the images. Only left projections are used for this step, since cvSBA does not allow users to establish custom restrictions between camera frames. Nonetheless, left projections contain enough information to reconstruct the object (the stereo information will be used to scale the object after the optimization, as described later). It is assumed that all of the pictures from the dataset are arranged as they were captured. Then, the features are matched sequentially to obtain the inter-frame visibility of the features. With this step, we obtained the relations in sequential frames (Figure 3 shows sequential matches of features for the creation of the inter-frame visibility matrix).

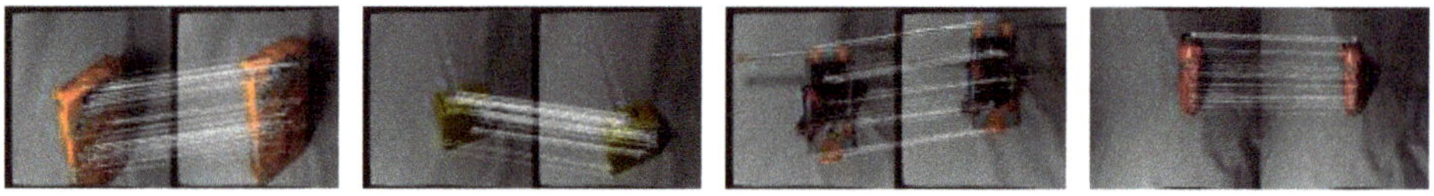

(a) **(b)** **(c)** **(d)**

Figure 3. Sequential association of features to compute their inter-frame visibility. (**a**) Whoopies box 640 × 480; (**b**) Gena box 640 × 480; (**c**) Drilling tool 640 × 480; (**d**) Coke 1280 × 720. Then, following Algorithm 1, the visibility between non-sequential frames is computed.

Nevertheless, it is necessary to obtain the remaining relations within all of the frames. Let us denote $P = \{p_k \forall k = 1...K\}$ a vector in which each p_k is a vector with the features on the frame k; M a matrix where each element m_{ij} contains a vector with the matches between the frame i and the frame j. All the elements $m_{i(i+1)}$ are filled from the sequential match of frames. Then, the rest of the elements of m_{ij} above the diagonal (i.e., $j > i$) can be filled using Algorithm 1. An improvement can be made in detecting loop closure; however, for most cases, this method is enough to reconstruct the object. A diagram of this visibility problem is shown in Figure 4.

Algorithm 1 Correlate back matches

for $offset = 2, offset < K$ **do**
  for $i = 0, i < K - offset$ **do**
    $j = i + offset$
    for match in $m_{i(j-1)}$ **do**
      if match is visible in $m_{(j-1)j}$ **then**
        add match in m_ij
      end if
    end for
  end for
end for

The BA consists of a global optimization using Levenberg and Marquardt's [38] algorithm to minimize the re-projection errors. Let there be a set of N 3D points, observed from K cameras (at T_k position and R_k orientation). Then, given the correlations between the projections of the 3D points into the cameras, an optimization is performed, minimizing the errors. Using the previously defined matrix M of matches between frames, it is possible to generate a unique list of 3D points and the inter-visibility of the points within the frames. Gordon and Lowe [3] mentioned that, in order to ensure the convergence of the BA, it is enough to place all the cameras at the same distance from the origin on the Z-axis and place all the projections in the XY-plane. It is important to highlight that it is necessary to keep track of the descriptors of the features, as they need to be stored with the 3D points as part of the model of the object. Figure 5 shows how the model has been iteratively constructed using the BA.

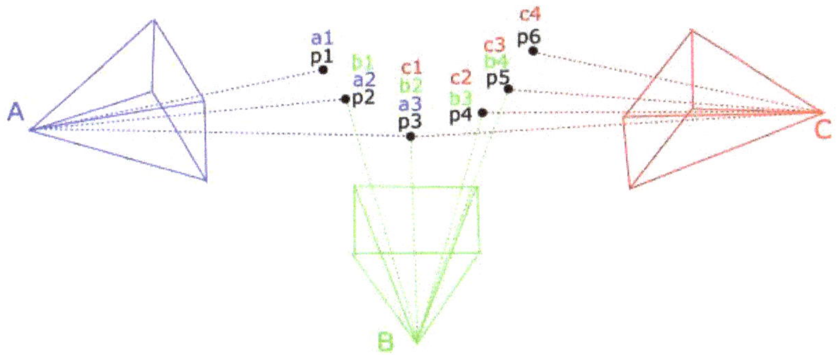

Figure 4. Diagram of elements in the Bundle Adjustment problem. A, B, and C represent the position of the camera from where the observations were taken. $p_i \forall i = 1...6$ are six features in the space and a_i, b_i, and c_i are the features observed by each of the positions.

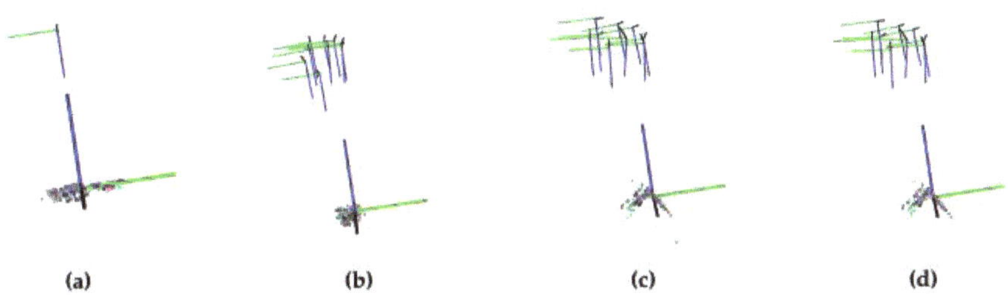

Figure 5. Different steps of the Bundle Adjustment optimization process. (**a**) Starting state; (**b**) First iteration; (**c**) Iteration 5; (**d**) Iteration 20.

Moreover, before computing the BA, an additional filtering can be done to improve the performance. Therefore, due to the fact that the inter-visibility matrix was obtained, we can compute the number of times that each point appears (i.e., in how many images each point is observed). Some of the features can be badly matched or just not matched. Hence, this could produce duplicated points that might complicate the convergence of the algorithm. To avoid this, removing points that appear in less than k images can improve and speed up the convergence of the BA. The minimum value for k is 2, as the points need to appear in at least two images to be able to "triangulate" it. On the other hand, increasing this parameter too much is not possible, because the SBA solver might not be able to solve the problem if the number of observations is lower than the number of variables in the problem. Therefore, k is set to 3.

Once the BA process is performed, we obtain a 3D model of the object. However, as described in [4], because of the optimization algorithm, the points are not scaled according to the real size. Authors in [4] record an extra dataset in which the position and orientation of the object is known, then a second optimization algorithm is performed to obtain the correct model scale. In contrast, this extra dataset is not needed when using stereo cameras. As the correlation of all the points is known, it is possible to get the projections on both left and right images at each frame. P^m are the points obtained from the BA, and P^t is a cloud reconstructed from features of a frame k using the known stereo geometry. It is possible to estimate the transformation T between them using a SVD (Singular

Value Decomposition) based estimator. k is the current frame, N_k is the number of points seen on that frame, p_i^m is the point i on the model, and p_i^t is the triangulated point from the stereo pair; the score of each transformation is computed as

$$score = \sum_{i=1...N_k} (\|p_i^m - p_i^t\|)/N_k$$

the transformation that produces the minimum score is used to scale the model to the real-world size. Being

$$T = \begin{bmatrix} a_{11} & a_{12} & a_{13} & t_x \\ a_{21} & a_{22} & a_{23} & t_y \\ a_{31} & a_{32} & a_{33} & t_z \\ 0 & 0 & 0 & 1 \end{bmatrix}$$

the scale factor can be computed as

$$s = [s_x, s_y, s_z] = \begin{cases} s_x = \|[a_{11}, a_{21}, a_{31}]\| \\ s_y = \|[a_{11}, a_{21}, a_{31}]\| \\ s_z = \|[a_{11}, a_{21}, a_{31}]\| \end{cases}$$

Eventually, this model does not contain information about how to grasp it. This is done now manually in order to ensure a correct manipulation. As the object modeling is performed offline, it is realistic to choose it manually at this stage. Nevertheless, the detection of the grasping points can be analyzed depending on the manipulator and the geometry of the object using diverse quality metrics [2]—this is beyond the scope of this paper.

3.3. Finding Object in a Scene

In this subsection, the online detection of the object in new images and the position estimation is described. First of all, using the camera calibration, the acquired images are undistorted. The same feature detector and descriptor as the one used in the modeling stage is used to extract features in both input images. Then, the features in the pair of images are matched. As described before, the known parameters of the stereo calibration are used to filter the outliers. Hence, the remaining points are stronger as they appear in both cameras and they are easy to match.

At this point, there is a set of point candidates on the scene that become part of the object. To detect it and estimate its position, a PnP (Perspective-n-Points) formulation is used. $P = \{[x_i, y_i, z_i], \forall i = 1...N\}$ is a set of 3D points, and $U = \{[u_i, v_i], \forall i = 1...N\}$ is their projection on the camera plane. The objective is to find the rotation R and translation T of the object in the camera's coordinates (knowing the calibration parameters of the camera), minimizing the re-projection error of the points. Particularly, a RANSAC [39] implementation is used. It computes randomly possible solutions using the data matched between the scene and the model. Then, the matched points that lie far from the model are considered as outliers (i.e., rejected). In conclusion, it is less sensitive to local minima, and more robust to outliers.

Now, we need to match the features in the scene with the model of the object to be able to start the PnP problem. In order to do that, each descriptor is matched with the points in the scene, and then filtered to remove outliers. The inliers are used in the PnP problem to detect the position of the object. Figure 6 shows screen-shots of results outdoors with a featured floor.

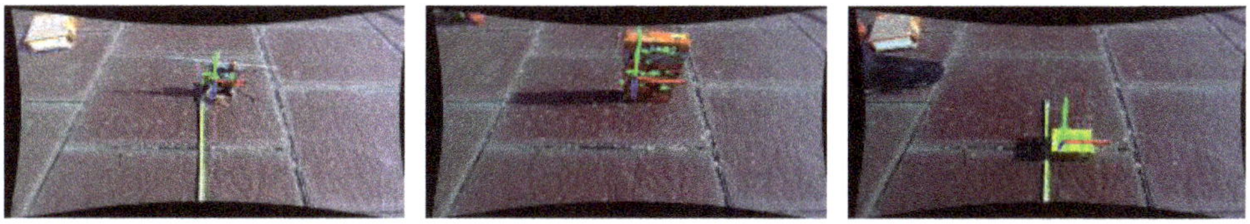

(a) (b) (c)

Figure 6. Examples of detection and position estimation of objects outdoors. White thin circles are candidate features in the scene. Green thick circles are the features assigned to the object, and the coordinate system is the representation of the position of the object. It depends on the coordinate system chosen at the modeling stage. (**a**) Drilling tool; (**b**) Whoopies box; (**c**) Gena box.

Nevertheless, during each step, several features belonging to the background are detected and described, which slows down the algorithm and increases the possibility of bad matches. In order to speed up the online stage, a moving window tracker was implemented. In the beginning, the algorithm searches for the objects over the whole image. However, if the confidence of the result is larger than a threshold, the expected portion of the next image in which the object appears is computed. As the pixel area of the following images are reduced, the amount of features computed decreases, and consequently, the algorithm runs faster. Algorithm 2 summarizes the process in the online stage.

Algorithm 2 Online stage for finding learned objects.

```
searchWindow ← size(images)
while images ← camera do
    Compute features on pair of images
    Filter features
    Match scene features with model features
    Estimate object relative pose
    if Number inliers threshold then
        Found object
        searchWindow ← boundBox(inliers)
        move arm to relative pose
    else
        searchWindow ← Size(images)
    end if
end while
```

Finally, once the object is detected and its pose estimated, the algorithm sends the desired relative position to the robot arm and orients the gripper using the estimated orientation. This only happens if the number of detected inliers is higher than a manually set threshold. If the number of inliers decreased, the robot is returned to a safe position to start again when the object is detected again. Therefore, the gripper closes when the error between the estimated position of the object and the end-effector (using the direct-kinematic of the arm) is lower than a threshold. As mentioned before, the grasp position is determined in the offline stage, ensuring it is graspable for the designed arm.

4. Experimental Validation

4.1. Hardware Setup

In order to test the algorithm, a hexacopter was built and equipped with an on-board computer, the ZED cameras, and a 3DOF robotic arm (also designed by the authors). The UAV uses a F550 frame, and the engines were chosen to have a maximum thrust of 6 kg. The whole system (including batteries) weighs 4 kg. The arm is capable of lifting up to 500 g. The inertial measurement unit (IMU) and controller of the hexacopter is the well known 3DR PIXHAWK [40]. To ensure enough computational power, the computer that was used was an INTEL NUC5i7RYH [24]. This compact computer has a CPU i7 3.1 GHz and 8 GB of RAM. An Arduino Uno [41] board was added as an interface between the computer and the manipulator.

The specifications of the robotic arm are: to be lightweight, have large range operation, and 3-DOF to accomplish the grasping task. Figure 7a shows a simplified model for the kinematics of the arm. Joints are represented in blue, and its variables in red. The inverse kinematic of the arm is governed by:

$$[\theta_1, \theta_2, \theta_3] = F(x, y, z) = \begin{cases} \theta_1 = \text{atan}(y/x) \\ \theta_2 = \text{acos}((l_2^2 - d^2 - l_1^2)/(-2 \times d \times l_1)) \\ \theta_3 = \text{acos}((d^2 - l_1^2 - l_2^2)/(-2 + l_1 \times l_2)) \end{cases}$$

being, $d = \sqrt[2]{p^2 + z^2}$ and $p = \sqrt[2]{x^2 + y^2}$.

Figure 7b shows the CAD design of the parts, which were built by 3D printing. Finally, Figure 7c shows the whole structure that we built. The arm is attached to the bottom part of the drone (centered) with a custom piece that screws to the base of the robot and to the arm.

(a) (b) (c)

Figure 7. Model of robotic arm designed for the aerial robot. (**a**) Simplified model of the arm; (**b**) CAD design; (**c**) Final built-in platform.

Additionally, a transformation is needed between the coordinate system of the camera and the coordinate system of the arm. This transformation is composed of a translation between the centers of the coordinates and a simple spin on the X axis:

$$T^{C \leftarrow A} = \begin{bmatrix} 1 & 0 & 0 & t_x \\ 0 & \cos(\alpha) & -\sin(\alpha) & t_y \\ 0 & \sin(\alpha) & \cos(\alpha) & t_z \\ 0 & 0 & 0 & 1 \end{bmatrix}$$

The parameters of the transformation were experimentally obtained as $\alpha = 30^{o}$, $t_x = 0.06$ m, $t_y = 0.1$ m, $t_z = 0$.

4.2. Validation Tests

Figure 8 shows pictures of the testing environment. As mentioned, the UAV was hung on a structure for security reasons (since its control is not the target of this article). The objects were placed in its workspace so it could detect them and grasp them. The drone was controlled in loiter mode, describing up and down movements.

Figure 8. Robot grasping an object. The camera's view is given in the inset. The green rectangle is the tracked moved window.

As mentioned in Section 3, the performance of the system depends on the election of the features extractor and descriptors. Table 1 summarizes the results for different combinations of detectors and descriptors at different resolutions in an outdoor environment. Results in indoor environments with poor light conditions are usually faster, generally because fewer features are detected so fewer features need to be described. Table 1 presents the performance of the algorithm for total image resolution without using the window tracker (i.e., before the object is tracked).

Table 1. Average computational times for the feature detection, matching, and stereo filtering for different feature detectors and descriptors. Using different image resolutions (640 × 480 and 1280 × 720). FAST: features from accelerated segment test; SIFT: scale-invariant feature transform; SURF: speeded up robust features; BRIEF: binary robust independent elementary features; rBRIEF: rotated BRIEF; DAISY: an efficient dense descriptor applied for wide baseline stereo.

	FAST Detector		**SIFT Detector**		**SURF Detector**	
	640 × 480	**1280 × 720**	**640 × 480**	**1280 × 720**	**640 × 480**	**1280 × 720**
SIFT descriptor	0.318 s	0.739 s	0.510 s	1.299 s	1.532 s	2.412 s
BRIEF descriptor	0.042 s	0.214 s	0.250 s	0.660 s	0.235 s	1.012 s
rBRIEF descriptor	0.045 s	0.229 s	0.237 s	0.715 s	0.256 s	1.100 s
SURF descriptor	0.074 s	0.215 s	0.380 s	0.986 s	0.368 s	1.098 s
DAISY descriptor	0.319 s	0.876 s	0.523 s	1.516 s	0.489 s	1.421 s

The algorithm's execution time is divided mainly into two processes; the first is the feature detection and description (Table 1), matching, and filtering, and the second one is the PnP solving method. If the object is not on the scene, the PnP solver takes longer due to the fact that it does not converge, and it performs all the defined number of iterations. On the other hand, the time for the first stage is usually stable. This only depends on the choice of the detector and descriptor, and on the smoothness of the image.

The PnP process was analyzed regarding the confidence parameter and the reprojection error parameter. These two parameters affect the performance of the algorithm in both time and pose estimation. To give a numerical idea of the influence of the parameters, Table 2 summarizes the average time for the algorithm, varying the parameters using FAST and SIFT. The reprojection error is the maximum allowed "projection" error for the inliers, and the confidence parameter's influence on the quality of the result.

Table 2. Computation times of the PnP (Perspective-n-Points) algorithm varying the confidence parameter and the reprojection error.

	Reprojection Error			
	3 pxs.	5 pxs.	7 pxs.	8 pxs.
$confidence = \mathbf{0.99}$	0.031 s	0.028 s	0.025 s	0.024 s
$confidence = \mathbf{0.999}$	0.036 s	0.031 s	0.027 s	0.026 s
$confidence = \mathbf{0.9999}$	0.039 s	0.034 s	0.028 s	0.028 s

Increasing the reprojection error increases the speed, but as shown in Figure 9, results in worsening of the position. Similarly, decreasing the confidence parameter speeds up the PnP algorithm, but decreases the quality of the result. Figure 9 shows the estimated position of an object, varying the reprojection parameter. It can be seen that the estimation on the Z axis (forward direction of the camera) is worst when the reprojection error increases. As the projections of the points are allowed to fall further, larger errors in translation and rotation can be produced. Additionally, these computation times are reduced up to 75% thanks to the optimization described in Algorithm 2.

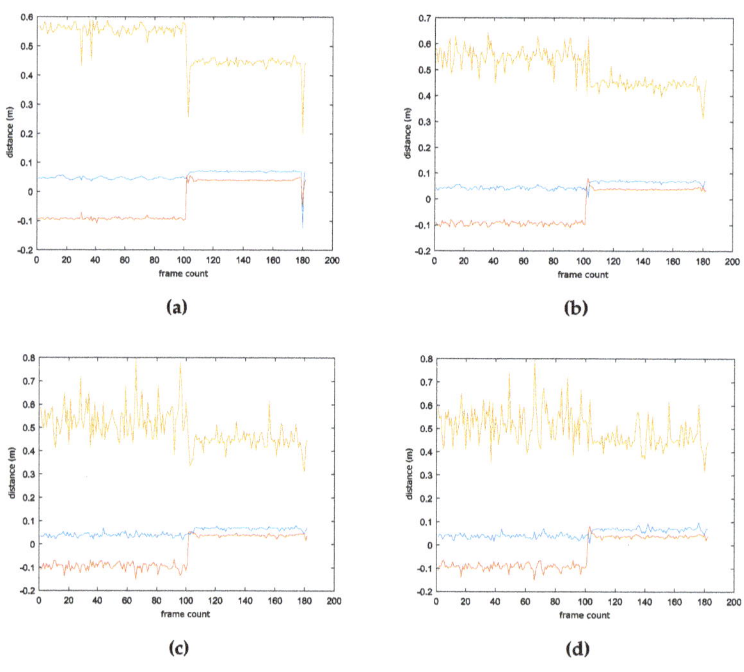

Figure 9. Result of pose estimation algorithm varying the reprojection error. Lines from top to bottom are: Z-coordinate (yellow), X-coordinate (blue), and Y-coordinate (red). Increasing the parameter decreases the quality of the results. However, as described in Table 2, it is slightly faster. (**a**) Reprojection error 3; (**b**) Reprojection error 5; (**c**) Reprojection error 7; (**d**) Reprojection error 8.

The algorithm is also proven to be robust to occlusions. The position of the object can be reconstructed with a small fraction of points of the model. Figure 10a,b shows the estimated position of an object occluded partially by a person. Additionally, in Figure 10c,d, one can see how the position of the object remains stable, even with the arm occluding the object during the grasping trajectory. It is noticeable that the position is more stable than the orientation against partial occlusions.

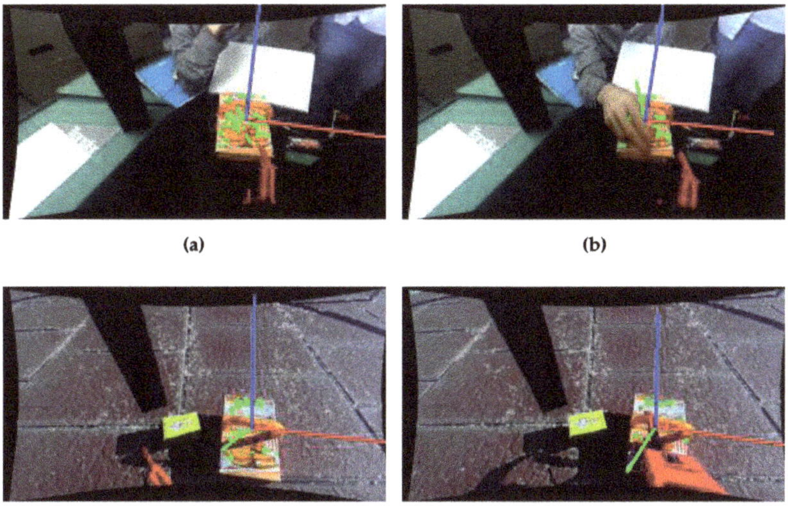

(a) (b) (c) (d)

Figure 10. Testing detection and position estimation with partial occlusions. (**a**) Non-occluded indoor; (**b**) Occluded indoor; (**c**) Non-occluded outdoor; (**d**) Occluded outdoor. The top figures show an indoor test, where the object is occluded by a human hand. In the bottom figures, the object is occluded by the arm of the UAV during the grasp process.

The performance of the FAST/SIFT features under different light conditions was also validated, as can be seen in Figure 11. This figure shows the result of the object detection and pose estimation with different types of lights and shadows (indoor, outdoor with shadows, and without shadows). The algorithm performs well in outdoor environments with and without shadows, as the camera automatically adjusts the exposure of the sensor. Indoors, the result is initially the same. However, as the exposure time of the camera increases, the image is more prone to have motion blur. This implies large variations in the feature descriptors. Because of this, the algorithm may lose the object tracking during fast movements.

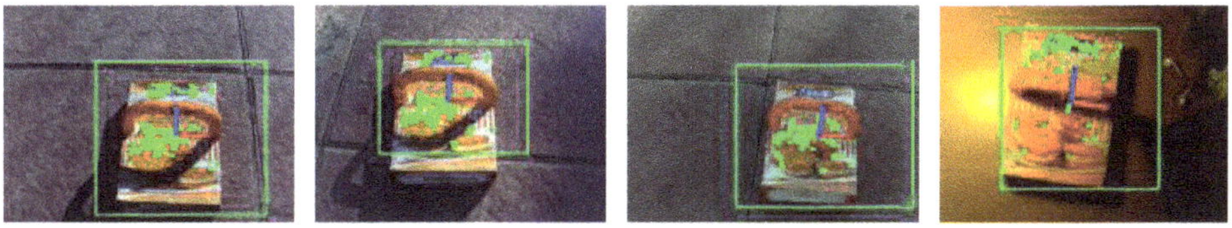

(a) (b) (c) (d)

Figure 11. Testing algorithm with different light conditions. (**a**) Without shadow; (**b**) Partial tree shadow; (**c**) Complete tree shadow; (**d**) Lamp light.

It was observed that the use of the FAST detector and SIFT descriptors produced the best results. In the learning stage, these features produced accurate models. Subsequently, the position estimation was recovered—in both indoor and outdoor environments—more easily than with the other feature descriptors. However, the computation time of this descriptor is too high. Indoors, a frame speed within 8 and 13 FPS was obtained without using Algorithm 2, and 15–19 FPS using it. Outdoors, however, due to the light conditions and the texture of the floor, the FPS decreased drastically to within

2 and 4 FPS without optimization and up to 11 FPS with the moving window. This happened because the feature detector detects more features outdoors, as images are sharper.

The second-best option is the use of the FAST detector and rBRIEF (rotated BRIEF). The computation time for this descriptor is significantly lower than SIFT, and the computation of distances in the PnP solver takes less time, as the descriptors are smaller. It works within 25–30 FPS indoors and reaches 25 outdoors thanks to the optimization. Nevertheless, this descriptor showed the worst behavior against variations in the scale.

Last but not least, the algorithm was tested with multiple objects at the same time in the scene. Figure 12 shows the results obtained by varying the chosen object model for the detection.

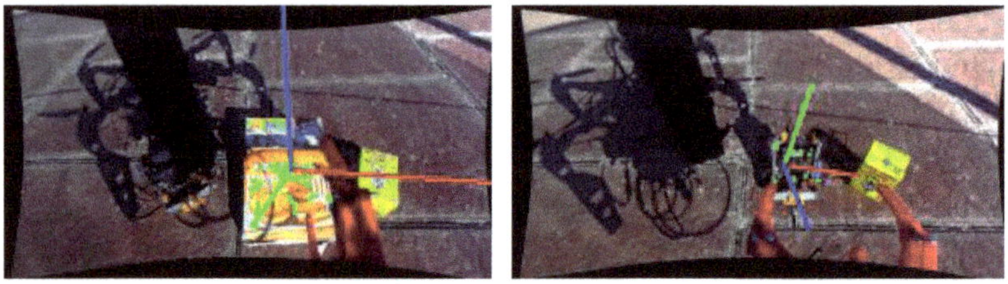

(a) (b)

Figure 12. Testing grasp with multiple objects. The algorithm is able to switch the targeted object according to any desired task if the model is learned. (**a**) Picking whoopies box; (**b**) Picking drilling tool.

5. Conclusions and Future Work

An on-board object detection method for an aerial robot that computes the needed information for the autonomous grasping of objects was developed. The algorithm was tested outdoors to test strong light conditions and its robustness against the vibrations generated by the UAV. The UAV was provided with a lightweight 3DOF arm for proof of concept of grasping objects.

In contrast to previous work, stereo cameras were chosen for two reasons: (1) to automate the learning process (the images are filtered using the stereo geometry, and the scale of the object is obtained automatically from the set of images without needing a manually calibrated dataset); and (2) for filtering bad features in the detection stage, making it more robust.

This can be used in several manipulation applications, such as inspection or maintenance of pipes or wind turbines. A drop-off/pick-up zone for objects (sensors, tools, etc.) can be selected, and the drone is able to pick up objects autonomously without requiring any information about the exact location. In contrast to RGB-D systems, the proposed method can be used robustly in outdoor environments. Furthermore, the method performs well under occlusions and the presence of outliers due to the feature-based modeling of the objects.

A speed comparison of different features has been made. This made it possible to choose the features that are better suited to the problem. As mentioned, the SIFT descriptors are more robust, as they perform well with different rotations and scales. However, this descriptor is slower than others, so if the UAV needs faster results, it is better to use other descriptors. rBRIEF (rotated BRIEF) is a good alternative. It is much faster than SIFT, and it is also invariant to rotations. Its main disadvantage is being less robust to scales.

As a future step, it might be interesting to compute the grasping points using quality metrics instead of choosing them manually at the learning stage. So far, all the tests have been performed by tying the UAV to a secure structure. The next step is to perform experiments while undertaking an autonomous flight. Finally, we want to speed up the feature detection using GPU to reduce the CPU computations and allow the UAV to perform more operations on the computer.

Acknowledgments: This work has been carried out in the framework of the AEROARMS (SI-1439/2015) EU-funded projects and the AEROMAIN (DPI2014-59383-C2-1-R) Spanish National Research project. The authors wish to thank to José María Aguilar and Javier Curado Soriano for the support given during the experiments.

Author Contributions: Pablo Ramon Soria, Begoña C. Arrue, Anibal Ollero conceived the methodology. Pablo Ramon Soria and Begoña C. Arrue designed the experiments; Pablo Ramon Soria performed the experiments; Pablo Ramon Soria and Begoña C. Arrue analyzed the data; Begoña C. Arrue and Anibal Ollero contributed materials and tools; Pablo Ramon Soria and Begoña C. Arrue wrote the paper.

Conflicts of Interest: The authors declare no conflict of interest.

References

1. Pajares, G. Overview and Current Status of Remote Sensing Applications Based on Unmanned Aerial Vehicles (UAVs). *Photogramm. Eng. Remote Sens.* **2015**, *81*, 281–329.
2. Roa, M.A.; Suárez, R. Grasp quality measures: Review and performance. *Auton. Robot.* **2015**, *38*, 65–88.
3. Gordon, I.; Lowe, D.G. What and Where: 3D Object Recognition with Accurate Pose. In *Toward Category-Level Object Recognition*; Ponce, J., Hebert, M., Schmid, C., Zisserman, A., Eds.; Springer: Berlin/Heidelberg, Germany, 2006; pp. 67–82.
4. Collet, A.; Berenson, D.; Srinivasa, S.S.; Ferguson, D. Object Recognition and Full Pose Registration from a Single Image for Robotic Manipulation. In Proceedings of the 2009 IEEE International Conference on Robotics and Automation, Kobe, Japan, 12–17 May 2009; pp. 3534–3541.
5. Lenz, I.; Lee, H.; Saxena, A. Deep Learning for Detecting Robotic Grasps. *Int. J. Robot. Res.* **2015**, *34*, 705–724.
6. Malisiewicz, T.; Gupta, A.; Efros, A.A. Ensemble of Exemplar-SVMs for Object Detection and Beyond. In Proceedings of the 2011 IEEE International Conference on Computer Vision (ICCV), Barcelona, Spain, 6–13 November 2011.
7. Rai, A.; Patchaikani, P.K.; Agarwal, M.; Gupta, R.; Behera, L. Grasping Region Identification in Novel Objects Using Microsoft Kinect. In *Neural Information Processing*; Huang, T., Zeng, Z., Li, C., Leung, C.S., Eds.; Springer: Berlin/Heidelberg, Germany, 2012; pp. 172–179.
8. Roy, N.; Newman, P.; Srinivasa, S. Recognition and Pose Estimation of Rigid Transparent Objects with a Kinect Sensor. In *Robotics: Science and Systems VIII*; MIT Press: Los Angeles, CA, USA, 2013.
9. Zhu, M.; Derpanis, K.G.; Yang, Y.; Brahmbhatt, S.; Zhang, M.; Phillips, C.; Lecce, M.; Daniilidis, K. Single image 3D object detection and pose estimation for grasping. In Proceedings of the 2014 IEEE International Conference on Robotics and Automation (ICRA), Hong Kong, China, 31 May–7 June 2014; pp. 3936–3943.
10. Marchand, E.; Bouthemy, P.; Chaumette, F.; Moreau, V. Robust real-time visual tracking using a 2D-3D model-based approach. In Proceedings of the Seventh IEEE International Conference on Computer Vision, Kerkyra, Greece, 20–27 September 1999; Volume 1, pp. 262–268.
11. Klingbeil, E.; Rao, D.; Carpenter, B.; Ganapathi, V.; Ng, A.Y.; Khatib, O. Grasping with application to an autonomous checkout robot. In Proceedings of the 2011 IEEE International Conference on Robotics and Automation (ICRA), Shanghai, China, 9–13 May 2011; pp. 2837–2844.
12. Saxena, A.; Driemeyer, J.; Ng, A.Y. Robotic Grasping of Novel Objects Using Vision. *Int. J. Robot. Res.* **2008**, *27*, 157–173.
13. Dragiev, S.; Toussaint, M.; Gienger, M. Uncertainty aware grasping and tactile exploration. In Proceedings of the 2013 IEEE International Conference on Robotics and Automation (ICRA), Karlsruhe, Germany, 6–10 May 2013; pp. 113–119.
14. Pelossof, R.; Miller, A.; Allen, P.; Jebara, T. An SVM learning approach to robotic grasping. In Proceedings of the 2004 IEEE International Conference on Robotics and Automation, New Orleans, LA, USA, 26 April–1 May 2004; Volume 4, pp. 3512–3518.
15. Pounds, P.E.I.; Bersak, D.R.; Dollar, A.M. Grasping from the air: Hovering capture and load stability. In Proceedings of the 2011 IEEE International Conference on Robotics and Automation (ICRA), Shanghai, China, 9–13 May 2011; pp. 2491–2498.
16. Spica, R.; Franchi, A.; Oriolo, G.; Bülthoff, H.H.; Giordano, P.R. Aerial grasping of a moving target with a quadrotor UAV. In Proceedings of the 2012 IEEE/RSJ International Conference on Intelligent Robots and Systems, Piscataway, NJ, USA, 7–12 October 2012; pp. 4985–4992.

17. Orsag, M.; Korpela, C.; Pekala, M.; Oh, P. Stability control in aerial manipulation. In Proceedings of the 2013 American Control Conference, Washington, DC, USA, 17–19 June 2011; pp. 5581–5586.
18. Danko, T.W.; Chaney, K.P.; Oh, P.Y. A parallel manipulator for mobile manipulating UAVs. In Proceedings of the 2015 IEEE International Conference on Technologies for Practical Robot Applications (TePRA), Woburn, MA, USA, 11–12 May 2015; pp. 1–6.
19. Fabresse, F.R.; Caballero, F.; Maza, I.; Ollero, A. Localization and mapping for aerial manipulation based on range-only measurements and visual markers. In Proceedings of the 2014 IEEE International Conference on Robotics and Automation (ICRA), Hong Kong, China, 31 May–7 June 2014; pp. 2100–2106.
20. Buonocore, L.R.; Cacace, J.; Lippiello, V. Hybrid visual servoing for aerial grasping with hierarchical task-priority control. In Proceedings of the 23th Mediterranean Conference on Control and Automation (MED), Torremolinos, Spain, 16–19 June 2015; pp. 617–623.
21. Laiacker, M.; Schwarzbach, M.; Kondak, K. Automatic aerial retrieval of a mobile robot using optical target tracking and localization. In Proceedings of the 2015 IEEE Aerospace Conference, Big Sky, MT, USA, 7–14 March 2015; pp. 1–7.
22. Thomas, J.; Loianno, G.; Sreenath, K.; Kumar, V. Toward Image Based Visual Servoing for Aerial Grasping and Perching. In Proceedings of the 2014 IEEE International Conference on Robotics and Automation (ICRA), Hong Kong, China, 31 May–7 June 2014.
23. Ramon Soria, P.; Bevec, R.; Arrue, B.C.; Ude, A.; Ollero, A. Extracting Objects for Aerial Manipulation on UAVs Using Low Cost Stereo Sensors. *Sensors* **2016**, *16*, 700.
24. Small Mini PC. Available online: http://www.intel.com/content/www/us/en/nuc/overview.html (accessed on 5 January 2017).
25. The World's First 3D Camera for Depth Sensing and Motion Tracking. Available online: https://www.stereolabs.com/ (accessed on 5 January 2017).
26. Lowe, D.G. Distinctive Image Features from Scale-Invariant Keypoints. *Int. J. Comput. Vis.* **2004**, *60*, 91–110.
27. Alhwarin, F.; Ristić-Durrant, D.; Gräser, A. VF-SIFT: Very Fast SIFT Feature Matching. In *Pattern Recognition*; Goesele, M., Roth, S., Kuijper, A., Schiele, B., Schindler, K., Eds.; Springer: Berlin/Heidelberg, Germany, 2010; pp. 222–231.
28. Grabner, M.; Grabner, H.; Bischof, H. Fast Approximated SIFT. In *Computer Vision – ACCV 2006*; Narayanan, P.J., Nayar, S.K., Shum, H.Y., Eds.; Springer: Berlin/Heidelberg, Germany, 2006; pp. 918–927.
29. Bay, H.; Tuytelaars, T.; Van Gool, L. SURF: Speeded up Robust Features. In *Computer Vision – ECCV 2006*; Leonardis, A., Bischof, H., Pinz, A., Eds.; Springer: Berlin/Heidelberg, Germany, 2006; pp. 404–417.
30. Rublee, E.; Rabaud, V.; Konolige, K.; Bradski, G. ORB: An Efficient Alternative to SIFT or SURF. In Proceedings of the 2011 International Conference on Computer Vision, Barcelona, Spain, 6–13 November 2011; pp. 2564–2571.
31. Tola, E.; Lepetit, V.; Fua, P. A Fast Local Descriptor for Dense Matching. In Proceedings of the IEEE Conference on Computer Vision and Pattern Recognition, Anchorage, AK, USA, 23–28 June 2008.
32. Calonder, M.; Lepetit, V.; Strecha, C.; Fua, P. BRIEF: Binary Robust Independent Elementary Features. In *Computer Vision—ECCV 2010*; Daniilidis, K., Maragos, P., Paragios, N., Eds.; Springer: Berlin/Heidelberg, Germany, 2010; pp. 778–792.
33. Rosten, E.; Porter, R.; Drummond, T. Faster and Better: A Machine Learning Approach to Corner Detection. *IEEE Trans. Pattern Anal. Mach. Intell.* **2010**, *32*, 105–119.
34. Muja, M.; Lowe, D.G. Fast Matching of Binary Features. In Proceedings of the 2012 Ninth Conference on Computer and Robot Vision (CRV), Toronto, ON, Canada, 28–30 May 2012; pp. 404–410.
35. Collet Romea, A.; Berenson, D.; Srinivasa, S.; Ferguson, D. Object Recognition and Full Pose Registration from a Single Image for Robotic Manipulation. In Proceedings of the IEEE International Conference on Robotics and Automation (ICRA '09), Kobe, Japan, 12–17 May 2009 .
36. Triggs, B.; McLauchlan, P.F.; Hartley, R.I.; Fitzgibbon, A.W. Bundle Adjustment—A Modern Synthesis. In *Vision Algorithms: Theory and Practice*; Springer: London, UK, 2000; pp. 298–372.
37. Lourakis, M.A.; Argyros, A. SBA: A Software Package for Generic Sparse Bundle Adjustment. *ACM Trans. Math. Softw.* **2009**, *36*, 1–30.
38. Marquardt, D.W. An Algorithm for Least-Squares Estimation of Nonlinear Parameters. *J. Soc. Ind. Appl. Math.* **1963**, *11*, 431–441.

39. Fischler, M.A.; Bolles, R.C. Random Sample Consensus: A Paradigm for Model Fitting with Applications to Image Analysis and Automated Cartography. *Commun. ACM* **1981**, *24*, 381–395.
40. Open-Hardware Autopilot. Available online: https://pixhawk.org/ (accessed on 5 January 2017).
41. Arduino. Available online: https://www.arduino.cc (accessed on 5 January 2017).

sensors

MDPI

Article

Small UAS-Based Wind Feature Identification System Part 1: Integration and Validation

Leopoldo Rodriguez Salazar *, Jose A. Cobano and Anibal Ollero

Robotics, Vision and Control Group, Universidad de Sevilla, 41092 Sevilla, Spain; jcobano@us.es (J.A.C.); aollero@us.es (A.O.)

* Correspondence: lrodriguez15@us.es; Tel.: +34-663-225-662

Academic Editors: Felipe Gonzalez Toro and Antonios Tsourdos

Received: 31 October 2016; Accepted: 19 December 2016; Published: 23 December 2016

Abstract: This paper presents a system for identification of wind features, such as gusts and wind shear. These are of particular interest in the context of energy-efficient navigation of Small Unmanned Aerial Systems (UAS). The proposed system generates real-time wind vector estimates and a novel algorithm to generate wind field predictions. Estimations are based on the integration of an off-the-shelf navigation system and airspeed readings in a so-called direct approach. Wind predictions use atmospheric models to characterize the wind field with different statistical analyses. During the prediction stage, the system is able to incorporate, in a big-data approach, wind measurements from previous flights in order to enhance the approximations. Wind estimates are classified and fitted into a Weibull probability density function. A Genetic Algorithm (GA) is utilized to determine the shaping and scale parameters of the distribution, which are employed to determine the most probable wind speed at a certain position. The system uses this information to characterize a wind shear or a discrete gust and also utilizes a Gaussian Process regression to characterize continuous gusts. The knowledge of the wind features is crucial for computing energy-efficient trajectories with low cost and payload. Therefore, the system provides a solution that does not require any additional sensors. The system architecture presents a modular decentralized approach, in which the main parts of the system are separated in modules and the exchange of information is managed by a communication handler to enhance upgradeability and maintainability. Validation is done providing preliminary results of both simulations and Software-In-The-Loop testing. Telemetry data collected from real flights, performed in the Seville Metropolitan Area in Andalusia (Spain), was used for testing. Results show that wind estimation and predictions can be calculated at 1 Hz and a wind map can be updated at 0.4 Hz. Predictions show a convergence time with a 95% confidence interval of approximately 30 s.

Keywords: wind prediction; wind estimation; UAS; wind shear; gust; multi-platform integration

1. Introduction

Current UAS technology has advanced in such a way that any unexperienced user is able to plan a route with relatively good accuracy. As the reliability has increased, applications using small UAS are growing rapidly. In addition, nonlinear natural effects, such as winds, can be compensated even with Commercial-Off-The-Shelf (COTS) components. Nevertheless, to compensate wind effects efficiently, the use of a sensor that can provide wind measurements is sometimes limited by the platform payload and the cost. This leads to inefficient attitude compensations, producing drift and sometimes missing waypoints, which may result likely into higher energy consumptions [1]. Currently, there are several research efforts to provide wind estimations without a direct measurement of the wind. Langelaan, et al. [2] proposed two ways of estimating the wind field, both using measurements from a standard sensor suite, i.e., Inertial Measurement Unit (IMU) and Global Navigation Satellite System (GNSS). The first method consists in a comparison between predictions generated with a dynamic model and actual measurements of the aircraft motion. The second one consists in the estimation on wind acceleration and its derivatives from the GNSS velocity, i.e., using the pseudorange rate of change together with direct measurements of the vehicle acceleration. Johansen, et al. [3] have developed a method in which the wind is estimated using an observer which leads to the calculation of sideslip and Angle-of-Attack (AOA). They estimate the wind from the difference between the platform velocity relative to the wind and the velocities in the body frame utilizing a Kalman Filter, which is a similar approach than the one presented in [4]. Other approaches, such as the one presented by Larrabee et al. [5] uses flow angle sensors and a Pitot tube with two Unscented Kalman Filters (UKF). This innovation compares information from different platforms in order to produce real time estimates. Neummann & Bartholmai [6] produced wind estimations with a quadcopter UAS without the use of any additional sensors rather than its standard sensor suite, even without a dedicated airspeed sensor, and/or anemometer, based mainly in the wind triangle and the vector difference between the ground speed and an estimated speed. Condomines, et al. [7] have published a set of results of a flight campaign with estimations of the wind field considering non-linear wind estimation with an square-root UKF in which the platform was equipped with an standard sensor suite which provides measurements that estimate angle of attack and sideslip.

Previously, as part of this research effort, the authors have introduced an algorithm that can estimate the wind field in such way that the different wind features (gust, shear, etc.) can be identified separately with a method that calculates statistical properties and based on distribution models of the wind, such as the $1-\cos$ model for gusts and the wind shear model [8,9]. Lawrance & Sukkarieh [4] propose a method that incorporates a Gaussian regression in order to predict within a limited amount of time (up to 10 s) the local wind field despite the feature that is present. The identification of features, such as shear, thermals [10], gusts is of particular importance in the so-called atmospheric energy harvesting [4]. On this field, several authors such as Cutler et al. [11] and Chakrabarty et al. [12] have published successful results on the generation of static soaring trajectories and others, such as Montella & Spletzer [13] and Bird et al. [14] have developed systems that produce and follow dynamic soaring trajectories. In addition, Bencatel et al. [15] have performed an analysis on necessary conditions for dynamic soaring and how this problem can be seen as a function of aircraft and environmental parameters. Despite the advances in the generation of the trajectories (Rucco et al. [16]), there are few methods for identifying wind features, and the creation of real-time algorithms for energy harvesting should be addressed and improved. A few authors have described the integration of such methods in a level of detail that can identify areas of opportunities in hardware selection, software architecture, computational time, etc.

This paper presents and describes the detailed integration at a hardware and software level of the system. This enables the estimation of the wind vector and identification of the features in real-time with a standard sensor suite. In the previous works of the authors [8,9], the identification system was firstly introduced. In the work presented, wind features (wind shear and discrete gusts) are identified separately based on statistical analysis by fitting wind estimates into a Weibull distribution. The wind identification system allows the generation of a 3-dimensional wind map with predictions of what the wind vector would be at a certain location. The system presents innovations regarding its architecture, and adds the capability for continuous gust identification. Preliminary results are presented in two stages: simulations of the different features and a Software-In-The-Loop (SITL) testbed fed up with previous flight information. The verification with actual experiments is going to be presented in a follow-up manuscript.

The paper is organized as follows: Section 2 presents a brief summary of the methods and statistical analysis utilized. Section 3 describes the hardware and software architecture of the system. Section 4 shows the validation results of the different components. Section 5 presents a discussion on the obtained results. Finally, conclusions are presented in Section 6.

2. Wind Field Estimation and Wind Field Prediction

The generation of the wind field considers both the estimation and prediction processes. Both are equally important, however they not necessary have to occur at the same time and rate. This section provides the insights of the selected methods for these operations.

2.1. Wind Field Estimation

The selected method for estimation process was originally presented in [2]. It estimates the wind without the use of an observer (Kalman or Particle filters) by using the velocity vector calculated by the GNSS module together with measurements of the vehicle acceleration and a portion of the state vector of the platform. The goal is to calculate the wind acceleration and velocity using the relationship between the GNSS velocity and the body-axis state from the COTS Autopilot Module (APM).

Consider a UAS located in $\mathbf{r}$ in the inertial frame $\mathbf{I}$. The unit vectors of this frame are defined as $(\hat{x}_\mathbf{I}, \hat{y}_\mathbf{I}, \hat{z}_\mathbf{I})$. Consider also a body frame $\mathbf{b}$ with unit vectors $(\hat{x}_\mathbf{b}, \hat{y}_\mathbf{b}, \hat{z}_\mathbf{b})$ with its origin at the center of mass of the vehicle. The wind vector $\mathbf{w}$ and the airmass-relative velocity v_a are illustrated in Figure 1.

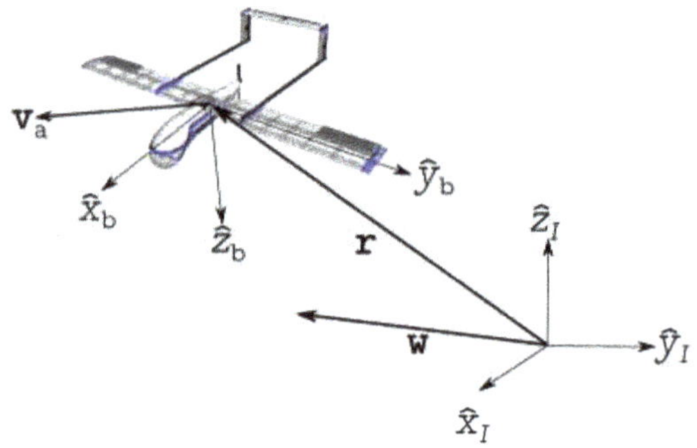

Figure 1. Reference frames utilized in the formulation of the identification of wind vector problem.

The velocity of the vehicle expressed in the inertial frame is.

$$\dot{\mathbf{r}} = \mathbf{v}_a + \mathbf{w} \tag{1}$$

From Equation (1), two relationships are used to characterize the instantaneous wind vector. Further details on the derivation of these relationships can be found in [2,8,9].

The first one indicates the correspondence between the vehicle kinematics and the GNSS velocity expressed with respect to the $\mathbf{I}$ frame:

$$\begin{bmatrix} w_x \\ w_y \\ w_z \end{bmatrix}_\mathbf{I} = \begin{bmatrix} \dot{x} \\ \dot{y} \\ \dot{z} \end{bmatrix}_{\mathbf{I}_{\text{GNSS}}} - \left(\mathbf{C}_\mathbf{I}^\mathbf{b}\right)^{-1} \begin{bmatrix} u \\ v \\ w \end{bmatrix}_\mathbf{b} \tag{2}$$

where $(w_x, w_y, w_z)_\mathbf{I}$ is the wind velocity vector, $(\dot{x}, \dot{y}, \dot{z})_{\mathbf{I}_{\text{GNSS}}}$ is the GNSS velocity vector and $(u, v, w)_\mathbf{b}$ are the components of the velocity with respect the air mass expressed in the body frame and assumed to be calculated by the autopilot. $\mathbf{C}_\mathbf{I}^\mathbf{b}$ is the Direction Cosine Matrix, which transforms a vector expressed in the inertial frame to one expressed in the body frame.

The second relationship aims to calculate the wind acceleration expressed in the body frame at the previous step $(k-1)$. Since the IMU body-axis accelerations expressed in the body frame are given by.

$$\begin{bmatrix} a_x \\ a_y \\ a_z \end{bmatrix}_{\mathbf{b}} = \begin{bmatrix} \dot{w}_x \\ \dot{w}_y \\ \dot{w}_z \end{bmatrix}_{\mathbf{b}} + \begin{bmatrix} \dot{u} \\ \dot{v} \\ \dot{w} \end{bmatrix}_{\mathbf{b}} + \begin{bmatrix} qw - rv + g\sin\theta \\ ru - pw - g\cos\theta\sin\phi \\ pv - qu - g\cos\theta\cos\phi \end{bmatrix} + \mathbf{b}_{\mathrm{imu}} + \mathbf{n}_{\mathrm{imu}} \quad (3)$$

where (a_x, a_y, a_z) is the body axis accelerations vector, (p, q, r) is the rotation rate vector, g is the gravity force, θ is the roll angle, ϕ is the pitch angle, $\mathbf{b}_{\mathrm{imu}}$ is the accelerometer bias and $\mathbf{n}_{\mathrm{imu}}$ is the white noise from the IMU.

Since the calculation of the rate of change of the velocity with respect the air mass can't be determined with the on-board sensors it can be estimated with a second order numerical differentiation. Therefore, the wind speed rate of change at the previous step $k-1$ was derived:

$$\begin{bmatrix} \dot{w}_x \\ \dot{w}_y \\ \dot{w}_z \end{bmatrix}_{\mathbf{b},k-1} = \begin{bmatrix} a_x \\ a_y \\ a_z \end{bmatrix}_{\mathbf{b},k-1} - \begin{bmatrix} b_x \\ b_y \\ b_z \end{bmatrix}_{k-1} + \begin{bmatrix} -g\sin\theta \\ g\cos\theta\sin\phi \\ g\cos\theta\cos\phi \end{bmatrix}_{k-1} - \begin{bmatrix} qw - rv \\ pw - ru \\ qu - pv \end{bmatrix}_{k-1} - \frac{1}{2\Delta t}\begin{bmatrix} u_k - u_{k-2} \\ v_k - v_{k-2} \\ w_k - w_{k-2} \end{bmatrix} \quad (4)$$

Equations (2) and (4) are necessary for trajectory planning. Using different sources of error increases the reliability of the solution. Moreover, the calculation of wind acceleration and velocity based on actual inertial and GNSS measurements ensures bounded errors which is a key advantage compared to other methods (e.g., the use of a dynamic model).

2.2. Wind Field Prediction

The wind field could be estimated at each time step from the data provided by the IMU, GNSS and the vehicle dynamics as shown in Section 2.1. Previous results [8,9] show that the estimation algorithm produce accurate results. However, these estimations are not sufficient if the information is going to be used for precise trajectory planning. Therefore, a prediction stage is needed so that the wind field could be inferred within a reasonable time window.

In this context, three models of different wind features have been selected: the wind shear model, the discrete gust model and the continuous (Dryden) wind turbulence model. These are widely used in the aerospace industry and are contained in the Military Specification MIL-F-8785C [17] and Military Handbook MIL-HDBK-1797 [18].

2.2.1. Wind Shear Model

The magnitude of the wind is modeled by the following equation:

$$W_{\mathrm{shear}} = W_{20}\frac{\ln\frac{h}{z_0}}{\ln\frac{6.096}{z_0}}, 1\,\mathrm{m} < h < 300\,\mathrm{m} \quad (5)$$

where W_{shear} is the mean wind speed, W_{20} is the wind speed at 20 ft (6.096 m) and z_0 varies depending on the flight phase. However, a value of 0.0457 m (0.15 ft) is selected due to the characteristics of the platform, i.e., flight below 1000 ft. Finally, h is the actual altitude of the vehicle.

The wind shear is illustrated in Figure 2.

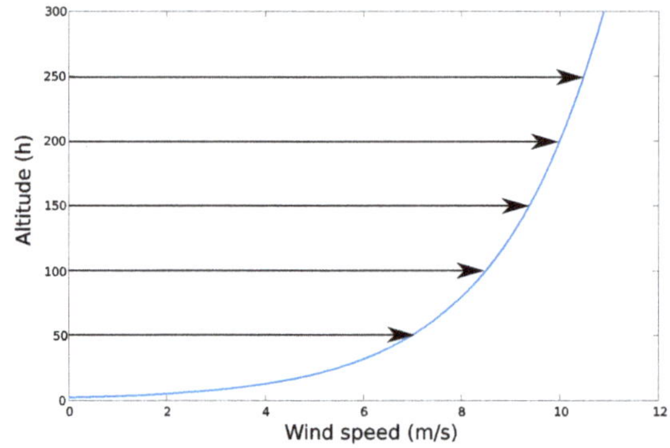

Figure 2. Typical shear profile that shows the increase of wind speed over as the altitude increases. The relationship is exponential between the two variables.

In order to characterize the wind shear, it is assumed that the wind varies with the altitude following the Prandtl Ratio based on an Empirical Power Law (EPL) [8]:

$$\frac{W_1}{W_2} = \left(\frac{h_1}{h_2}\right)^{\xi} \tag{6}$$

where ξ is the Prandtl coefficient that shapes the EPL function. (W_1, W_2) are two wind speeds and (h_1, h_2) are the corresponding altitudes.

2.2.2. Discrete Gust Model

This model uses the implementation of the $1 - \cos$ shape and its mathematical representation is as follows:

$$W_{\text{gust}} = \begin{cases} 0 & x < 0 \\ \frac{W_m}{2}(1 - \cos\frac{\pi x}{d_m}) & 0 \leq x \leq d_m \\ W_m & x > d_m \end{cases} \tag{7}$$

where W_m is the magnitude of the gust and d_m is the gust length and x is the distance traveled.

The discrete gust is illustrated in Figure 3.

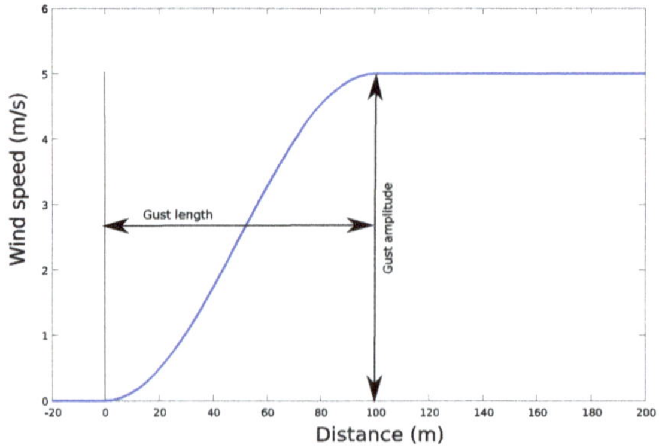

Figure 3. Typical discrete gust profile that shows a growth over the wind on a short period of time from the initial wind speed of the gust magnitude, and a permanent increase at the end of the gust length.

2.2.3. Continuous Gust Model

The selected model for continuous gust utilizes the Dryden spectral representation in which the turbulence is considered a stochastic process defined by velocity spectra. In [17–19], the power spectral densities are defined. Note that for simulation purposes the Low-Altitude scale lengths have been used.

The number of variables in the continuous gust model is vast. Therefore, inferring these values from actual wind measurements trough a regression is very complex. Thus, this model is used only as a simulation input.

Two methods are proposed for continuous gust identification in both short and long term. The first one incorporates a Standard Gaussian Process (GP) Regression [20].

Considering a set of vertical wind observations of size $\bar{M}$, $\hat{\mathbf{W}}_{\mathbf{z}} = \hat{W}_{z,i}|_{i=1}^{\hat{M}}$. The wind speed prediction $\bar{W}_p(x)$ at any location x can be expressed as:

$$\bar{W}_p(x) = \sum_{i=1}^{\bar{M}} k_i \hat{W}_{z,i} \tag{8}$$

in which k_i is the i-th coefficient of the linear combination of wind measurements $\hat{\mathbf{W}}_{\mathbf{z}}$. Based on the work presented by Park et al. [21], an optimal coefficient is determined by minimizing the prediction error.

$$\min_{\mathbf{k}} E\left[(\bar{W}_p(x) - W_p(x))^2\right] = \min_{\mathbf{k}}(k^T\left[\mathbf{Q}(\mathbf{X},\mathbf{X}) + \sigma_n^2\mathbf{I}\right]k - 2k^T\mathbf{q}(\mathbf{X},x) + q(x,x)) \tag{9}$$

which can be determined by calculating the covariance matrix $\mathbf{Q}(\mathbf{X},\mathbf{X})$ and the covariance vector $\mathbf{q}(\mathbf{X},x)$ between every two observations at locations $\mathbf{X}$ and x; finally $q(x,x)$ represents the covariance value . This leads to express standard GP regression of the linear predictor as:

$$\bar{p}(X) = \mathbf{k}^T\hat{\mathbf{W}}_{\mathbf{z}} = \mathbf{q}(x,\mathbf{X})\left[\mathbf{Q}(\mathbf{X},\mathbf{X}) + \sigma_n^2\mathbf{I}\right]^{-1}\hat{\mathbf{W}}_{\mathbf{z}} \tag{10}$$

and the covariance value $cov(\bar{p}(x)))$ can be expressed as:

$$cov(\bar{p}(x)) = q(x,x) - \mathbf{q}(x,\mathbf{X})\left[\mathbf{Q}(\mathbf{X},\mathbf{X}) + \sigma_n^2\mathbf{I}\right]^{-1}\mathbf{q}(x,\mathbf{X}) \tag{11}$$

where σ_n^2 is the measurement noise covariance.

An alternative approach can be used to perform long-term predictions by employing a non homogeneous regression prediction model. Lerch and Thoraninsdottir [22] have performed a comparison between three non-homogeneous regression model, which allows to produce predictions in day time window. Since the intention of the intended testing flight campaigns is to store data into a single database, a big amount of data can be utilized to perform the predictions with the selected regression model.

At this stage, the truncated normal model was selected as a form of wind estimation. Being W the wind speed and $X_1, \ldots, X_j$ the ensemble member forecasts, the predicted distribution of W can be approximated by a truncated normal distribution:

$$W|X_1, \ldots, X_j \sim \mathbf{N}_{[0,\infty)}(\mu, \sigma^2) \tag{12}$$

where the mean μ is an affine function of the ensemble forecast and the variance σ^2 is an affine function of the ensemble variance. If these exchange members are exchangeable [22], the distribution function of the Truncated Normal (TN) distribution $F(z)$ is given by:

$$F(z) = \Phi\left(\frac{\mu}{\sigma}\right)^{-1} \Phi\left(\frac{z-\mu}{\sigma}\right) \tag{13}$$

for $z > 0$, where Φ is the cumulative standard normal distribution.

This is indeed a simple non-homogeneous method. However, results indicate that the training period to produce accurate predictions in one-day ahead forecasts is of the equivalent 30 days of continuous measurements [22].

2.2.4. Weibull Distribution

The Weibull distribution is a key part of the research performed as many datasets, including wind speed have been proved to fit in. The Weibull distribution has three main parameters, the shaping factor κ, the scaling factor ν and the threshold. Given a dataset $\mathbf{W}= (W_1 ... W_n)$, the Weibull probability density function can be expressed as a function of a wind magnitude W [8]:

$$f(W) = \frac{\kappa}{\nu}\frac{W}{\nu}^{\kappa-1} e^{\frac{W}{\nu}\kappa} \tag{14}$$

From this function the most probable wind speed W_{mp} at a particular location can be expressed in terms of the Weibull parameters:

$$W_{\mathrm{mp}} = \nu(1 - \frac{1}{\kappa})^{\frac{1}{\kappa}} \tag{15}$$

A typical Weibull distribution is shown in Figure 4.

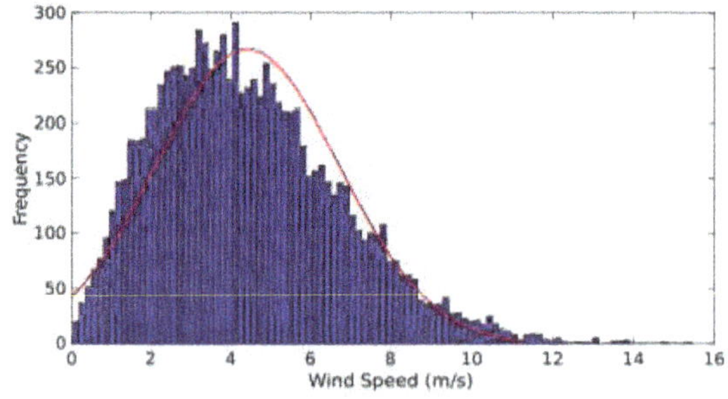

Figure 4. Typical weibull distribution from a National Oceanic and Atmospheric Administration (NOAA) measurement station at an altitude of 30 m.

2.2.5. Genetic Algorithm and the Weibull Distribution Parameters

Genetic Algorithm is a searching method that simulates the evolution theory. The method aims to generate possible random solutions (chromosomes) to a problem stated in a for of an objective function (fitness function). A given set of chromosomes is a population in a generation. Every one of them will produce evolved chromosomes based on three operations: reproduction, crossover and mutation. Details in the implementation of the GA can be found in [23].

In order to calculate the shaping parameter κ of a Weibull-distributed data set, one has to calculate the residual error ϵ between the measured mean and the standard deviation of the wind estimates (see Section 2.1) and a theoretical mean and standard deviation derived from the Weibull distribution moment, as stated in the following equation:

$$\epsilon = \sigma^2/\mu^2 - \frac{\Gamma(1+2/\kappa)+\Gamma^2(1+1/\kappa)}{\Gamma^2(1+1/\kappa)} \quad (16)$$

where Γ is the gamma function σ is the standard deviation of the wind estimates and μ is the mean of the wind estimates.

Once Equation (16) converges to a desired tolerance value, an acceptable κ value is obtained and the scaling parameter can be calculated based on the following equation:

$$\mu = v\Gamma(1+1/k) \quad (17)$$

2.3. Wind Mapping

In order to generate a full 3D wind map, a combination of methods is required. Initially, the work presented in [8] suggests the use of a Newton polynomial extrapolation in order to generate local values of the Prandtl coefficient, ξ, from which the shaping and scaling parameters can be calculated in order to obtain a local most probable wind speed W_{mp}. This is true in case of the presence of a wind shear. However, if a gusts is detected, the extrapolation will accumulate error, producing an inaccurate wind map.

Therefore, a 3D map can be generated based on the feature that is detected. The GP regression shown in Section 2.2.3 allows the generation of a wind map with predictions based on the estimates found at position X. These estimates carry information of the covariance which is continuously updated with the different feature detection algorithms. Details on the wind mapping algorithm and the results are to be found in the Part 2 of this research.

3. System Architecture

This section describes the hardware, software and communication architecture of the wind identification system.

The selected hardware takes mainly two COTS components in order to perform the estimation and the prediction of the wind field. The selected autopilot is the Pixhawk (3D Robotics, Berkeley, CA, USA) which is based on the PX4 open-hardware project. The characteristics of this module can be found in [24]. Given the processor characteristics, the wind estimation and wind prediction algorithms have to reside in a dedicated computer. The selected computer is the ODROID-C2 (Hardkernel, Anyang Gyeonggi-do, Korea) [25] that contains a quad-core processor at 2 Ghz at 64 bit. The main characteristics are enumerated in Table 1.

Table 1. ODROID-C2 Specifications from [25].

CPU	Amlogic S905 SoC, 4×ARM Cortex-A53 2 GHz, 64 bit ARMv8 Architecture @28 nm
RAM	2 GB 32 bit DDR3 912 MHz
Flash Storage	Micro-SD UHS-1 @83MHz/SDR50, eMMC5.0 storage option
ADC	10 bit SAR 2 channels
Size	85 mm × 56 mm (3.35 inch × 2.2 inch)
Weight	40 g (1.41 oz)

The required algorithms need an additional platform that shall do the data analysis of the stored variables. All the wind estimates and predictions are kept in a database. As more flights are to be performed as part of the validation, verification and other applications, the wind database will grow. Due to its size and for reliability, a ground station contains the wind prediction and estimation database. A PC with an ©Intel Core(TM) i755000U CPU (Seattle, WA, USA) at 2.4 GHz with 16 GB of RAM was used.

The software design has evolved deeply since its conception. Initially the system was created in a multi-platform way with different computing languages interacting at a very high level. The proposed architecture intends to minimize these interfaces at component-level in order to enhance maintainability and upgradeability of the system. In the architecture shown in Figure 5, the autopilot sends information from a request made by the communication module, this information is sent to the wind estimation algorithm that generates wind estimates that will go to the prediction block which uses information from the wind database and also calls the storage module once a prediction is performed.

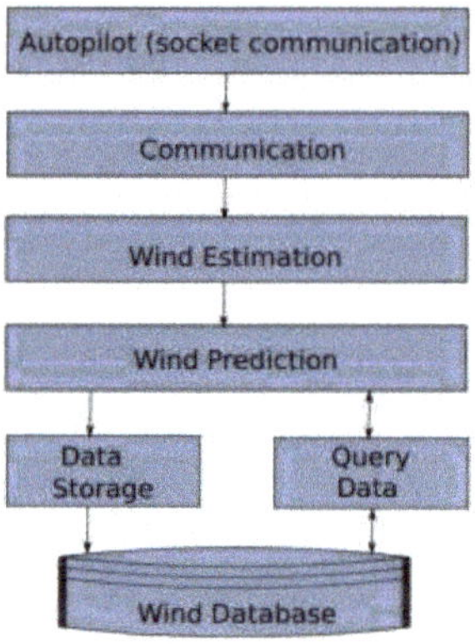

Figure 5. High level software architecture that shows the flow of information of the wind identification system.

As it was mentioned before, the designed architecture considered a diversity of programming languages and even various operating systems. The modules communication of this system was done in Linux with pymavlink (MAVLINK (Micro Air Vehicle Communication Protocol) is a communication library for UAS that can pack C-structures over a serial channel and send this packets with other modules. It was originally released in 2009 by Lorenz Meier with a GNU Lesser General Public Licence (LPGL). Pymavlink is a Python implementation of MAVLINK [26]). The wind estimations and the simulation test-bed with the models shown in Section 2 were done using MATLAB, Simulink® (The MathWorks, Inc., Natick, MA, USA). Finally, in order to generate a database, initially the idea was to create comma separated (*.csv) files, however, Structured Query Langate (SQL) was selected to be utilized for Database Accessing and Management, which required Java and C++ connector of SQL.

After observing problems in the synchronization of the systems, the solution was to migrate everything to C++ leaving only the database management in JAVA with the MySQL® (Oracle Corporation, Redwood Shores, CA, USA) C++ connector. The concept was to build a modular architecture that runs under a handler that manages the communication between the various modules that interact to identify the wind (see Figure 6).

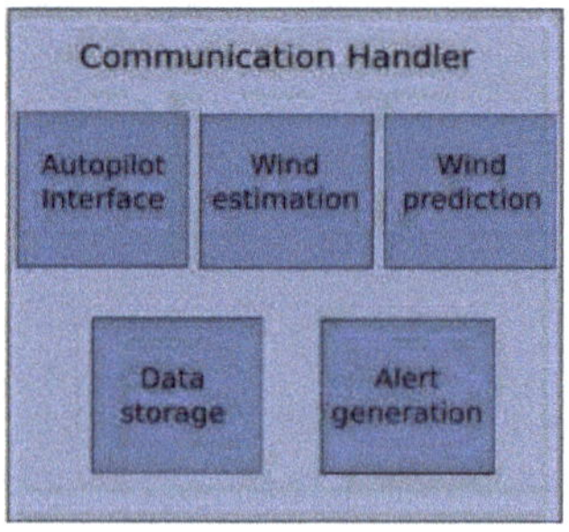

Figure 6. Architecture design with communications handler.

The modules are the same shown in Figure 5 plus the alerts generation.

An advantage of the modular implementation is that the system can be easily expanded to provide additional functionalities besides the wind identification system. This was thought in order to be able to integrate trajectory optimization functions and controlling modules to follow the desired trajectory.

The details on the implementation of the modules are described in Sections 3.1–3.4.

3.1. *Communication Block and Handler*

The explanation of the communications is divided in two parts. The first one, described in Section 3.1.1 analyzes the details of the communication between the three main hardware components: the ODROID, the Pixhawk and the PC with SQL. The second part, Section 3.1.2, explains the details of the communication between the functional software blocks.

3.1.1. Hardware Communication

The hardware communication is performed by a C++ Software implementation derived from the MAVCONN software created by Lorenz Meier as a complementary MAVLINK toolset [27]. The main characteristic is the low latency that allow the communication between processes approximately at 100 microseconds. The system was implemented asynchronously, allowing the data to be sent immediately after it is available. The asynchronous communication is an alternative solution to the widely use polling which is proven to require extra CPU resources because of the context switch. On the other hand, asynchronous design requires minimum CPU resources. Nevertheless, it needs a multi-threaded implementation which is computationally more complex. The ODROID computer allows this type of implementation. Further details of this implementation can be found in [28].

3.1.2. Module Communication

The communication between modules is managed by a handler (see Figure 6). Each module publishes its information at a certain order based on a request and the importance of the information. Therefore, if a module requires priority information the framework will designate this request over others. Table 2 shows the selected requirements in terms of communication rate and an assigned priority based on the importance of its information to other subsystems.

Table 2. Communication Scheme of Handler. The publishing rate which was determined based on the results found in [9] and the priority based on the system requirements.

Function	Publishing Rate	Priority
APM Comm Request	2 Hz	1
Wind Estimation	1 Hz	1
Wind Prediction	0.25 Hz *	2
Database Management	0.2 Hz	3
Database Search	0.25 Hz	2
Alert Generation	0.1 Hz	4

* This rate was selected due to the current time required to scan the wind database. Future work will optimize this rate allowing the generation of predictions at a lower rate.

The main advantage of this system is the modularity, since the intention is to have total independence between systems. If there is any communication problem, or the data is proved to be corrupted, this is handled directly by the communication handler which will continue to serve the other functions to preserve the overall integrity.

The processes with highest priority of publishing are the communication request between the ODROID and the PIXHAWK and the wind estimation processes (see Table 2). The first one was based on the publishing rate of the information available from the User Datagram Protocol (UDP) connection with MAVLINK. The second one was based on the computational time that requires the prediction which was subject of previous study in [8,9].

3.2. Wind Prediction

The main part of the system consists in a prediction algorithm that is able to recognize wind features (gust and shear) separately. The algorithm performs a statistical analysis to wind velocity estimates in order to determine if a feature is present. First, the module requests a wind estimation to the communication handler. Once it is requested, it stores the data into a temporary database that is going to be used for analysis.

If there are sufficient estimates from the current flight, the system starts a feature detection process by ordering the wind database with respect the UAS altitude. Since the altitude reading vary a lot with time, even in small amounts, the estimates are grouped according to a reference altitude by selecting those altitudes that are close within a given tolerance. Normally the references altitudes are integer numbers and the groups are conformed by those readings between a $\pm 1\,\mathrm{m}$ tolerance. At this point the module calls the communication handler in order to request additional measurements. These measurements may introduce significant noise to the system. Therefore, the conditions for the selection of previous measurements include date, time, location, altitude and some weather information. The database query instructions may vary from flight to flight, therefore, the specific conditions and the tolerances value can be specified on a flight-to-flight basis.

For those grouped wind estimates, the module tries to find the corresponding Weibull parameters using GA. If the system finds the Weibull parameters, a most probable wind speed at the corresponding reference altitude is generated. The process is repeated until the local maximum altitude is reached.

At first, the system performs an analysis to determine the presence of a shear, which is a very common feature [23,29]. The wind prediction module tries to find a Prandtl coefficient that minimizes the error between the most probable wind speeds for the reference altitudes. If the estimates are distributed according to the Weibull distribution and there is a Prandtl coefficient ξ that produces an acceptable error into the system (during the testing, the Prandtl coefficient was selected when the average error among the different altitudes was $\epsilon \leq 5\,\mathrm{m/s}$). Then, and alert is triggered and the system recognizes the presence of a shear. Afterwards, the system performs a statistical analysis to determine anomalies (significant jumps) in consecutive wind estimates. These were performed by looking for sudden increases into the running standard deviation of the wind estimates. If there is a sudden increase an initial alert is generated that potentially a discrete gust is identified. If the system is not capable to determine accurately the Weibull parameters of the system, most probable wind speeds cannot be fitted into a shear, and/or the running standard deviation presents drastic changes, i.e., there are continuous increases in the running standard deviation, the system assumes the presence of a continuous gust which triggers a short term Gaussian Regression process in order to characterize the feature. Algorithm 1 describes the insights of the prediction algorithm.

Algorithm 1 Wind prediction algorithm.

```
procedure REQUESTWINDESTIMATION
    WindEst = CommHandler.Request.CurrentWindVel          ▷ Request a wind speed estimation
  (see Equation (2)).
    WDb(CommHandler.Request.WVelCount++)= WindEst;         ▷ Store WindEst to Database.
end procedure

procedure DETECTFEATURE(WDb)   ▷ Requires Wind Database (WDb) with at least 30 elements
    Start=False;
    if WDb.Size ≥ 30 then
        Start = True;                                      ▷ Start detection of features.
    else
        CommHandler.Alert = InsufficienElements;     ▷ Wait until DB has sufficient elements.
    end if
    if Start==True then
        WDb = OrderAltitudes(WDb);                         ▷ Order WDb based on altitude.
        for i = 1 ←AltMax do                 ▷ Check for altitudes 1 m to maximum altitude .
            NearAlts = FindNearAltitudes(WDb,i,thres);
            AdNearAlts = CommHandler.RequestDb(i); ▷ Additional WDb elements to master Db.
            NearAlts = [NearAlts:AdNearAlts];                            ▷ Group elements.
            WindVelMP = FindMPWVel(NearAlts) ▷ Find most probable wind speed at altitude i m.
            MPS(i) = Store(WindVelMP);                ▷ Store the most probable wind speeds.
        end for
        Prandtl = CalcPrandtl(MPS)          ▷ Calculate Prandtl coefficient from Equation (6).
        if Prandtl.Exist = True then
            ξ = Prandtl;
            CommHandler.Alert = ShearDetected;
        end if
        i++;
        if Exist(Prandtl) = False then
            DetectJumps(WDb.Velocity,Std(WDb.Velocity))   ▷ Look for jumps in running std. dev.
        end if
        if CommHandler.Request.Alert.JumpDetected = True then
            JumpCounter++;
        end if
        if JumpCounter≥threshold then
            Commhandler.Alert = ContGustDetected;                  ▷ Is a continuous gust.
        else
            Commhandler.Alert = DiscGustDetected;                    ▷ Is a discrete gust.
        end if
        if CommHandler.Request.Alert.DiscGustDetected = True then
            Gust = DetectJumps.Jumpsize
        else if CommHandler.Request.Alert.DiscGustDetected then
            ContGust = PerformGaussianRegressionWDb                         ▷ See note **.
        end if
    else
        CommHandler.Alert = NoFeatureDetected;                  ▷ No feature was detected.
    end if
end procedure
```

** The system may perform a long-term and a short term prediction. For this research activity only the short-term which is a Standard GP regression. The non-homogeneous GP regression requires a vast amount of information which is part of future activities.

The algorithm that is used to group the altitudes based on a reference is shown in Algorithm 2.

Algorithm 2 Grouping near altitudes algorithm.

```
procedure FINDNEARALTITUDES(WDb,alt,thres)          ▷ Find altitudes in WDb close to alt.
    Counter = 0;
    for i = 1 ←WDb.Size do
        if alt-thres≤WDb(i).Altitude≤alt+thres then
            NearAlts(Counter++) = WDb(i);           ▷ Store whole WDb.
        end if
    end for
    return NearAlts;
end procedure
```

The determination of the most probable wind speed at a given altitude is described in Algorithm 3.

Algorithm 3 Weibull parameter calculation algorithm.

```
procedure FINDMPWVEL(NearAlts)                      ▷ Find altitudes.
    κ = CalcKappa(NearAlts)                         ▷ Calculate shaping parameter using GA.
    ν = 1/Mean(NearAlts) Γ(1 + 1/κ)                 ▷ Calculate scaling parameter from Equation (17).
end procedure

procedure CALCKAPPA(Altitudes)                      ▷ GA Implementation (see note***).
    PopulationSize = 50;
    FunctionTolerance = 1 × 10^-3;
    MaxGenerations = 100;
    CrossOverFraction = 0.8;
    StdAlt = Std(NearAlts);                         ▷ Calculate standard deviation.
    MeanAlt = Mean(NearAlts);                       ▷ Calculate mean.
    PopKappa == rand(PopulationSize);               ▷ Initialize with random population.
    while ε > FunctionTolerance do
        for j = 1 ←PopKappa.Size do
            Results(j) = ObjFunc(PopKappa(j),StdAlt,MeanAlt);   ▷ Evaluate Objetive Function.
        end for
        Parents = Selection(Results,PopKappa);      ▷ Selection of elements for newGeneration
  Equation (16)).
        Reproduction(Parents,PopKappa,MaxGenerations;)   ▷ Creation of new population.
        Crossover(CrossOverFraction);               ▷ Scattered crossover function.
        Migration();                                ▷ Gaussian Mutation function.
    end while
end procedure
```

*** The selected parameters were the same ones utilized in previous implementations [8,9].

Algorithm 4 describes the calculation of the Prandtl coefficient once the system detects a stable running standard deviation.

Algorithm 4 Wind Prediction Algorithm.

```
procedure CALCPRANDLT(WindSpeeds)                         ▷ Determine Prandtl Coefficient.
    Prandtl.Exist = False;                                ▷ Initialize values.
    Prandtl.Value = 0;
    for m = 0.01 ← 1 do                                   ▷ Evaluate potential Prandtl coefficient.
        for l = 1 ←MaxAlt do                              ▷ Evaluate for altitudes in MPWS.
            CalError = ComparePrandtlValues
        end for
        if Mean(Error)≤thres and Std(erorr)≤thres then
            Prandtl.Exist = True;                         ▷ If coefficient gives a minimum average error.
            Prandtl.Value = m;                            ▷ Prandtl coefficient is m.
            break;
        end if
    end for
    return Prandtl
end procedure
```

Algorithm 5 is used for detection of anomalies in the running standard deviation.

Algorithm 5 Jump detection algorithm.

```
procedure DETECTJUMPS(WindSpeeds)                         ▷ Look for jumps in running std dev.
    PrevStd = Std(WindSpeeds(k − 1));                     ▷ Look for previous std. dev.
    DiffStd = PrevStd-Std(WindSpeeds)                     ▷ Difference between std. deviations.
    AcumDiffStd(count + 1) = DisffStd
    if DiffStd ≥ thres then                               ▷ If error is bigger than threshold.
        CommHandler.Alert = JumpDetected
        JumpSize = Mean(AcumDiffStd)                      ▷ Estimate the size of the jump.
    end if
end procedure
```

3.3. Data Storage and Wind Database

An important part of the designed system is the storage and management of the information. This information is the one generated by the estimation and the prediction modules, and also the one generated by the autopilot (vehicle state: position, velocity, acceleration).

SQL is selected as a means of the generation, storage and management of the database. This was because SQL is a standardized language for database management. SQL is a language by itself, therefore, it requires an interface with the wind identification system. The SQL system that is selected for this research is MySQL® and the interfacing between the database and the wind identification comm-handler is done trough the MySQL® C++ connector. This allows the generation of C++ commands that will read and write information from any SQL database.

The communication scheme is shown in Figure 7.

Figure 7 illustrates the two modules that are required to interact with wind database. One is the MySQL C++ connector and the other the MySQL system, which have to be compatible with the used operating system.

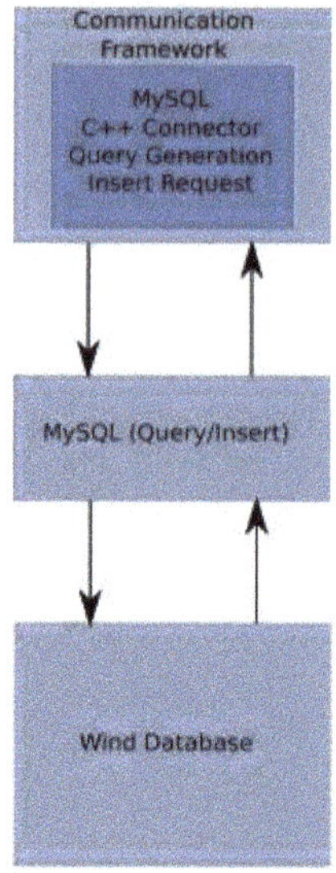

Figure 7. Database interfacing with communication handler.

The algorithm for accessing the database, perform a query of the useful data and write the generated data for the modules consists in a series of calls to the SQL connector which needs to open a Connection to the SQL server and then to execute an update or a query to the database based on the parameters that are needed. This is described in Algorithm 6.

Algorithm 6 Database Access, Query and Writing Algorithm.

```
procedure WAITFORREQUEST(DBReq)
    if DBReq = Write then
        WriteDB(DB);
    else if DBReq = Query then
        Query(DB,Cond);
    else
        TriggerException;
    end if
end procedure
procedure WRITEDB(DB,WindVector)
    Con→ CreateDriver();                          ▷ Create Database Driver
    Con→GetDriverInstance();     ▷ Used to get the Driver Instance and Load the DB.
    Con→setSchema(DB)                             ▷ Set the DB to write to
    Stmt→WindVector
end procedure
procedure QUERYDB(DB,Cond) State Con→ CreateDriver();
    Con→GetDriverInstance();
    Con→setSchema(DB);
    execute;
    stmt→executeQuery(Condition);                 ▷ See note below.
end procedure
```

In the previous algorithm, the query is based on the location the time of the year. Since the location of a typical mission may not vary the data should be valid, however a future step is to include

Meteorological Terminal Aviation Routine Weather Report (METAR) weather reports to the query so that it only looks for wind predictions and estimations performed in similar meteorological conditions.

3.4. Alert Generation Module

A complementary part of the estimation module is the alert generation algorithm. This part illustrates what kind of alerts need to be triggered internally to the system and to the user so that it can take a decision. These alerts are related to the detection and the uncertainty of a feature . In addition, there are different alerts that are generated inside each module related to the information that each module produces, including the generation of software exceptions.

Table 3 shows the main alerts that are generated once a feature is detected, the variable type of the alert and the priority.

Table 3. Alerts, data types and the priority values.

Alert Type	Data Type	Priority Value
Feature Is Present	Boolean	1
Wind Shear Detected	Boolean	1
Discrete Gust Detected	Boolean	2
Continous Gust Detected	Boolean	3
Prediction Time Window	Integer	2
Uncertainty Level	Double	4

The alert priority value aids on determining how often an alert is generated. The goal of the system is to alert to other modules the presence of a feature and to display these alerts in the ground station.

The prediction time window (τ) requires additional computational resources. If a discrete gust of a shear is detected, and the running standard deviation remains stable, a prediction time window alert is not required (for computation purposes is considered as infinite). However, if a continuous gust is detected the time window of the prediction goes critical depending to the behavior of the difference between the prediction and the estimation. If the running standard deviation of this difference is bounded, the prediction window can be slightly increased. If there is an unbounded behavior, then the system stays at its initial value ($\tau_0 = 5\,\mathrm{s}$), based on the results published by Passner et al. [30].

4. Simulation and Experimental Results

This section presents the preliminary validation results of the system obtained with simulations and Autopilot/Framework SITL experiments with real telemetry data obtained in four flights which took place in the Seville Metropolitan Area in Andalusia (Spain).

4.1. Simulation Test Bed Description

MATLAB® and Simulink® has been utilized for simulation. The Aerospace toolbox contains wind-model blocks of shear, discrete and continuous gusts. In addition the AeroSim® blockset has been utilized to generate 6DOF model of a small UAS.

The 6DOF model utilized together with the wind dynamic model blocks are shown in Figure 8.

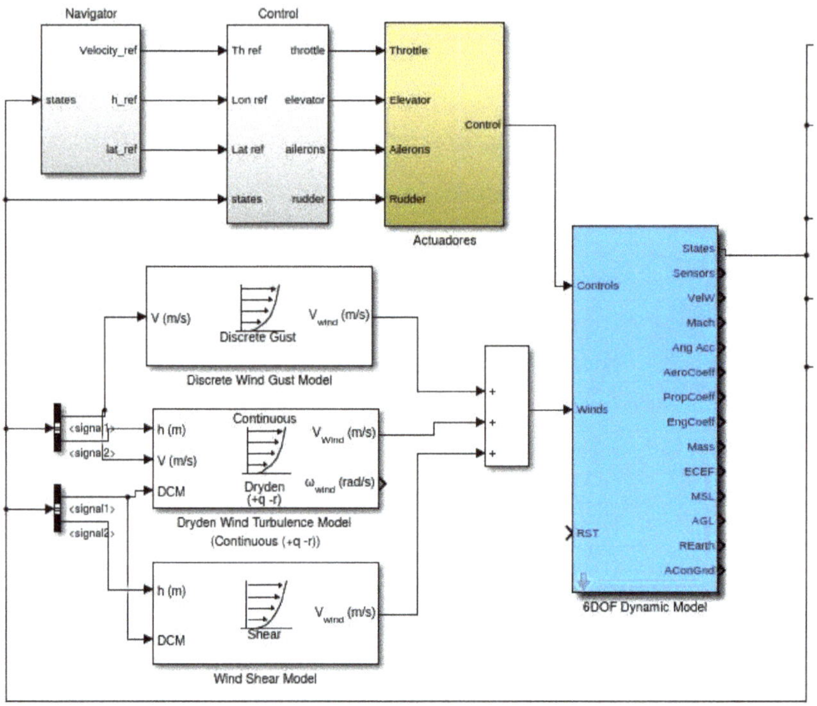

Figure 8. Simulink model of the simulation environment for the wind identification system.

The blue block shows the 6 DOF dynamic model and the white blocks show the wind dynamic model. In addition, there is an actuator block that corresponds the dynamic model of the actuators. There are two additional blocks that show the navigation and the control modules which are a series of nested Proportional-Integral-Derivative (PID) controllers.

Table 4 show the characteristics of the computer used in order to perform the simulations.

Table 4. Simulation computer relevant characteristics.

Component	Specification
CPU	Intel Core i7-5500U CPU 2.40 GHz × 4
RAM	15.6 GiB
Graphics	Intel HD Graphics 5500 (Broadwell GT2)
OS Type	64-Bit
OS	Ubuntu 16.04 lts

The corresponding trimming parameters for a typical flight condition [31] are utilized in the simulation are shown in Table 5.

Table 5. Selected Trimming Parameters.

Parameter	Value
Trim airspeed	25 m/s
Trim altitude	150 m
Trim bank angle	0°
Fuel mass	2 kg
Flap setting	0

The scenario considers a planned helix flight ascending trajectory. Once the vehicle starts its flight, the trajectory is under the influence of different wind types. Two scenarios are considered. The first one considers each feature separately (shear, discrete gust, continuous gust) and the second one considers all features at the same time. The purpose of this simulation is to prove the ability of the system in controlled conditions of detecting the features separately. Figure 9 depicts the wind effects on the trajectory.

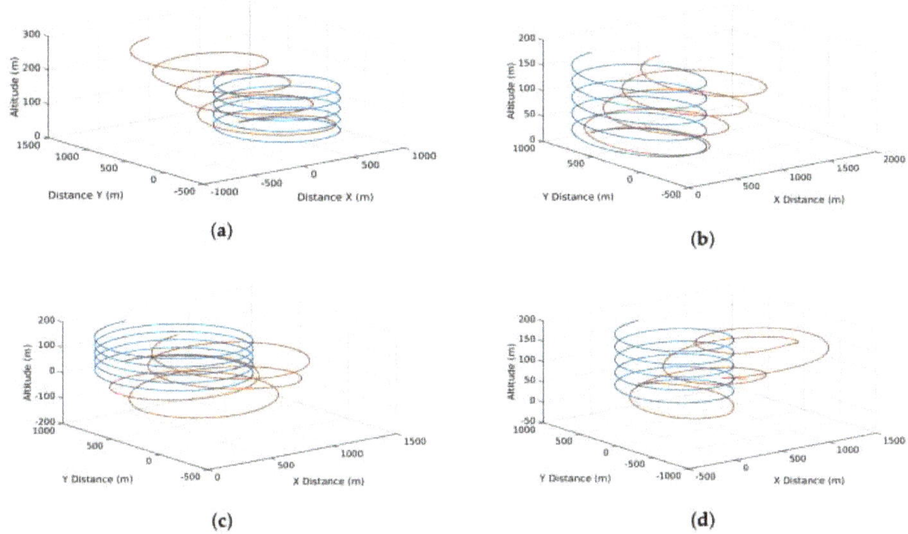

Figure 9. Effects of different simulated features on the vehicle trajectory: (**a**) the effect of a shear wind with increasing deviation as altitude rises; (**b**) the effects of a discrete gust with a constant deviation on the trajectory in a single direction; (**c**) a chaotic deviation due to the effects of a continuous gust; and finally (**d**) the total effects of the wind present at the same time.

4.2. Simulation Results

The detection capabilities of the system are illustrated Figure 10. It shows the information that feeds up the system and how it will detect and identify the different features.

Figure 10d shows two trends (vertical successions of wind velocity points). One in which the wind speeds are distributed uniformly across altitudes with a mean value of approximately 3 m/s. At 100 m one can observe another succession of points with a mean value of approximately 8.2 m/s. This produces a sudden increase (jump of aproximately 5 m/s) in the standard deviation which triggers an alert of gust detected and forces the system to characterize two separate distributions, one after and one before the gust.

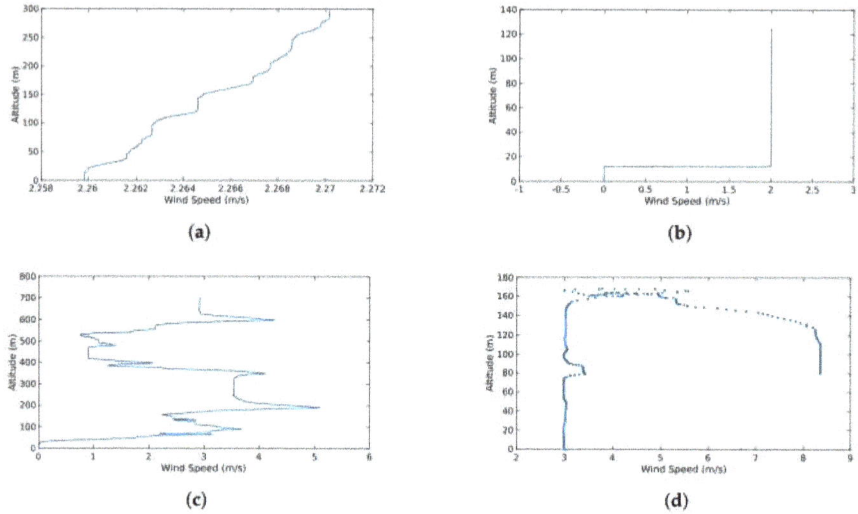

Figure 10. Wind Speed/Altitude maps of the different simulation scenarios: (**a**) the wind shear as an increase of wind speed with altitude; (**b**) a sudden increase in wind speed at a certain altitude (discrete gust); (**c**) a continuous gust with a chaotic effect and rapid increases and decreases of wind speed; and (**d**) the sum of the three effects.

The results of the wind estimates and predictions of the wind are show in Figure 11.

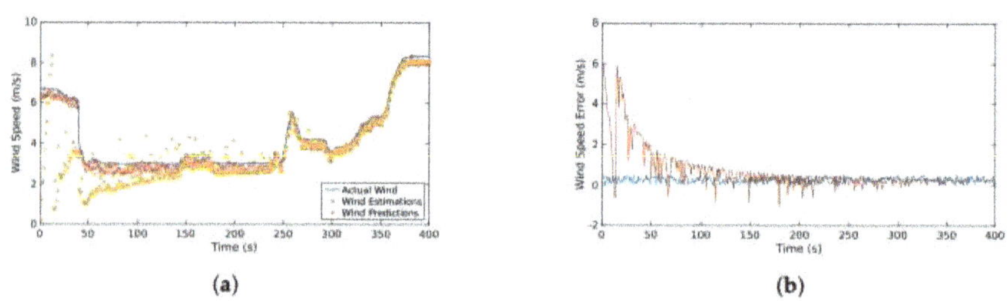

Figure 11. Actual, estimated an predicted wind speed (**a**) and wind speed error (**b**) in the considered scenario. The predicted wind starts with high dispersion, however, it converges to the actual value within 100 s.

In this flight, an alert of a continuous gust detected was triggered almost immediately (at approximately 20 s). In addition, there was an alarm of two detected gusts: one occurred at approximately 40 s and the other occurred at 250 s. This coincides with the jumps, abrupt changes of the wind speed, that can be seen in Figure 11.

4.3. *Software-in-the-Loop Experiments*

The wind identification system has been functionally tested with real telemetry data. The data were fed into the system using Mavlink interfacing with a Ground Control Station as in [32]. The sensor information was transmitted to the wind identification system at a rate of 0.5 Hz. Nevertheless the communication framework demands varied in frequency due to the asynchronous scheme. The transmission to the system does not match the actual duration of the telemetry log, it was truncated once the platform had landed.

The test-bed architecture uses a MATLAB®- Mavlink interface implemented in the Robotic Operating System (ROS). The wind identification system interfaces with MATLAB® through a series of S-Functions. This concept is illustrated in Figure 12.

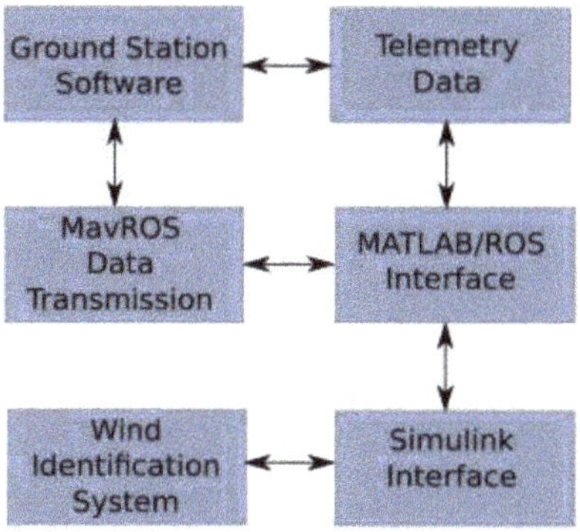

Figure 12. Information flow for on-the-loop experiments of the wind identification system. The blocks show the multi-platform interfaces that allowed the validation tests.

The platform and the airspeed sensor in which the experiments were performed is shown in Figure 13a.

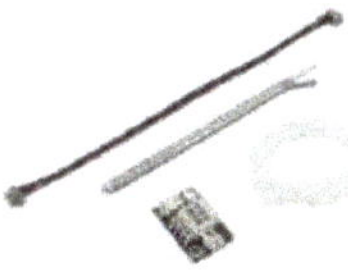

(a) (b)

Figure 13. Sensor and UAS platform utilized in experiments. (**a**) UAS SkyWalker X8, (SkyWalker Technology Co., Ltd, Wuhan, China) with carbon fiber frame equipped with a 12 × 6 prop with 2 × 20 g servomotor; (**b**) Digital airspeed sensor utilized in experiments which contains a 4525DO sensor (TE Connectivity Ltd., Schaffhausen, Switzerland) which enables a resolution of 0.84 Pa [33].

Table 6 presents the characteristics of the platform shown in Figure 13a.

Table 6. Skywalker Characteristics.

Parameter	Value
Wing Span	2122 mm
Wing Area	80 dm^2
Max Payload	2 kg
Center of Gravity	435 mm away from nose

The vehicle was equipped with an APM2.6 autopilot (3D Robotics, Berkeley, CA, USA) with the airspeed sensor illustrated in Figure 13b.

4.4. Software-in-the-Loop Experiments Results

The information of the flights performed in the Seville Metropolitan Area (Brenes) is shown in Table 7.

Table 7. Experiments information.

	Flight 1	Flight 2	Flight 3	Flight 4
Duration	521 s	315 s	631 s	749 s
Distance Traveled	5.1 km	3.7 km	6.3 km	7.4 km
Maximum Altitude	179 m	125 m	134 m	146 m

Figure 14 depicts the flight trajectories of the scenarios described in Table 7. The first flight shows 8 maneuvers performed at different altitudes. The other three flights consisted on takeoff, several spirals at a target altitude and then the descent and landing.

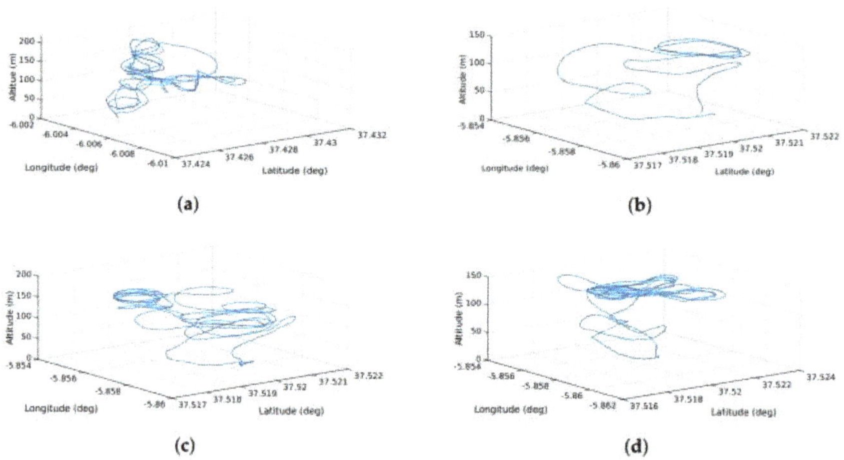

Figure 14. UAS Trajectory for validation of the wind information system. (**a**) shows a medium altitude with few spirals; (**b**) shows the shortest flight; (**c**) shows a flight with spirals performed at an altitude of 120 m; and (**d**) shows a flight with wide spirals at an altitude of 120 m.

Figure 15 depicts the results obtained from Experiment 1:

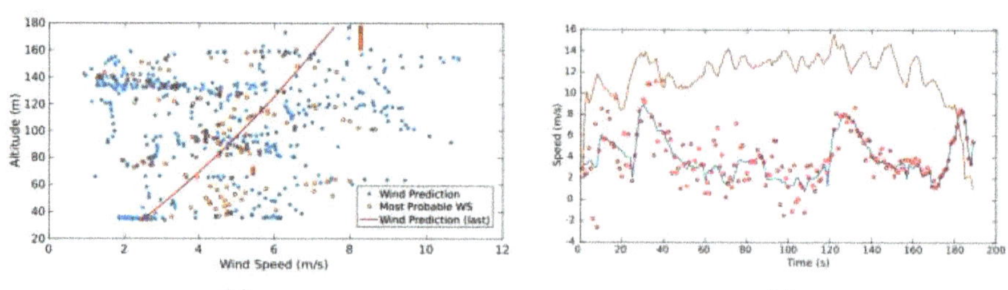

Figure 15. (**a**) shows the estimation, most probable wind speeds and last shear wind prediction generated throughout the flight; (**b**) indicates the wind speed estimation (blue line), wind speed prediction (red dots) and airspeed (orange line).

The red dots in Section 4.4 indicate the predicted wind speed. The blue line represents the estimations which were obtained with the direct computation method presented in [2,8,9].

A continuous gust alarm was generated almost at the beginning of the flight due to the continuous changes in wind speed over time.

The second experiment (see Figure 16) shows a significant decrease of the estimates estimation as the UAS reaches its maximum altitude. The system interpreted this tendency as a negative gust, i.e., a sudden reduction of the wind speed. Once the system generates the corresponding alarm and the running standard deviation of the estimates stops growing, the system starts characterizing a second shear which is represented by the purple line.

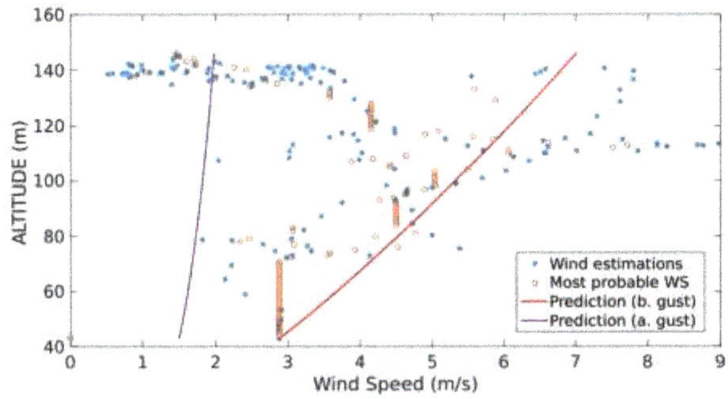

Figure 16. Estimation, most probable wind speeds and last wind prediction generated throughout the flight shown in Figure 14b.

Experiments 3 and 4 (see Figures 17 and 18) show a very similar behavior. The wind speed estimates show higher values as the altitude grows. The high concentration of estimations of the wind speed between 120 m and 140 m altitude show big dispersion which suggests that the UAS maneuvers affect the speed reading as the accelerometers and the GNSS speed readings affect the computation of the estimates. More testing is required to support this hypothesis.

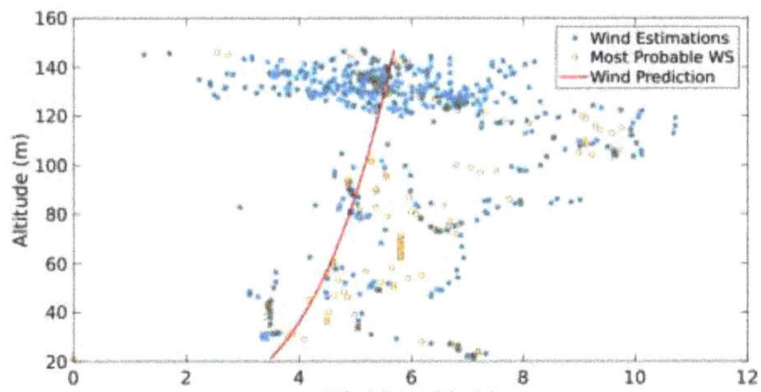

Figure 17. Estimation, most probable wind speeds and last wind prediction generated throughout the flight illustrated in Figure 14c.

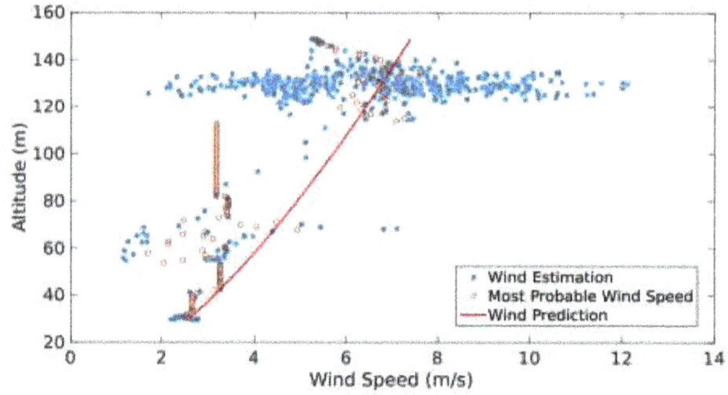

Figure 18. Estimation, most probable wind speeds and last wind prediction generated throughout the flight depicted in Figure 14c.

The average Weibull shaping parameter, κ, Weibull scaling parameter, ν, and the calculated Prandtl coefficient, ξ, are shown in Figure 14d.

Table 8. Main SITL outputs (single run).

Scenario	$\mu(\kappa)$	$\mu(\nu)$	$\mu(\xi)$
1	4.24	1.9825	0.6628
2	1.1579 & 3.1425	0.4531 & 1.6671	0.5314 & 0.6628
3	2.9820	0.8349	0.3124
4	5.7425	0.9623	0.7614

Note that in scenario 2 of Table 8 two values of shaping parameter, scaling parameter and Prandtl coefficient appear for scenario 2. They correspond to two different shear characteristics detected before and after the presence of a discrete gust.

Figure 19 shows the difference comparison between the estimations and the predictions of the wind speed magnitude. It is important to consider that it is the difference, therefore, it cannot be considered as an absolute error. However it aims to prove that this difference is bounded since it considers local measurements.

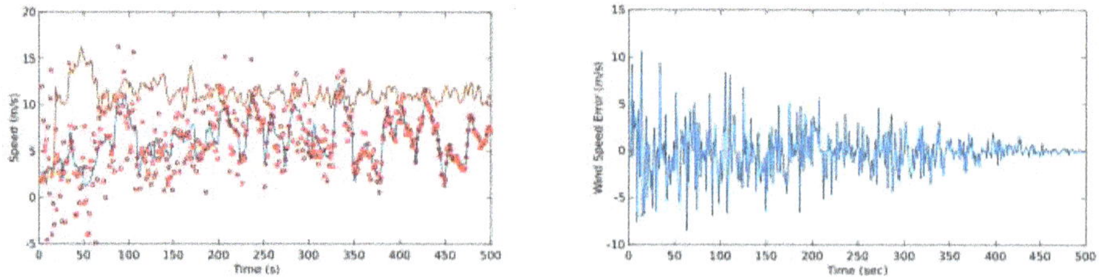

(a) (b)

Figure 19. Predicted (red circles) vs. Estimated wind speed (blue line): (**a**) shows the difference between the estimation and the prediction of the wind speeds with the airspeed (orange) as a reference; (**b**) shows the actual difference between these two quantities which appears to be bounded and shows a normal behavior.

The difference analysis between all the flights together showing the mean $\mu(W_e)$ and the standard deviation $\sigma(W_e)$ are shown in Table 9.

Table 9. Mean and standard deviation of difference between the wind speed estimations and the wind speed predictions.

Flight	$\mu\,(W_e)$	$\sigma\,(W_e)$
1	1.985 m/s	2.54 m/s
2	1.197 m/s	1.21 m/s
3	0.932 m/s	0.74 m/s
4	0.854 m/s	0.27 m/s

5. Results Discussion

In the simulation results, the effects of the wind features (continuous and discrete gusts and shear) are shown in Figure 9. It is observed that single or multiple features can affect the trajectory without any sort of drifting compensation. Figure 10 illustrates the effects of the features in wind speed/altitude charts. The system is able to identify this feature from the summed effects plot shown in Figure 10d. However, at this stage noise is not considered and it can have a significant impact to the plot, so the estimation process has to be accurate enough since it is the only input to the wind identification system. The case shown in Figure 10d is analyzed in detail since the effects of each feature separately were analyzed in [8,9].

The results plotted in Figure 11 show the behavior of the estimation and prediction processes. The estimation process considered a slight Gaussian noise which is typical for airspeed sensors [2]. The error of the estimation process is bounded as observed in Figure 11b which is indeed the expected behavior and is consistent with the result presented in [2,8,9]. On the other hand, the prediction shows a different behavior. At the beginning and up to 70 s there is a considerable dispersion of the prediction due to the assumption that only a shear feature is present since the beginning. Then, the system starts identifying other features. At 40 s a rapid change is observed which triggers a discrete gust alarm. Up to that point the system starts converging and the predictions with a variable window start happening. Since there is a continuous gust during the entire scenario, the system utilizes Gaussian regression to start predicting the behavior of the plot. Even though there are rapid changes at some parts, the identification of the discrete gust minimizes the effect in the Gaussian regression. The prediction error shows a gradual decrease up to the point that it follows a similar behavior than the estimation. This concludes that the prediction error tends to be bounded as more data is fed to the system.

The SITL testing illustrates four scenarios with actual airspeed measurements. Figure 14 shows different paths with different maneuvers at various altitudes. These scenarios are very helpful to comprehend how the noise of the airspeed measurements affects the estimation. However, these results have to be treated carefully since there is no ground-truth and intend to validate the functionality of the system only. Full validation of the results needs to come from full validation and verification with Software, and Hardware-In-The-Loop and extensive flight testing in different conditions. The intention of the upcoming testing activities is to prove every aspect of the system and to analyze how the obtained results support the hypotheses on the wind identification problem. Nevertheless, current preliminary results prove that the wind estimates behave statistically as expected, in cases of shear and discrete gusts, estimations and predictions are Weibull distributed and they keep following the Prandtl law with a low increasing running standard deviation. In the case of continuous gusts, the short-term regression has proved to be accurate, keeping the running standard deviation of the predicted wind bounded throughout the flight.

In the scenario shown in Figure 14a the results of the prediction and estimation process show a very dispersed behavior (see Figure 15a). The system triggers a continuous gust alarm and the prediction process employs a GP Regression. Note that a shear is identified (red line), however, this is not taken into account in the prediction, since the system always assumes that once there are sufficient wind estimates there is a shear. Once the continuous gust is detected and during the computation of the GP regression, the system stops calculating the shear characteristics releasing computational load as no GA is performed.

The scenario shown in Figure 14b shows two shear features that are identified due to a sudden change of wind speed that occurs at 40 m. Since a discrete gust was detected the system tries to identify this two features. It is observed that there is are three points that the most probable wind speeds looks constant. This is due to the lack of measurements possibly by a communication error between the system and the ground station.

The remaining two flights (see Figure 14c,d) show a very similar behavior at high altitude. Nevertheless in the fourth flight there is a lack of airspeed measurements and the most probable wind speed is assumed to be constant. However the actual prediction (red line) shows what in reality the airspeed has to behave. Since a lot of spiral maneuvers were performed, the system has a vast amount of estimations over an altitude of 120 m, however, there is a substantial dispersion that follows the Weibull distribution allowing the generation of coherent most probable wind speeds and to treat this measurements as part of a wind shear feature.

The prediction models illustrated in Figure 16 show two characterized shear predictions as a results of the identification of a gust. The system performs this interpretation as the estimates from 40 m to 120 m show a tendency of a sustained growth, however when the vehicle passes 130 m most of the estimates go back to values around 2 m/s.

In Figure 17 one can observe dispersion in the wind estimates from the lowest altitude up to 120 m even with a few measurements at some altitudes, e.g., between 50 m and 80 m. However, the system was able to identify a Prandtl coefficient to characterize an average shear. The last prediction (red line) is moved ot the left as most of the available estimates were above 120 m. In higher altitudes, the measurements are dispersed (2 m/s to 8 m/s), however, the alerts generated indicate that the system was able to find Weibull parameters for these measurements. This indicates that the system might require an adjustment of the tolerances for continuous gust detection.

Figure 18 depicts a more stable behavior across different altitudes. Most of the measurements are concentrated above 120 m, however, with the measurements below those altitudes the system is able to produce a solution in which a wind shear is identified. The measurements above 120 m show big dispersion, possibly due to sensor noise, however these were proved to be Weibull-distributed, hence, the system was able to produce a set of most probable wind speeds and a prediction tendency.

For the last scenario, a comparison between estimation and the prediction can be observed in Figure 19. The error between the predictions and estimation is bounded since the very beginning, mainly to the absence of continuous gust features. This results can be confirmed with the study of the dispersion of this difference that is shown in Table 9.

6. Conclusions and Future Work

This paper presents the integration of a wind identification system using small UAS. It describes the high and low level architecture and provides a initial validation with simulations and software-in-the-loop testing.

The system architecture integrates different components at various levels, and presents significant advances from the previous research activities presented in [8,9]. In terms of hardware, the proposed system uses COTS components which help on cost efficiency without the sacrifice of functionality or reliability mainly due to the system characteristics and current state-of-the-art. On the other hand, the software has a decentralized integration at system and component level due to the development of a communication handler. This manages the information exchange between the different modules of the system. The main advantages of this communication scheme are the separation between different functional modules, which ensures the upgradeability and the module dependencies. Also the possibility can be easily expanded by adding other functional modules, such as trajectory generation/optimization. Another factor considered in the design was the possibility of asynchronous communication between blocks. This is an important requirement due to the possible variation on the processing time for different modules regardless if the variation is generated at a software or hardware level.

The core function of the system which is the prediction module is described and presents significant improvements from the previous research activities. The algorithm has unique way to characterize the features as it intends to find statistical key values that will lead to the identification of a feature. The system now triggers alarms to the communication handler and the sub-procedures were clearly defined and tested. Other modules such as the communication handler and the database management and query were also analyzed. The implemented algorithms work asynchronously and even though the computational demand may be significant to the use of nested loops and complex algorithms such as GA, they do not impact the prediction computation since the algorithms intend to find a minimum of variables. The most costly algorithm is the database management as it intends to do a smart search of accumulated data from previous flights. This is not an issue since it is done in a separate dedicated computer. Even though there is no information from the wind database, the system is able to produce results as it depends only in the current estimates.

In terms of the wind speed estimation and wind speed prediction validation, the system was tested with both simulations and software-in-the-loop. In the simulations, results indicate that the wind identification system is capable of identifying the different features and eventually converges to the actual wind field within variable time windows. However, the accuracy varies according to the identification of other features. Therefore, as clearer features are identified, the convergence time is reduced together with the error magnitude and dispersion. In the SITL testing, the system exhibits dispersion on the wind estimates which is mainly attributable to the noise from the airspeed sensor. The system estimates were Weibull-distributed in altitudes on which the aircraft remains for longer periods and presented inaccurate predictions at altitudes in which the aircraft has low or null density of measurements. The presented results lead to the conclusion that the system fulfills the design requirements and provides the identification of separated wind features which could be really useful for trajectory planning and optimization. The novelty of the system relies in two main aspects: first the architecture with a upgradeable system with minimum module dependency and secondly the information that the system generates since the identification of separated wind field features could easily be used for efficient trajectory planning, for instance in dynamic soaring.

Future work includes details on the generation of 3D wind maps and a complete validation and verification of the system at system and component levels, as well as on-board/real-time testing of the system. This paper intends to present a detailed description and the initial stages of validation and verification of the system. The full testing, including hardware-in-the-loop and on-board testing activities and the integration of the mapping feature will be subject of a further publication (Part 2) which will provide results at system and component level in terms of accuracy and reliability and a detail analysis of the computational cost of the different methods. In addition, upcoming work includes the integration of a trajectory generation module and the generation of control commands to follow the wind-efficient trajectories which ultimately is derived from the objective of increasing substantially the flight duration in a given mission.

Acknowledgments: This work has been supported by the MarineUAS project, funded by the European Comission under the H2020 Programme (MSCA-ITN-2014-642153) and the AEROMAIN Project (DPI2014-5983-C2-1-R), funded by the Science and Innovation Ministery of the Spanish Government.

Author Contributions: Leopoldo Rodriguez have designed the overall system, Jose Antonio Cobano has contributed in the architecture conception and the experiments, Anibal Ollero revised and edit the paper.

Conflicts of Interest: The authors declare no conflict of interest.

Abbreviations

The following abbreviations are used in this manuscript:

ADC	Analog to Digital converter
AOA	Angle of Attack
APM	Autopilot Module
COTS	Commercial-Off-The-Shelf
CPU	Central Processing Unit
DB	Database
EKF	Extended Kalman Filter
GA	Genetic Algorithm
GEV	Generalized Extreme Value
GNSS	Global Navigation Satellite System
GP	Gaussian Proccess
IMU	Inertial Measurement Unit
LPGL	GNU Lesser General Public License
PID	Proportional, Integral, Derivative
RAM	Random Access Memory
SITL	Software-In-The-Loop
SQL	Structured Query Language
TN	Truncated Normal (Distribution)
UAV	Unmanned Aerial Vehicle
UDP	User Datagram Protocol
UKF	Unscented Kalman Filter

References

1. Biradar, A.S. Wind Estimation and Effects of Wind on Waypoint Navigation of UAVs. Master's Thesis, Arizona State University, Tempe, AZ, USA, 2014.
2. Langelaan, J.W.; Alley, N.; Neidhoefer, J. Wind Field Estimation for Small Unmanned Aerial Vehicles. *J. Guid. Control Dyn.* **2011**, *34*, 109–117.
3. Johansen, T.; Crisofaro, A.; Sorensen, K.; Fossen, T. On estimation of wind velocity, angle-ofattack and sideslip angle of small UAVs using standard sensors. In Proceedings of the International Conference on Unmanned Aircraft Systems, Denver, CO, USA, 9–12 June 2015; pp. 510–519.
4. Lawrance, N.; Sukkarieh, S. Simultaneous Exploration and Exploitation of a Wind Field for a Small Gliding UAV. In Proceedings of the AIAA Guidance, Navigation and Control Conference, Toronto, ON, Canada, 2–5 August 2010.
5. Larrabee, T.; Chao, H.; Gu, Y.; Napolitano, M. Wind field estimation in UAV formation flight. In Proceedings of the American Control Conference, Portland, OR, USA, 4–6 June 2014; pp. 5408–5413.
6. Neumann, P.; Bartholmai, M. Real-time wind estimation on a micro unmanned aerial vehicle using its inertial measurement unit. *Sens. Actuators A. Phys.* **2015**, *235*, 300–310.
7. Condomines, J.F.; Bronz, M.; Erdely, J.F. Experimental Wind Field Estimation and Aircraft Identification. In Proceedings of the International Micro Air Vehicles Conference and Flight Competition, Aachen, Germany, 15–16 September 2015.
8. Rodriguez, L.; Cobano, J.; Ollero, A. Wind Characterization and Mapping Using Fixed-Wing Small Unmanned Aerial Systems. In Proceedings of the International Conference on Unmanned Aircraft Systems, Arlington, VA, USA, 7–13 June 2016; pp. 178–184.
9. Rodriguez, L.; Cobano, J.; Ollero, A. Wind Fiel Estimation and Identification Having Shear Wind and Discrete Gust Features with a Small UAS. In Proceedings of the IEEE/RSJ International Conference on Intelligent Robots and Systems, Daejon, Korea, 9–14 October 2016.
10. Cobano, J.A.; Alejo, D.; Vera, S.; Heredia, G.; Sukkarieh, S.; Ollero, A. Distributed Thermal Identification and Exploitation for Multiple Soaring UAVs. In *Human Behavior Understanding in Networked Sensing: Theory and Applications of Networks of Sensors*; Spagnolo, P., Mazzeo, L.P., Distante, C., Eds.; Springer: Cham, Switzerland, 2014; pp. 359–378.

11. Cutler, M.J.; McLain, T.W.; Beard, R.W.; Capozzi, B. Energy Harvesting and Mission Effectiveness for Small Unmanned Aircraft. In Proceedings of the AIAA Guidance, Navigation and Control Conference, Toronto, ON, Canada, 2–5 August 2010.
12. Chakrabarty, A.; Langelaan, J.W. Flight Path Planning for UAV Atmospheric Energy Harvesting Using Heuristic Search. In Proceedings of the AIAA Guidance, Navigation and Control Conference, Toronto, ON, Canada, 2–5 August 2010.
13. Montella, C.; Spletzer, J.R. Reinforcement Learning for Autonomous Dynamic Soaring in Shear Winds. In Proceedings of the IEEE/RSJ International Conference on Intelligent Robots and Systems, Chicago, IL, USA, 14–18 September 2014.
14. Bird, J.; Langelaan, J.; Montella, C.; SPletzer, J.; Grenestedt, J. Closing the Loop in Dynamic Soaring. In Proceedings of the AIAA Guidance, Navigation, and Control Conference, National Harbor, MD, USA, 13–17 January 2014.
15. Bencatel, R.; Kabamba, P.; Girard, A. Perpetual Dynamic Soaring in Linear Wind Shear. *J. Guid. Control. Dyn.* **2014**, *37*, 1712–1716.
16. Rucco, A.; Aguiar, P.A.; Lobo Pereira, F.; Borges de Sousa, J. A Predictive Path-Following Approach for Fixed-Wing Unmanned Aerial Vehicles in Presence of Wind Disturbances. In *Robot 2015 Second Iberian Robotics Conference*; Springer: Cham, Switzerland, 2015; Volume 417, pp. 623–634.
17. U.S. Department of Defense. *Military Specification, Flying Qualities of Piloted Airplanes MIL-F-8785C*; U.S. Department of Defense: Washington, DC, USA, 1980.
18. U.S. Department of Defense. *Handbook, Flying Qualities of Piloted Airplanes MIL-F-8785C*; U.S. Department of Defense: Washington, DC, USA, 1997.
19. The MathWorks, Inc. *MATLAB Reference Pages, Dryden Wind Turbulence Model (Continuous)*; The MathWorks, Inc.: Natick, MA, USA, 2010.
20. Gao, C. Autonomous Soaring and Surveillance in Wind Fields with an Unmanned Aerial Vehicle. Ph.D. Thesis, University of Toronto, Toronto, ON, Canada, 2015.
21. Park, C.; Huang, J.; Ding, Y. Domain decomposition for fast Gaussian process regression. *J. Mach. Learn. Res.* **2011**, *12*, 1697–1728.
22. Lerch, S.; Thorarinsdottir, T.L. Comparison of nonhomogeneous regression models for probabilistic wind speed forecasting. *Tellus A* **2013**, *65*, 21206.
23. Liu, F.J.; Chen, P.H.; Kuo, S.S.; Su, D.C.; Chang, T.P.; Yu, Y.H.; Lin, T.C. Wind characterization analysis incorporating genetic algorithm: A case study in Taiwan Strait. *Energy* **2011**, *36*, 2611–2619.
24. 3DR. Pixhawk Autopilot, Quick Start Guide, 2014. Available online: https://goo.gl/lhi0jx (accessed on 1 August 2016).
25. Hardkernel Co., LTD. Odroid Platforms, ODROID-C2, 2013. Available online: http://goo.gl/8nQVKO (accessed on 1 August 2016).
26. Control, Q. MAVLink Micro Air Vehicle Communication Protocol, 2016. Available online: http://qgroundcontrol.org/mavlink/start (accessed on 15 September 2016).
27. Meier, L. Mavconn- Micro Air Vehicle Middleware, 2014. Available online: http://goo.gl/P82kbG (accessed on 15 September 2016).
28. Pixhawk, Computer Vision on Autonomous Aerial Robots. MAVCONN- MICRO AIR VEHICLE MIDDLEWARE, 2014. Available online: http://goo.gl/P82kbG (accessed on 15 September 2016).
29. Tar, K. Some statistical characteristics of monthly average wind speed at various heights. *Renew. Sustain. Energy Rev.* **2008**, *12*, 1712–1724.
30. Passner, J.; Knapp, I.; Ding, Y. Using Wrf-Arw Data to Forecast Turbulence at Small Scales. In Proceedings of the 13th Conference on Aviation, Range, and Aerospace Meteorology, Waltham, MA, USA, 17–20 April 2007; Volume 13.
31. Dynamics, U. AeroSim Blockset Version 1.01 User's Guide. Available online: https://goo.gl/gWNWhC (accessed on 5 December 2016).

32. Team, A.D. SITL Simulator (Software in the Loop), 2016. Available online: https://goo.gl/thkBTu (accessed on 18 November 2016).
33. JDrones. Digital Airspeed Sensor, 2016. Available online: https://goo.gl/1T73k2 (accessed on 1 August 2016).

sensors

MDPI

Article

Augmented Reality Tool for the Situational Awareness Improvement of UAV Operators

Susana Ruano [1,*], Carlos Cuevas [1], Guillermo Gallego [2] and Narciso García [1]

1 Grupo de Tratamiento de Imágenes (GTI), Information Processing and Telecommunications Center (IPTC) and ETSI Telecomunicación, Universidad Politécnica de Madrid (UPM), ES-28040 Madrid, Spain; ccr@gti.ssr.upm.es (C.C.); narciso@gti.ssr.upm.es (N.G.)

2 Robotics and Perception Group, University of Zurich, CH-8050 Zurich, Switzerland; guillermo.gallego@ifi.uzh.ch

* Correspondence: srs@gti.ssr.upm.es; Tel.: +34-91-336-7353

Academic Editors: Felipe Gonzalez Toro and Antonios Tsourdos

Received: 31 December 2016; Accepted: 2 February 2017; Published: 6 February 2017

Abstract: Unmanned Aerial Vehicles (UAVs) are being extensively used nowadays. Therefore, pilots of traditional aerial platforms should adapt their skills to operate them from a Ground Control Station (GCS). Common GCSs provide information in separate screens: one presents the video stream while the other displays information about the mission plan and information coming from other sensors. To avoid the burden of fusing information displayed in the two screens, an Augmented Reality (AR) tool is proposed in this paper. The AR system has two functionalities for Medium-Altitude Long-Endurance (MALE) UAVs: route orientation and target identification. Route orientation allows the operator to identify the upcoming waypoints and the path that the UAV is going to follow. Target identification allows a fast target localization, even in the presence of occlusions. The AR tool is implemented following the North Atlantic Treaty Organization (NATO) standards so that it can be used in different GCSs. The experiments show how the AR tool improves significantly the situational awareness of the UAV operators.

Keywords: situational awareness; augmented reality; unmanned aerial vehicle; tool; ground control station

---

1. Introduction

The use of Unmanned Aerial Vehicles (UAVs) has become increasingly popular in recent years, especially because the majority of them are equipped with at least an Electro-Optical (EO) sensor. Consequently, the use of these platforms has been considered for applications that traditionally use aerial imagery, such as surveillance [1] and remote sensing [2]. Additionally, UAVs have enabled the use of aerial images to improve tasks in which they were not previously used, such as structure inspection [3].

The research work involving the use of UAVs is oriented to solve problems that are directly related to the type of used platforms. This is because the quality and accuracy of the sensors, as well as the mission specifications, are different for each one. There are studies for the autonomous navigation of micro aerial vehicles (MAV) [4], while others focus on target detection with platforms that can reach higher altitudes and can fly during longer periods of time [5].

MAVs can be operated by people without previous experience piloting aircrafts. On the contrary, Medium-Altitude Long-Endurance (MALE) UAVs are usually operated by pilots of traditional manned aerial platforms that have adapted their expertise to control the aerial systems from a Ground Control Station (GCS). These MALE UAVs are usually operated in automatic or semi-automatic mode and are used in Intelligence, Surveillance, Target Acquisition and Reconnaissance (ISTAR) missions. Therefore, the operator is simultaneously supervising the flight and manually operating the payload to fulfill the specific mission requirements. The GCS has at least two different screens. One of them shows the mission plan, as it is represented in Figure 1a, and the other screen shows the video stream captured by the drone, as depicted in Figure 1b. In the example, the operator must accomplish a task as fast as possible: to determine the region of the images captured by the UAV (Figure 1b) where the target of interest is placed (illustrated in Figure 1c). However, such a task can be tedious and sometimes leads to misinterpretations when the targets are occluded. To this end, we propose to use augmented reality (AR) to help the UAV operator.

(**a**) screen where the mission planning is shown during the operation. In a standard GCS, the route (in green), the waypoints (in red), the targets (in blue) and the UAV (aircraft icon) position are shown

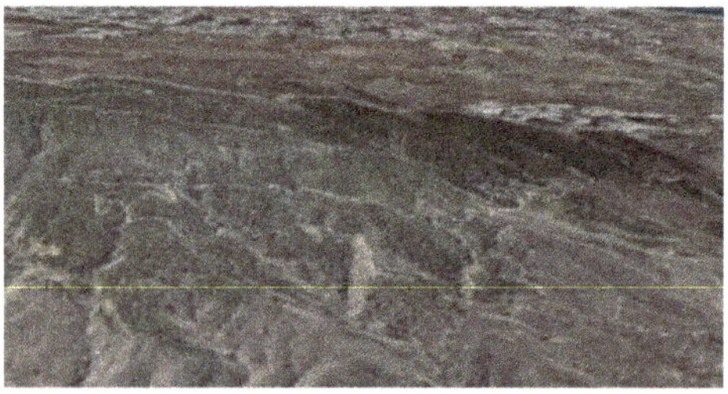

(**b**) raw video stream from the UAV, without any virtual aid

(**c**) augmented video: video stream from the UAV camera, augmented with the proposed Augmented Reality (AR) tool

Figure 1. Information available in the screens of the Ground Control Station (GCS) of the Unmanned Aerial Vehicles (UAVs).

AR systems have numerous possibilities and can be used to obtain information about the environment [6]. This can be done by mixing virtual objects and annotations in the user's view of the surroundings, increasing the meaningful semantic context of the world. In addition, if we take into account that, besides the sensed information by the UAV (such as a video stream), we have as input some additional geographic information, research related to integrating this kind of information is paramount. In [7], an example of the importance of integrating geographical information is given. They remark that one of the main challenges in this kind of integration is the establishment of a link between the geographic information system (GIS) data and the real world context (i.e., comprehensibility of the AR visualization and the geographic database). The goals they want to solve with their work are: (*i*) easy interpretation of the visualized information; (*ii*) understanding between the spatial location of virtual elements and the real world; and (*iii*) consistency in AR visualization when the GIS dataset is modified. However, they do not show results with aerial data, which is the type of information that is considered in this work.

The situational awareness problem has been addressed in the past [8–10]; some solutions proposed the enhancement of the acquired video stream with virtual elements to provide the operator with additional information. In this research direction, some techniques of AR have been used for outdoor elements annotation [8] of the objects seen by the user. A study of the optimal representation of occluded elements has been discussed in [9]. Finally, an AR system to improve the situational awareness and depth perception of UAV operators is proposed in [10], with a focus on small platforms.

Examples of how AR can be useful for UAV operations are given in [11,12]. The AR system in [11] increases the safety during take-off and landing UAV operations under harsh conditions such as fog, or at night. However, the tests are not carried out for MALE UAVs, do not follow standards and are not oriented to reconnaissance missions. In [12], a vision-based navigation method for UAVs using AR is proposed. The idea is to develop a system that can be used to superimpose the virtual elements over the video captured by the UAV. However, they only show experiments detecting markers in a controlled environment.

The purpose of this paper is to improve the situational awareness of UAV operators by providing a tool with AR capabilities valid for MALE UAVs in reconnaissance missions, and to make the tool available to the public on a website. Our system avoids the burden of fusing information that comes from different sources and is displayed on separate screens. Additionally, the tool complies with North Atlantic Treaty Organization (NATO) standards, which increases its usability: the application can be executed in different GCSs that follow the same standards. The contribution of this work is shown by the development of two different AR capabilities:

- Enhancement of the video stream with the UAV flight route.
- Identification of targets (as illustrated in Figure 1c) and viewpoint-based classification of targets according to occlusion with respect to geographical information.

The paper is organized as follows. Section 2 describes the main functionalities of the AR tool. Section 3 describes the structure of the proposed AR system. Input data processing is explained in Section 3.1. Then, the AR solution is presented in Section 3.2 and the augmented video that is shown to the operators is detailed in Section 3.3. Finally, results and conclusions are presented in Sections 4 and 5, respectively.

2. Functionalities of the Augmented Reality (AR) Tool

The functionalities of the AR solution are oriented to improve the situational awareness of the UAV operators when they have to accomplish a specific mission. The proposed tool provides two different functionalities: route orientation and target identification.

2.1. Route Orientation

MALE UAVs are usually equipped with a gimbal camera sensor payload. The sensor can operate in automatic or manual mode. When the manual mode is activated, the operator can steer the UAV camera with a joystick and point it in any direction. This freedom of movement gives flexibility to explore the world because the sensor is not limited to the flight or the downward-looking directions, but it can also lead to disorientation of the operator. After managing the payload for a while, it may be difficult for the operator to distinguish if the camera is aligned with the flight direction. The proposed AR solution overcomes that problem by superimposing the flying route on the video stream.

The visual assistance given to the operators allows their situational awareness without having to establish correspondences between the 2D map mission information and the video stream, which are shown on different screens. Operators become aware of the camera orientation, not in a global frame but with respect to the flying route at a glance. This can benefit the world exploration because they can infer the remaining time to visit a waypoint and what the next movements of the UAV will be.

2.2. Target Identification

During a mission, the UAV operator could be in charge of identifying some strategic positions. The strategic positions can be targets that are detected by additional sensors (e.g., radar) or a list of targets that are known in advance and should be monitored. Sometimes, it can be difficult for the operators to distinguish the exact position of these targets in the video stream if they are carrying out manual inspection operations. If the targets are far from the UAV, the operators may not distinguish them easily unless they use the camera zoom (if available). This situation, i.e., looking for a target with a close-up view, reduces the situational awareness of the operators. The proposed AR tool overcomes this problem by superimposing virtual beacons on the acquired video stream, even in the presence of occlusions, thus improving target identification.

The functionalities included in the proposed AR tool give assistance to the operators to easily determine where the targets are in the video stream. Additionally, the tool informs the operators about the visibility of the targets with respect to the terrain. This reduces the impact of using the camera zoom because the operators can distinguish if the target is going to be visible when zooming in or if it is occluded by the terrain (e.g., by a mountain) and they should wait for a more appropriate viewpoint along the UAV trajectory.

3. Structure of the AR Tool

The system is structured in three different modules, as it is shown in Figure 2. One of the main parts of the system is that of input data processing; this module encompasses the processing of data coming from the UAV and the information available in the GCS. The AR solution module comprises the information of both real and virtual worlds as well as the establishment of the relationship between them to achieve visual coherence of the whole. Finally, the augmented video module is where the semantic meaning of the virtual elements is explained.

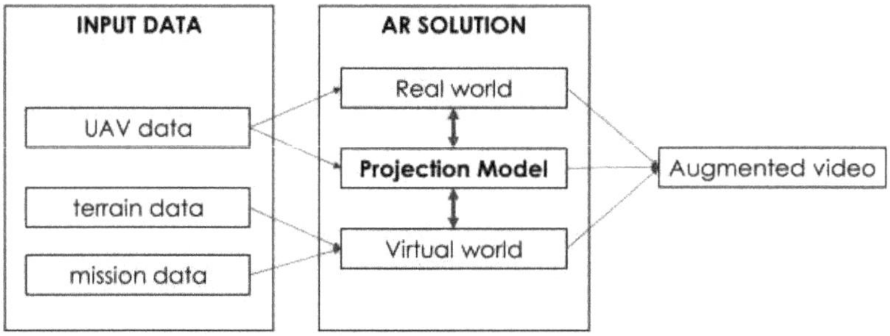

Figure 2. Principal modules of the AR tool: input data processing, AR solution and augmented video. The input data module encompasses the processing of data coming from the UAV and the GCS (terrain and mission). The AR solution module is responsible for achieving coherence of real and virtual worlds. Finally, the augmented video module manages the information shown to the UAV operator.

3.1. Input Data Processing

The input data of the AR tool is provided by the GCS. Two types of data are distinguished according to their availability: (*i*) mission planning data and (*ii*) mission execution data.

3.1.1. Mission Planning Data

Mission planning data are available before the flight is carried out. During mission planning, the route and the flying parameters are settled to allow the autonomous flight of the UAV. The proposed AR tool uses the route path and the digital terrain model from the GIS.

The route information is obtained from the mission plan, exported in an Extensible Markup Language (XML) file using the Common Route Definition (CRD) that is defined according to the standard interfaces of the UAV Control System for NATO UAV interoperability. The latitude and the longitude of each waypoint are obtained from the XML file. The XML file also encapsulates the information about the order in which the waypoints will be visited and the flight altitude when the UAV visits them. The terrain data available in the GCS are the Digital Terrain Elevation Data (DTED). This data consists of a grid of square pixels that contain height information as well as a header with geographic information used to geo-reference the data. The DTED used in the performed experiments is DTED-Level 2, which has a Ground Sampling Distance (GSD) of 30 meters.

3.1.2. Mission Execution Data

Mission execution data provide the information that the GCS receives from the UAV payload (i.e., camera, Global Positioning System–GPS, Inertial Measurement Unit–IMU) after the mission starts. During the flight, the GCS collects information through the datalink established with the UAV.

The proposed AR tool uses the payload data provided by a Motion Imagery Standards Board (MISB) stream file constructed from technologies approved by the MISB and referenced in the NATO Motion Imagery Standard STANAG 4609 [13]. Any Motion Imagery Standards Profile (MISP) compliant file for Full Motion Video must have three components [14]: (*i*) motion imagery (MI); (*ii*) metadata; and (*iii*) a media container. The MI is the essence of the file. It corresponds to the imagery obtained from the Electro-Optical (EO) capture device. The imagery can be in compressed or uncompressed format. The metadata contains the information coming from other sensors (e.g., GPS, IMU) and is encapsulated in Key-Length-Value (KLV) form. Finally, the media container carries the MI and the metadata in two possible ways: using MPEG-2 transport stream (TS) or using a Real-Time Transport Protocol (RTP). The former is the one used in the proposed AR tool.

The metadata is usually collected by the mission computer, although on some occasions more information such as an operator command can be included. In the end, the source is not relevant until the information is interpreted. For this reason, the most important thing is to know how the metadata is encoded. All the MISB metadata is encoded following the Society of Motion Picture and Television Engineers (SMPTE) KLV with the specifications given in MISB STD 0902.1 [15]. This document specifies the Minimum Metadata Set of metadata elements to enable the functionality required for the Situational Awareness Product for Intelligence, Surveillance and Reconnaissance (ISR) missions.

Each metadata package associated to an image has the same scheme, which is illustrated in Figure 3. The set starts with the 16-bytes Universal Label (UL) key (in green), it is followed by the length of the KLV packet (in purple) and a sequence of Tag-Length-Value (TLV) encoded data. The TLV consists of a byte with the metadata tag (in cyan), the length of the metadata package (in magenta), and the value itself depending on the tag data type (in orange). All the metadata as well as the bytes, are represented using big-endian encoding (with the most significant bit first). The information relative to the interpretation of each TLV is given in the document MISB Standard 0601.2 [16]. The tag of the TLV is a unique identification of the type of encapsulated metadata, and the interpretation is reported in the first two columns of Table 1. The rest of the columns correspond to the values and interpretation of an example of KLV packet.

The KLV packets contain:

- information that remains constant throughout a mission: mission identification (tag 3), the type of image sensor (tag 11) and the coordinate system (tag 12).
- information that changes in time: the UNIX Time Stamp (tag 2), the camera sensor, the platform position and orientation.

This information is used to obtain the intrinsic and extrinsic camera parameters that are needed to create the projection matrix. The intrinsic camera parameters are obtained from the Horizontal and Vertical Field of Views (FOVs). They correspond to the metadata with tags 16 and 17, respectively.

Concerning extrinsic parameters, the position of the sensor in the world and the pose are needed. The former is given by the Sensor Latitude (tag 13), Sensor Longitude (tag 14) and the Sensor True Altitude (tag 15), and the latter is calculated from the Platform Heading Angle (tag 5), Platform Pitch Angle (tag 6), Platform Roll Angle (tag 7), Sensor Relative Azimuth Angle (tag 18), Sensor Relative Elevation Angle (tag 19) and Sensor Relative Roll Angle (tag 20).

```
06 0E 2B 34 02 0B 01 01 0E 01 03 01 01 00 00 00
81 AE 02 08 00 04 60 50 58 4E 01 80 03 0A 4D 69
73 73 69 6F 6E 20 31 32 05 02 71 C2 06 02 FD 3D
07 02 08 B8 0A 08 50 72 65 64 61 74 6F 72 0B 07
45 4F 20 4E 6F 73 65 0C 0E 47 65 6F 64 65 74 69
63 20 57 47 53 38 34 0D 04 55 95 B6 6D 0E 04 5B
53 60 C4 0F 02 C2 21 10 02 CD 9C 11 02 D9 17 12
04 72 4A 0A 20 13 04 87 F8 4B 86 14 04 00 00 00
00 15 04 03 83 09 26 16 02 12 81 17 04 F1 01 A2
29 18 04 14 BC 08 2B 19 02 34 F3 30 1C 01 01 01
02 01 07 03 05 2F 2F 55 53 41 0C 01 07 0D 06 00
55 00 53 00 41 16 02 04 01 41 01 02 01 02 29 72
```

Figure 3. Example of a metadata Key-Length-Value (KLV) packet. It is formed by a key (in green), the length of the whole packet (in purple), and a sequence of metadata. Each metadata is identified by a tag (in cyan), the length of the data (in magenta) and the information itself (in orange). Grid patterned colors have the same meaning as the solid colors [15].

Table 1. Example of Tag-Length-Value (TLV) packets contained in a Key-Length-Value (KLV) packet. The table shows: the TLV hexadecimal value (last column), the tag (first column) of the metadata (second column) and its value (third column), and the interpretation of the specific value (fourth column).

Tag	Name	Value	Interpretation	TLV Hex Bytes
2	UNIX Time Stamp	1,231,798,102,000,000 ms	Mon Jan 12 2009 22:08:22 (UTC)	02 08 00 04 60 50 58 4E 01 80
3	Mission ID	Mission 12	Mission 12	03 0A 4D 69 73 73 69 6F 6E 20 31 32
5	Platform Heading Angle	0x71C2	159.9744 Degrees	05 02 71 C2
6	Platform Pitch Angle	0xFD3D	−0.4315251 Degrees	06 02 FD 3D
7	Platform Roll Angle	0x08B8	3.405814 Degrees	07 02 08 B8
11	Image Source Sensor	EO Nose	EO Nose	0B 07 45 4F 20 4E 6F 73 65
12	Image Coordinate System	Geodetic WGS84	Geodetic WGS84	0C 0E 47 65 6F 64 65 74 69 63 20 57 47 53 38 34
13	Sensor Latitude	0x5595B66D	60.17682296 Degrees	0D 04 55 95 B6 6D
14	Sensor Longitude	0x5B5360C4	128.42675904 Degrees	0E 04 5B 53 60 C4
15	Sensor True Altitude	0xC221	14190.72 Meters	0F 02 C2 21
16	Sensor Horizontal FoV	0xCD9C	144.5713 Degrees	10 02 CD 9C
17	Sensor Vertical FoV	0xD917	152.6436 Degrees	11 02 D9 17
18	Sensor Rel. Azimuth Angle	0x724A0A20	160.71921147 Degrees	12 04 72 4A 0A 20
19	Sensor Rel. Elevation Angle	0x87F84B86	−168.79232483 Degrees	13 04 87 F8 4B 86
20	Sensor Rel. Roll Angle	0x00000000	0.0 Degrees	14 04 00 00 00 00

The video stream with the motion imagery data should be parsed, decoded, and decompressed. We have used the Fast Forward MPEG (FFMPEG) libraries to read a User Datagram Protocol (UDP) stream, demultiplex the MPEG-2 TS into video and metadata, and perform video decoding. However, it is not possible to decode the KLV metadata format with them, so we have implemented a decoder to collect the information associated with each frame.

3.2. AR Solution Module

The AR solution module encloses the procedures to achieve visual coherence of the whole scene, so that virtual elements can be included in the video stream in a natural way. Three essential elements of an AR system are distinguished: the real world information, the virtual world, and the relation between the real and virtual worlds.

3.2.1. Real World

Real world information is obtained from the UAV in real-time through the GCS encapsulated in the MPEG-2 TS, as it is explained in Section 3.1.2. MI that is captured by the gimbal camera of the drone is the canvas where the virtual elements will be displayed. In addition, the information corresponding to the acquisition of the images and the UAV position and orientation is needed to build the appropriate projection model.

3.2.2. Projection Model: Conversion between Coordinate Systems

The inclusion of virtual elements in a scene is done by rendering the projection of such elements on the images: virtual elements are defined in the same reference system as the scene and are then rendered on the image using a perspective projection model.

Let us consider that the scene is given in a Cartesian world reference system called Earth-Centered Earth-Fixed frame (ECEF) (Figure 4a). In this system, the Earth's center corresponds to the origin of the ECEF frame, the x-axis points to the intersection of the prime meridian with the Equator (point at (0° latitude, 0° longitude)); the y-axis points to (0° latitude, 90° longitude), and the z-axis points to 90° latitude along the Earth axis of rotation.

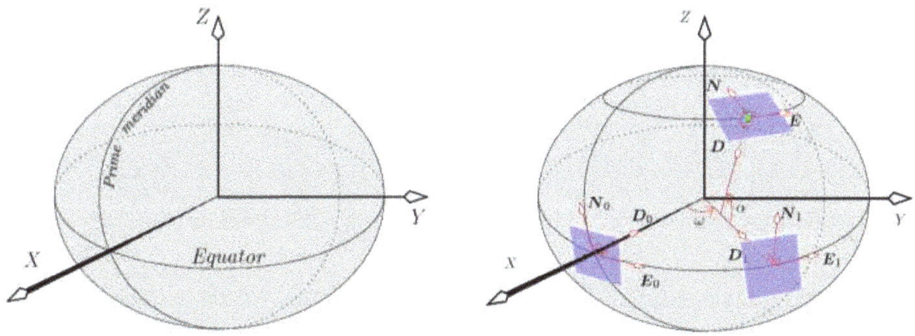

(**a**) Earth-Centered Earth-Fixed reference system (ECEF) (**b**) local geographic frame, North-East-Down (NED)

Figure 4. Geographic coordinate systems used: ECEF and NED [17].

The perspective projection is given by the pinhole camera model [18], which may include lens distortion parameters. In this work, we consider that the lens distortion is negligible so that the projection model is solely described by a 3×4 projection matrix $\mathtt{P}$. In practice, the optics of the camera may be calibrated using an algorithm such as [19], so that lens distortion can be considered compensated. In homogeneous coordinates [18], a world point $\mathbf{X} = (X, Y, Z, 1)^\top$ projects onto the image point $\mathbf{x} = (x, y, 1)^\top$ according to $\mathbf{x} \sim \mathtt{P}\mathbf{X}$, where $\sim$ means an equality up to a non-zero scale factor. The projection matrix $\mathtt{P} = \mathtt{K}[\mathtt{R}|\mathbf{t}]$ consists of the intrinsic ($\mathtt{K}$) and the extrinsic camera parameters ($\mathtt{R}, \mathbf{t}$).

The intrinsic parameter matrix

$$\mathtt{K} = \begin{pmatrix} f_x & 0 & x_0 \\ 0 & f_y & y_0 \\ 0 & 0 & 1 \end{pmatrix} \tag{1}$$

comprises the focal lengths in horizontal and vertical directions (f_x and f_y, respectively) and the principal point (x_0, y_0), assumed to be at the center of the image. For an image of size $w \times h$ pixels (width $\times$ height), the principal point is at $(x_0, y_0) = (w/2, h/2)$. The focal lengths, f_x and f_y, may be calculated from the horizontal and vertical FOVs (FoV_x and FoV_y) and the image size as follows: $f_x = (w/2)/\tan(FoV_x/2)$ and $f_y = (h/2)/\tan(FoV_y/2)$.

The extrinsic camera parameters (translation $\mathbf{t}$ and rotation $\mathtt{R}$) provide the position and orientation (i.e., the pose) of the camera in the world, but they are not as straightforward to set as the intrinsic parameters. Since the experiments performed are based on simulated data (see Section 4), the metadata used to build the camera orientation and position matrix are error free. However, when working with real data, the provided extrinsic parameters will be subject to error, which could lead to virtual objects not being projected on the desired image location. Accurate pose estimation is a different problem from the one tackled here. The proposed AR tool assumes that the pose provided in the metadata is accurate enough for a correct projection of virtual elements.

The camera pose in the world reference system is obtained by interpreting the metadata, which is given in a different coordinate system: World Geodetic System 1984 (tag 12 in Table 1). This standard states that a world location is specified by its latitude, longitude and height with respect to an oblate spheroid that models the shape of the Earth. Thus, the position of the camera in the scene coordinate system is obtained by converting the sensor geodetic coordinates given in the metadata (latitude α, longitude ω, and true altitude h—tags 13–15 in Table 1) to the ECEF reference system:

$$\begin{aligned} X &= (N+h)\cos(\alpha)\cos(\omega), \\ Y &= (N+h)\cos(\alpha)\sin(\omega), \\ Z &= (N(1-ec^2)+h)\sin(\alpha), \end{aligned} \tag{2}$$

where ec and N are the eccentricity and prime vertical radius of curvature of the spheroid, respectively. Such parameters are given by

$$ec = \frac{a^2-b^2}{a^2}, \qquad N = \frac{a}{\sqrt{1-ec^2\sin(\alpha)}}, \tag{3}$$

where $a = 6,378,137$ and $b = 6,356,752.3142$ are the lengths, in meters, of the semi-mayor and semi-minor axes of the spheroid.

The orientation of the camera is obtained by concatenating the rotations that define (i) the orientation of the UAV's local navigation frame (NED) with respect to the world (ECEF) frame; (ii) the orientation of the UAV with respect to its NED frame; and (iii) the orientation of the camera with respect to the UAV. These three changes of coordinates are specified next.

The local geographic reference system used by the UAV is defined by the gravity direction and its perpendicular plane (i.e., plane parallel to the ground). It is called the North-East-Down (NED) system and is shown in Figure 4b. As it can be observed, a plane tangent to the Earth spheroid contains the North and East directions, and the Down (i.e., gravity) is normal to this plane. The center of this local geographic reference system coincides with the position of the sensor in the world. As illustrated in Figure 4b, the NED frame is constructed in two steps, by applying the longitude (ω) and latitude (α) rotations to the $N_0E_0D_0$ frame defined by the vectors

$$\mathbf{n}_0 = (0,0,1)^\top, \quad \mathbf{e}_0 = (0,1,0)^\top, \quad \mathbf{d}_0 = (-1,0,0)^\top. \tag{4}$$

Recall that a frame is rotated by multiplying each of its basis vectors by the rotation matrix. Let the matrix describing the rotation of a point by an angle θ around axis $\mathbf{n}$ (in a right-handed way) be $\mathtt{R}_{\mathbf{n}}(\theta)$, which is given by Rodrigues' formula ([18], p. 585):

$$\mathtt{R}_{\mathbf{n}}(\theta) = (1-\cos(\theta))\mathbf{n}\mathbf{n}^\top + \cos(\theta)\mathbf{I}_3 + \sin(\theta)[\mathbf{n}]_\times, \tag{5}$$

where $\mathbf{n}$ has unit norm, $\mathbf{I}_3$ is the 3×3 identity matrix, and $[\mathbf{n}]_\times$ is the cross-product matrix associated with $\mathbf{n}$. Hence, the NED frame $(\mathbf{n},\mathbf{e},\mathbf{d})$ is obtained from the $N_0E_0D_0$ frame (4) by concatenating two rotations (Figure 4b): $(\mathbf{n}_1,\mathbf{e}_1,\mathbf{d}_1) = \mathtt{R}_{\mathbf{n}_0}(\omega)\,(\mathbf{n}_0,\mathbf{e}_0,\mathbf{d}_0)$, and $(\mathbf{n},\mathbf{e},\mathbf{d}) = \mathtt{R}_{-\mathbf{e}_1}(\alpha)\,(\mathbf{n}_1,\mathbf{e}_1,\mathbf{d}_1)$.

Next, the rotation of the UAV with respect to the NED system is performed (Tags 5–7 in Table 1). The rotation angles of the platform are illustrated in Figure 5. First, the heading is rotated around **d**, $(\mathbf{x}_0, \mathbf{y}_0, \mathbf{z}_0) = \mathrm{R}_\mathbf{d}(heading)\,(\mathbf{n}, \mathbf{e}, \mathbf{d})$; then, the pitch is rotated around the rotated **e**, $(\mathbf{x}_1, \mathbf{y}_1, \mathbf{z}_1) = \mathrm{R}_{\mathbf{y}_0}(pitch)(\mathbf{x}_0, \mathbf{y}_0, \mathbf{z}_0)$, and, lastly, the roll angle is rotated around the latest rotated **n** (i.e., **x**), yielding $(\mathbf{x}_2, \mathbf{y}_2, \mathbf{z}_2) = \mathrm{R}_{\mathbf{x}_1}(roll)(\mathbf{x}_1, \mathbf{y}_1, \mathbf{z}_1)$.

Finally, the rotation of the sensor with respect to the platform is applied (Tags 18–19 in Table 1). A rotation of the sensor relative azimuth angle, $(\mathbf{x}_a, \mathbf{y}_a, \mathbf{z}_a) = \mathrm{R}_{\mathbf{z}_2}(azimuth)\,(\mathbf{x}_2, \mathbf{y}_2, \mathbf{z}_2)$, is followed by a rotation of the relative elevation angle, $(\mathbf{x}_e, \mathbf{y}_e, \mathbf{z}_e) = \mathrm{R}_{\mathbf{y}_a}(elevation)\,(\mathbf{x}_a, \mathbf{y}_a, \mathbf{z}_a)$.

In summary, the camera rotation matrix is $\mathrm{R} = (\mathbf{x}_e, \mathbf{y}_e, \mathbf{z}_e)$, and the camera translation is $\mathbf{t} = -\mathrm{R}\mathbf{C}$, where the optical center of the camera (**C**) is the position of the camera in ECEF coordinates (2).

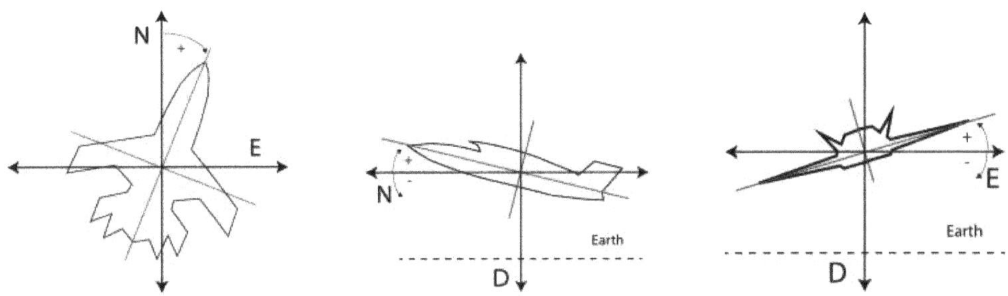

(**a**) heading (**b**) pitch (**c**) roll

Figure 5. On the left, the heading angle of the platform in the plane *N*–*E*. In the middle, the pitch angle of the platform with respect to the plane *D*–*N*. On the right, the roll angle of the platform in the *D*–*E* plane.

3.2.3. Virtual World

The proposed AR tool has been developed with OpenSceneGraph [20] as a 3D rendering engine for the virtual world. This is an open source 3D graphics engine written in C++ that acts as an object-oriented wrapper for OpenGL. The virtual world is defined according to the WSG84 Earth model but in the ECEF reference system. Therefore, to place virtual elements in correspondence with their real world positions, the latitudes, longitudes and altitudes are converted to the ECEF reference system [21].

3.2.4. Flying Route

The flying route is composed of two different entities: the waypoints and the legs (i.e., the part of the route between two waypoints). The positions of the waypoints are defined in the mission planning step, obtained from the XML file, and then transformed from WGS84 to the ECEF reference system. Each waypoint is represented by a semi-transparent sphere and its correspondent leg (a semi-transparent cylinder that extends from one waypoint to another). The route can be interpreted as a directed graph $G = (W, L)$ where $W = \{w_0, w_1, ..., w_i, ..., w_{n-1}\}$ denotes the set of waypoints and $L = \{l_{0,1}, l_{1,2}, ..., l_{i-1,i}, ..., l_{n-2,n-1}\}$ denotes the directed legs (i.e., $l_{i-1,i}$ is the directed leg from w_{i-1} to w_i). The colors of the waypoints and legs change according to the UAV position during flight to give additional information to the operators. Three different colors are used: green, gray and white. They correspond to three different states of the waypoints and legs, respectively: *upcoming*, *visited*, and *not visited*.

Two different situations are considered in the algorithm that controls the waypoint state: (*i*) initialization and (*ii*) standard use case. The initialization step is not crucial if the video stream is received before take-off, but it becomes essential if the datalink is lost and the connection is re-established at any time during the flight. The route information is also taken into consideration to distinguish the conectivity among waypoints. Although a more complex color code could have been chosen, the one selected is a trade-off between giving enough relevant information to the operators and not overloading them.

Figure 6 illustrates the initialization process. The first step in this stage is determining the closest waypoint (highlighted with a dash red circle), $w_{closest}$, to the current position of the UAV (in blue). Once this is known, the next step is distinguishing whether the UAV is getting closer or getting further away from $w_{closest}$. If the UAV is getting closer, as it is shown in Figure 6a, $w_{closest}$ is the *upcoming* one, so $w_{closest}$ and the leg $l_{closest-1,closest}$ are rendered in green. Otherwise, if the UAV is getting further away, as it is illustrated in Figure 6b, $w_{closest}$ is set as *visited* (in gray) and the next one, $w_{closest+1}$, is set as the *upcoming* one, so that the leg shown in green is $l_{closest,closest+1}$. Finally, when the *upcoming* waypoint is identified (i.e., the green one is selected), the previous ones in the route are settled as *visited* (in gray) and the others as *non-visited* (in white). During the standard case of use, a waypoint changes from *upcoming* to *visited* when the UAV is in a sphere of a predefined radius centered in the waypoint.

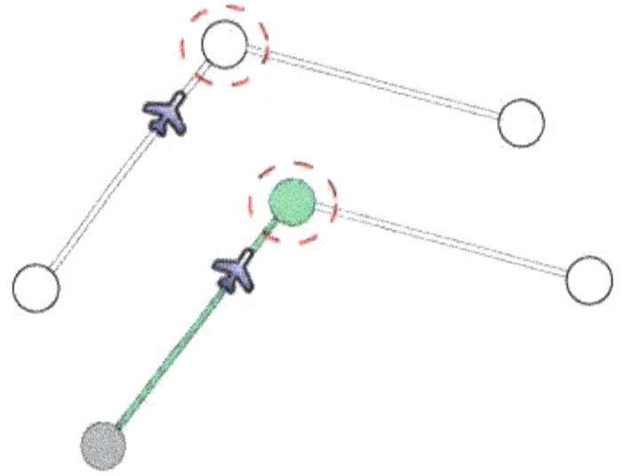

(**a**) UAV approaching *closest* waypoint. The *closest* waypoint coincides with the *upcoming* waypoint.

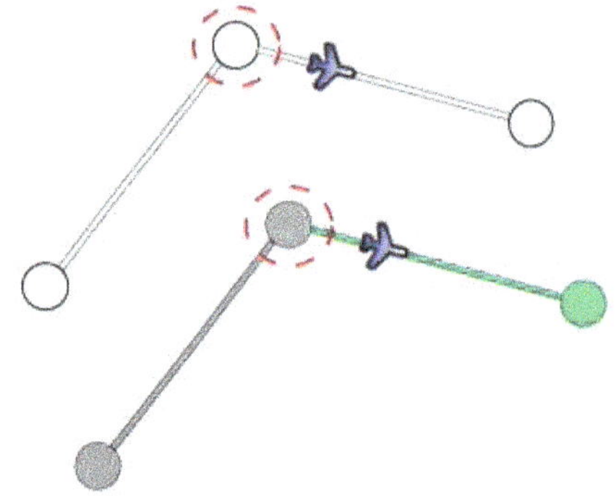

(**b**) UAV getting further away from the *closest* waypoint; the next waypoint is the *upcoming* one.

Figure 6. Initialization process of the flying route. The top of each subfigure shows the UAV situation with respect to the *closest* waypoint (in red). The bottom of each subfigure shows the color coding of the legs and waypoints that will be presented to the operator. The *upcoming* waypoint and the currently flown leg are always displayed in green.

3.2.5. Localization and Visualization of Targets

The targets are represented, as it is illustrated in Figure 7, using four different virtual entities: a cylinder, a cube, a label and a semi-transparent sphere. The cylinder represents a post that starts at the target position and ends at the same latitude and longitude but with a higher altitude. At the top of the post, a cube is added to highlight where the targets are. Both the cylinder and the cube are colored in blue. Additionally, next to the cube, a label that contains information of the specific target is shown in red. This allows the operator not only to distinguish where the targets are, but also to identify them. Finally, a semi-transparent red sphere is displayed at the bottom of the post. This sphere encloses the part of the image where the target should appear, and it is used to represent a region of uncertainty around the target.

The positions of the targets are obtained from an XML file and then transformed from the WGS84 to the ECEF reference system. The use of an XML provides the possibility to exchange information between different sensor types regardless of the specific format used in each one of them.

To achieve a correct visualization of the targets and therefore improve the situational awareness of the operators, it is paramount to take into account occlusions. To this end, the proposed AR tool incorporates the terrain information into the virtual world. The terrain is built from the DTED-Level 2 information of the area that is going to be flown over during the mission, which is established in the mission planning stage. Occlusions are computed on the fly using the 3D rendering engine. The terrain information is stored in a polygon mesh defined with pointers to a vertex list. In particular, we have chosen Wavefront Object (OBJ) format [22], a useful standard for representing polygonal data in the ASCII form that is widely used in computer graphics. This format is chosen because it is not limited to a specific terrain model and it provides the possibility to add terrain models built using different methods such as computer vision techniques, like those in [23]. In addition to the terrain, the use of the OBJ format allows the incorporation of other modeled elements such as buildings, which can improve the situational awareness when the mission requires target identification in an urban environment.

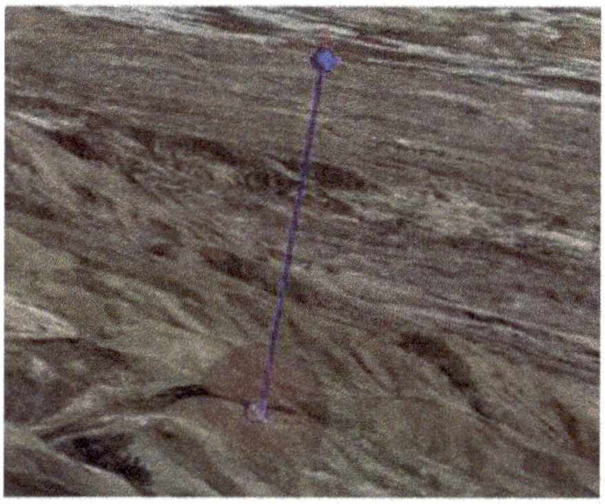

Figure 7. Virtual target beacon—composed by a blue post with a cube at the top, a red label, and a semi-transparent red sphere at the bottom.

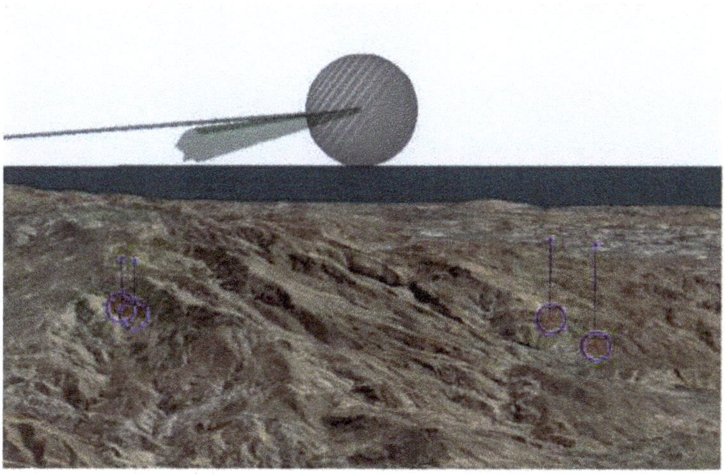

Figure 8. Augmented video with highlighted route (waypoint and legs) and four targets. Same notation for virtual targets as in Figure 7. Same notation for waypoints and legs as in Figure 6: the camera is looking at the last visited waypoint (hence, it is colored in grey, as in Figure 6b).

3.3. Augmented Video

The result of fusing the real world images and the virtual world elements with the correct projection model, as explained in Section 3.2, is an augmented video stream, as illustrated in Figure 8. The UAV operators reduce their workload by visualizing the information contained in the video: they can distinguish the route that the UAV is going to follow and the targets that must be monitored.

The route has three different colors: green, for the upcoming waypoint, gray for the visited ones, and white for the non visited. The operators can easily infer from the colors the direction of flight. Indeed, they can also infer when a waypoint is going to be reached because the closer you get to the waypoint, the bigger it is displayed on the screen.

The targets can be identified even if the UAV is flying far away from them. The operators can obtain information about the target positions quickly by observing the video. Then, they can identify them with the labels.

4. Results

The proposed AR tool for improving drone operations has been tested in a GCS demonstrator at Airbus facilities in Getafe, Madrid, Spain. The input data of the application is a synthetic video and metadata stream following the NATO standard 4609 [13] transmitted through UDP protocol. An XML file containing mission planning information, following the common route definition standards, is also provided. Additionally, digital terrain information and a list of targets that should be identified by the operator are given.

The AR tool has been tested during a mission that takes place in the south of Spain. The objective of the mission was the identification of several targets that were reported to the operator. The targets chosen for the test were buildings, and the operator had to check if the targets were actually present in the indicated locations. This assignment was framed in a reconnaissance procedure. The UAV followed a route that was predefined according to several restrictions (e.g., non-flying zones) during mission planning. The UAV is flown with an automated control system and the operator is responsible for the supervision the flight, the alerts and the payload. The operator can control the camera sensor manually with a joystick. Several tests were carried out with different operators and some representatives moments of the mission are discussed below. Additional material, including the proposed tool, are publicly available (http://www.gti.ssr.upm.es/data/) [24].

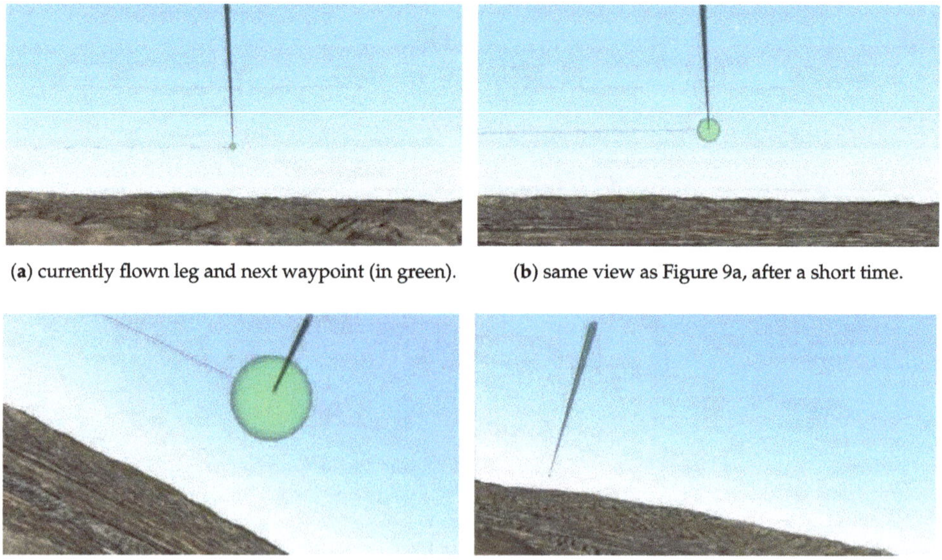

(**a**) currently flown leg and next waypoint (in green). (**b**) same view as Figure 9a, after a short time.

(**c**) next waypoint: the UAV starts turning left. (**d**) the leg has changed the color after the UAV has turned to the left.

Figure 9. Route orientation. Four different moments of the video stream augmented with the proposed AR tool during a mission.

4.1. Results on Route Orientation

Figure 9 shows four different frames of the video stream, augmented with the route during the flight. In Figure 9a, the next waypoint of the route as well as the leg that is being currently followed are shown in green. The operator can easily infer from the image that there is enough time to inspect the environment and the UAV is then going to turn left. Therefore, it is important to explore the zone to the right of the UAV. In Figure 9b, the same waypoint is shown after approximately half a minute. It can be seen that the waypoint is bigger than before, thus indicating that it is going to be reached shortly. Figure 9c shows the waypoint when the UAV is turning left. It can be seen that the UAV is turning left because of the white leg in the image. Finally, Figure 9d shows, in green, the leg that is being currently followed. This leg corresponds to the one shown in white in Figure 9c. This illustrates that, as the UAV progresses, the color information changes accordingly to the situation, thus improving the situational awareness of the operator.

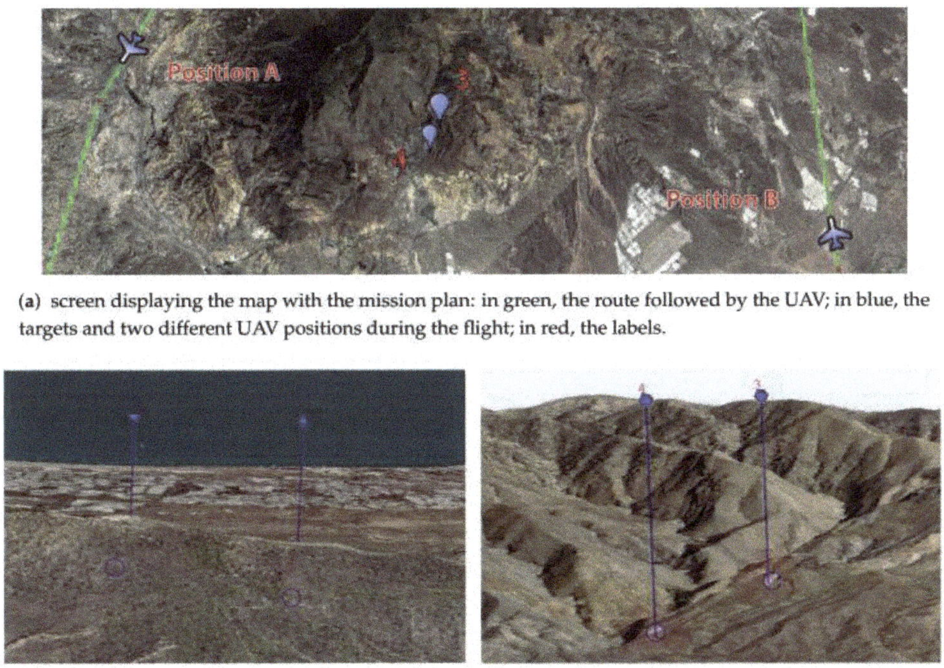

(a) screen displaying the map with the mission plan: in green, the route followed by the UAV; in blue, the targets and two different UAV positions during the flight; in red, the labels.

(b) augmented video (AR tool) from position A. Occluded targets are easy to locate using the virtual beacons.

(c) augmented video from position B. Targets are not occluded. Beacons highlight location and uncertainty of the targets.

Figure 10. Target identification. Results of the AR tool, displayed on the screens of the GCS. A blue circle is surrounding the targets to mark the true position.

4.2. Results on Target Identification

During the mission, the operator is in charge of identifying four targets, which correspond to four buildings. Figure 10 illustrates how some of such targets are seen from different positions along the UAV route. In Figure 10a, the map with the route (in green) is shown. In the middle of the map, two targets labeled with "3" and "4" are depicted in red. The operator should find these targets with the camera. However, in this example, the targets are placed over a mountainous terrain. Thus, an appropriate point of view should be found. The results from two different UAV positions are shown in Figure 10b,c. The targets seen from "position A" are displayed on Figure 10b: the operator can infer from the image that the targets are occluded by the terrain because the post beacons do not end in semi-transparent red spheres; they end with the shape of the mountain. With this information, the operator is aware of the fact that the camera zoom is not going to improve the visibility, hence another point of view is needed. In contrast, the augmented video observed from "position B", displayed in Figure 10c, allows the operator to note that such a position is appropriate to distinguish the targets since they are not occluded. In such a case, the semi-transparent red spheres are shown at the end of the posts, which indicate that the targets are visible from that point of view. Therefore, the target inspection can be carried out from that perspective.

Finally, Figure 11 illustrates the difference between having the raw video stream and the augmented one. The images in this figure show how difficult it is for the operator to know if there are visible targets in the video stream. As it can be seen, the proposed tool significantly benefits the target search. The benefits of the tool have been validated by the GCS experts of Airbus. Their comments have been considered in order to improve the tool because they are aware of the operator needs. For further validation, the AR tool could be tested by UAV operators under some experimental cases in order to provide objective time search measurements.

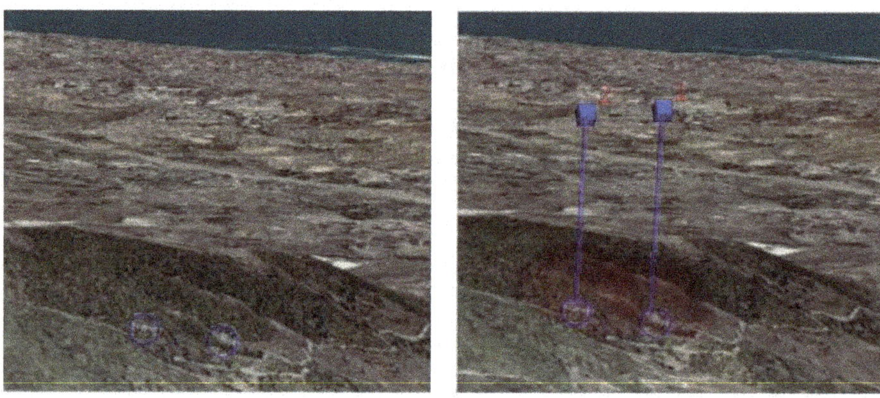

(**a**) Raw video stream (**b**) Augmented video

Figure 11. Difference between the raw video stream (**a**) and the augmented with the proposed AR tool (**b**) for distinguishing buildings in reconnaissance missions. A blue circle surrounding the targets has been superimposed on both images to mark the true position.

5. Conclusions

An AR tool to improve the situational awareness of UAV operators during ISTAR missions with MALE UAVs has been presented. The tool is available online with test data in a public website. The AR system provides information about the flying path, letting the operator know the direction of flight and the next waypoints to visit. Additionally, the targets are highlighted, allowing the operator to easily identify them. Moreover, the presence of occlusions is taken into account so that the operator can reduce the time to find them and prevent the use of the camera zoom when it is not necessary. The usability of the proposed AR tool is assured by the adoption of NATO standards for motion imagery, KLV for metadata and CRD for mission plans. Therefore, the tool is valid for any GCS that follows the same standards. The performance of the AR tool has been tested in an Airbus GCS demonstrator, where it has been shown how the enhancement of the video stream with virtual elements avoids the burden of fusing information displayed in separate screens and improves the situational awareness of the UAV operators.

Acknowledgments: This work has been partially supported by the Ministerio de Economía, Industria y Competitividad of the Spanish Government under projects TEC2013-48453 (MR-UHDTV) and TEC2016-75981 (IVME) and by Airbus Defence and Space under project SAVIER, Open Innovation Program. The authors would also like to acknowledge the help of Álvaro Valverde-Grimaldi.

Author Contributions: All authors conceived the methodology. S.R. designed the augmented reality tool and performed the experiments. S.R. and C.C. analyzed the data. All authors wrote the paper.

Conflicts of Interest: The authors declare no conflict of interest. The founding sponsors had no role in the design of the study; in the collection, analyses, or interpretation of data; in the writing of the manuscript, and in the decision to publish the results.

Abbreviations

The following abbreviations are used in this manuscript:

AR	Augmented Reality
CRD	Common Route Definition
DTED	Digital Terrain Elevation Data
ECEF	Earth-Centered Earth-Fixed Coordinate system
FOV	Field of View
GCS	Ground Control Station
GPS	Global Positioning System
GSD	Ground Sampling Distance
IMU	Inertial Measurement Unit
ISTAR	Intelligence, Surveillance, Target Acquisition and Reconnaissance
KLV	Key-Length-Value
MALE	Medium-Altitude Long-Endurance
MAV	Micro Aerial Vehicles
MI	Motion Imagery
NATO	North Atlantic Treaty Organization
NED	North-East-Down Coordinate System
OBJ	Wavefront Object
SMPTE	Society of Motion Picture and Television Engineers
TLV	Tag-Length-Value
UAV	Unmanned Aerial Vehicle
UL	Universal Label
WGS84	World Geodetic System 1984
XML	Extensible Markup Language

References

1. Acevedo, J.J.; Arrue, B.C.; Maza, I.; Ollero, A. Cooperative large area surveillance with a team of aerial mobile robots for long endurance missions. *J. Intell. Robot. Syst.* **2013**, *70*, 329–345.
2. Colomina, I.; Molina, P. Unmanned aerial systems for photogrammetry and remote sensing: A review. *ISPRS J. Photogramm. Remote Sens.* **2014**, *92*, 79–97.
3. Máthé, K.; Buşoniu, L. Vision and control for UAVs: A survey of general methods and of inexpensive platforms for infrastructure inspection. *Sensors* **2015**, *15*, 14887–14916.
4. Schauwecker, K.; Zell, A. On-board dual-stereo-vision for the navigation of an autonomous MAV. *J. Intell. Robot. Syst.* **2014**, *74*, 1–16.
5. Oliveira, T.; Aguiar, A.P.; Encarnação, P. Moving path following for unmanned aerial vehicles with applications to single and multiple target tracking problems. *IEEE Trans. Robot.* **2016**, *32*, 1062–1078.
6. Chen, J.; Cao, R.; Wang, Y. Sensor-Aware Recognition and Tracking for Wide-Area Augmented Reality on Mobile Phones. *Sensors* **2015**, *15*, 31092–31107.
7. Zollmann, S.; Schall, G.; Junghanns, S.; Reitmayr, G. Comprehensible and interactive visualizations of GIS data in augmented reality. In *International Symposium on Visual Computing*; Springer: Berlin, Heidelberg, Germany, 2012; pp. 675–685.

8. Wither, J.; DiVerdi, S.; Höllerer, T. Annotation in outdoor augmented reality. *Comput. Graph.* **2009**, *33*, 679–689.
9. Zollmann, S.; Grasset, R.; Reitmayr, G.; Langlotz, T. Image-based X-ray visualization techniques for spatial understanding in Outdoor Augmented Reality. In Proceedings of the 26th Australian Computer-Human Interaction Conference on Designing Futures: The Future of Design, New York, NY, USA, 2–5 December 2014; pp. 194–203.
10. Zollmann, S.; Hoppe, C.; Langlotz, T.; Reitmayr, G. FlyAR: Augmented Reality Supported Micro Aerial Vehicle Navigation. *IEEE Trans. Vis. Comput. Graph.* **2014**, *20*, 560–568.
11. Cai, Z.; Chen, M.; Yang, L. Multi-source information fusion augmented reality benefited decision-making for unmanned aerial vehicles: A effective way for accurate operation. In Proceedings of the 2011 6th IEEE Conference on Industrial Electronics and Applications, Beijing, China, 21–23 June 2011; pp. 174–178.
12. Wu, H.; Cai, Z.; Wang, Y. Vison-based auxiliary navigation method using augmented reality for unmanned aerial vehicles. In Proceedings of the IEEE 10th International Conference on Industrial Informatics, Beijing, China, 25–27 July 2012; pp. 520–525.
13. STANAG 4609 Ed.3, NATO Motion Imagery. Available online: http://www.gwg.nga.mil/misb/docs/nato_docs/STANAG_4609_Ed3.pdf (accessed on 3 February 2017).
14. MISB TRM 0909.3, Constructing a MISP Compliant File/Stream. Available online: http://www.gwg.nga.mil/misb/docs/trm/TRM0909.3.pdf (accessed on 3 February 2017).
15. MISB STD 0902.1 Motion Imagery Sensor Minimum Metadata Set. Available online: http://www.gwg.nga.mil/misb/docs/standards/ST0902.1.pdf (accessed on 3 February 2017).
16. MISB Standard 0601.2, UAS Datalink Local Metadata Set. Available online: http://www.gwg.nga.mil/misb/docs/standards/ST0601.2.pdf (accessed on 3 February 2017).
17. Koks, D. *Using Rotations to Build Aerospace Coordinate Systems*; Technical Report; DSTO Systems Sciences Laboratory: Edinburgh, Australia, 2008.
18. Hartley, R.; Zisserman, A. *Multiple View Geometry in Computer Vision*; Cambridge University Press: Cambridge, UK, 2003.
19. Zhang, Z. A flexible new technique for camera calibration. *IEEE Trans. Pattern Anal. Mach. Intell.* **2000**, *22*, 1330–1334.
20. OpenSceneGraph. Available online: http://www.openscenegraph.com/ (accessed on 3 February 2017).
21. *Department of Defense World Geodetic System 1984, Its Definition and Relationships With Local Geodetic Systems, NIMA TR8350.2*; Technical Report; National Imagery and Mapping Agency: St. Louis, MO, USA, 1984.
22. WavefrontOBJ. Available online: http://www.fileformat.info/format/wavefrontobj/egff.htm (accessed on 3 February 2017).
23. Ruano, S.; Gallego, G.; Cuevas, C.; García, N. Aerial video georegistration using terrain models from dense and coherent stereo matching. In Proceedings of the International Society for Optics and Photonics, SPIE Defense + Security, Baltimore, MD, USA, 5–9 May 2014.
24. Augmented Reality Tool for the Situational Awareness Improvement of UAV Operators. Avalilable online: http://www.gti.ssr.upm.es/data/ (accessed on 3 February 2017).

sensors

MDPI

Article

Cramer-Rao Lower Bound Evaluation for Linear Frequency Modulation Based Active Radar Networks Operating in a Rice Fading Environment

Chenguang Shi [1,2], Sana Salous [2], Fei Wang [1] and Jianjiang Zhou [1,*]

1 Key Laboratory of Radar Imaging and Microwave Photonics, Ministry of Education, Nanjing University of Aeronautics and Astronautics, Nanjing 210016, China; scg_space@163.com (C.S.); wangxiaoxian@nuaa.edu.cn (F.W.)

2 School of Engineering and Computing Sciences, Durham University, Durham DH1 3DE, UK; salous.durham@gmail.com

* Correspondence: zjjee@nuaa.edu.cn; Tel.: +86-25-8489-2838

Academic Editors: Felipe Gonzalez Toro and Antonios Tsourdos

Received: 17 September 2016; Accepted: 28 November 2016; Published: 6 December 2016

Abstract: This paper investigates the joint target parameter (delay and Doppler) estimation performance of linear frequency modulation (LFM)-based radar networks in a Rice fading environment. The active radar networks are composed of multiple radar transmitters and multichannel receivers placed on moving platforms. First, the log-likelihood function of the received signal for a Rician target is derived, where the received signal scattered off the target comprises of dominant scatterer (DS) component and weak isotropic scatterers (WIS) components. Then, the analytically closed-form expressions of the Cramer-Rao lower bounds (CRLBs) on the Cartesian coordinates of target position and velocity are calculated, which can be adopted as a performance metric to access the target parameter estimation accuracy for LFM-based radar network systems in a Rice fading environment. It is found that the cumulative Fisher information matrix (FIM) is a linear combination of both DS component and WIS components, and it also demonstrates that the joint CRLB is a function of signal-to-noise ratio (SNR), target's radar cross section (RCS) and transmitted waveform parameters, as well as the relative geometry between the target and the radar network architectures. Finally, numerical results are provided to indicate that the joint target parameter estimation performance of active radar networks can be significantly improved with the exploitation of DS component.

Keywords: Cramer-Rao lower bound (CRLB); Fisher information matrix (FIM); joint parameter estimation; linear frequency modulation (LFM) signal; Rician target; active radar networks

1. Introduction

1.1. Related Works and Motivation

With widely separated transmitters and receivers, the distributed radar networks, also known as spatial distributed multiple-input multiple-output (MIMO) radar systems [1–3], can view the target from different aspect angles and provide spatial and signal diversities. To be specific, for a distributed radar network system with M transmitters and N receivers, the various transmitter-receiver pairs observe the different aspects of the target. In this way, we can obtain the equivalent of MN radars by optimizing the selection of the transmitted signals from different transmitters. However, the conventional radar observes only single aspect of the target. As we can conclude in [4], the advantage of the radar network is that the average received energy across all the transmitter-receiver pairs is approximately constant, and it overcomes deep fades other than the conventional systems. Therefore, the radar network systems have attracted considerable attention and on a path from theory to practice [4–10].

The Cramer-Rao lower bound (CRLB) is an important tool for analyzing the performance of radar networks, which can provide the smallest variance estimates for any unbiased estimation [11,12]. The mean-square error (MSE) of the maximum likelihood estimator (MLE) is close to the CRLB when the high signal-to-noise ratio (SNR) is satisfied. It is also worth mentioning that the performance of multiple signal classification (MUSIC) in computational time-reversal (TR) applications is studied in [13,14], where the closed-form MSE matrix of TR-MUSIC is calculated for the single-frequency case in multistatic co-located and non co-located scenarios. Simulation results show that TR-MUSIC can predict a more accurate MSE than CRLB, while it is a sub-optimal estimator since it does not asymptotically achieve the CRLB as the MLE. In the last couple of years, there is a growing interest on the CRLB studies for the target estimation performance of distributed radar networks [11–19]. The authors in [11] derive the analytical expressions of CRLB for both noncoherent mode and coherent mode in MIMO radar systems, which shows that the CRLB is inversely proportional to the carrier frequency and signals averaged effective bandwidth. In [12], the problem of target parameter estimation for noncoherent MIMO radar is addressed, and the joint CRLB of target position and velocity is computed. Reference [15] further extends the results in [12] to a multiple targets scenario. Later, He et al. investigate the coherent MIMO radar performance when the oscillators at each transmitter and receiver are aligned in phase [16]. The work in [17] studies the target localization accuracy for MIMO radar systems with static phase errors. In [18], the CRLBs of the joint time delay and Doppler shift estimation are derived for an extended target, and the effects of transmitted waveform parameters on the CRLBs are analyzed. Assuming that the approximation state of the target is unknown without previous target detection, a generalized CRLB for distributed active and passive radar networks is calculated in [19].

Recently, the CRLB has been investigated and applied to passive radar systems that employ signals of opportunity as illuminators for target detection, estimation and tracking [20–24]. Since passive radar does not use its own transmitter to radiate electromagnetic wave, it has been a potential technology for low cost, low probability of intercept (LPI) [25–27], antijamming and other advantages. The authors in [20] present the CRLB analysis for the joint target estimation of position and velocity in a frequency modulation (FM) based passive radar networks. In [21–24], the modified CRLB (MCRLB) is employed as a good alternative to the classical CRLB due to the presence of random parameters in the transmitted waveforms, which has been shown to offer a looser bound in practical applications. The target estimation performance of a universal mobile telecommunications systems (UMTS)-based passive multistatic radar and an orthogonal frequency-division multiplexing (OFDM)-based passive radar network in a line-of-sight (LoS) environment is analyzed in [23,24] respectively, where the Rician target model is composed of two components, that is, fixed amplitude or dominant scatterer (DS) and weak isotropic scatterers (WIS). It is shown that the target estimation accuracy will be increased with an increase in reflection coefficient, number of transmitter-receiver pairs, the choice of the transmitter-receiver pairs and duration time. Furthermore, the work in [28] proposes two transmitter of opportunity selection algorithms for FM-based passive radar network systems, which are formulated as knapsack problems (KPs) and tackled with greedy selection approaches. On the basis of the research mentioned above, almost all of previous works focus on stationary platforms. The CRLB analysis for joint moving target position and velocity estimation of linear frequency modulation (LFM) based active radar networks with sensors placed on moving platforms operating in a Rice fading environment, which has not been considered, needs to be investigated.

1.2. Major Contributions

The major contributions of this paper are fourfold:

(1) We formulate the linear frequency modulation (LFM) signal model and derive the log-likelihood function of the received signal for a Rician target. The Rician target model is composed of DS component and WIS components, which are the signals received after striking from the target [21,22]. It is worth pointing out here that [15,16,27] only study the target parameter estimation accuracy limits either when the target' radar cross section (RCS) observes as a Rayleigh model in a non-coherent scenario or the target is modeled as a point target in a coherent scenario for all the transmitter-receiver pairs [23]. While utilizing a Rician target model, the estimation performance can be generalized and evaluated when the target has different RCS models for different transmitter-receiver pairs.
(2) On the basis of the previous works [15–24,29], almost all the studies concentrate on stationary platforms. In this paper, we build an LFM-based active radar network configuration and extend it to a more general case, which consist of multiple radar transmitters and multichannel receivers placed on moving platforms. On the other hand, only the CRLBs for LFM-based bistatic radar channels are computed in [29]. To the best of our knowledge, the CRLB for an LFM-based radar network has not been derived. Thus, the joint CRLB for position and velocity estimation of a Rician target in LFM-based radar networks is computed, where we assume that the signals scattered off the target due to different radar transmitters can be received and separated at the receivers. The cumulative Fisher information matrix (FIM) can be factored into two terms: one term accounting for the effect of the DS component, and another incorporating the effect of the WIS components.
(3) Simulation results have shown that the DS component can be exploited to decrease the target parameter estimation errors, which is due to the fact that the reception of DS component increases the received SNR at the radar receiver. Previous results in [20,29] only show that the CRLB is a function of the signal parameters as well as the geometry between the target and the radar network architecture. In this paper, the effects of SNR and target's RCS on the target parameter estimation performance are also analyzed. It is demonstrated that the joint CRLB is not only a function of SNR, target's RCS and transmitted waveform parameters, but also a function of the geometry between the target and the active radar network systems.
(4) The closed-form expressions of CRLB can be used as a performance metric to access the target estimation performance for LFM-based active radar networks in a Rice fading environment. Since the DS component can be exploited to increase the received SNR at the receiver, the geometry-dependent CRLB analysis will open up a new dimension for active radar network systems by aiding the optimal power allocation for radar networks to achieve a given estimation requirement with the minimum system cost.

1.3. Outline of the Paper

The rest of the paper is organized as follows. Section 1 describes the signal model for LFM-based radar networks. In Section 2, the joint CRLB is computed for target position and velocity estimation by deriving the closed-form expressions of FIM. The numerical simulations are provided to demonstrate our analytical results in Section 3. Finally, conclusion remarks are drawn with potential future work in Section 4.

Notation: The superscript T represents the transpose operator; $\mathbb{E}\{\cdot\}$ and $(\cdot)^*$ represent the expectation and conjugation operators, respectively. $|\cdot|$ denotes the absolute value, $\Re\{\cdot\}$ is the real part, and $\Im\{\cdot\}$ is the imaginary part. $S_i(f)$ denotes the Fourier transform of $s_i(t)$.

2. Signal Model

Consider a active radar network architecture comprising of N_T radar transmitters and N_R multichannel receivers. Let the ith radar transmitter and the jth receiver be located at $\overrightarrow{\mathbf{p}_i^t} = [x_i^t, y_i^t]$ and $\overrightarrow{\mathbf{p}_j^r} = [x_j^r, y_j^r]$ respectively, in a 2-dimensional Cartesian coordinate system for simplicity. The target position and velocity are supposed to be deterministic unknown and denoted by $\overrightarrow{\mathbf{p}} = [x, y]$ and $\overrightarrow{\mathbf{v}} = [v_x, v_y]$. We define the unknown target state vector:

$$\mathbf{U} = [x, y, v_x, v_y]^T. \tag{1}$$

Without loss of generality, we will concentrate on a single target scenario. However, the results can be extended to multiple targets.

Let τ_{ij} represent the bistatic time delays corresponding to the path between the ith radar transmitter, moving target, and the jth radar receiver, which is a function of the unknown target position $\overrightarrow{\mathbf{p}} = [x, y]$:

$$\begin{aligned} \tau_{ij} &= \frac{\sqrt{(x-x_i^t)^2+(y-y_i^t)^2}+\sqrt{(x-x_j^r)^2+(y-y_j^r)^2}}{c} \\ &= \frac{\left\|\overrightarrow{\mathbf{p}}-\overrightarrow{\mathbf{p}_i^t}\right\|+\left\|\overrightarrow{\mathbf{p}}-\overrightarrow{\mathbf{p}_j^r}\right\|}{c}, \end{aligned} \tag{2}$$

where c is the speed of light, $\left\|\overrightarrow{\mathbf{p}}-\overrightarrow{\mathbf{p}_i^t}\right\|$ denotes the distance from the ith radar transmitter to the target and $\left\|\overrightarrow{\mathbf{p}}-\overrightarrow{\mathbf{p}_j^r}\right\|$ denotes the distance from the target to the jth receiver, respectively. In this paper, the ith radar transmitter and the jth multichannel receiver are moving with velocities $\overrightarrow{\mathbf{v}_i^t} = [v_{x,i}^t, v_{y,i}^t]$ and $\overrightarrow{\mathbf{v}_j^r} = [v_{x,j}^r, v_{y,j}^r]$, respectively. With the aforementioned positions/velocities of the target, the radar transmitters and receivers, the Doppler shift of the moving target corresponding to the ijth path is the time rate of change of the total ijth path length:

$$f_{D_{ij}} = \frac{1}{\lambda}\left[\frac{\partial\left\|\overrightarrow{\mathbf{p}}-\overrightarrow{\mathbf{p}_i^t}\right\|}{\partial t} + \frac{\partial\left\|\overrightarrow{\mathbf{p}}-\overrightarrow{\mathbf{p}_j^r}\right\|}{\partial t}\right], \tag{3}$$

where λ denotes the carrier wavelength, $\frac{\partial\left\|\overrightarrow{\mathbf{p}}-\overrightarrow{\mathbf{p}_i^t}\right\|}{\partial t}$ and $\frac{\partial\left\|\overrightarrow{\mathbf{p}}-\overrightarrow{\mathbf{p}_j^r}\right\|}{\partial t}$ are the relative velocities for the ith radar transmitter and the jth receiver, respectively. Then, we have:

$$\begin{aligned} f_{D_{ij}} &= \tfrac{1}{\lambda}\left[v_x\left(\frac{x-x_i^t}{\left\|\overrightarrow{\mathbf{p}}-\overrightarrow{\mathbf{p}_i^t}\right\|}+\frac{x-x_j^r}{\left\|\overrightarrow{\mathbf{p}}-\overrightarrow{\mathbf{p}_j^r}\right\|}\right)\right]+\tfrac{1}{\lambda}\left[v_y\left(\frac{y-y_i^t}{\left\|\overrightarrow{\mathbf{p}}-\overrightarrow{\mathbf{p}_i^t}\right\|}+\frac{y-y_j^r}{\left\|\overrightarrow{\mathbf{p}}-\overrightarrow{\mathbf{p}_j^r}\right\|}\right)\right] \\ &+\tfrac{1}{\lambda}\left[v_{x,i}^t\frac{x-x_i^t}{\left\|\overrightarrow{\mathbf{p}}-\overrightarrow{\mathbf{p}_i^t}\right\|}+v_{y,i}^t\frac{y-y_i^t}{\left\|\overrightarrow{\mathbf{p}}-\overrightarrow{\mathbf{p}_i^t}\right\|}\right]+\tfrac{1}{\lambda}\left[v_{x,j}^r\frac{x-x_j^r}{\left\|\overrightarrow{\mathbf{p}}-\overrightarrow{\mathbf{p}_j^r}\right\|}+v_{y,j}^r\frac{y-y_j^r}{\left\|\overrightarrow{\mathbf{p}}-\overrightarrow{\mathbf{p}_j^r}\right\|}\right], \end{aligned} \tag{4}$$

which is a function of the unknown target position $\overrightarrow{\mathbf{p}} = [x, y]$ and velocity $\overrightarrow{\mathbf{v}} = [v_x, v_y]$.

The LFM signal transmitted by the ith radar transmitter is given by [29]:

$$s_i(t) = \frac{1}{\sqrt{N}}\sum_{n=0}^{N-1} u_i(t-nT_R), \tag{5}$$

where

$$u_i(t) = \begin{cases} \frac{1}{T}e^{j\pi k t^2}, & |t| \le \frac{T}{2}, \quad (6a) \\ 0, & \text{elsewhere} \quad (6b) \end{cases}$$

N is the number of subpulses for each transmitted burst, T_R is the pulse repetition interval (PRI) and T is the duration of each pulse, such that $T < T_R/2$. Moreover, $kT^2 = BT$ represents the effective time-bandwidth product of the signal and B denotes the total frequency derivation. Note that each transmitter-receiver pair has its own angle of view for the target because of the widely spaced antennas that leads to different attenuation factors [20]. It is assumed that the signals from different radar transmitters are supposed to be received and processed at the multichannel radar receivers. For a Rician target, it consists of a DS and many independent WIS. In this paper, by utilizing the Rician target model [23], the reflection coefficient ζ_{ij} is modeled as a complex Gaussian random variable with mean d_{ij} and variance σ^2, i.e., $\zeta_{ij} \sim \mathcal{CN}(d_{ij}, \sigma^2)$. Then, the signal from the ith radar transmitter arriving at the jth receiver can be expressed as:

$$r_{ij}(t) = \zeta_{ij} s_i(t - \tau_{ij}) e^{j2\pi f_{D_{ij}}(t-\tau_{ij})} + w_{ij}(t), \tag{7}$$

where $w_{ij}(t)$ denotes the additive zero-mean white Gaussian noise of variance corresponding to the ijth path, i.e., $w_{ij} \sim \mathcal{CN}(0, \sigma_n^2)$, which is independent to ζ_{ij}. We assume that the parameters d_{ij}, σ^2 and σ_n^2 are deterministic and known.

Following the concepts and derivations in [12,23], the likelihood ratio of the ijth transmitter-receiver pair can be given by:

$$\begin{aligned} \Lambda(r_{ij}(t); \mathbf{U}) &= \exp\Bigg\{ \tfrac{\sigma^2}{\sigma^2+\sigma_n^2} \left| \int_{-\infty}^{+\infty} r_{ij}(t) s_i^*(t-\tau_{ij}) e^{-j2\pi f_{D_{ij}}(t-\tau_{ij})} dt \right|^2 \\ &- \tfrac{1}{\sigma^2+\sigma_n^2} \left| \int_{-\infty}^{+\infty} a_{r_{ij}} s_i^*(t-\tau_{ij}) e^{-j2\pi f_{D_{ij}}(t-\tau_{ij})} dt \right|^2 \\ &+ \tfrac{2}{\sigma^2+\sigma_n^2} \Re \left[\int_{-\infty}^{+\infty} r_{ij}(t) s_i^*(t-\tau_{ij}) e^{-j2\pi f_{D_{ij}}(t-\tau_{ij})} dt \right. \\ &\left. \times d_{r_{ij}}^* s_i(t-\tau_{ij}) e^{j2\pi f_{D_{ij}}(t-\tau_{ij})} dt \right] \Bigg\} + \left(\tfrac{\sigma_n^2}{\sigma^2+\sigma_n^2} \right), \end{aligned} \tag{8}$$

where $d_{r_{ij}}$ represents the mean of the received signal $r_{ij}(t)$, i.e., $d_{r_{ij}} = d_{ij} s_i(t-\tau_{ij}) e^{j2\pi f_{D_{ij}}(t-\tau_{ij})}$. Furthermore, the log-likelihood ratio is written as:

$$\begin{aligned} L(r_{ij}(t); \mathbf{U}) &= \tfrac{\sigma^2}{\sigma^2+\sigma_n^2} \left| \int_{-\infty}^{+\infty} r_{ij}(t) s_i^*(t-\tau_{ij}) e^{-j2\pi f_{D_{ij}}(t-\tau_{ij})} dt \right|^2 \\ &- \tfrac{1}{\sigma^2+\sigma_n^2} \left| \int_{-\infty}^{+\infty} d_{r_{ij}} s_i^*(t-\tau_{ij}) e^{-j2\pi f_{D_{ij}}(t-\tau_{ij})} dt \right|^2 \\ &+ \tfrac{2}{\sigma^2+\sigma_n^2} \Re \left\{ \int_{-\infty}^{+\infty} r_{ij}(t) s_i^*(t-\tau_{ij}) e^{-j2\pi f_{D_{ij}}(t-\tau_{ij})} dt \right. \\ &\left. \times d_{r_{ij}}^* s_i(t-\tau_{ij}) e^{j2\pi f_{D_{ij}}(t-\tau_{ij})} dt \right\} + \ln \left(\tfrac{\sigma_n^2}{\sigma^2+\sigma_n^2} \right). \end{aligned} \tag{9}$$

Due to the fact that the radar transmitters and receivers are widely separated, the received signals $r_{ij}(t)$ are mutually independent for different transmitter-receiver pairs. Therefore, the joint log-likelihood ratio across all the transmitter-receiver pairs can be written as the sum of all single transmitter-receiver pair log-likelihood ratios:

$$\begin{aligned} L(\mathbf{r}(t); \mathbf{U}) &= \textstyle\sum_{i=1}^{N_T} \sum_{j=1}^{N_R} L(r_{ij}(t); \mathbf{U}) \\ &= \textstyle\sum_{i=1}^{N_T} \sum_{j=1}^{N_R} (\Gamma_{ij}^1 - \Gamma_{ij}^2 + \Gamma_{ij}^3) + \mathbb{C}, \end{aligned} \tag{10}$$

where $\mathbf{r}(t) = \left[r_{11}(t), r_{12}(t), \cdots, r_{N_T N_T}(t) \right]^T$ is the observed signals from the entire set of the receivers. Γ_{ij}^1, Γ_{ij}^2 and Γ_{ij}^3 denote the first, second and third terms in (9), respectively. The constant

$\mathbb{C} = \sum_{i=1}^{N_T}\sum_{j=1}^{N_R} \ln\left(\frac{\sigma_n^2}{\sigma^2+\sigma_n^2}\right)$ is independent of the target state vector **U**. Therefore, the MLE of the unknown target state vector **U** can be expressed as:

$$\begin{aligned} \hat{\mathbf{U}}_{\mathrm{ML}} &= \arg\max_{\mathbf{U}} L(\mathbf{r}(t);\mathbf{U}) \\ &= \arg\max_{\mathbf{U}} \textstyle\sum_{i=1}^{N_T}\sum_{j=1}^{N_R} L(r_{ij}(t);\mathbf{U}) \\ &= \arg\max_{\mathbf{U}} \textstyle\sum_{i=1}^{N_T}\sum_{j=1}^{N_R} (\Gamma_{ij}^1 - \Gamma_{ij}^2 + \Gamma_{ij}^3), \end{aligned} \tag{11}$$

where $\hat{\mathbf{U}}_{\mathrm{ML}}$ represents the MLE of the unknown parameter vector **U**.

3. Derivation of Joint Cramer-Rao Lower Bound

It is discussed in [12,16] that the CRLB indicates the smallest variance estimate of any unbiased estimate, which can be adopted as a performance metric in parameter estimation problems because that the CRLB is close to the MSE of the MLE when the high SNR is satisfied. Using the derivations in [12,21], the FIM is a 4×4 matrix related to the second-order derivatives of the joint log-likelihood function:

$$\begin{aligned} \mathbf{J}(\mathbf{U}) &= (\bigtriangledown_{\mathbf{U}}\mathbf{Q}^T)\mathbf{J}(\mathbf{Q})(\bigtriangledown_{\mathbf{U}}\mathbf{Q}^T)^T \\ &= (\bigtriangledown_{\mathbf{U}}\mathbf{Q}^T)\left(-\mathbb{E}_{\mathbf{r}(t);\mathbf{U}}\{\bigtriangledown_{\mathbf{Q}}[\bigtriangledown_{\mathbf{Q}}L(\mathbf{r}(t);\mathbf{U})]^T\}\right)(\bigtriangledown_{\mathbf{U}}\mathbf{Q}^T)^T, \end{aligned} \tag{12}$$

where we define **Q** as an alternative representation of the unknown parameter vector:

$$\mathbf{Q} = \left[\tau_{ij}, f_{D_{ij}}\right]^T (\forall i,j). \tag{13}$$

We first derive the Jacobian matrix $(\bigtriangledown_{\mathbf{U}}\mathbf{Q}^T)$, whose entries can be obtained by taking the first-order derivatives of the time-delays in (2) and the Doppler shifts in (4) with respect to target positions:

$$\frac{\partial\tau_{ij}}{\partial x} \equiv \frac{1}{c}\left(\frac{x - x_i^t}{\left\|\overrightarrow{\mathbf{p}} - \overrightarrow{\mathbf{p}_i^t}\right\|} + \frac{x - x_j^r}{\left\|\overrightarrow{\mathbf{p}} - \overrightarrow{\mathbf{p}_j^r}\right\|}\right), \tag{14}$$

$$\frac{\partial\tau_{ij}}{\partial y} \equiv \frac{1}{c}\left(\frac{y - y_i^t}{\left\|\overrightarrow{\mathbf{p}} - \overrightarrow{\mathbf{p}_i^t}\right\|} + \frac{y - y_j^r}{\left\|\overrightarrow{\mathbf{p}} - \overrightarrow{\mathbf{p}_j^r}\right\|}\right), \tag{15}$$

$$\begin{aligned} \frac{\partial f_{D_{ij}}}{\partial x} \equiv \frac{1}{\lambda}\Bigg\{ & v_x\left[\frac{(y-y_i^t)^2}{\left\|\overrightarrow{\mathbf{p}}-\overrightarrow{\mathbf{p}_i^t}\right\|^3} + \frac{(y-y_j^r)^2}{\left\|\overrightarrow{\mathbf{p}}-\overrightarrow{\mathbf{p}_j^r}\right\|^3}\right] + v_y\left[-\frac{(x-x_i^t)(y-y_i^t)}{\left\|\overrightarrow{\mathbf{p}}-\overrightarrow{\mathbf{p}_i^t}\right\|^3} - \frac{(x-x_j^r)(y-y_j^r)}{\left\|\overrightarrow{\mathbf{p}}-\overrightarrow{\mathbf{p}_j^r}\right\|^3}\right] \\ & + \left[v_{x,i}^t\frac{(y-y_i^t)^2}{\|\overrightarrow{\mathbf{p}}-\overrightarrow{\mathbf{p}_i^t}\|^3} - v_{y,i}^t\frac{(x-x_i^t)(y-y_i^t)}{\|\overrightarrow{\mathbf{p}}-\overrightarrow{\mathbf{p}_i^t}\|^3}\right] + \left[v_{x,j}^r\frac{(y-y_j^r)^2}{\|\overrightarrow{\mathbf{p}}-\overrightarrow{\mathbf{p}_j^r}\|^3} - v_{y,j}^r\frac{(x-x_j^r)(y-y_j^r)}{\|\overrightarrow{\mathbf{p}}-\overrightarrow{\mathbf{p}_j^r}\|^3}\right]\Bigg\}, \end{aligned} \tag{16}$$

$$\begin{aligned} \frac{\partial f_{D_{ij}}}{\partial y} \equiv \frac{1}{\lambda}\Bigg\{ & v_x\left[-\frac{(x-x_i^t)(y-y_j^r)}{\left\|\overrightarrow{\mathbf{p}}-\overrightarrow{\mathbf{p}_i^t}\right\|^3} - \frac{(x-x_j^r)(y-y_j^r)}{\left\|\overrightarrow{\mathbf{p}}-\overrightarrow{\mathbf{p}_j^r}\right\|^3}\right] + v_y\left[\frac{(x-x_i^t)^2}{\left\|\overrightarrow{\mathbf{p}}-\overrightarrow{\mathbf{p}_i^t}\right\|^3} + \frac{(x-x_j^r)^2}{\left\|\overrightarrow{\mathbf{p}}-\overrightarrow{\mathbf{p}_j^r}\right\|^3}\right] \\ & + \left[-v_{x,i}^t\frac{(x-x_i^t)(y-y_i^t)}{\|\overrightarrow{\mathbf{p}}-\overrightarrow{\mathbf{p}_i^t}\|^3} + v_{y,i}^t\frac{(x-x_i^t)^2}{\|\overrightarrow{\mathbf{p}}-\overrightarrow{\mathbf{p}_i^t}\|^3}\right] + \left[-v_{x,j}^r\frac{(x-x_j^r)(y-y_j^r)}{\|\overrightarrow{\mathbf{p}}-\overrightarrow{\mathbf{p}_j^r}\|^3} + v_{y,j}^r\frac{(x-x_j^r)^2}{\|\overrightarrow{\mathbf{p}}-\overrightarrow{\mathbf{p}_j^r}\|^3}\right]\Bigg\}, \end{aligned} \tag{17}$$

Similarly, the derivatives with respect to the target velocities can be calculated as:

$$\frac{\partial\tau_{ij}}{\partial v_x} \equiv 0, \tag{18}$$

$$\frac{\partial \tau_{ij}}{\partial v_y} \equiv 0, \tag{19}$$

$$\frac{\partial f_{D_{ij}}}{\partial v_x} \equiv \frac{1}{\lambda}\left(\frac{x - x_i^t}{\left\|\overrightarrow{\mathbf{p}} - \overrightarrow{\mathbf{p}_i^t}\right\|} + \frac{x - x_j^r}{\left\|\overrightarrow{\mathbf{p}} - \overrightarrow{\mathbf{p}_j^t}\right\|}\right), \tag{20}$$

$$\frac{\partial f_{D_{ij}}}{\partial v_y} \equiv \frac{1}{\lambda}\left(\frac{y - y_i^t}{\left\|\overrightarrow{\mathbf{p}} - \overrightarrow{\mathbf{p}_i^t}\right\|} + \frac{y - y_j^r}{\left\|\overrightarrow{\mathbf{p}} - \overrightarrow{\mathbf{p}_j^t}\right\|}\right). \tag{21}$$

After lengthy algebraic derivations, the FIM $\mathbf{J}(\mathbf{Q})$ can be correspondingly expressed by:

$$\begin{aligned} \mathbf{J}(\mathbf{Q}) &= -\mathbb{E}_{\mathbf{r}(t);\mathbf{U}}\{[\nabla_{\mathbf{Q}} L(\mathbf{r}(t);\mathbf{U})][\nabla_{\mathbf{Q}} L(\mathbf{r}(t);\mathbf{U})]^T\} \\ &= -\mathbb{E}_{\mathbf{r}(t);\mathbf{U}}\{\nabla_{\mathbf{Q}}[\nabla_{\mathbf{Q}} L(\mathbf{r}(t);\mathbf{U})]^T\} \\ &= \sum_{i=1}^{N_T}\sum_{j=1}^{N_R} \frac{8\pi^2\sigma^4}{\sigma_n^2(\sigma^2+\sigma_n^2)}\left[1 + 2h_{ij} + \frac{2h_{ij}}{(\sigma^2/\sigma_n^2)}\right] \times \begin{bmatrix} \varepsilon_i & \gamma_{ij} \\ \gamma_{ij} & \eta_{ij} \end{bmatrix}, \end{aligned} \tag{22}$$

where the terms ε_i, η_{ij}, and γ_{ij} are dependent on the radar waveforms, which can be calculated as:

$$\begin{aligned} \varepsilon_i &\equiv \int_{-\infty}^{+\infty} f^2 \left|S_i(f)\right|^2 df - \left|\int_{-\infty}^{+\infty} f \left|S_i(f)\right|^2 df\right|^2 \\ &= \tfrac{1}{3}\pi^2 k^2 T^2, \end{aligned} \tag{23}$$

$$\begin{aligned} \eta_{ij} &\equiv \int_{-\infty}^{+\infty} t^2 \left|s_i(t)\right|^2 df - \left|\int_{-\infty}^{+\infty} t \left|s_i(t)\right|^2 df\right|^2 \\ &= \tfrac{1}{12}T^2 + \tfrac{1}{12}T_R^2(N^2 - 1), \end{aligned} \tag{24}$$

$$\begin{aligned} \gamma_{ij} &\equiv \Im\left\{\int_{-\infty}^{+\infty} t s_i^*(t)\frac{\partial s_i(t)}{\partial t}dt\right\} - \int_{-\infty}^{+\infty} t|s_i(t)|^2 dt \int_{-\infty}^{+\infty} s_i(t)\frac{\partial s_i^*(t)}{\partial t}dt \\ &= -\tfrac{1}{6}k\pi T^2. \end{aligned} \tag{25}$$

The derivation of $\mathbf{J}(\mathbf{Q})$ is provided in Appendix A. Then, we can write the final expression for total FIM across all the transmitter-receiver pairs as:

$$\mathbf{J}(\mathbf{U}) = \sum_{i=1}^{N_T}\sum_{j=1}^{N_R} \frac{8\pi^2\sigma^4}{\sigma_n^2(\sigma^2+\sigma_n^2)}\left[1 + 2h_{ij} + \frac{2h_{ij}}{(\sigma^2/\sigma_n^2)}\right]\mathbf{J}_{ij}(\mathbf{U}). \tag{26}$$

The expressions for the elements of the bistatic FIM $\mathbf{J}_{ij}(\mathbf{U})$ corresponding to the ijth transceiver pair are given in Appendix B. The CRLB for the joint position and velocity estimation of a Rician target can be obtained by taking inverse of FIM in (26), i.e.,

$$\mathrm{CRLB}(\mathbf{U}) = \mathbf{J}^{-1}(\mathbf{U}). \tag{27}$$

Remark 1. *It is obvious that the final expression of FIM in (26) is a linear combination of FIMs due to DS component and WIS components [23]. In this paper, one of our goals is to increase the SNR value at the radar receiver by employing the DS component, which leads to lower radar transmit power and better target estimation performance.*

Remark 2. *From (26) and (27), we can observe that the MCRLB depends on a number of factors. It not only depends on the relative geometry between the target and the radar networks, but also depends on the transmitted LFM waveform parameters such as the duration of each pulse and bandwidth. In addition, it shows dependence on the target's RCS and the SNR.*

4. Simulation Results

In the following, numerical results are dedicated to compute the joint CRLB for active active radar networks as well as reveal the effects of several factors on the CRLB.

For numerical simulations, we consider a radar network with five active radar transmitters and an equal number of multichannel receivers, i.e., $N_T = 5$ and $N_R = 5$. The Cartesian coordinates of their positions are provided in Figure 1. The position/moving parameters of the radar transmitters are shown in Table 1. The receivers are co-located with the corresponding transmitters and have the same velocities. It is assumed that the target is located at $[6000, 6000]$ m with velocity $[30, 50]$ m/s. For simulation parameters, we set the LFM signal parameters as follows [29]: the number of subpulses $N = 256$, the bandwidth $B = 50$ MHz, the duration of each pulse $T = 1$ µs, the PRI $T_R = 0.1$ ms, and the carrier wavelength $\lambda = 0.03$ m.

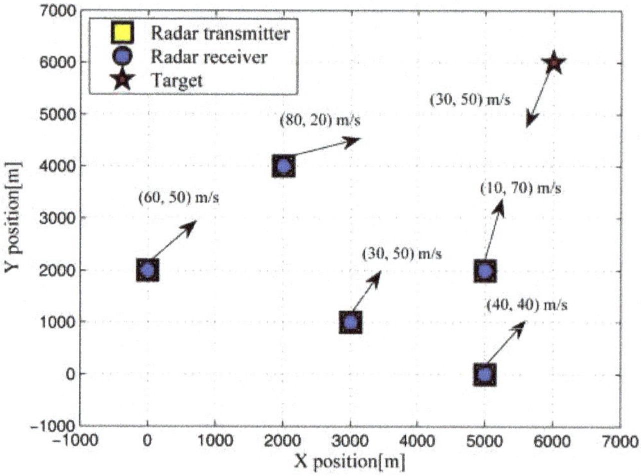

Figure 1. Target and radar networks configuration used in the numerical simulations.

Table 1. Location and Moving Parameters of the Radar Transmitters.

Transmitter Index	Locations [m]	Velocities [m/s]
Transmitter 1	[3000, 1000]	[30, 50]
Transmitter 2	[5000, 2000]	[10, 70]
Transmitter 3	[2000, 4000]	[80, 20]
Transmitter 4	[0, 2000]	[60, 50]
Transmitter 5	[5000, 0]	[40, 40]

Define the SNR as:

$$\mathrm{SNR} = 10\log\left(\frac{\sigma^2}{\sigma_n^2}\right). \tag{28}$$

Without loss of generality, we assume that the reflection coefficients are the same for all transmitter-receiver pairs, i.e., $h_{ij} = h$. In Figures 2 and 3, the MSE curves are plotted versus SNR in the x-dimension and y-dimension of target position with different h. Solid and dashed curves show the CRLBs and the MSE curves of the ML estimation, respectively. As indicated in [12], it can be observed that the MSE is close to the CRLB in value and slope at an SNR threshold, see the green arrows in the figures. Similarly, we depict the MSE curves of target velocity against SNR in Figures 4 and 5.

From Figures 2–5, we can notice that as the value of SNR goes up, the MSE decreases for both target position and velocity estimates.

In addition, it should be pointed out that the CRLB decreases significantly with an increase in h. This is due to the fact that an increase in h provides a rise in target RCS [23], which leads to the increase in the received SNR at the radar receiver. The CRLB will achieve a maximum value when DS component does not exist, i.e., $h = 0$, where the target RCS follows Rayleigh fluctuations in a non-coherent mode for all the transmitter-receiver pairs. In contrast, the CRLB will be minimum at an asymptotic limit, i.e., $h \to \infty$, and the target is idealistically a point target in a coherent mode, which has a fixed amplitude RCS value for all the transmitter-receiver pairs. For the rest of the other cases, the CRLB lies in between these two values [23].

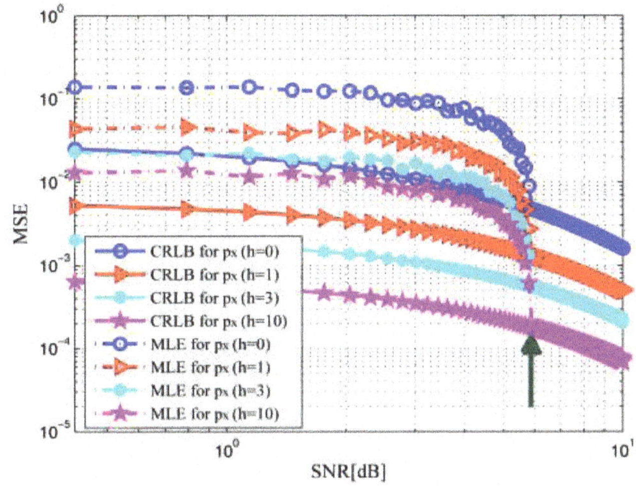

Figure 2. MSE versus SNR for x-dimension of target position with different h.

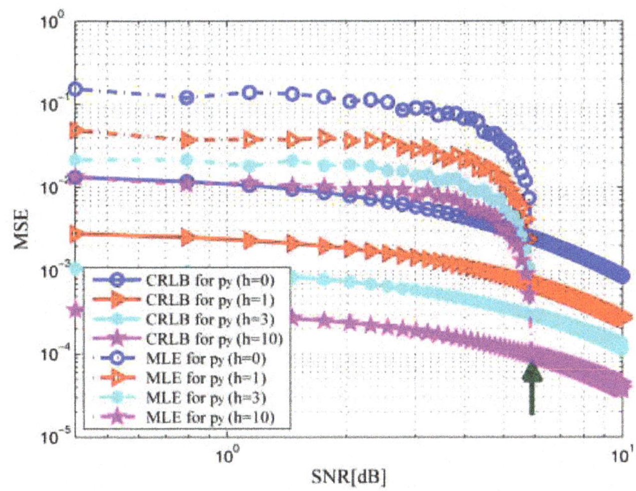

Figure 3. MSE versus SNR for y-dimension of target position with different h.

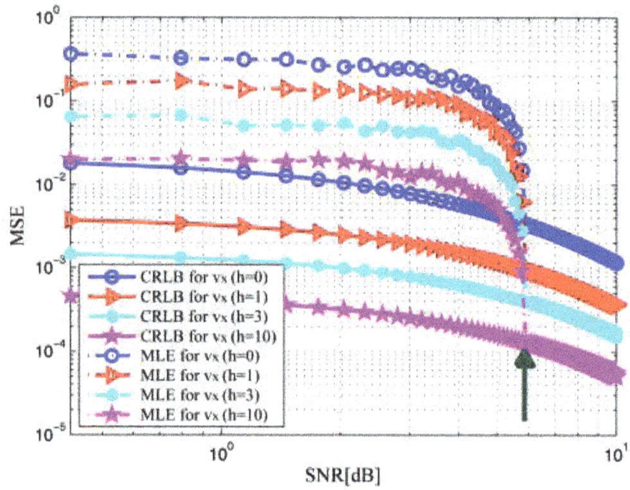

Figure 4. MSE versus SNR for x-dimension of target velocity with different h.

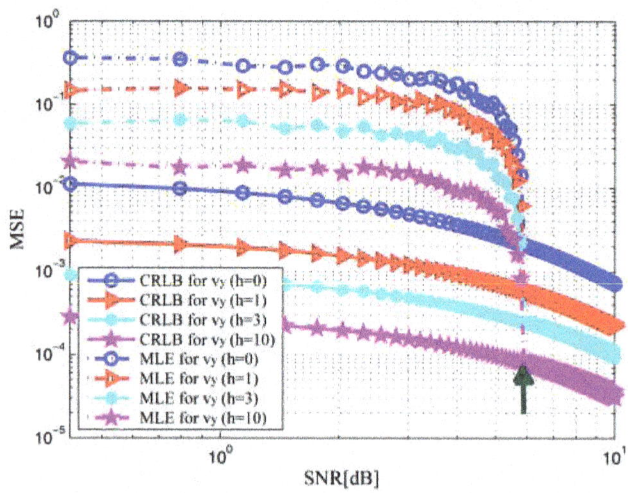

Figure 5. MSE versus SNR for y-dimension of target velocity with different h.

Furthermore, we change the location of the target to different positions to investigate the effects of the geometry between the target and the radar networks. In Figures 6–9, we show the CRLBs for both target position and velocity in different position when SNR = 0 dB, $h = 3$. From these results, we can observe that the CRLBs on the Cartesian coordinates of target position and velocity are different when the target is in different positions. This is because the geometry between the target and the radar network systems impacts the derivatives of the delay-Doppler terms with respect to the Cartesian coordinates significantly [20,23].

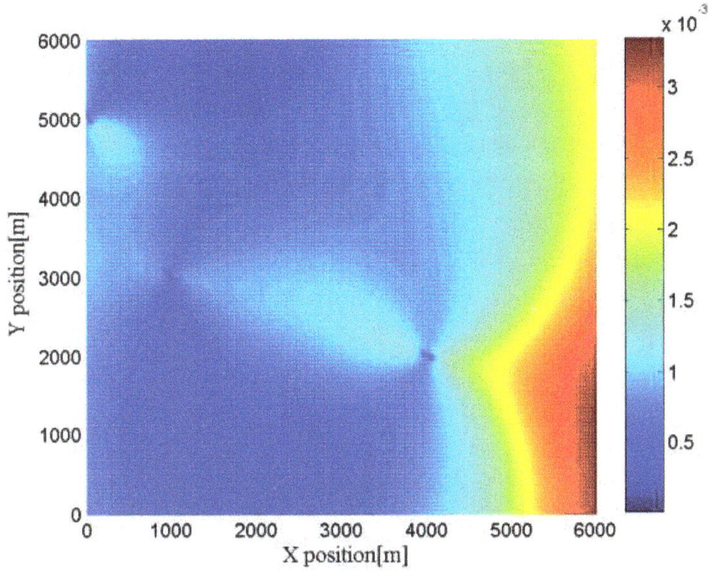

Figure 6. CRLB for x-dimension of target position in different position when SNR = 0 dB, $h = 3$.

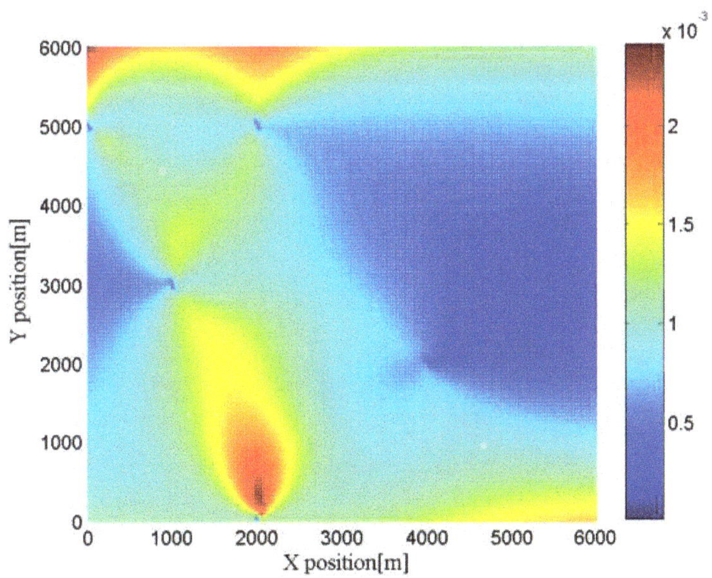

Figure 7. CRLB for y-dimension of target position in different position when SNR = 0 dB, $h = 3$.

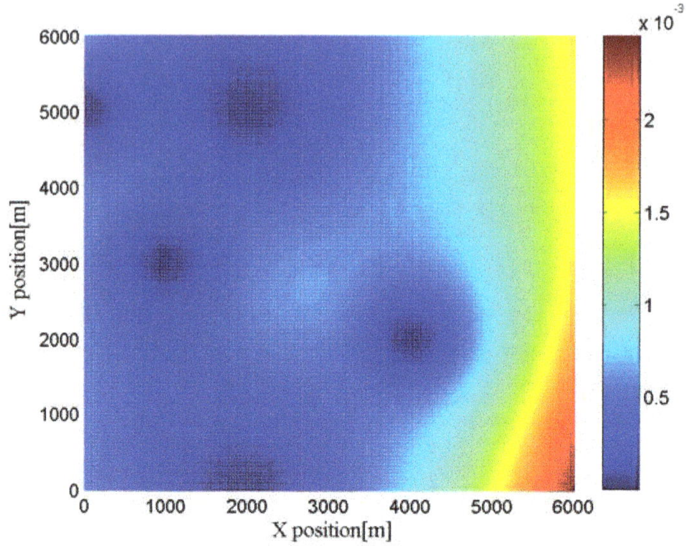

Figure 8. CRLB for x-dimension of target velocity in different position when SNR = 0 dB, $h = 3$.

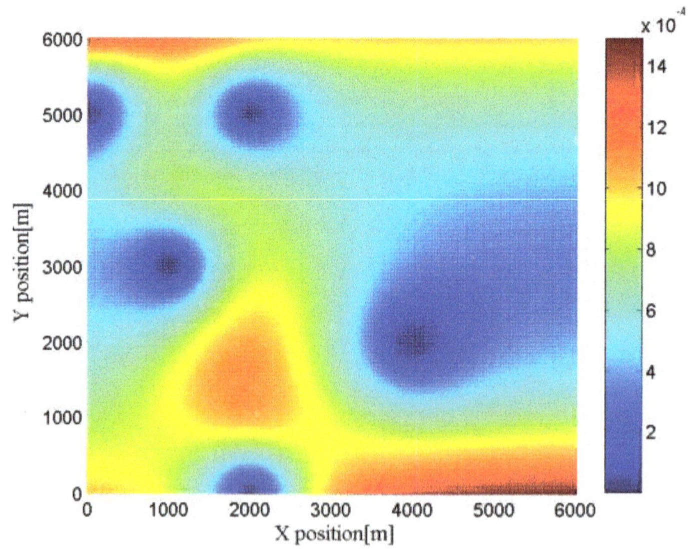

Figure 9. CRLB for y-dimension of target velocity in different position when SNR = 0 dB, $h = 3$.

In Figure 10, we depict the square root of CRLBs (RCRLBs) for target position coordinates against the duration time of each pulse T and bandwidth B at 0 dB with different h. One can notice that the RCRLBs reduce as the waveform parameters increase, confirming that a waveform with a larger time-bandwidth product can provide better target estimation performance. It is worth mentioning that the figures of target velocity are omitted for the sake of brevity, which are similar to the figures of target position. Overall, it can be concluded that the CRLB shows dependence on the SNR, target's RCS, geometry and waveform parameters.

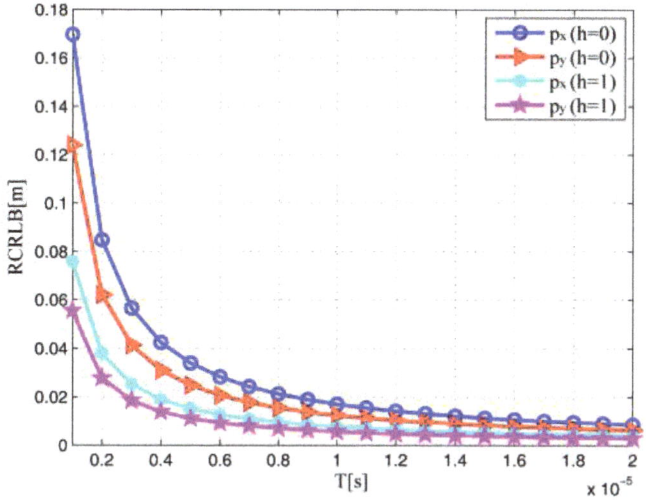

(**a**)

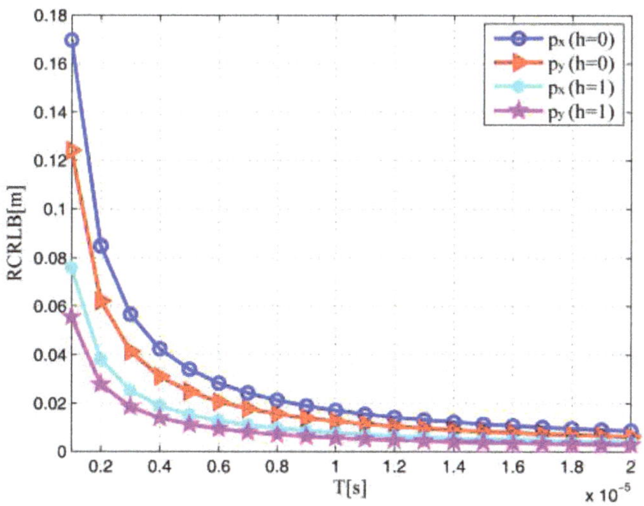

(**b**)

Figure 10. RCRLB in the target position dimensions versus waveform parameters when SNR = 0 dB with different h: (**a**) T; (**b**) B.

5. Conclusions

In this paper, we examined the problem of moving target parameter estimation for active radar network systems with sensors on moving platforms in a Rice fading environment, which consist of multiple radar transmitters and multichannel receivers. The CRLB for joint position and velocity estimation of a Rician target has been derived. It should be noted that the cumulative FIM is a linear combination of both DS component and WIS components. Numerical examples have been provided to demonstrate that the joint target parameter estimation accuracy of active radar networks can be significantly improved with the exploitation of the DS component. Furthermore, it is shown that the joint CRLB is a function of the transmitted waveforms as well as the geometry between the target and the radar networks. Also, it depends on the SNR and target's RCS. In future work, we will utilize this framework to investigate the problem of optimal power allocation of the radar networks in a Rice fading environment.

Acknowledgments: This research is supported by the National Natural Science Foundation of China (Grant No. 61371170, No. 61671239), the Fundamental Research Funds for the Central Universities (Grant No. NS2015038, No. NP2015404), the National Aerospace Science Foundation of China (Grant No. 20152052028), the Priority Academic Program Development of Jiangsu Higher Education Institutions (PADA) and Key Laboratory of Radar Imaging and Microwave Photonics (Nanjing Univ. Aeronaut. Astronaut.), Ministry of Education, Nanjing University of Aeronautics and Astronautics, Nanjing, 210016, China.

Author Contributions: C.G. Shi, S. Salous and F. Wang conceived and designed the experiments; C.G. Shi, S. Salous and F. Wang performed the experiments; C.G. Shi and J.J. Zhou analyzed the data; C.G. Shi wrote the paper; S. Salous and J.J. Zhou contributed to data analysis revision and English language correction. All authors of article provided substantive comments.

Conflicts of Interest: The authors declare no conflict of interest.

Appendix A. The Derivation of FIM $\mathbf{J}(\mathbf{Q})$

We have:

$$\Gamma_{ij}^1 = \frac{\sigma^2}{\sigma^2+\sigma_n^2}\left|\int_{-\infty}^{+\infty} r_{ij}(t)s_i^*(t-\tau_{ij})e^{-j2\pi f_{D_{ij}}(t-\tau_{ij})}dt\right|^2. \tag{A1}$$

Let us suppose $\xi_{ij} = \int_{-\infty}^{+\infty} r_{ij}(t)s_i^*(t-\tau_{ij})e^{-j2\pi f_{D_{ij}}(t-\tau_{ij})}dt$, then the first derivative with respect to τ_{ij} is given by:

$$\begin{aligned}\frac{\partial\Gamma_{ij}^1}{\partial\tau_{ij}} &= \frac{\sigma^2}{\sigma^2+\sigma_n^2}\left(\xi_{ij}^*\frac{\partial\xi_{ij}}{\partial\tau_{ij}} + \xi_{ij}\frac{\partial\xi_{ij}^*}{\partial\tau_{ij}}\right)\\ &= \frac{2\sigma^2}{\sigma_n^2(\sigma^2+\sigma_n^2)} \times \Re\left(\xi_{ij}\int_{-\infty}^{+\infty} r_{ij}^*(t)\frac{\partial s_i(t-\tau_{ij})}{\partial\tau_{ij}}e^{j2\pi f_{D_{ij}}(t-\tau_{ij})}\right).\end{aligned} \tag{A2}$$

To compute the expectation with respect to the second-order derivative, we have:

$$\begin{aligned}-\mathbb{E}\left(\frac{\partial^2\Gamma_{ij}^1}{\partial\tau_{ij}^2}\right) &= -\frac{2\sigma^2(\sigma^2+d_{ij}^2)}{\sigma^2+\sigma_n^2} \times \mathbb{E}\left[\Re\left(\left|\int_{-\infty}^{+\infty} s_i(t-\tau_{ij})\frac{\partial s_i^*(t-\tau_{ij})}{\partial\tau_{ij}}dt\right|^2\right.\right.\\ &\left.\left.+\int_{-\infty}^{+\infty}|s_i(t-\tau_{ij})|^2dt\int_{-\infty}^{+\infty} s_i^*(t-\tau_{ij})\frac{\partial^2 s_i(t-\tau_{ij})}{\partial\tau_{ij}^2}dt\right)\right].\end{aligned} \tag{A3}$$

With the derivations in [29,30], hence we obtain:

$$-\mathbb{E}\left(\frac{\partial^2\Gamma_{ij}^1}{\partial\tau_{ij}^2}\right) = \frac{8\pi^2\sigma^4}{\sigma_n^2(\sigma^2+\sigma_n^2)}(1+2h_{ij})\varepsilon_i, \tag{A4}$$

where $h_{ij} = |d_{ij}|^2/(2\sigma^2)$.

To calculate the expectation with respect to other second-order derivatives, we follow the same procedure and arrive at the following closed form expression:

$$-\mathbb{E}\left(\frac{\partial^2\Gamma_{ij}^1}{\partial f_{D_{ij}}^2}\right) = \frac{8\pi^2\sigma^4}{\sigma_n^2(\sigma^2+\sigma_n^2)}(1+2h_{ij})\eta_{ij}. \tag{A5}$$

The off-diagonal terms can be obtained as:

$$-\mathbb{E}\left(\frac{\partial^2\Gamma_{ij}^1}{\partial\tau_{ij}\partial f_{D_{ij}}}\right)=\frac{8\pi^2\sigma^4}{\sigma_n^2(\sigma^2+\sigma_n^2)}(1+2h_{ij})\gamma_{ij}. \tag{A6}$$

Following the same lines, we can get the expectation of second-order derivatives of Γ_{ij}^2 and Γ_{ij}^3 as follows:

$$-\mathbb{E}\left(\frac{\partial^2\Gamma_{ij}^2}{\partial\tau_{ij}^2}\right)=\frac{8\pi^2\sigma^4}{\sigma_n^2(\sigma^2+\sigma_n^2)}\left(-\frac{2h_{ij}}{\sigma^2/\sigma_n^2}\right)\varepsilon_i, \tag{A7}$$

$$-\mathbb{E}\left(\frac{\partial^2\Gamma_{ij}^2}{\partial f_{D_{ij}}^2}\right)=\frac{8\pi^2\sigma^4}{\sigma_n^2(\sigma^2+\sigma_n^2)}\left(-\frac{2h_{ij}}{\sigma^2/\sigma_n^2}\right)\eta_{ij}, \tag{A8}$$

$$-\mathbb{E}\left(\frac{\partial^2\Gamma_{ij}^2}{\partial\tau_{ij}\partial f_{D_{ij}}}\right)=\frac{8\pi^2\sigma^4}{\sigma_n^2(\sigma^2+\sigma_n^2)}\left(-\frac{2h_{ij}}{\sigma^2/\sigma_n^2}\right)\gamma_{ij}, \tag{A9}$$

$$-\mathbb{E}\left(\frac{\partial^2\Gamma_{ij}^3}{\partial\tau_{ij}^2}\right)=\frac{8\pi^2\sigma^4}{\sigma_n^2(\sigma^2+\sigma_n^2)}\left(\frac{4h_{ij}}{\sigma^2/\sigma_n^2}\right)\varepsilon_i, \tag{A10}$$

$$-\mathbb{E}\left(\frac{\partial^2\Gamma_{ij}^3}{\partial f_{D_{ij}}^2}\right)=\frac{8\pi^2\sigma^4}{\sigma_n^2(\sigma^2+\sigma_n^2)}\left(\frac{4h_{ij}}{\sigma^2/\sigma_n^2}\right)\eta_{ij}, \tag{A11}$$

$$-\mathbb{E}\left(\frac{\partial^2\Gamma_{ij}^3}{\partial\tau_{ij}\partial f_{D_{ij}}}\right)=\frac{8\pi^2\sigma^4}{\sigma_n^2(\sigma^2+\sigma_n^2)}\left(\frac{4h_{ij}}{\sigma^2/\sigma_n^2}\right)\gamma_{ij}. \tag{A12}$$

Appendix B. The Elements of FIM $\mathbf{J}_{ij}(\mathbf{U})$

The elements of the symmetric FIM $\mathbf{J}_{ij}(\mathbf{U})$ corresponding to the ijth transmitter-receiver pair are given by:

$$J_{ij}^{11}(\mathbf{U})=\frac{k^2\pi^2T^2}{3}\left(\frac{\partial\tau_{ij}}{\partial x}\right)^2-\frac{k\pi T^2}{3}\left(\frac{\partial\tau_{ij}}{\partial x}\right)\left(\frac{\partial f_{D_{ij}}}{\partial x}\right)+\frac{1}{12}\left[T^2+T_R(N^2-1)\right]\left(\frac{\partial f_{D_{ij}}}{\partial x}\right)^2, \tag{B1}$$

$$\begin{aligned}J_{ij}^{12}(\mathbf{U})&=J_{ij}^{21}(\mathbf{U})=\left\{\tfrac{k^2\pi^2T^2}{3}\left(\tfrac{\partial\tau_{ij}}{\partial x}\right)-\tfrac{k\pi T^2}{6}\left(\tfrac{\partial f_{D_{ij}}}{\partial x}\right)\right\}\left(\tfrac{\partial\tau_{ij}}{\partial y}\right)\\&+\left\{-\tfrac{k\pi T^2}{6}\left(\tfrac{\partial\tau_{ij}}{\partial x}\right)+\tfrac{1}{12}\left[T^2+T_R(N^2-1)\right]\left(\tfrac{\partial f_{D_{ij}}}{\partial x}\right)\right\}\left(\tfrac{\partial f_{D_{ij}}}{\partial y}\right),\end{aligned} \tag{B2}$$

$$J_{ij}^{13}(\mathbf{U})=J_{ij}^{31}(\mathbf{U})=\left\{-\frac{k\pi T^2}{6}\left(\frac{\partial\tau_{ij}}{\partial x}\right)+\frac{1}{12}\left[T^2+T_R(N^2-1)\right]\left(\frac{\partial f_{D_{ij}}}{\partial x}\right)\right\}\left(\frac{\partial f_{D_{ij}}}{\partial v_x}\right), \tag{B3}$$

$$J_{ij}^{14}(\mathbf{U})=J_{ij}^{41}(\mathbf{U})=\left\{-\frac{k\pi T^2}{6}\left(\frac{\partial\tau_{ij}}{\partial x}\right)+\frac{1}{12}\left[T^2+T_R(N^2-1)\right]\left(\frac{\partial f_{D_{ij}}}{\partial x}\right)\right\}\left(\frac{\partial f_{D_{ij}}}{\partial v_y}\right), \tag{B4}$$

$$J_{ij}^{22}(\mathbf{U})=\frac{k^2\pi^2T^2}{3}\left(\frac{\partial\tau_{ij}}{\partial y}\right)^2-\frac{k\pi T^2}{3}\left(\frac{\partial\tau_{ij}}{\partial y}\right)\left(\frac{\partial f_{D_{ij}}}{\partial y}\right)+\frac{1}{12}\left[T^2+T_R(N^2-1)\right]\left(\frac{\partial f_{D_{ij}}}{\partial y}\right)^2, \tag{B5}$$

$$J_{ij}^{23}(\mathbf{U}) = J_{ij}^{32}(\mathbf{U}) = \left\{ -\frac{k\pi T^2}{6}\left(\frac{\partial \tau_{ij}}{\partial y}\right) + \frac{1}{12}\left[T^2 + T_R(N^2-1)\right]\left(\frac{\partial f_{D_{ij}}}{\partial y}\right)\right\}\left(\frac{\partial f_{D_{ij}}}{\partial v_x}\right), \tag{B6}$$

$$J_{ij}^{24}(\mathbf{U}) = J_{ij}^{42}(\mathbf{U}) = \left\{ -\frac{k\pi T^2}{6}\left(\frac{\partial \tau_{ij}}{\partial y}\right) + \frac{1}{12}\left[T^2 + T_R(N^2-1)\right]\left(\frac{\partial f_{D_{ij}}}{\partial y}\right)\right\}\left(\frac{\partial f_{D_{ij}}}{\partial v_y}\right), \tag{B7}$$

$$J_{ij}^{33}(\mathbf{U}) = \frac{1}{12}\left[T^2 + T_R(N^2-1)\right]\left(\frac{\partial f_{D_{ij}}}{\partial v_x}\right)^2, \tag{B8}$$

$$J_{ij}^{34}(\mathbf{U}) = J_{ij}^{43}(\mathbf{U}) = \frac{1}{12}\left[T^2 + T_R(N^2-1)\right]\left(\frac{\partial f_{D_{ij}}}{\partial v_x}\right)\left(\frac{\partial f_{D_{ij}}}{\partial v_y}\right), \tag{B9}$$

$$J_{ij}^{44}(\mathbf{U}) = \frac{1}{12}\left[T^2 + T_R(N^2-1)\right]\left(\frac{\partial f_{D_{ij}}}{\partial v_y}\right)^2. \tag{B10}$$

References

1. Li, J.; Stoica, P. *MIMO Radar Signal Processing*; Wiley: Hoboken, NJ, USA, 2009; pp. 348–381.
2. Pace, P.E. *Detecting and Classifying Low Probability of Intercept Radar*; Artech House: Boston, MA, USA, 2009; pp. 319–378.
3. Haimovich, A.M.; Blum, R.S.; Cimini, L.J., Jr. MIMO radar with widely separated antennas. *IEEE Signal Process. Mag.* **2008**, *25*, 116–129.
4. Fisher, E.; Haimovich, A.; Blum, R.S.; Cimini, L.J.; Chizhik, D.; Valenzuela, R. Spatial diversity in radars-models and detection performance. *IEEE Trans. Signal Process.* **2006**, *54*, 823–836.
5. Naghsh, M.M.; Mahmoud, M.H.; Shahram, S.P.; Soltanalian, M.; Stoica, P. Unified optimization framework for multi-static radar code design using information-theoretic criteria. *IEEE Trans. Signal Process.* **2013**, *61*, 5401–5416.
6. Niu, R.X.; Blum, R.S.; Varshney, P.K.; Drozd, A.L. Target localization and tracking in noncoherent multiple-input multiple-output radar systems. *IEEE Trans. Aerosp. Electron. Syst.* **2010**, *48*, 1466–1487.
7. Godrich, H.; Tajer, A.; Poor, H.V. Distributed Target Tracking in Multiple Widely Separated Radar Architectures. In Proceedings of the 2012 7th IEEE Sensor Array and Multichannel Signal Processing Workshop (SAM), Hoboken, NJ, USA, 17–20 June 2012; pp. 153–156.
8. Chen, Y.F.; Nijsure, Y.; Yuen, C.; Chew, Y.H.; Ding, Z.G. Adaptive distributed MIMO radar waveform optimization based on mutual information. *IEEE Trans. Aerosp. Electron. Syst.* **2013**, *49*, 1374–1385.
9. Godrich, H.; Petropulu, A.P.; Poor, H.V. Sensor selection in distributed multiple-radar architectures for localization: A knapsack problem formulation. *IEEE Trans. Signal Process.* **2012**, *60*, 247–259.
10. Song, X.F.; Willett, P.; Zhou, S.L. Optimal Power Allocation for MIMO Radars with Heterogeneous Propagation Losses. In Proceedings of the IEEE International Conference on Acoustics, Speech and Signal Processing (ICASSP), Kyoto, Japan, 25–30 March 2012; pp. 2465–2468.
11. Godrich, H.; Haimovich, A.M.; Blum, R.S. Target localization accuracy gain in MIMO radar-based systems. *IEEE Trans. Inf. Theory* **2010**, *56*, 2783–2803.
12. He, Q.; Blum, R.S.; Haimovich, A.M. Noncoherent MIMO radar for location and velocity estimation: More antennas means better performance. *IEEE Trans. Signal Process.* **2010**, *58*, 3661–3680.
13. Ciuonzo, D.; Romano, G.; Solimene, R. On MSE Performance of Time-Reversal MUSIC. In Proceedings of the 2014 IEEE 8th Sensor Array and Multichannel Signal Processing Workshop (SAM), A Coruna, Spain, 22–25 June 2014; pp. 13–16.
14. Ciuonzo, D.; Romano, G.; Solimene, R. Performnce analysis of time-reversal MUSIC. *IEEE Trans. Signal Process.* **2015**, *63*, 2650–2662.

15. Wei, C.; He, Q.; Blum, R.S. Cramer-Rao Bound for Joint Location and Velocity Estimation in Multi-Target Non-Coherent MIMO Radars. In Proceedings of the 2010 44th IEEE Annual Conference on Information Sciences and Systems (CISS), Princeton, NJ, USA, 17–19 March 2010; pp. 1–6.
16. He, Q.; Blum, R.S. Noncoherent versus coherent MIMO radar: Performance and simplicity analysis. *Signal Process.* **2012**, *92*, 2454–2463.
17. He, Q.; Blum, R.S. Cramer-Rao bound for MIMO radar target localization with phase errors. *IEEE Signal Process. Lett.* **2010**, *17*, 83–86.
18. Zhao, T.; Huang, T.Y. Cramer-Rao lower bounds for the joint delay-Doppler estimation of an extended extended target. *IEEE Trans. Signal Process.* **2016**, *64*, 1562–1573.
19. Hu, J.B.; He, Q.; Blum, R.S. Comparing the Cramer-Rao Bounds for Distributed Radar with and without Previous Detection Information. In Proceedings of the IEEE China Summit and International Conference on Signal and Information Processing (ChinaSIP), Chengdu, China, 12–15 July 2015; pp. 334–338.
20. Shi, C.G.; Wang, F.; Zhou, J.J. Cramer-Rao bound analysis for joint target position and velocity estimation in FM-based passive radar networks. *IET Signal Process.* **2016**, *10*, 780–790.
21. He, Q.; Blum, R.S. The significant gains from optimally processed multiple signals of opportunity and multiple receive stations in passive radar. *IEEE Signal Process. Lett.* **2014**, *21*, 180–184.
22. Filip, A.; Shutin, D. Cramer-Rao bounds for L-band digital aeronautical communication system type 1 based passive multiple-input multiple-output radar. *IET Radar Sonar Navig.* **2016**, *10*, 348–358.
23. Javed, M.N.; Ali, S.; Hassan, S.A. 3D MCRLB evaluation of a UMTS-based passive multistatic radar operating in a line-of-sight environment. *IEEE Trans. Signal Process.* **2016**, *64*, 5131–5144.
24. Shi, C.G.; Salous, S.; Wang, F.; Zhou, J.J. Modified Cramer-Rao lower bounds for joint position and velocity estimation of a Rician target in OFDM-based passive radar networks. *Radio Sci.* **2016**, submitted.
25. Shi, C.G.; Wang, F.; Zhou, J.J.; Zhang, H. Security Information Factor Based Low Probability of Identification in Distributed Multiple-Radar System. In Proceedings of the IEEE International Conference on Acoustics, Speech and Signal Processing (ICASSP), Brisbane, Australia, 19–24 April 2015; pp. 3716–3720.
26. Shi, C.G.; Zhou, J.J.; Wang, F. LPI Based Resource Management for Target Tracking in Distributed Radar Network. In Proceedings of the IEEE Radar Conference (RadarConf), Philadelphia, PA, USA, 2–6 May 2016; pp. 1–5.
27. Shi, C.G.; Wang, F.; Sellathurai, M.; Zhou, J.J.; Zhang, H. Robust transmission waveform design for distributed multiple-radar systems based on low probability of intercept. *ETRI J.* **2016**, *38*, 70–80.
28. Shi, C.G.; Wang, F.; Sellathurai, M.; Zhou, J.J. Transmitter subset selection in FM-based passive radar networks for joint target parameter estimation. *IEEE Sens. J.* **2016**, *16*, 6043–6052.
29. Greco, M.S.; Stinco, P.; Gini, F.; Farina, A. Cramer-Rao bounds and selection of bistatic channels for multistatic radar systems . *IEEE Trans. Aerosp. Electron. Syst.* **2011**, *47*, 2934–2948.
30. Dogandzic, A.; Nehorai, A. Cramer-Rao bounds for estimating range, velocity, and direction with an active array. *IEEE Trans. Signal Process.* **2001**, *49*, 1122–1137.

sensors

MDPI

Technical Note

Low Cost and Flexible UAV Deployment of Sensors

Lars Yndal Sørensen [1], Lars Toft Jacobsen [2] and John Paulin Hansen [1,*]

1 Management Engineering, Technical University of Denmark, Diplomvej, Building 372, 2800 Kgs. Lyngby, Denmark; larynd@dtu.dk
2 IT University of Copenhagen, Rued Langgaards Vej 7 2300 Copenhagen S, Denmark; latj@itu.dk
* Correspondence: jpha@dtu.dk; Tel.: +45-4046-9626

Academic Editors: Felipe Gonzalez Toro and Antonios Tsourdos
Received: 18 August 2016; Accepted: 10 January 2017; Published: 14 January 2017

Abstract: This paper presents a platform for airborne sensor applications using low-cost, open-source components carried by an easy-to-fly unmanned aircraft vehicle (UAV). The system, available in open-source , is designed for researchers, students and makers for a broad range of exploration and data-collection needs. The main contribution is the extensible architecture for modularized airborne sensor deployment and real-time data visualisation. Our open-source Android application provides data collection, flight path definition and map tools. Total cost of the system is below 800 dollars. The flexibility of the system is illustrated by mapping the location of Bluetooth beacons (iBeacons) on a ground field and by measuring water temperature in a lake.

Keywords: UAV; drone; monitoring; multisensor; platform; software framework; beacons

1. Introduction

The use of civilian unmanned aircraft vehicles (UAVs, or simply, drones) by professionals, researchers and hobbyists alike, has increased over the recent years. Commonly, the drones are used for aerial photography, inspection and surveys. Possible applications extend to any conceivable task where an aerial presence with minimal environmental impact brings value or new insights [1]. What sets a modern UAV platform apart from earlier flying robots is that enhancements in technology allow for the UAV to carry a networked computer payload. Consequently, the UAVs now have the ability to collect, process, store and relay data on their own and in real time.

UAVs afford three-dimensional, spatial coverage, which makes them particularly suitable for airborne sensor deployment in, for instance, ecology research and disaster management [2–4]. The most common deployed sensors are cameras sensitive to different light spectra. They are most useful for manual as well as automated inspection, and are often assisted by computer vision to identify relevant conditions or objects on the ground. Single drones can be programmed to follow predefined paths, covering a particular area of interest, or are flown manually either in line-of-sight, if possible, or guided by GPS or other sensory feedback. Multiple drones may be deployed in swarms, e.g., to map out and track pollution [5] or to make up a grid network and relay the data over large distances.

This paper proposes an airborne sensor platform based on commercial off-the-shelf components that provides a modularized sensor system and data acquisition infrastructure. The drone is a commercial quad-copter that allows for attaching external sensors and relaying the data back to a ground station using a telemetry communications link. The platform supports a simple and expandable interface for attaching custom sensors to the UAV, overcoming the limitation of single-purpose platforms which are costly to convert for other tasks. Since the sensor system is modularized, sensors can be exchanged rapidly. Besides the hardware platform, which easily integrates sensors, we provide the possibility to use an established open-source infrastructure to collect and visualize sensor data from the drone in real time. This allows for both autonomous path generation and following (see Figure 1) and operator-assisted (see Figure 2) flights towards areas of interest based on sensor feedback — and beyond line-of-sight if needed (The drone needs not be visible to the operator, but some line-of-sight is, however, required in the sense that telemetry signals may not be blocked, as this would result in lost data points).

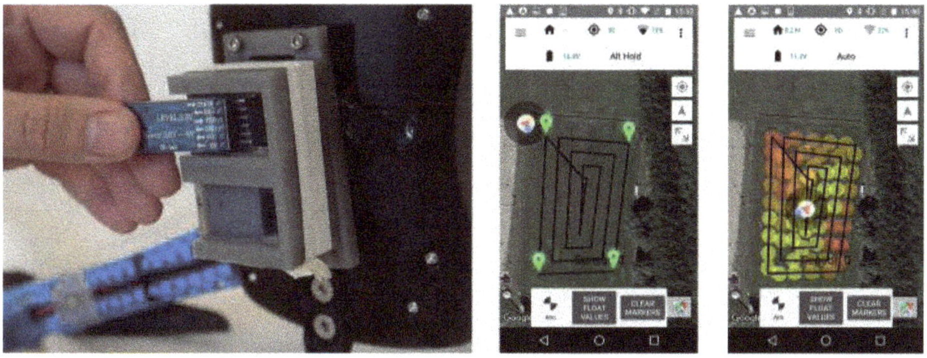

(a) (b) (c)

Figure 1. For autonomous flight: (**a**) Install sensor (Bluetooth in this scenario) and write a small sketch in Arduino, (**b**) mark the area to cover and (**c**) take-off and interpret the data (see "Supplementary files" for the data collected from this flight).

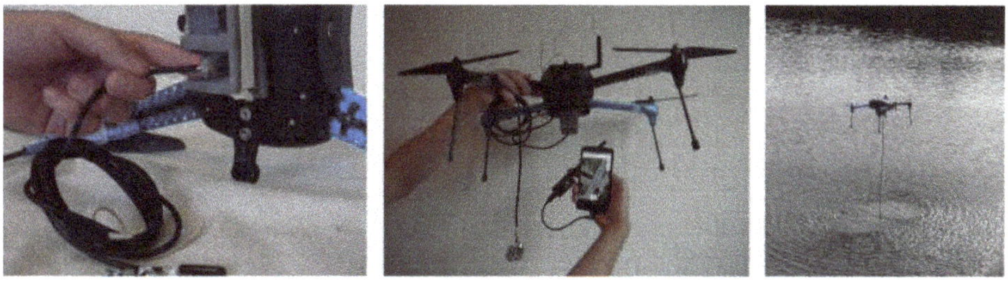

(a) (b) (c)

Figure 2. Operator-assisted flight: (**a**) A sensor (i.e., water thermometer) is mounted in the multi-socket and a small sketch in Arduino is written, (**b**) entire system with drone, sensor, smart-phone and antenna (**c**) flying manually over the lake to measure water temperature.

Our choice of a multi-rotor aircraft delivers desirable operational parameters in some respect (i.e., agility, vertical take-off and landing, hovering and low-altitude performance), but sacrifices on other parameters like range, speed, and altitude. However, the proposed design targets makers, students and researchers that intend to conduct experiments with custom sensors and only need data collected within the boundaries of unlicensed operation. To that end, our system may be considered a prototyping platform for airborne sensor deployments, although it provides a way to rapidly implement a concrete application in its own right. In particular, we hope it may become a flexible first-step learning tool for students within engineering and environmental programmes.

While designing the system, four primary design objectives were kept in mind. The system should be: (1) low-cost; (2) easy to use; (3) modularized, for fast development and deployment of sensors; and (4) provide real-time data.

The following sections will first outline some usage scenarios. A more detailed system description follows with its technical implementation presented and illustrated with a field demonstration. Finally, we discuss shortcomings and future improvements.

2. Previous Work

Low-cost, light-weight UAVs have been used extensively within research for more than a decade to measure, for instance, air quality [6,7], mapping of 3D geodata [8–10] and remote sensing within agriculture [1,11–13]. These UAVs were build with a single purpose and with one particular sensor on board. While relatively cheap in comparison with other means for airborne measurements (e.g., airplanes, helicopters or high-grade UAVs), they cost several thousand dollars.

The use of open source UAV software is promoted by several research teams, (e.g., [10,14]). Aside from research, amateurs with an interest in UAVs have formed communities at places like diydrones.com to show their prototypes and discuss further developments. Some of the drones are made of parts from electronics outlets, while others are built using open-source Arduino parts intended for aviation (e.g., "Arducopters"). Anderson [15], who initiated the diydrones community, sees this as an example of amateur makers potentially revolutionizing the industry by sharing, for instance, code for a 3D-printout of a construction part. When cheap, consumer-grade drones became available, their flight computers were hacked almost immediately (e.g., [16]). Even with this "maker" movement, modifying low-cost UAVs for remote sensing is still difficult, which may prevent people with limited technical skills in computers and embedded control systems to take advantage of them.

Sensor measurements from UAVs may benefit teaching, for example, in engineering and environmental disciplines. Jung et al. [17] developed a low-cost UAV test-bed based on a model airplane, with in-house -design and -assembling of most system parts. They argue that a way to provide interdisciplinary skills for UAV development is to promote educational projects on UAV technologies. Recently, Eriksen et al. [18] and Mathias [19] augmented low-cost drones with the same educational purpose.

3. Usage Scenarios

The following fictitious scenarios guided our implementation. All address the collection of real-time data from the sensor attached to the drone.

- **Mapping WiFi coverage in outdoor areas**

  The task is to map out wireless network (WiFi) coverage in 2D or 3D for a large outdoor area, for instance a festival venue. A technician sets up access points and mounts a sensor on the drone that collects signal strength (i.e., received signal strength indicator (RSSI) values). A telemetry radio dongle is connected to the technicians' smart phone on which a flight path covering the entire area is marked out using an Android ground station application. The drone will then fly between the set way-points at regular intervals. For each flight, it sends timestamped and absolute positioned sensor values back via the telemetry link. The Android app generates a heat map in real time, allowing the technician to quickly assess the coverage without waiting for the entire flight path to complete. Offline, the technician further analyses the data collected.

- **Detecting radio beacons**

  A simple Bluetooth 4.0 (BLE) module is used to detect and track radio tags, i.e., iBeacons [20]. A beacon is attached to a key-chain that has been lost on a field. The drone flies over the field and maps signal strengths, guiding the search for the key-chain.
  A signal map can be updated by flying over the area multiple times, for example, when tracking objects in motion. A zoologist tags a beacon to a badger cub. In the following weeks, the drone tracks the cub when outside its cave, mapping how its territory expands day by day.

- **Tracking pollution sources**

  An agriculturalist detects an airborne aggressive pollutant on field crops by manual inspection. The source of it needs to be tracked down quickly, and the agriculturalist, therefore, mounts a sensor for the drone platform that measures the concentration of the pollutant in the air. The drone is instructed to follow an inwards spiraling pattern from the point of detection. The agriculturalist receives values sent from the sensor as the drone continues its flight path. A heatmap pattern starts to emerge on his mobile phone, indicating a stronger concentration of pollutants in a certain direction. The agriculturalist can choose to manually alter the flight path or specify a concentration threshold for the drone sensor on the ground station software.

4. System Description

Our system is meant for attaching arbitrary low-cost sensors to an open-source robotics vehicle platform. Figure 3 provides a conceptual overview of the entire platform. The sensors are, in part, a piece of hardware that attaches to the drone. A simple protocol based on the common I2C peripheral protocol is used for interfacing with the flight computer. As long as the sensor modules conform to our hardware and software specifications, any low-weight sensor can be used.

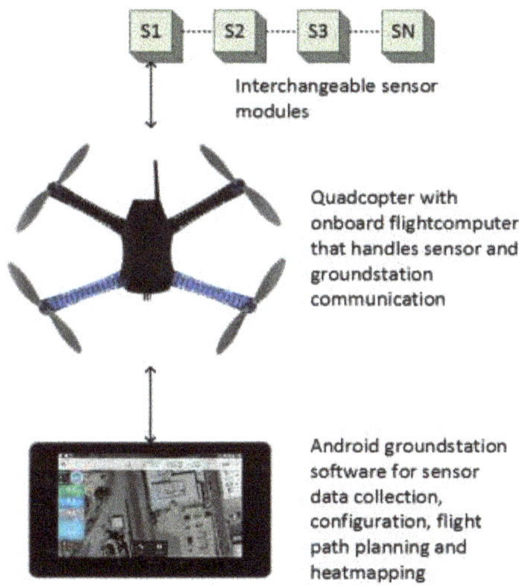

Figure 3. System overview. Interchangeable sensor modules attach to the quadcopter platform. Sensor data is relayed through the flight computer via a wireless telemetry link to the Android ground station application. The ground station software stores sensor data and provides intelligent survey flight path planning, heat-maps and platform configuration.

The UAV, a 3D Robotics IRIS+ [21], which can be classified as an MAV (Micro Arial Vehicle) and a VTOL (Vertical Take-Off and Landing) according to [1], is a common commercially available drone. It was selected from a set of criteria that met our needs:

- *Cost and availability*. The drone is relatively low-cost, produced in high quantities and obtainable from resellers in both North America and Europe. Total cost for the system described herein runs just short of $800.
- *Spare-parts and repairability*. Except for the body, no legacy components are used; all spare-parts are low-cost and available from alternative brands.
- *Open-source hardware and software*. The drone is equipped with a Pixhawk flight controller [22,23]. This is a spin-off from an ETH Zürich research project [24], where the software and hardware are open-source, and the system is well-documented and supported through a community effort.
- *Telemetry options*. The drone ships with long-range radio telemetry modules. One is attached to the Pixhawk, and the other can be tethered to a computer or smart phone using a USB cable.

The radio module firmware is optimized for communication using the open standard MAVLink protocol [25].

- *Attaching payloads*. The drone has mounting holes underneath the body for attaching a camera gimbal. This makes it easy to make a custom bracket for attaching other hardware without altering the body.

We prototyped a casing that attaches to a bracket that fits in the gimbal mounting points (see the first image at Figure 1). The housing contains an Arduino-based microcontroller that is powered by and interfaces with the flight computer. Smaller housings containing the actual sensors can then be attached to the microcontroller unit.

The readings picked up by the sensor module are relayed to a ground station using the featured telemetry modules; on the drone, the radio link module is connected directly to the flight computer (the Pixhawk/PX4). For this to work seamlessly with the drone setup, modifications have been made to the open-source flight computer firmware and ground station. Firstly, drivers and user space code on the Pixhawk computer interface with the sensor module and communicate with the ground station using custom messages defined as part of an existing open protocol format. Extensive modifications have gone into the code base of an open-source ground station application for Android. This allows the user to configure flight paths for surveying and retrieving the sensor data stream that, in turn, are seen in real time and are visualized on the map for every 10th reading—used later for offline analysis (see "Supplementary file" for example of collected data).

In addition, built-in safety features in the flight computer manage potentially dangerous situations; the UAV will automatically land when the battery is running low or when crossing a geofence boundary. In its current state, the drone is operational for ∼25 min on a single battery, while the Android phone (the used Android phone is a Google Nexus 5 running Android 6.0) is able to operate for ∼1 h. The Pixhawk flight computer is not solely intended for air vehicles. It may just as well be applied for autonomous path following in rovers or other ground vehicles, and our sensor platform may be ported to these vehicles without changing any parts in the system.

5. Technical Implementation

Our design goals call for a "full stack" implementation, touching components from sensors to the ground station. This includes an IRIS+ drone, a Pixhawk flight computer, an ArduPilot flight stack and a microcontroller driving the sensors of choice.

Substantial previous work [22,23,26,27] has gone into creating the flight computer, its firmware and the ground station software, thus enabling us to focus on integrating the features supporting the usage scenarios. Our original contributions include building and programming sensor modules, developing drivers and communication extensions for the flight computer, and building system specific features into the Android ground station application. Figure 4 shows the overall system components and the software packages that our project has contributed.

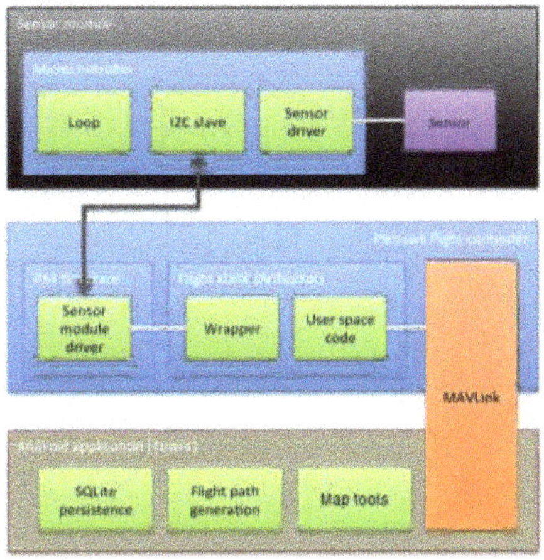

Figure 4. General package diagram of software involved in the system. The green packages are contributions of our package, including map tools, i.e., heat-map visualization.

5.1. *Sensor Modules*

To facilitate easy module installation, a bracket has been designed [28] to fit two mounting points underneath the drone, as originally intended for a camera gimbal. A sensor module can then be fitted to the bracket in various ways. It will not interfere with the structural integrity of the drone as long as the module is kept within certain physical limits: The IRIS+ specifications list a maximum payload of 400 g, and, to minimize handling interference, we suggest a sensor module design *no larger* than approximately the volume of the body of the drone (L 200 $\times$ W 120 $\times$ H 70 mm). The design of the module can be seen in image 1 of Figure 1: three parts are 3D printed to act as (i) base; (ii) container for micro controller and (iii) sensor mount. The sensor mount is for convenience to easily replace a sensor, while the base and container for the micro controller are the main parts of the system. By 3D-printing the mounts ourselves, we can comply with a broad range of sensors within a very short production time (see Figure 5 for examples), at a low cost, and just adding a few extra grams to the payload of the drone.

The Pixhawk flight controller exposes several pins useful for connecting peripherals. Besides digital GPIO (general purpose input/output) pins, there are dedicated connectors for UARTs (universal asynchronous receiver/transmitter) and SPI (serial peripheral interface)/I2C (inter-integrated circuit) buses. Since most UARTs are used for other peripherals and SPI requires a dedicated slave-select signal for each connected device, the I2C bus [29] was chosen for its availability and the possibility to daisy-chain multiple devices using just two signal wires.

The choice of I2C also means that a cable with just four wires (+5 V, Ground, Data and Clock) needs to be brought out from the flight computer to the underside of the chassis where the sensor module attaches to the mounting bracket. Power is provided by the Pixhawk as a regulated 5 V source, but the I2C clock and data lines are designed for 3.3 V logic only. The sensor module therefore must be able to accept 5 V power while honoring the 3.3 V levels on the bus lines, a common feature for several modern microcontrollers.

Although many sensor ICs support I2C directly, these are meant to be connected to an intermediary, such as a microcontroller for offloading communication and to perform preliminary data processing. This is also our strategy: a sensor module must contain at least a microntroller or embedded processor with a hardware I2C interface and support for the addressing and register scheme that we outline. The embedded computer can then connect to and retrieve readings from any number of sensor peripherals and relay the data back to the flight computer upon request in a standardized format.

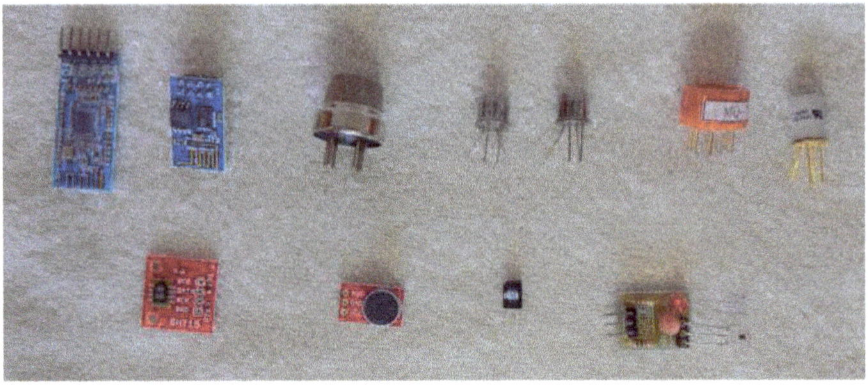

Figure 5. Examples of 11 low-cost sensors that can be integrated with the Arduino based micro controller. From top left are the following modules/sensors: **HM-10** (Bluetooth), **ESP8266** (WiFi), **MQ-135** (air quality: NH3, NOx, alcohol, benzene, smoke and CO_2), **Figaro TGS2600** (air quality: methane, CO, ethanol, hydrogen and iso-butane), **Figaro TGS2602** (air quality: ammonia, hydrogen sulfide, toluene, ethanol and hydrogen), **MQ-7** (air quality: CO), **Figaro TGS2442** (air quality: CO), **SparkFun 13683** (humidity and temperature), **SparkFun 12758** (electric microphone), **IR (infrared light) receiver** and **IR receiver module**. The weight and cost of the sensors can be seen in Table 1.

At this point, we only support one sensor module at a time. This limitation is a result of not being able to handle requests for more than one value at a time from the sensor module. The module must respond when addressed using the address reserved for this purpose, and it must implement responses to the virtual registers that make up a simple protocol on top of the I2C layer (see GitHub repository [30] for details of the protocol).

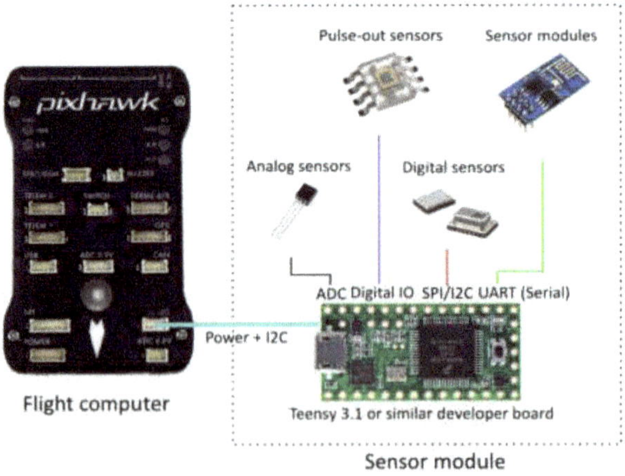

Figure 6. Electronic components on the drone. Basically, any type of sensor peripheral can be attached to the development board as long as it can be connected to the the Pixhawk via the I2C + power interface and implements the I2C register semantics (see code repository [30]). Photo by PixHawk/CC.

The two sensor modules we developed use a Teensy 3.1 development board [31] that contains an ARM Cortex-M4 processor running at 3.3 V. The board supports a 5V power input. Combined with a small footprint, the Teensy board is well suited for this application. The module design only occupies one I2C interface and all remaining pins (GPIO and DAC (digital to analog converter)) and bus interfaces can be used to connect the actual sensors. Figure 6 shows the electronic components involved on the drone.

If a new sensor is required, it can be implemented within hours, as the Arduino related world allows for rapidly prototyping new sensors — they just need to comply with the simple requirement that the interface is I2C at a 3.3v logic level, which many MCUs (micro controllers units) support, e.g., the entire Arduino world. If a specific MCU is needed, a voltage regulator makes it possible to use the drone's battery as a power source, while having a logical converter to bridge between the device and the Pixhawk.

Table 1 provides cost and weight information for some of the sensors that may be attached to the drone within a few hours, while Figure 5 provides a visual of those sensors.

Table 1. Examples of sensors, including cost and weight, which may be used with the proposed system. The sensors can be seen in Figure 5.

Sensor	Price	Weight	Notes
HM-10	2.92 USD	4 g	**Bluetooth** Low Energy (BLE) 4.0 module for iBeacon detection
ESP8266	1.90 USD	2 g	**WiFi** module, that i.e., can measure signal strength
MQ-135	1.70 USD	4 g	For **air quality**: NH3, NOx, alcohol, benzene, smoke and CO_2
Figaro TGS2442	18.19 USD	2 g	For **air quality**: CO
Figaro TGS2600	15.71 USD	2 g	For **air quality**: Mehtane, CO, ethanol, hydrogen and iso-butane
Figaro TGS2602	15.52 USD	2 g	For **air quality**: Ammonia, hydrogen sulfide, toluene, ethanol and hydrogen
MQ-7	7.25 USD	2 g	For **air quality**: CO
SparkFun 13683	41.95 USD	2 g	**Humidity and temperature**
SparkFun 12758	5.95 USD	2 g	Electric **microphone** (noise filtering *will* be needed)
IR receiver	2.04 USD	2 g	Just the **IR** component. (Price includes emitter)
IR receiver module	1.29 USD	2 g	**IR** light module

5.2. Flight Computer Firmware

In order to communicate with the sensor module, a driver for the Pixhawk firmware is needed. The software that runs on the flight computer consists of firmware on top of a real-time OS (RTOS) and a flight stack. The RTOS, based on NuttX [32], is responsible for scheduling and generally adding deterministic behavior to the timing of critical paths in the flight control software, i.e., position and attitude control. The firmware takes care of hardware abstraction, inter-module communication, and drivers. The flight stack is a set of modules or processes that perform specific tasks. These modules cooperate to get the wanted behavior from the drone. At its core, the flight stack uses inertial sensors and precise motor drivers to maintain attitude and position while responding correctly to user RC (radio controlled) inputs. These tasks have the highest priority and update frequency. Other modules with lower priority provide telemetry communication, waypoint management and more. Currently, two open-source flight stacks exist for the Pixhawk:

1. *PX4*: a highly modularized flight stack where each flight function runs in separate threads. It builds on the current NuttX and PX4 firmware development efforts.
2. *ArduPilot*: flight control runs as one large program on top of NuttX and the PX4 firwmare. The reason for this is legacy considerations. The ArduPilot flight stack can also be compiled for

older integrated flight computers. In addition, it has a huge code base that supports casual and light commercial flying very well.

We have chosen the ArduPilot flight stack because it provides good support for the drone and the software ecosystem around it. From a software engineering perspective, the monolithic structure is not ideal. However, since we are running it on the Pixhawk hardware, most of the ArduPilot codes are thin wrappers around the PX4 firmware layer.

The driver that we developed for the sensor module is a native PX4 firmware driver. It is loaded by the firmware, and its update cycle is scheduled directly in NuttX. Hence, it runs autonomously from the ArduPilot flight stack, and, to access it, a wrapper driver is needed to communicate with the native driver using an *ioctl* (input/output control) like interface. This wrapper driver can then be used in the Ardupilot flight stack to read and configure the sensor module and relay data to the ground station using its own telemetry transport.

To recap, our modifications to the flight computer to support the external sensor modules are:

- *PX4 firmware native driver*: a low-level driver that accesses the I2C peripheral directly and exposes a device interface for configuration and reading data. The driver runs at a user-defined interval where it commands the sensor module to convert readings and read the actual values after an appropriate amount of time depending on the sensor.
- *Wrapper driver for the ArduPilot flight stack*: an adapter class that wraps the ioctl interface of the native driver and exposes methods that can be used directly in Ardupilot user space code.
- *ArduPilot user space code*: minor additions to the ArduPilot handles messaging to and from the sensor module drivers using a radio telemetry link.

5.3. MAVLink Protocol Messages

Micro Air Vehicle Link (MAVLink) is *"a very lightweight, header-only message marshalling library for micro air vehicles"* [25]. It defines an extensible set of messages and mechanisms to transfer data such as streams. The IRIS+ drone comes with MAVLink optimized radio telemetry modules, and we take advantage of this feature by describing our own MAVLink messages for sending and receiving sensor data. Because the active set of MAVLink messages must be identical on both the drone flight computer and the ground station, all messages are defined in a platform-agnostic XML file that can be used in different development projects to automatically generate the necessary message code. The code generation is done by utilities provided by the MAVlink project.

We have added two new message types to the hundreds of existing messages. The messages contain GPS position, timestamp and sensor values. Even though GPS coordinates and timestamps are sent via other messages, it is important that the sensor values are reliably correlated to the position and time of conversion. At this point, the two custom messages are just placeholders for simple single integer or floating point values. The basic structure of the sensor messages, which may be e-mailed post-flight, are: latitude, longitude, altitude in cm, sensor value, and time. e.g., 55.4868037, 12.1692884, 461.0, 27, Fri Jun 24 15:33:31 GMT+02:00 2016.

To illustrate how data collection may be used for educational purposes, a student assignment could be to conduct a test of the accuracy by which the Bluetooth sensor locates a beacon on the ground (cf. Figure 2). The students would then be asked to place two beacons somewhere on a field and collect data from e.g. a 2 m altitude. First, students should make a graph of signal strengths (i.e., RSSI-values) from their test flight (cf. Figure 7). The next assignment could then be to compare these results with recordings made when flying at, for example, 10 m, to conduct low-pass filtering of the raw data [33], and, finally, to depict the relation between true distance to the beacons and the RSSI-values.

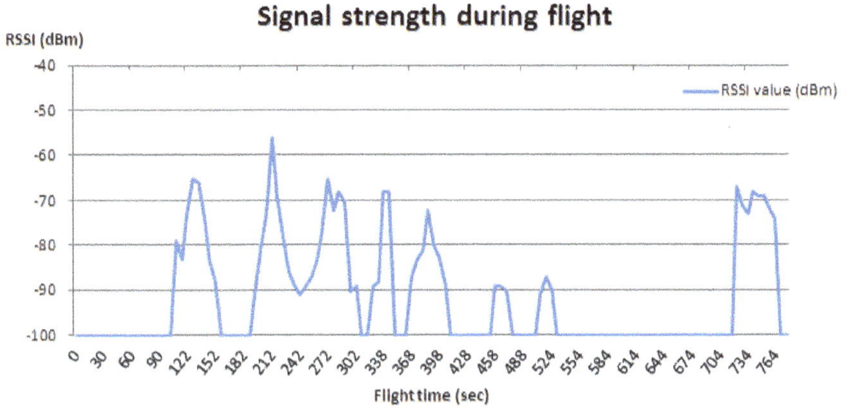

Figure 7. Signal strength measures during a flight above two beacons on a field. These data were collected with the drone flying at an altitude between 2 to 3 m.

5.4. Android Ground Station Application

The Android application is divided into two parts because of the modularized APIs provided by 3D Robotics. The first part is the 3DR Services Library [26]. Its responsibility is to transmit and receive data between the device itself and the drone via the included radio telemetry module. This part is acting as a service on the device, providing a pipeline to other applications that need to communicate with the drone. The service will copy and store the latest of each data type coming from the drone, i.e., altitude, battery level, etc, just before notifying all subscribing applications. The subscribers may then collect the latest data in a way that prevents jeopardizing the stability of the 3DR Services Library. Our contribution to this part of the system is the ability to handle sensor values and provide these to other applications.

The second part of the Android system is the user interface (UI) named *Tower* [27], which allows the user to interact with the drone by using the 3DR Services Library. We have extended this to include the possibility to collect the sensor values and store these in a local database (the SQLite implementation for Android) from where they can be forwarded by email in a CSV (comma separated values) format. This is done pre-flight by defining the flight parameters in terms of, for example, frequency of measurements, how long each measurement is estimated to take, altitude, etc. After the flight, the email address may be defined, if not already done, and the data are then forwarded. Another feature is the possibility to show the 300 latest sensor values as a heat map, either by defining the area to be observed in the UI and letting the autopilot handle the rest, or by manually controlling the drone. In both cases, real-time data are displayed in terms of the measured value, while a graphical map is updated with every 10th data point.

If an area is marked for observation, the UI will show the flight path based on an algorithm that creates a spiral going outside-in (see Figure 1 for the process of using the drone autonomously in a scenario to locate two iBeacons in a soccer field). It is possible to easily mark an area and scan it, while still being able to override the autopilot at any given point—or not use the autopilot at all, but instead operate it manually (see Figure 2).

6. Discussion

The main contribution of this paper is the extensible architecture that facilitates modularized airborne sensor deployment and real-time data feedback. We expand and augment a proven embedded robotics research platform and provide a foundation for continued work on airborne sensor prototyping.

Most previous research concerning drone measurements of the environment have used computer vision (CV). Our motivation went towards a more generic sensor approach, contributing to a system that may be applied on both multi-copters, fixed-wings, rovers, and other vehicles as well. We developed a platform for of-the-shelf sensors, based on the PX4 driver, as this supports a great variety of vehicle platforms and is open-source. The PX4 system by Meier et al. [22] was intended not just for aerial vehicles but for any novel vehicle platform, and we have deliberately chosen a commercial drone based on the PX4 platform. This allows for taking our contribution to the PX4 middleware and using it in application where other types of vehicles may be more suited, be it in the air, under water or on the ground. Villa et al. [7] points to a number of limitations for the use of small lightweight UAVs in research that is critical for our system, namely short range operation, low payload capacity, and sensitivity limitations of smaller sensors. Undoubtedly, there will be research projects requiring far more than this system supports. However, the area of UAV deployment of sensors is rapidly evolving, and we believe it will take some years to sort out when to use what equipment—just like land transportation conducted with a range of vehicles, from mini-vans to long-haul trucks. Our present system offers some of the benefits that Villa et al. [7] mentions: cost effectiveness, flexibility, short time for set-up, high repeatability of data collection and safety in operation.

The used drone from 3D Robotics, which is based on the PixHawk flight computer, comes with two default telemetry systems: one for communicating with the remote control and one for communicating with other systems which can interface with the USB antenna, i.e., a computer or a smartphone. Throughout the testing and usage of the proposed system, we experienced that the USB signal was rather weak, compared to the remote control signal. At a distance of roughly 100 m, the signal strength was occasionally too weak.

In its current state, the proposed system does not support the recovery of a missed data package. Thus, a measurement may be permanently lost if the drone moves out of the telemetric field of the ground station, which also is prohibited in several countries.

A key aspect for the project was uncovering the benefits and possibilities of the system. We devised three usage scenarios that the platform should ultimately support. A weakness of this approach is the lack of proper validation of the scenarios. Are they credible to potential users and how should they be evaluated under real task conditions?

The community around research and DIY drones is thriving, and open designs and software for UAVs are becoming increasingly available. This makes it harder to maintain a plug-and-play solution for a particular brand of drone. One could picture at least two paths for future efforts: one solution is to simply ignore the drone and flight systems and make a go for an isolated and autonomous payload. Another path is to support a complete open design for a research drone system that is tailored for carrying generic sensor modules. It is our hope and ambition that we have contributed to the latter approach by the work presented in this paper and by the available source code.

7. Future Work

To prevent data from getting lost, the firmware on the drone should be improved to expect a confirmation message from the ground station upon retrieval of a measurement. In case the confirmation signal is absent, the measurement should be stored until the connection is re-established.

At the moment, the system is unable to accurately track a moving source in real time. However, since the system can collect and store all measured values, with some further work, it might eventually track down a source in motion. This requires an algorithm for determining the vehicles' next most optimal heading, while the system already supports a change of way-points in mid-air.

The current usage scenarios are outdoor. However, we foresee a range of indoor sensing tasks in which drones—flying or driving—may do well, for instance locating gas leaks or mapping signal strengths in a building. Our future research will address indoor navigation by use of various non-vision sensors, e.g., ultra-sound, IR-beacons and by fingerprinting ubiquitous radio waves. In order to do this, we need a platform that can change sensors easily and quickly. Eventually, the current platform should be improved in order to carry more than one sensor at a time.

8. Conclusions

We have demonstrated the feasibility of turning a commercial drone into an extensible airborne sensor platform by adding new functionality to the stack of components in a drone system, from the sensor attachment to the user interface in the ground station software. The one important premise for this to succeed, however, is the availability of well-documented open-source or open-API software (and hardware for that matter) in all of the subsystems.

Acknowledgments: John Paulin Hansen's contribution was supported by the Bevica foundation.

Author Contributions: Lars Yndal Sørensen contributed with concept and design of the technical platform, acquisition of data, drafting and revising the manuscript. Lars Toft Jacobsen contributed with concept and design of the technical platform, and drafting of the manuscript. John Paulin Hansen contributed with data interpretation and revising the manuscript.

Conflicts of Interest: The authors declare no conflict of interest.

References

1. Watts, A.C.; Ambrosia, V.G.; Hinkley, E.A. Unmanned Aircraft Systems in Remote Sensing and Scientific Research: Classification and Considerations of Use. *Remote Sens.* **2012**, *4*, 1671–1692.
2. Erman, A.T.; Hoesel, L.; Havinga, P.; Wu, J. Enabling mobility in heterogeneous wireless sensor networks cooperating with UAVs for mission-critical management. *IEEE Wirel. Commun.* **2008**, *15*, 38–46.
3. Kerle, N.; Heuel, S.; Pfeifer, N. *Real-Time Data Collection and Information Generation Using Airborne Sensors;* Taylor & Francis/Balkema: Leiden, The Netherlands, 2008.
4. Quaritsch, M.; Kruggl, K.; Wischounig-Strucl, D.; Bhattacharya, S.; Shah, M.; Rinner, B. Networked UAVs as aerial sensor network for disaster management applications. *e & i Elektrotech. Inform.* **2010**, *127*, 56–63.
5. White, B.A.; Tsourdos, A.; Ashokaraj, I.; Subchan, S.; Zbikowski, R. Contaminant cloud boundary monitoring using network of UAV sensors. *IEEE Sens. J.* **2008**, *8*, 1681–1692.
6. Alvarado, M.; Gonzalez, F.; Fletcher, A.; Doshi, A. Towards the development of a low cost airborne sensing system to monitor dust particles after blasting at open-pit mine sites. *Sensors* **2015**, *15*, 19667–19687.
7. Villa, T.F.; Gonzalez, F.; Miljievic, B.; Ristovski, Z.D.; Morawska, L. An Overview of Small Unmanned Aerial Vehicles for Air Quality Measurements: Present Applications and Future Prospectives. *Sensors* **2016**, *16*, 1072.
8. Bendea, H.; Chiabrando, F.; Tonolo, F.G.; Marenchino, D. Mapping of archaeological areas using a low-cost UAV. The Augusta Bagiennorum test site. In Proceedings of the XXI International CIPA Symposium, Athens, Greece, 1–6 October 2007; Volume 1.
9. Neitzel, F.; Klonowski, J. Mobile 3D mapping with a low-cost UAV system. In Proceedings of the International Archives of the Photogrammetry, Remote Sensing and Spatial Information Sciences, Vol. XXXVIII-1/C22 UAV-g 2011, Conference on Unmanned Aerial Vehicle in Geomatics, Zurich, Switzerland, 14–16 September 2011.
10. Niethammer, U.; Rothmund, S.; Schwaderer, U.; Zeman, J.; Joswig, M. Open source image-processing tools for low-cost UAV-based landslide investigations. In Proceedings of the International Archives of the Photogrammetry, Remote Sensing and Spatial Information Sciences, Vol. XXXVIII-1/C22 UAV-g 2011, Conference on Unmanned Aerial Vehicle in Geomatics, Zurich, Switzerland, 14–16 September 2011.
11. Gonzalez, F.; Castro, M.P.; Narayan, P.; Walker, R.; Zeller, L. Development of an autonomous unmanned aerial system to collect time-stamped samples from the atmosphere and localize potential pathogen sources. *J. Field Robot.* **2011**, *28*, 961–976.
12. Xiang, H.; Tian, L. Development of a low-cost agricultural remote sensing system based on an autonomous unmanned aerial vehicle (UAV). *Biosyst. Eng.* **2011**, *108*, 174–190.

13. Bareth, G.; Aasen, H.; Bendig, J.; Gnyp, M.L.; Bolten, A.; Jung, A.; Michels, R.; Soukkamäki, J. Low-weight and UAV-based hyperspectral full-frame cameras for monitoring crops: Spectral comparison with portable spectroradiometer measurements. *Photogramm.-Fernerkund.-Geoinform.* **2015**, *2015*, 69–79.
14. Jang, J.S.; Liccardo, D. Automation of small UAVs using a low cost MEMS sensor and embedded computing platform. In Proceedings of the 2006 IEEE/AIAA 25TH Digital Avionics Systems Conference, Portland, OR, USA, 15–18 October 2006; pp. 1–9.
15. Anderson, C. *Makers: The New Industrial Revolution*; Crown Business: Dewsbury, UK, 2013.
16. Childers, B. Hacking the Parrot A.R. Drone, 2014. Available online: http://dl.acm.org/citation.cfm?id=2631585.2631586 (accessed on 18 March 2015).
17. Jung, D.; Levy, E.; Zhou, D.; Fink, R.; Moshe, J.; Earl, A.; Tsiotras, P. Design and development of a low-cost test-bed for undergraduate education in UAVs. In Proceedings of the 44th IEEE Conference on Decision and Control, Seville, Spain, 12–15 December 2005; pp. 2739–2744.
18. Eriksen, C.; Ming, K.; Dodds, Z. Accessible Aerial Robotics. *J. Comput. Sci. Coll.* **2014**, *29*, 218–227.
19. Mathias, H.D. An Autonomous Drone Platform for Student Research Projects. *J. Comput. Sci. Coll.* **2016**, *31*, 12–20.
20. Cavallini, A. iBeacon Bible 2.0. Available online: https://meetingofideas.files.wordpress.com/2014/06/ibeacon-bible-2-0.pdf (accessed on 14 May 2015).
21. IRIS + Drone Specifications. Available online: https://store.3drobotics.com/products/iris (accessed on 23 February 2015).
22. Meier, L.; Honegger, D.; Pollefeys, M. PX4: A Node-Based Multithreaded Open Source Robotics Framework for Deeply Embedded Platforms. In Proceedings of the 2015 IEEE International Conference on Robotics and Automation (ICRA), Seattle, WA, USA, 26–30 May 2015.
23. Meier, L.; Tanskanen, P.; Fraundorfer, F.; Pollefeys, M. PIXHAWK: A system for autonomous flight using onboard computer vision. In Proceedings of the 2011 IEEE International Conference on Robotics and Automation (ICRA), Shanghai, China, 9–13 May 2011; pp. 2992–2997.
24. Pixhawk project at ETH Zürich. Available online: https://pixhawk.ethz.ch/ (accessed on 4 March 2015).
25. MAVLink Project Website. Available online: http://qgroundcontrol.org/mavlink/start (accessed on 17 April 2015).
26. 3DR-Services-Library. Available online: https://github.com/ne0fhyk/3DR-Services-Library (accessed on 22 February 2015).
27. DroidPlaner-Tower. Available online: https://github.com/DroidPlanner/Tower (accessed on 17 February 2015).
28. Bracket for mounting the sensor module on the drone. Available online: http://www.thingiverse.com/thing:435675 (accessed on 4 April 2015).
29. Semiconductors, N. I2C-bus specification and user manual. *Rev* **2007**, *3*, 19.
30. Complete Source Code for the Project Found on Github. Available online: https://github.com/Yndal/ArduPilot-SensorPlatform (accessed on 14 January 2017).
31. Teensy Development Board. Available online: https://www.pjrc.com/teensy/index.html (accessed on 9 February 2015).
32. NuttX Real-Time Operating System. Available online: http://nuttx.org/ (accessed on 21 March 2015).
33. Jung, J.; Kang, D.; Bae, C.; Personal Computing Platform Research Team Distance Estimation of Smart Device Using Bluetooth. In Proceedings of the ICSNC 2013: The Eighth International Conference on Systems and Networks Communications, Venice, Italy, 27 October–1 November 2013; pp. 13–18.

sensors

MDPI

Article

A Camera-Based Target Detection and Positioning UAV System for Search and Rescue (SAR) Purposes

Jingxuan Sun, Boyang Li, Yifan Jiang and Chih-yung Wen *

Department of Mechanical Engineering, The Hong Kong Polytechnic University, Hong Kong, China; jingxuan.j.sun@connecy.polyu.hk (J.S.); boyang.li@connect.polyu.hk (B.L.); jiang.uhrmacher@connect.polyu.hk (Y.J.)

* Correspondence: cywen@polyu.edu.hk; Tel.: +852-2766-6644

Academic Editors: Felipe Gonzalez Toro and Antonios Tsourdos

Received: 30 August 2016; Accepted: 19 October 2016; Published: 25 October 2016

Abstract: Wilderness search and rescue entails performing a wide-range of work in complex environments and large regions. Given the concerns inherent in large regions due to limited rescue distribution, unmanned aerial vehicle (UAV)-based frameworks are a promising platform for providing aerial imaging. In recent years, technological advances in areas such as micro-technology, sensors and navigation have influenced the various applications of UAVs. In this study, an all-in-one camera-based target detection and positioning system is developed and integrated into a fully autonomous fixed-wing UAV. The system presented in this paper is capable of on-board, real-time target identification, post-target identification and location and aerial image collection for further mapping applications. Its performance is examined using several simulated search and rescue missions, and the test results demonstrate its reliability and efficiency.

Keywords: unmanned aerial vehicle (UAV); wilderness search and rescue; target detection

1. Introduction

Wilderness search and rescue (SAR) is challenging, as it involves searching large areas with complex terrain for a limited time. Common wilderness search and rescue missions include searching and rescuing injured humans and finding broken and lost cars in deserts, forests or mountains. Incidents of commercial aircraft disappearing from radar, such as the case in Indonesia in 2014 [1–3], also entail a huge search radius and search timeliness is critical to "the probability of finding and successfully aiding the victim" [4–7]. This research focuses on applications common in eastern Asian locations such as Hong Kong, Taiwan, the southeastern provinces of mainland China, Japan and the Philippines, where typhoons and earthquakes happen a few times annually, causing landslides and river flooding that result in significant damage to houses, roads and human lives. Immediate assessment of the degree of damage and searching for survivors are critical requirements for constructing a rescue and revival plan. UAV-based remote image sensing can play an important role in large-scale SAR missions [4–6,8,9].

With the development of micro-electro-mechanical system (MEMS) sensors, the use of small UAVs (with a wing-span of under 10 m) is a promising platform for conducting search, rescue and environmental surveillance missions. UAVs can be equipped with various remote sensing systems, such as powerful tools for observing disaster mitigation, including rapid all-weather flood and earthquake damage assessment. Today, low price drones allow people to quickly develop small UAVs, which have the following specific advantages:

- Can loiter for lengthy periods at preferred altitudes;

- Produce remote sensor data with better resolution than satellites, particularly in terms of image quality;
- Low cost, rapid response;
- Capable of flying below normal air traffic height;
- Can get closer to areas of interest.

Applying UAV technology and remote sensing to search, rescue and environmental surveillance is not a new idea. Habib et al. stated the advantages of applying UAV technologies to surveillance, security and mission planning, compared with the normal use of satellites, and various technologies and applications have been integrated and tested on UAV-assisted operations [9–13].

A fact people cannot ignore when applying UAV-assisted SAR is the number of required operators. It is claimed that at least two roles are required: one pilot who flies, monitors, plans and controls the UAV, and a second pilot who operates the sensors and information flow [14]. Practically, these two roles can be filled by a single operator, yet studies on ground robots have also suggested that a third person is recommended to monitor and protect the operator(s). Researchers have also studied the human behavior involved in managing multi UAVs, and have found that "the span of the human control is limited" [4,14,15]. As a result, a critical challenge of applying multiple UAVs in SAR is simultaneously monitoring information-rich data streams, including flight data and aerial video. The possibility of simplifying the human roles by optimizing information presentation and automatizing information acquisition was also explored [4], in which a fixed-wing UAV was used as a platform, and they analyzed and compared three computer vision algorithms to improve the presentation.

To automatize the information acquisition, it has been suggested that UAV systems integrate target-detection technologies for detecting people, cars or aircraft. A common method of observing people is the detection of heat features, which can be achieved by applying infrared camera technology and specifically developed algorithms. In 2005, a two-stage method based on a generalized template was presented [16]. In the first stage, a fast screening procedure is conducted to locate the potential person. Then, the hypothesized location of the person is examined by an ensemble classifier. In contrast, human detection based on color imagery has also been studied for many years. The research on developing a human detection method was conducted, which uses background subtraction, but pre-processing is required before a search mission [17]. Another method of human detection was presented that uses color images and models the human/flexible parts, then detects the parts separately [18]. A combination of both thermal and color imagery for human detection was also studied in [19].

To enhance information presentation and support humanitarian action, geo-referenced data from disaster-affected areas is expected to be produced. Numerous different technologies and algorithms for generating geo-referenced data via UAV have been studied and developed. A self-adaptive, image-matching technique to process UAV video in real-time for quick natural disaster response was presented in [20]. A prototype UAV and a geographical information system (GIS) by applying the stereo-matching method to construct a three-dimensional hazard map was also developed [21]. Scale Invariant Features Transform (SIFT) algorithms was improved in [22] by applying a simplified Forstner operator. Rectifying images on pseudo center points of auxiliary data were proposed in [23].

The aim of this study is to build an all-in-one camera-based target detection and positioning system that integrates the necessary remote sensors for wilderness SAR missions into a fixed-wing UAV. Identification and search algorithms were also developed. The UAV system can autonomously conduct a mission, including auto-takeoff and auto-landing. The on-board searching algorithm can report victims or cars with GPS coordinates in real-time. After the mission, a map of the hazard area can be generated to facilitate further logistics decisions and rescue troop action. Despite their importance, the algorithms for producing the hazard map are beyond the scope of this paper. In this work, we focus on the possibility of using a UAV to simultaneously collect geo-referenced data and detect victims. A hazard map and points are generated by the commercial software Pix4Dmapper^{TM} (Pix4Dmapper Discovery version 2.0.83, Pix4D SA, Lausanne, Switzerland).

Figure 1 provides a mission flowchart. Once a wilderness SAR mission is requested to the Ground Control System (GCS), the GCS operator designs a flight path that covers the search area and sends the UAV into the air to conduct the mission. During the flight, the on-board image processing system is designed to identify targets such as cars or victims, and to report possible targets with the corresponding GPS coordinates to the GCS within 60 m accuracy. These real-time images and generalized GPS help the immediate rescue action including directing the victim to wait for rescue at the current location and delivering emergency medicine, food and water. Meanwhile, the UAV is transmitting real-time video to the GCS and recording high-resolution aerial video that can be used, once the UAV lands, in post-processing tasks such as target identification and mapping the affected area. The post-target identification is designed to report victims' accurate locations within 15 m, and the map of the affected area can be used to construct a rescue plan.

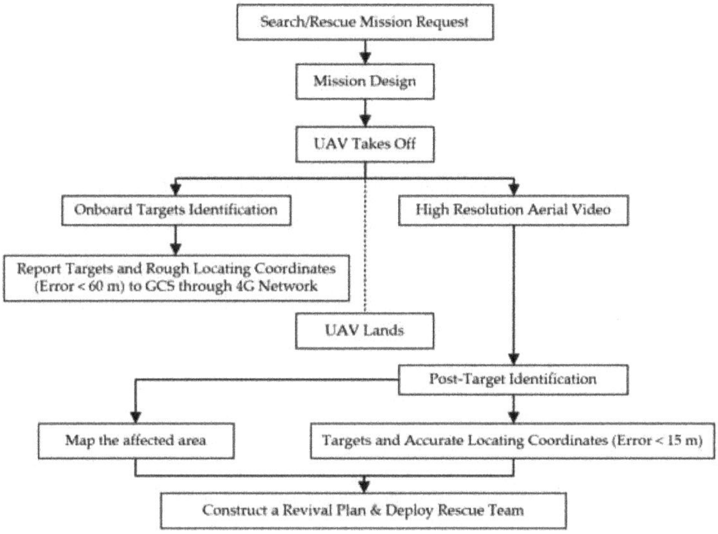

Figure 1. Flowchart of a wilderness SAR mission using the all-in-one UAV.

The remainder of this paper is organized as follows. Section 2 describes the details of the UAV system. Section 3 presents the algorithm and the implementation. Section 4 presents the tests and results, and Section 5 concludes the paper.

2. Experimental Design

The all-in-one camera-based target detection and positioning UAV system integrates the UAV platform, the communication system, the image system, and the GCS. The detailed hardware construction of the UAV is introduced in this section.

2.1. System Architecture

The purpose of the UAV system developed in this study was to find targets' GPS coordinates within a limited amount of time. To achieve this, a suitable type of aircraft frame was needed. The aircraft had to have enough fuselage space to accommodate the necessary payload for the task. The vehicle configuration and material had to exhibit the good aerodynamic performance and reliable structural strength needed for long-range missions. The propulsion system for the aircraft was calculated and selected once the UAV's configuration and requirements were known.

Next, a communication system, including a telemetry system, was used to connect the ground station to the UAV. After adding the flight control system, the aircraft could take off and follow the designed route autonomously. Finally, with the help of the mission system (auto antenna tracker (AAT), cameras, on-board processing board Odroid and gimbal), targets' and their GPS coordinates could be found. Figure 2 shows the UAV system's systematic framework, the details of which are explained in the following sub-sections. The whole system weighs 3.35 kg and takes off via hand launching.

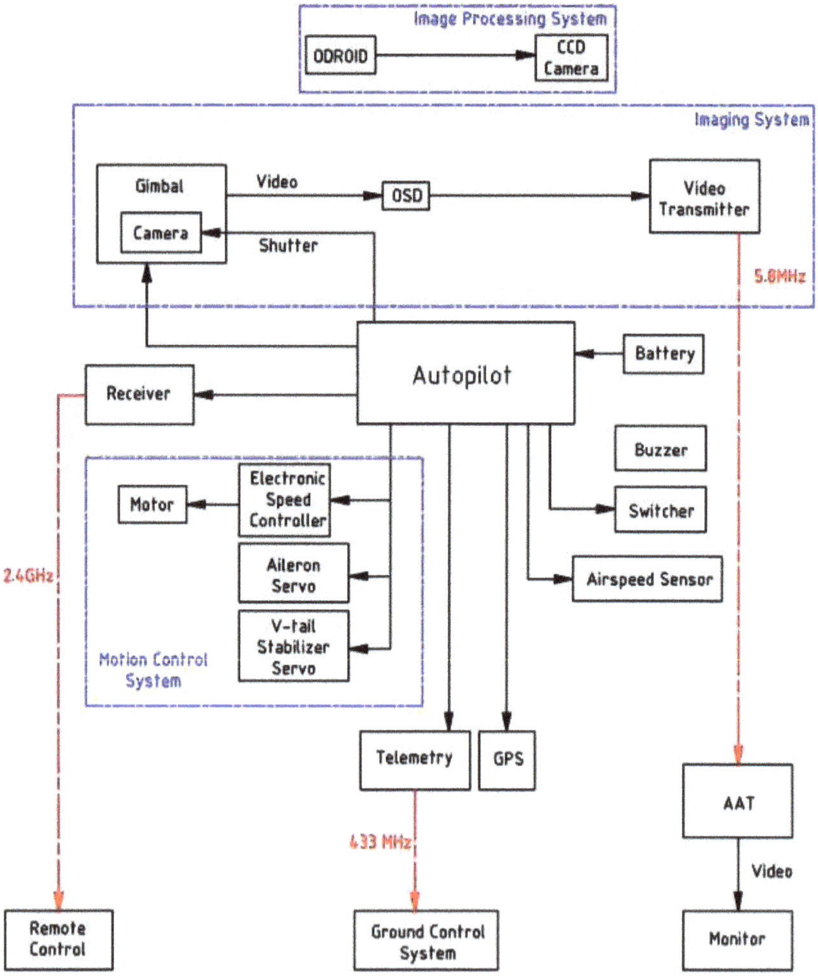

Figure 2. Systematic framework of the UAV system.

2.2. *Airframe of the UAV System*

The project objective was to develop a highly integrated system capable of large-area SAR missions. Thus, the flight vehicle, as the basic platform of the whole system, was chosen first. Given the prerequisites of quick response and immediate assessment capabilities, a fixed-wing aircraft was chosen for its high speed cruising ability, long range and flexibility in complex climatic conditions. To shorten the development cycle and improve system maintenance, an off-the-shelf commercial UAV platform "Talon" from X-UAV company was used (Figure 3). The wingspan of the Talon is 1718 mm and the wing area is 0.06 m^2. The take-off weight of this airframe can reach 3.8 kg.

Figure 3. Overall View of X-UAV Talon [24].

2.3. Propulsion System

The UAV uses Sunnysky X-2820-5 motor works in conjunction with an APC 11X5.5EP propeller. A 10,000 mAh Lipo 4-cell 20 C battery was used and this propulsion system provides a maximum cruse time of approximately 40 min at an airspeed of 18 m/s.

2.4. Navigation System

The main component of the navigation system is the Pixhawk flight controller running the free ArduPilot Plane firmware, equipped with GPS and compass kit, airspeed sensor and a sonar for measuring the height below 7 m. The airplane with this navigation system can conduct a fully autonomous mission, including auto take-off, cruise via waypoints, return to home position and auto landing, with enhanced fail-safe protection.

2.5. GCS and Data Link

The GCS works via a data link that enables the researcher to monitor or interfere with the UAV during an auto mission. Mission Planner, an open-source ground station application compatible with Windows, was installed on the GCS laptop for mission design and monitoring. An HKPilot 433 Mhz 500 Mw radio transmitter and receiver was installed on the GCS laptop, along with a Pixhawk flight controller. An auto antenna tracker (AAT) worked in conjunction with a 9 dBi patch antenna to provide a reliable data link within a 5-km range.

2.6. Post-Imaging Processing and Video Transmission System

The UAV system is designed with a fixed-wing aircraft flying at airspeeds ranging from 15 to 25 m/s for quicker response times on SAR missions. The ground speed may reach 40 m/s in extreme weather conditions. A GoPro HERO 4 was installed in the vehicle after considering the balance between its weight and image quality capabilities. In a searching and mapping mission, the aerial image always faces the ground. During flight, some actions such as rolling, pitching or other unexpected vibrations can disrupt the camera's stability, which may lead to unclear video. A Mini 2D camera gimbal produced by Feiyu Tech Co., Ltd. (Guilin, China), powered by two brushless motors, was used to stabilize the camera (Figure 4). The camera (GoPro HERO 4, GoPro, Inc., San Mateo, CA, USA) was set to video mode with a 1920 × 1080 pixel resolution in a narrow field of view (FOV) at 25 frames per second. During the flight, an analog image signal is sent to an on-screen display (OSD) and video transmitter. With a frequency of 5.8 GHz, the aerial video can be visualized by GCS in real-time as the high-resolution video is rerecorded for use during post-processing.

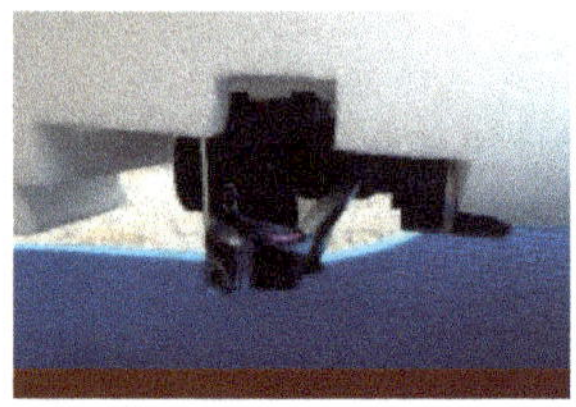

Figure 4. GoPro HERO 4 attached to the camera gimbal.

2.7. On-Board, Real-Time Imaging Process and Transmission System

A real-time imaging process and transmission system was setup on the UAV. The "oCam," (shows in Figure 5) a 5-mega pixel charge-coupled device (CCD) camera was chosen as the image source for the on-board target identification system. The focal length of the camera is 3.6 mm and it has a field of view of 65°. It weighs 37 g and has a 1920 × 1080 pixel resolution with 30 frames per second. The development of the on-board image processing was based on the Odroid XU4 (Hardkernel co., Ltd., GyeongGi, South Korea) (Figure 5b), which is a light, small, powerful computing device equipped with a 2-GHz core CPU and 2 Gbyte LPDDR3 Random-Access Memory (RAM). It also provides USB 3.0 interfaces that increase transfer speeds for high-resolution images. The Odroid XU4 used on the UAV in this system runs Ubuntu 14.04. The details of the algorithm and implementation will be discussed in Section 3. The Odroid board was connected to a 4th Generation (4G) cellular network via a HUAWEI (Shenzhen, China) E3372 USB dongle. Once the target is identified by the Odroid XU4, that particular image is transmitted through the 4G cellular network to the GCS.

(a) **(b)**

Figure 5. (**a**) oCam [25] and (**b**) Odroid XU4.

3. Algorithm for and Implementation of Target Identification and Mapping

The target identification program was implemented using an on-board micro-computer (Odroid XU4,) and the ground control station. The program can automatically identify and report cars, people and other specific targets.

3.1. Target Identification Algorithm

The mission is to find victims who need to be rescued, crashed cars or aircraft. The algorithm approaches these reconnaissance problems by using the color signature. These targets create a good contrast with the backgrounds due to their artificial colors. Figure 6 shows the flowchart of the reconnaissance algorithm. The aerial images are in YUV rather than RGB color space to identify the color signatures [26]. This progress can be achieved by calling back the function provided by OpenCV libraries. Both blue and red signatures are examined.

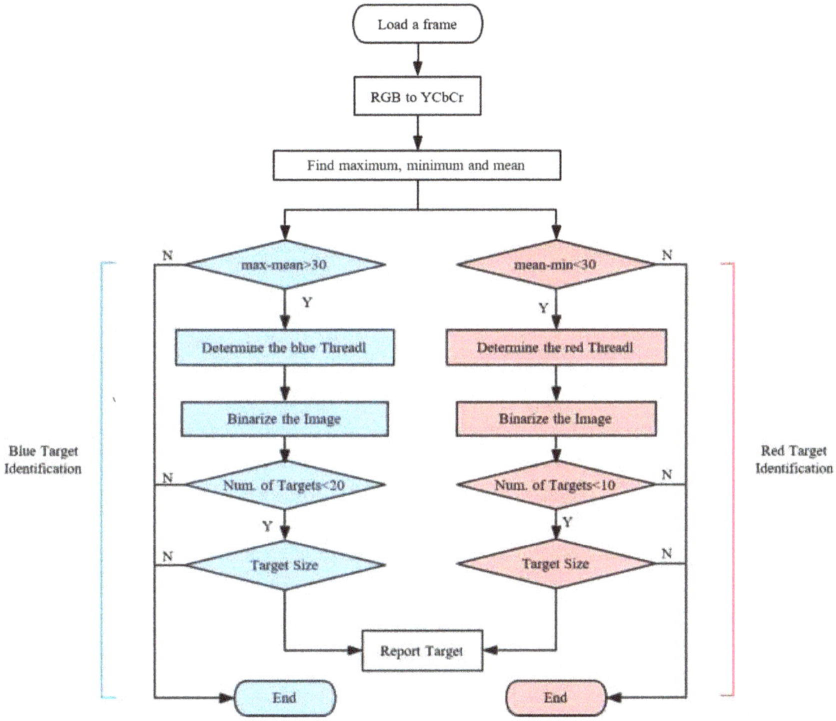

Figure 6. Flowchart of the identification algorithm.

The crucial step of the algorithm is to find an appropriate value of *Threadl*. A self-adapting method was applied to the reconnaissance program. The identification included the following steps.

Step 1: Read the blue and red chrominance values (Cb and Cr layers) of the image, and determine the maximum, minimum and mean values of the chrominance matrix. These values are then used to adapt the threshold.

Step 2: Distinguish whether existing objects are in great contrast. The distinction is processed by comparing the maximum/minimum and mean values of the chrominance. Introducing this step improves the efficiency with which the aerial video is processed, because the relevant identification is skipped if the criteria are not met. The criteria are expressed in Equation (1):

$$\begin{aligned} \max - \text{mean} > 30 \\ \text{mean} - \min < 30 \end{aligned} \quad (1)$$

Step 3: Determine the appropriate value of the threshold, which is determined by Equation (2), where the threshold with subscripts b and r donate blue and red, respectively. K_s is the sensitivity factor, and the program becomes more sensitive as it increases. K_s also changes with different cameras, and was set as 0.1 for the GoPro HERO 4 and 0.15 for the oCam in this study.

$$\begin{aligned} Threadl_b &= \max - (\max - \text{mean}) * K_s \\ Threadl_r &= (\text{mean} - \min) * K_s + \min \end{aligned} \tag{2}$$

Step 4: Binarize the image with the threshold.

$$f(p) = \begin{cases} 0; (p < Threadl) \\ 255; (p > Threadl) \end{cases} \tag{3}$$

where 0 represents the black color and 255 represents the white color.

Step 5: Examine the number of targets and their sizes. The results are abandoned if there are too many targets (over 20) in a single image because such results are typically caused by noise at the flight height of 80 m. The amount criterion is used because it is rare for a UAV to capture over 20 victims or cars in a single image in the wilderness. When examining the size of the targets, the results are abandoned if the suspected target only has a few or too many pixels. The criterion for the number of pixels is determined by the height of the UAV and the size of the target.

Step 6: The targets are marked with blue or red circles on the original image and reported to the GCS.

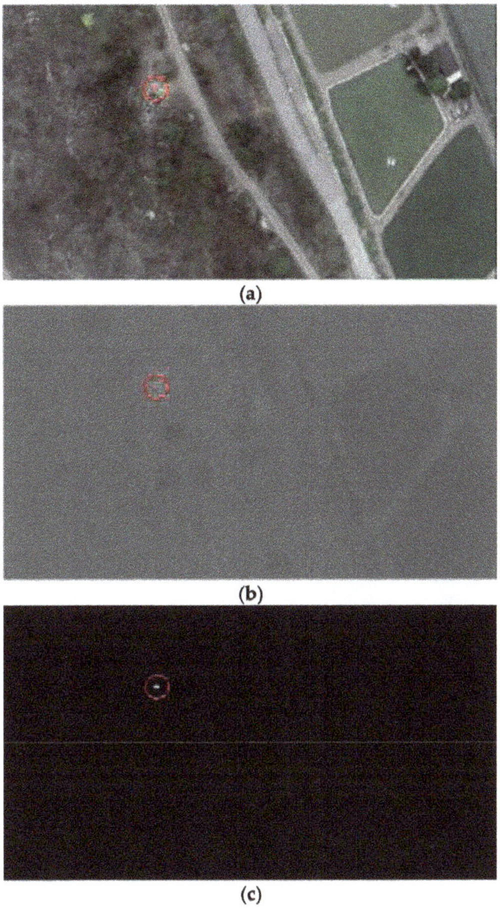

Figure 7. (a) The original image with red target in RGB color space; (b) the Cr layer of the YCbCr color space and (c) the binarized image with threshold.

Figure 7 demonstrates a test of the target identification algorithm using an aerial image with a tiny red target. Figure 7a is the original image captured from the aerial video with the target circled for easy identification. The Cr data were loaded for red color, as shown in Figure 7b. Figure 7c shows the results of the binarized image with a threshold of 0.44 (the white spot in the upper left quadrant).

3.2. On-Board Target Identification Implementation

Before developing the on-board system for identifying targets, the method used to report the targets and their locations to the GCS must be determined. Considering all of the subsystems on the vehicle and the frequencies used for the data link (433 MHz), live video transmission (5.8 GHz) and remote controller (2.4 GHz), the on-board target identification system is designed to connect to the base station of a cellular network, 800–900 MHz in the proposed testing area (Hong Kong and Taiwan). The results are then uploaded to the Dropbox server. Consequently, the on-board target identification system consists of four modules: Odroid as the core hardware, an oCam CCD camera, a GPS module and a dongle that connects to the 4G cellular network and provides it for the Odroid. The workflow of the on-board target identification system, designed as shown in Figure 8, includes three functions: Self-starting, identification and target reporting.

The self-starting is achieved via a Linux shell script. The program runs automatically when Odroid is powered on. The statuses of the camera, the Internet and the GPS module are checked. After successfully connecting all of the modules, the identification program runs on a loop until the Odroid is powered off. The identification program usually conducts four frames in a second.

During the flight, the GPS coordinates of the aircraft are directly treated as the location of the targets, because the rapid report is preferable to taking the time to get a highly accurate report during flight. The accurate locations of the targets are discovered post-flight using the high-resolution aerial video taken by the GoPro camera.

When reporting, the system scans the resulting files every 30 s and packs the new results, which are uploaded as a package instead of as frames to limit time consumption, because the Dropbox server requires verification for each file. The testing results show that uploading a package every 30 s is faster than uploading frame by frame. The reporting results include the images of the marked target and a text file of the GPS coordinates. These files are then stored in an external SD card that allows the GCS to quickly check the results post-flight. Figure 9 shows a truck reported by the on-board target identification system.

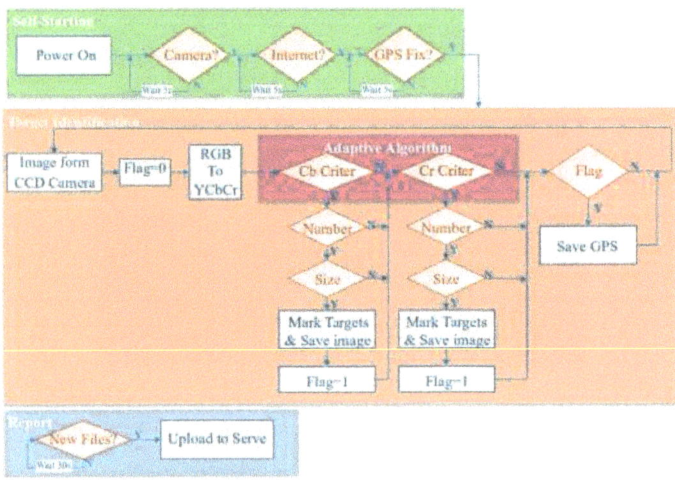

Figure 8. Flowchart of the on-board target identification system.

Figure 9. A blue truck reported by the on-board target identification system, marked by the identification program with a white circle.

3.3. Post-Target Identification Implementation via Aerial Video and Flight Log

Post-target identification is conducted using the high-resolution aerial video taken by the GoPro camera and stored in the SD card, and the flight data log from the flight controller to capture all possible targets to be rescued and obtain their accurate locations. In this section, the technical details of post-target identification are discussed.

3.3.1. Target Identification

The altitude of the flight path is carefully determined during the flight tests via the inertial-measurement unit and GPS data in the flight controller. Any targets coated with artificial colors of or larger than the estimated image size (15 × 15 pixels), calculated according to the height of the UAV and the target's physical dimensions, should be reported.

Figure 10 shows an aerial image of a 0.8 m × 0.8 m blue board with a letter 'Y' on it from flight heights of 50 m, 80 m and 100 m. The height of the flight path for the later field test was determined to be lower than 80 m accordingly, otherwise, the targets would only be several pixels in the image and might be treated as noise.

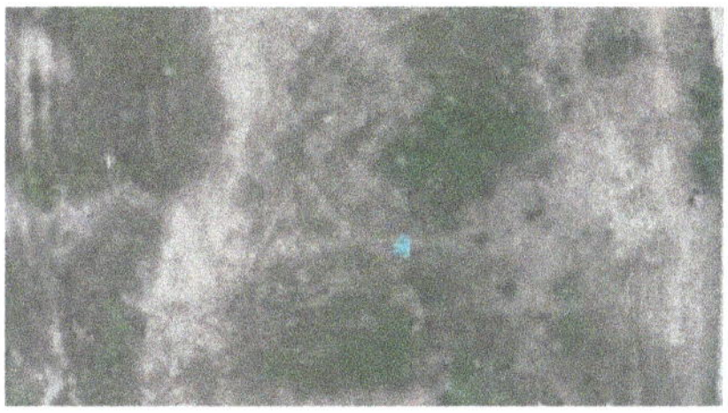

(a)

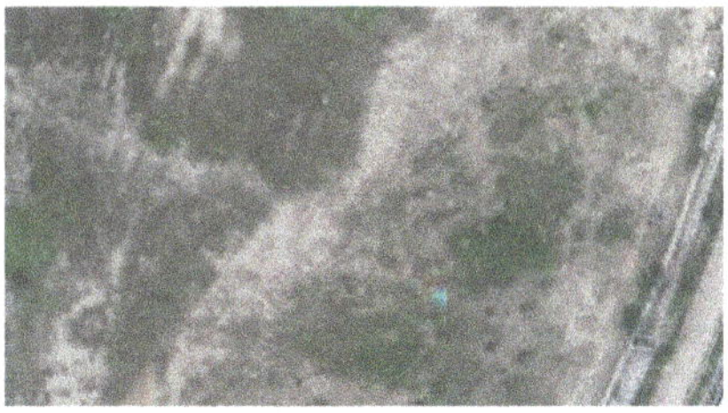

(b)

Figure 10. *Cont.*

(c)

Figure 10. The results of altitude tests with the vehicle cruising at (**a**) 50 m; (**b**) 80 m and (**c**) 100 m.

The main loop of the post-identification program was developed in the OPENCV environment. Similar to on-board target identification, the post-identification program loads the aerial video file and runs the algorithm in a loop with each frame. The targets are marked for the GCS operator, who engages in efficient confirmation. The flight data log and the aerial video are simultaneously synchronized to determine the reference frame number and reference shutter time. The technical details of this step are discussed in Section 3.3.2. The target image is saved as a JPEG file and named with its frame number. Figure 11 shows a red target board and a green agricultural net reported by the post-identification program. This JPEG file is sent to the GPS transformation program discussed in Section 3.3.3 to better position the target.

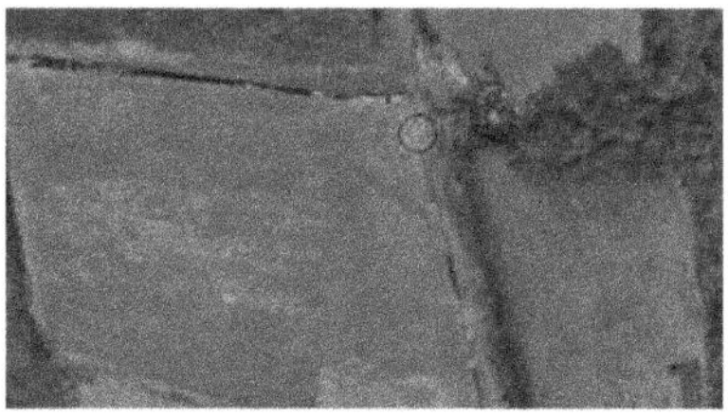

Figure 11. A red target board and a green agricultural net reported by the post-identification program, with both the red and blue targets marked with circles in corresponding colors.

To determine the image's frame number, we assume that the GoPro HERO 4 camera records the video with a fixed frame rate of 25 frames per second (FPS) in this study. Thus, the time interval (TI) of the target frame **F** in the aerial video and the reference frame can be determined by

$$TI = (Frame\ Number - Reference\ Frame\ No.) \times 40\ \mathrm{ms} \tag{4}$$

and the GPS time of **F** is

$$GPSTime = Reference\ GPS\ Time + TI \tag{5}$$

where the $Reference\ Frame\ No.$ and $Reference\ GPS\ Time$ are determined during synchronization, as discussed in Section 3.3.2.

Once the GPS time of the target frame is determined, the altitude and GPS coordinates of the camera are determined. The yaw angle Ψ is recorded as part of the Attitude messages in the flight data log, and the corresponding Attitude message can be searched via GPS time. The update frequencies of the Attitude messages come from an inertial-measurement unit IMU sensor, and the GPS messages are different. These two types of messages cannot be recorded simultaneously due to the control logic of the flight board. However, the updating frequency of the Attitude message is much higher than that of the GPS messages, thus the attitude message that is closest to the GPS time is treated as the vehicle's current attitude.

3.3.2. Synchronization of the Flight Data and Aerial Video

During the flight, the aerial video and flight data are recorded by the GoPro HERO 4 camera and flight controller, respectively. It is crucial to synchronize the flight data and the aerial video to obtain the targets' geo-information for the identification and mapping of the affected areas in a rescue mission.

Camera trigger distance (DO_SET_CAM_TRIGG_DIST), a camera control command provided by ArduPlane firmware, was introduced to synchronize the aerial video and the flight data log. DO_SET_CAM_TRIGG_DIST sets the distance in meters between camera triggers, and the flight control board logs the camera messages, including GPS time, GPS location and aircraft altitude when the camera is triggered. Compared with commercial quad-copters, fixed-wing UAVs fly at higher airspeeds. The time interval between two consecutive images should be small enough to meet the overlapping requirement for further mapping. However, the normal GoPro HERO 4 cannot achieve continuous photo capturing at a high frequency (5 Hz or 10 Hz) for longer than 30 s [27]. Thus, the GoPro was set to work in video recording mode with a frame rate of 25 FPS. The mode and shutter buttons were modified with a pulse width modulation (PWM)-controlled relay switch, as shown in Figure 12, so that the camera can be controlled by the flight controller. The shutter and its duration are configured in the flight controller.

Figure 12. Modification of the GoPro buttons to PWM-controlled relay switch.

The camera trigger distance can be set to any distance that will not affect the GoPro's video recording. A high-frequency photo capturing command will lead to video file damage. In this study, the flight controller sends a PWM signal to trigger the camera and record the shutter times and positions of the camera messages. However, the Pixhawk records the time that the control signal is sent out, and there is a delay between the image's recorded time and its real shutter time. This shutter delay was measured to be 40 ms and was introduced to the synchronization process.

The synchronization process shown in Figure 13 is conducted after the flight. The synchronization process shown in Figure 13 is conducted after the flight. The comparison process started with reading the aerial video and the photograph saved in GoPro's SD card. The original captured photo was resized to 1920 × 1080 pixels because the GoPro photograph was of a nonstandard size of 2016 × 1128 pixels. During the comparison process, both the video frames and photograph were treated as a matrix with a size of 1920 × 1080 × 3, where the number 3 denotes the 3 layers of RGB color space. The difference ε between the video frame and the photo was determined by the mean-square deviation value of $(\mathbf{Matrix_{photo}} - \mathbf{Matrix_{frame}})$. The video frame with minimum value of ε was considered the same as the original aerial photo (Figure 14) and the number of this video frame was recorded as the Reference Frame No (RFN). The recorded GPS time of sending the aerial photo triggering command was named as the Reference GPS time (RGT). Considering the above-mentioned 40 ms delay between sending out the command and capturing the photo the frame at RFN was taken at the time of (RGT + 40 ms delay time). Therefore, the video is combined with the flight log.

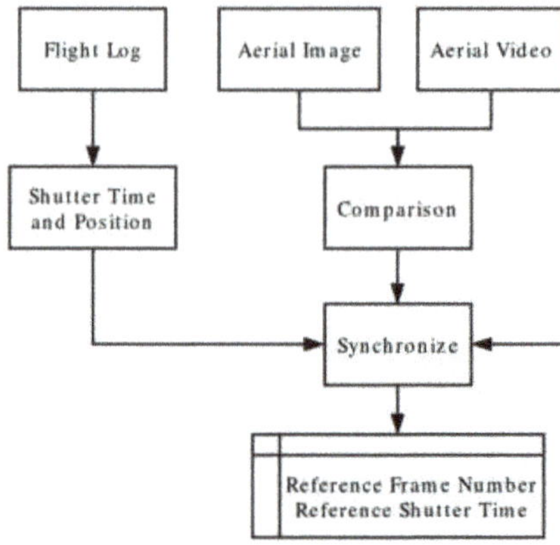

Figure 13. Flowchart for the synchronization of the aerial video and the flight data log.

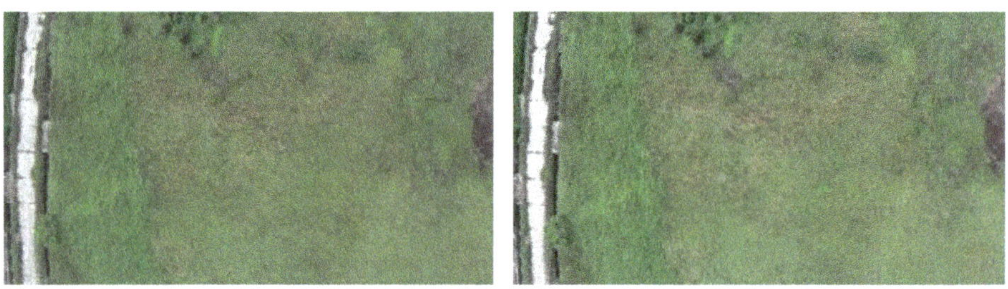

(**a**) (**b**)

Figure 14. Comparison results in synchronization process (**a**) the original photo taken by Gopro camera and (**b**) video frame captured by synchronization program.

3.3.3. GPS Transformation to Locate Targets

Once a target with its current aircraft position is reported to the GCS, an in-house MatLab locating program is used to report the target's GPS coordinates. In this study, the position of the aircraft is assumed to be at the center of the image, because the GPS module is placed above the camera.

The coverage of an image can be estimated using the camera's field of view (FOV) [28], as shown in Figure 15. The distances in the x and y directions are estimated using Equation (6).

$$a = \frac{2h}{\cos(\frac{\mathrm{FOV_X}}{2})} \tag{6}$$

$$b = \frac{2h}{\cos(\frac{\mathrm{FOV_Y}}{2})}$$

The resolution of the video frame is set to be 1920 × 1080 pixels. The scale between the distance and pixels is assumed to be a linear relationship, and is presented in Equation (7) as:

$$scale_x = \frac{a}{1920} = \frac{2h}{1920\left(\frac{\mathrm{FOV_X}}{2}\right)} \tag{7}$$

$$scale_y = \frac{b}{1080} = \frac{2h}{1080\left(\frac{\mathrm{FOV_Y}}{2}\right)}$$

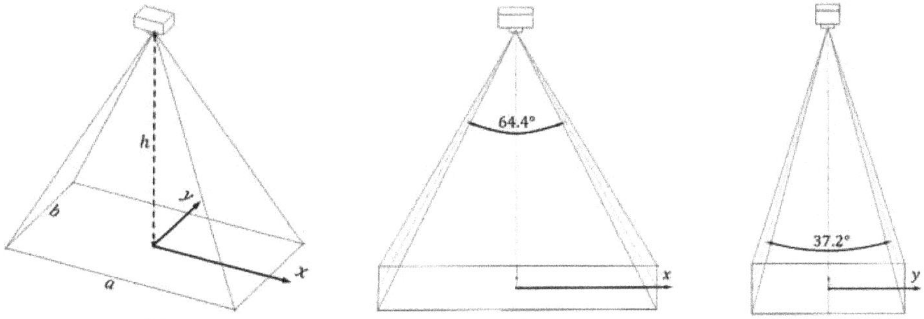

(a) (b) (c)

Figure 15. Camera and world coordinates.

As Figure 16 shows, a target is assumed to be located on the (x, y) pixel in the photo, and the offset of the target from the center of the picture is

$$\mathbf{offset_{target}} = \begin{bmatrix} scale_x \cdot x \\ scale_y \cdot y \end{bmatrix} (\text{m}) \tag{8}$$

For the transformation of a north-east (NE) world-to-camera frame with the angle of the Ψ, the rotation matrix is defined as

$$\mathbf{R_W^C} = \begin{bmatrix} \cos(\Psi) & -\sin(\Psi) \\ \sin(\Psi) & \cos(\Psi) \end{bmatrix} \tag{9}$$

where Ψ is the yaw angel of the aircraft. Thus, the position offset in the world frame can be solved with

$$P = \mathbf{R_W^C}^{\mathrm{T}} \mathbf{offset_{target}} = \begin{bmatrix} P_E \\ P_N \end{bmatrix} \tag{10}$$

Therefore, the target's GPS coordinates can be determined using

$$\boldsymbol{GPS_{target}} = \boldsymbol{GPS_{cam}} + \begin{bmatrix} P_E / f_x \\ P_N / f_y \end{bmatrix} \tag{11}$$

where f_x and f_y denote the distances represented by one degree of longitude and latitude, respectively.

A graphical user interface was designed and implemented in the MatLab environment to transform the coordinates with a simple 'click and run' function (Figure 17). The first step is opening the image containing the targets. The program automatically loads the necessary information for the image, including the frame number (also the image's file name), current location, camera attitude and yaw angle of the plane. The second step is to click the 'GET XY' button and use the mouse to click the target in the image. The program shows the coordinates of the target in this image. Finally, clicking the 'GET GPS' button provides the GPS coordinates reported by the program.

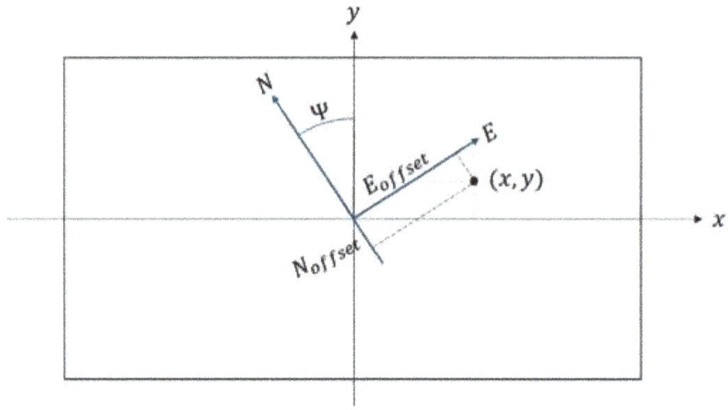

Figure 16. Coordinates of the camera and world frames.

Figure 17. Graphical user interface for the GPS transformation that allows end users to access a target's GPS coordinates using simple buttons.

3.4. Mapping the Searched Area

During rescue missions following landslides or floods, the terrain features can change significantly. After target identification, the local map must be re-built to guarantee the rescue team's safety and shorten the rescue time. In this study, we provide a preliminary demonstration of a fixed-wing UAV used to assist in post-disaster surveillance. Mapping algorithms are not discussed in this paper. The commercial software Pix4D was used to generate orthomosaic models and point clouds.

To map the disaster area, a set of aerial photos and their geo-information are applied to the commercial software, Pix4D. There should be at least 65% overlap between consecutive pictures, but aiming for 80% or higher is recommended. The distance between two flight paths should be smaller than a, and estimation Equation (6) can be found in Section 3.3.3. A mapping image capture program is shown in Figure 18.

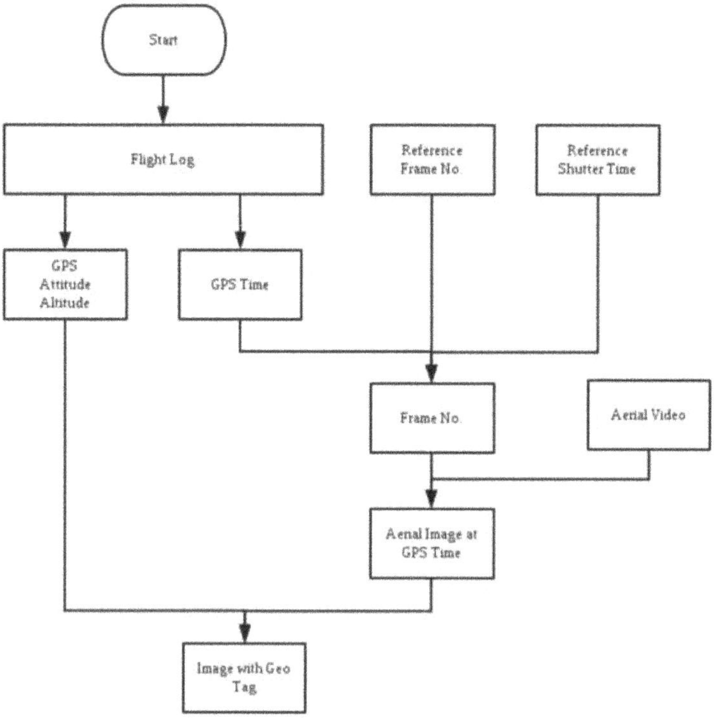

Figure 18. Flowchart of the mapping image capture program.

The mapping image capture program starts with GPS messages from the flight data log with reference frame numbers and shutter times generated by the synchronization step discussed in Section 3.2. The program loads the GPS times of all of the GPS messages in the loop and calculates the corresponding frame number N in the aerial video, which equals

$$N = \frac{GPS\ Time - Reference\ GPS\ Time}{40\ \text{ms}} + Reference\ Frame\ No.$$

Then, the mapping image capture program loads the Nth frame of the aerial video and saves it to the image file.

Once the mapping image capture program is complete, a series of photos and a text file containing the file names, longitude, latitude, altitude, roll, pitch and yaw are generated. The Pix4D then produces the orthomosaic model and point clouds using these two file types.

4. Blind Tests and Results

To test the all-in-one camera-based target detection and positioning system, a blind field test was designed. A drone, a 2 m × 2 m blue or red square board and a 0.8 m × 0.8 m blue or red square board were used to simulate a crashed airplane, broken cars and injured people, respectively (Figure 19a–c).

The flight tests were conducted at two test sites, the International Model Aviation Center (22°24′58.1″ N 114°02′35.4″ E) of the Hong Kong Model Engineering Club, Ltd. in Yuen Long town, Hong Kong and the Zengwun River (23°7′18.03″ N 120°13′53.86″ E) in the Xigang District, of Tainan city, Taiwan. Given concerns with the limited flying area in Hong Kong, the preliminary in-sight tests were conducted in Hong Kong and the main blind out-of-sight tests were conducted in

Taiwan. The flight test information is listed in Table 1. Only post-identification tests were conducted in Hong Kong. In Taiwan, no after-flight mapping was done for the first two tests (Tests 3 and 4).

(a) (b) (c)

Figure 19. (**a**) The drone simulated a crashed airplane, (**b**) the 2 m × 2 m blue or red target boards represented broken cars and (**c**) the 0.8 m × 0.8 m blue or red targets boards represented injured people to be rescued.

Table 1. Basic information for flight tests.

Flight Test	Test Site	Flight Time (min)	Testing Function		
			Real-Time Identification	Post-Identification	Mapping
Test 1	Hong Kong	15:36	×	√	×
Test 2	Hong Kong	3:05	×	√	×
Test 3	Taiwan	13:23	√	√	×
Test 4	Taiwan	17:41	√	√	×
Test 5	Taiwan	17:26	√	√	√
Test 6	Taiwan	16:08	√	√	√
Test 7	Taiwan	16:23	√	√	√
Test 8	Taiwan	17:56	√	√	√

Figure 20a,b shows the search site and its schematic in Hong Kong. The search path repeated the square route due to the limited flight area. The yellow path in Figure 20a is the designed mission path and the purple line indicates the real flight path of the vehicle. For the tests in Taiwan, there were two main search areas (A and B) along the bank of the Zengwun River in the Xigang District of Tainan city, Taiwan, as shown in Figure 20c. The schematics of the designed search route and areas are depicted in Figure 20d. The maximum communication distance was 3 km and the width of the flight corridor was 30 m. This width was intended to test the stability of the UAV and the geo-fencing function of the flight controller. If the UAV flies outside the corridor, it is considered to have crashed. After the flight performance tests, the UAV flew inside the corridor and was proven stable. An unknown number of targets were placed in search areas A and B by an independent volunteer before every test. The search team then conducted the field tests and tried to find the targets. The test results are discussed in the following sections.

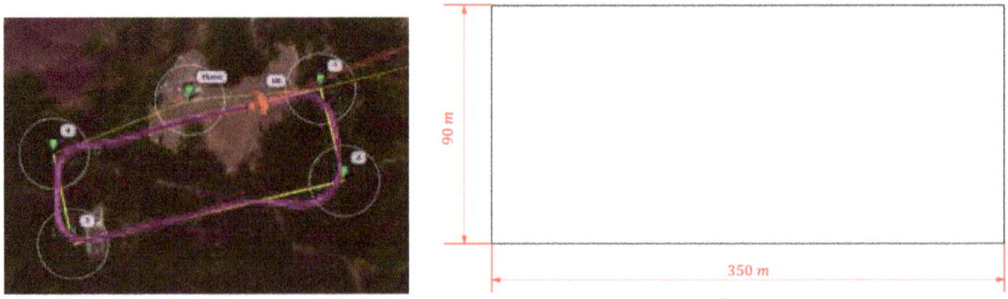

(a) (b)

Figure 20. *Cont.*

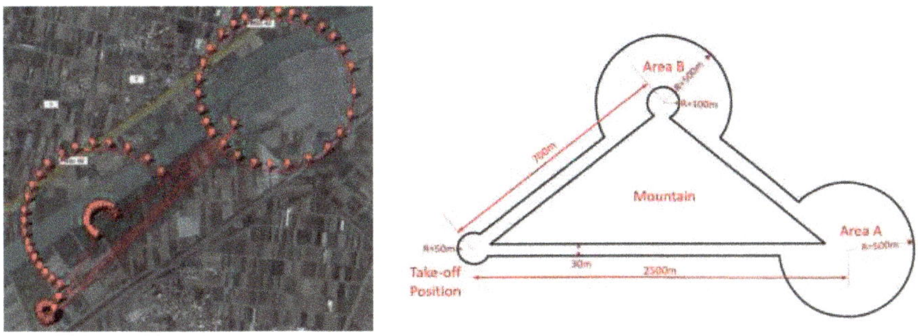

(c) (d)

Figure 20. (**a**) Test route in Hong Kong; (**b**) schematics of the designed route in Hong Kong; (**c**) search areas A and B for blind tests in Taiwan and (**d**) schematics of the designed search route and areas in Taiwan.

4.1. Target Identification and Location

Post-target identification processing was conducted in all eight flight test to assess the identification algorithm. The post-identification program ran on a laptop equipped with Intel Core i5-2430M CPU and 8 Gb RAM. The testing results are shown in Table 2. Note that the post-identification program only missed two targets for all of the tests.

Table 2. Post-target identification results.

Flight Test	Resolution	Flying Altitude	Flight Time (min)	Targets	Identified Targets	Total Post-Target Identification Time (min)
Test 1	1920 × 1080	80	15:36	3	2	11:08.6
Test 2	1920 × 1080	80	3:05	2	2	02:46.3
Test 3	1920 × 1080	80	13:23	3	3	12:57.1
Test 4	1920 × 1080	80	17:41	3	2	14:04.6
Test 5	1920 × 1080	80	17:26	3	3	13:23.9
Test 6	1920 × 1080	80	16:08	3	3	11:45.1
Test 7	1920 × 1080	80	16:23	6	6	12:16.9
Test 8	1920 × 1080	75	17:56	6	6	14:18.3

Taking test 7 as an example, 6/6 targets were found by the identification system, as shown in Figure 21, including a crashed aircraft, two crashed cars and three injured people. Note that in Figure 21g the target board, representing the injured people, was folded by gusts of wind to the extent that it is barely recognizable. Nevertheless, the identification system still reported this target, confirming its reliability. The locating error of 5 targets was less than 15 m as shown in Table 3 (having met the requirements discussed in Section 1). The targets and their locations were reported in 15 min.

Table 3. Locating results of flight test 7.

Target	Red Z	Red Plane	Blue I	Blue V	Blue J	Red Q
Latitude (N)	23.114536°	23.111577°	23.110889°	23.113637°	23.122189°	23.117840°
Longitude (E)	120.213111°	120.211898°	120.210819°	120.210463°	120.223206°	120.225225°
Error	2.8 m	13.9 m	1.6 m	0.8 m	11.3 m	4.8 m

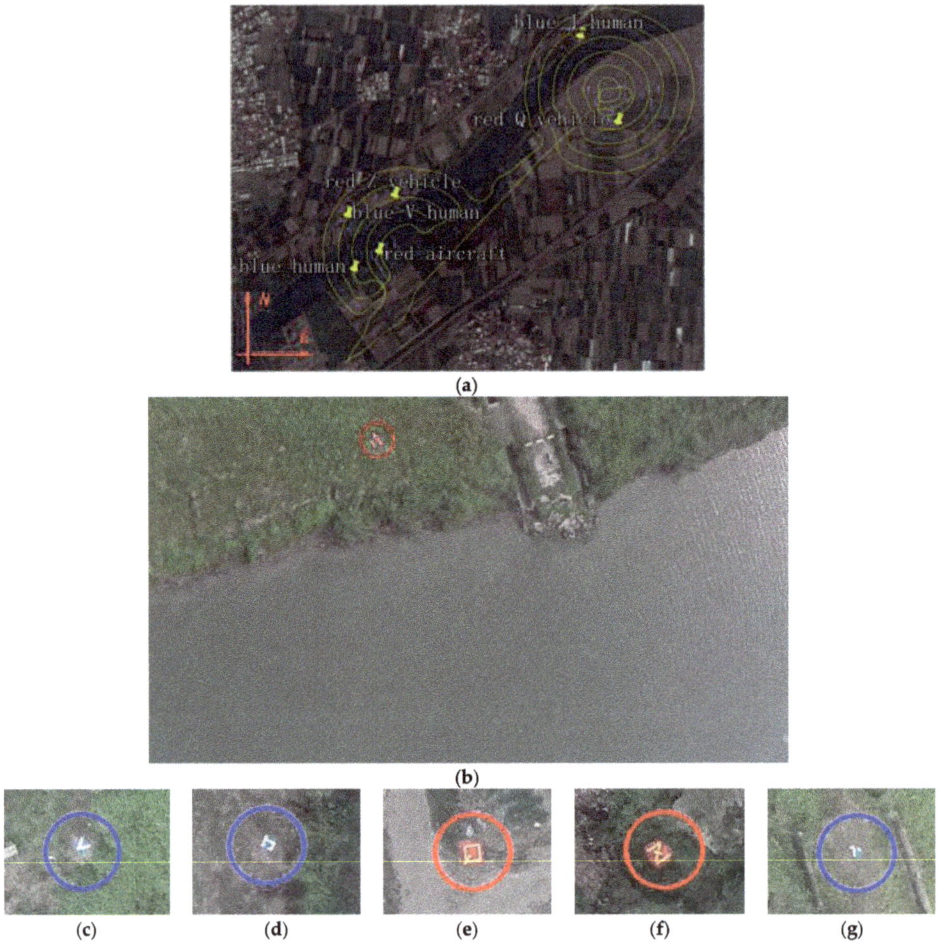

Figure 21. (**a**) The locations of six simulated targets; (**b**) the original image saved by the identification program with target drone; (**c**) designed target (blue board with letter V) represents an injured person; (**d**) designed target (blue board with letter J) represents an injured person; (**e**) designed target (red board with letter Q) represents a crashed car; (**f**) designed target (red board with letter Z) represents a crashed car and (**g**) designed target (small blue board) represents an injured person. The board was blown over by the wind.

In addition to the designed targets, the identification program reported real cars/tracks, people, boats and other objects. The percentages of each type of target are shown in Figure 22. The large amount of other targets is due to the nature of the search area. The testing site is a large area of cropland near a river, and the local farmers use a type of fertilizer that is stored in blue buckets and they use green nets to fence in their crops. These two item types were reported, as shown in Figure 23. However, these results can be quickly sifted through by the GCS operator. The identification program still reduces the operator's work load, and the search mission was successfully completed in 40 min, beginning when the UAV took off and ending when all of the targets had been reported.

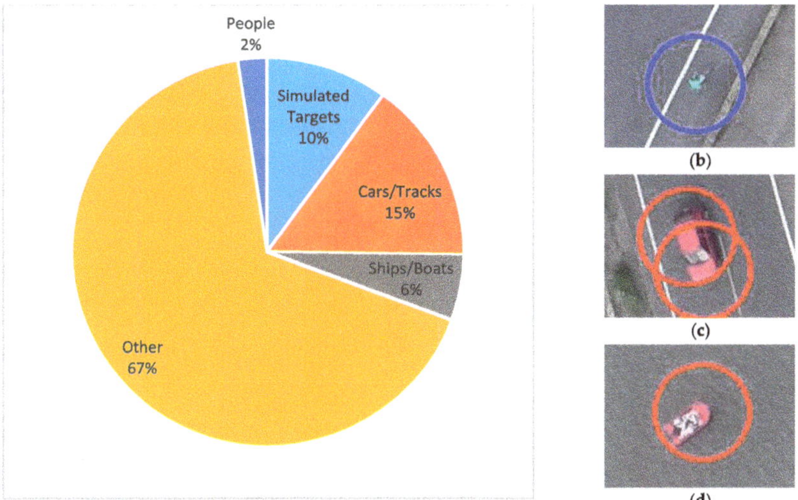

(a) (b) (c) (d)

Figure 22. (**a**) Composition of reporting targets; (**b**) a person on the road; (**c**) a red car and (**d**) a red boat reported by the identification program.

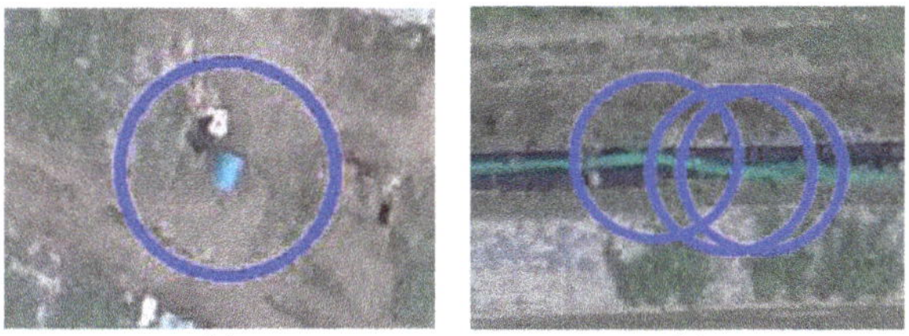

(a) (b)

Figure 23. The other reporting targets: (**a**) A blue bucket and (**b**) green nets.

In tests 3–8, both on-board real-time processing and post-processing were conducted and the results are shown in Figure 24. Note that the performance of the post-target identification is better than that of real-time onboard target identification, due to the higher resolution of the image source. Nevertheless, the on-board target identification system still reported more than 60% of the targets and provided an efficient real-time supplementary tool for the all-in-one rescue mission. A future study will be conducted to improve the success rates of on-board target identification systems.

4.2. Mapping

To cover the whole search area, the flight plan was designed as shown in Figure 25. The distance between 2 adjacent flight paths is 80 m. The total distance of flight plan is 20.5 km with a flight time of 18 min. The turning radius of the UAV was calculated, and it is 50 m for bank angles no larger than 35°. Thus, as shown in Figure 25b, the flight plan was designed with a 160-m turning diameter while the gap between the two flight paths remained 80 m to ensure overlapping and complete coverage.

After the flight, the mapping image capture program developed in this study was applied to capture the images from the high-resolution video and process the flight data log. A total of 2200 photos were generated and applied to Pix4D, and the resulting orthomosaic model and point clouds are shown in Figure 26. The missing part is due to the strong reflection on the water's surface resulting in mismatched features.

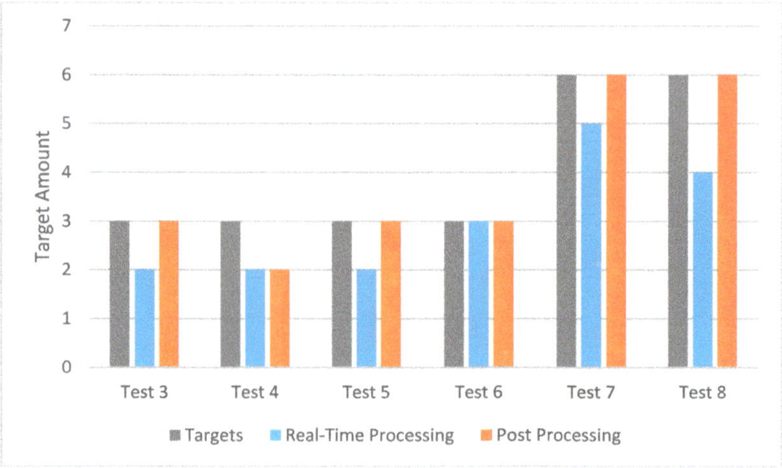

Figure 24. Target identification results of real-time processing and post-processing.

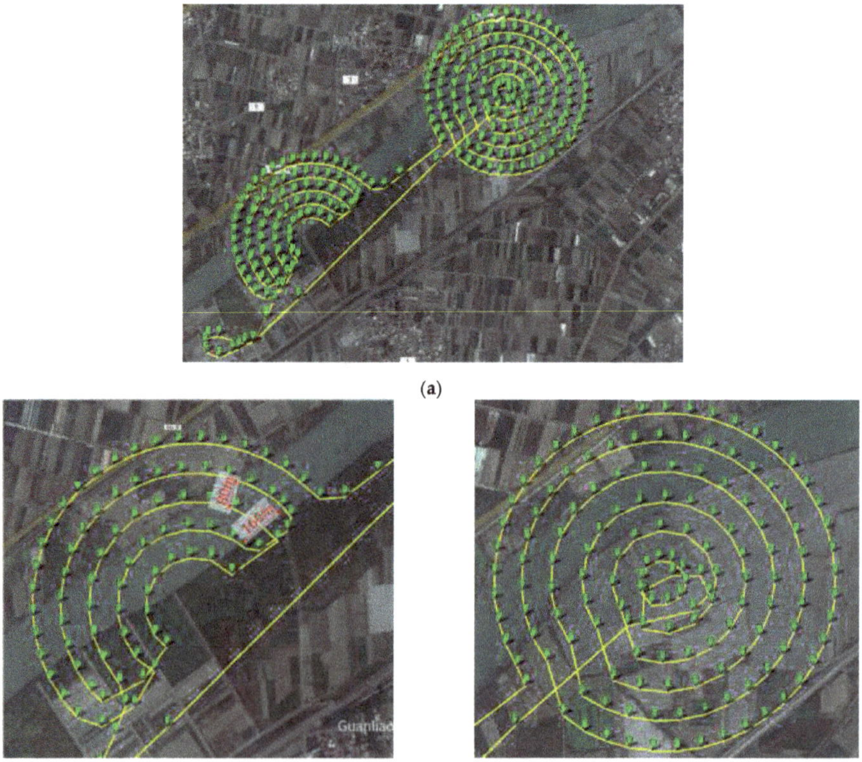

(a)

(b)

(c)

Figure 25. (**a**) Overall flight plan for the search mission, (**b**) flight plan for search area B (the turning diameter reaches 160 m to ensure the flight performance while the distance between the two flight paths remains 80 m, guaranteeing full coverage and overlap) and (**c**) flight plan for search area A.

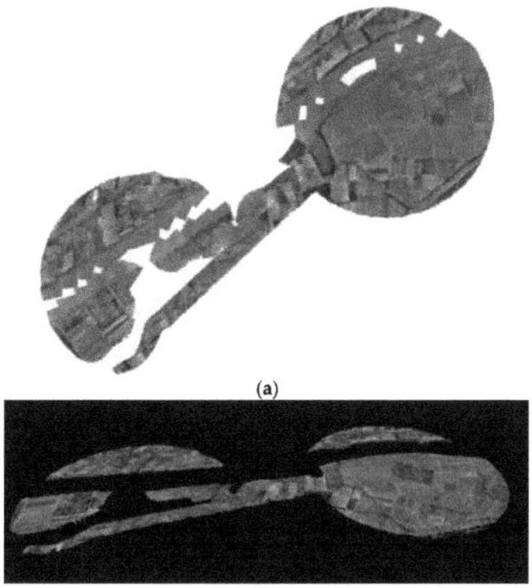

Figure 26. (**a**) Orthomosaic model of the testing area and (**b**) point clouds of the search area.

5. Conclusions

In this study, a UAV system was developed, and its ability to assist in SAR missions after disasters was demonstrated. The UAV system is a data acquisition system equipped with various sensors to realize searching and geo-information acquisition in a single flight. The system can reduce the cost of large-scale searches, improve the efficiency and reduce end-users' workloads.

In this paper, we presented a target identification algorithm with a self-adapting threshold that can be applied to a UAV system. Based on this algorithm, a set of programs was developed and tested in a simulated search mission. The test results demonstrated the reliability and efficiency of this new UAV system.

A further study will be conducted to improve the image processing in both onboard and post target identification, focusing on reducing the unexpected reporting targets. A proposed optimization method is to add an extra filtration process to the GCS to further identify the shape of the targets. This proposed method will not increase the computational time of the onboard device significantly. It is a simple but effective method concerning the limited CPU capability of an on-board processor. Generally speaking, most commercial software is too comprehensive to be used in the on-board device. Notably, the limitation of the computing power becomes a minor consideration during post-processing since powerful computing devices can be used at this stage. To evaluate and improve the performance of targets' identification algorithm in post-processing, further study will be conducted, including the application of the parallel computing technology and comparison with the advanced commercial software.

In this study, the scales of the camera and world coordinates were assumed to be linear. This assumption can result in target location errors. We tried to reduce the error by selecting the image with the target near the image center. Although the error of the current system is acceptable for a search mission, we will conduct a further study to improve the location accuracy. Lidar will be installed to replace the sonar, and more accurate relative vehicle height will be provided for auto-landing. Also, in the future, the vehicle will be further integrated to realize the 'Ready-to-Fly' stage for quick responses in real applications.

Supplementary Materials: The following is available online at https://www.youtube.com/watch?v=19_-RyPp93M. **Video S1:** A Camera-Based Target Detection and Positioning System for Wilderness Search and Rescue using a UAV. https://github.com/jingego/UAS_system/tree/master/Image%20Processing. **Source Code 1:** Matlab Code of targets identification. https://github.com/jingego/UAS_system/blob/master/Mapping_pre-process/CAM_clock_paper_version.m. **Source Code 2:** MatLab Code of synchronization.

Acknowledgments: This work is sponsored by Innovation and Technology Commission, Hong Kong under Contract No. ITS/334/15FP. Special thanks to Jieming Li for his help in building the image identification algorithm of this work.

Author Contributions: Jingxuan Sun and Boyang Li designed the overall system. In addition, Boyang Li developed the vehicle platform and Jingxuan Sun developed the identification algorithms, locating algorithms and post image processing system. Yifan Jiang developed the on-board targets identification. Jingxuan Sun and Boyang Li designed and performed the experiments. Jingxuan Sun analyzed the experiment results and wrote the paper. Chih-yung Wen is in charge of the whole project management.

Conflicts of Interest: The authors declare no conflict of interest.

Abbreviations

The following abbreviations are used in this manuscript:

UAV:	Unmanned Aerial Vehicle
SAR:	Search and Rescue
GCS:	Ground Control System
AAT:	Auto Antenna Tracker
OSD:	On Screen Display
FOV:	Field of view
FPS:	Frame Per Second

References

1. Indonesia Airasia Flight 8501. Available online: https://en.wikipedia.org/wiki/Indonesia_Air-Asia_Flight_8501 (accessed on 20 October 2016).
2. Qz8501: Body of First Victim Identified. Available online: http://english.astroawani.com/airasia-qz8501-news/qz8501-body-first-victim-identified-51357 (accessed on 20 October 2016).
3. Airasia Crash Caused by Faulty Rudder System, Pilot Response, Indonesia Says. Available online: https://www.thestar.com/news/world/2015/12/01/airasia-crash-caused-by-faulty-rudder-system-pilot-response-indonesia-says.html (accessed on 20 October 2016).
4. Goodrich, M.A.; Morse, B.S.; Gerhardt, D.; Cooper, J.L.; Quigley, M.; Adams, J.A.; Humphrey, C. Supporting wilderness search and rescue using a camera-equipped mini uav. *J. Field Robot.* **2008**, *25*, 89–110. [CrossRef]
5. Goodrich, M.A.; Cooper, J.L.; Adams, J.A.; Humphrey, C.; Zeeman, R.; Buss, B.G. Using a mini-uav to support wilderness search and rescue: Practices for human-robot teaming. In Proceedings of the 2007 IEEE International Workshop on Safety, Security and Rescue Robotics, Rome, Italy, 27–29 September 2007.
6. Goodrich, M.A.; Morse, B.S.; Engh, C.; Cooper, J.L.; Adams, J.A. Towards using unmanned aerial vehicles (UAVs) in wilderness search and rescue: Lessons from field trials. *Interact. Stud.* **2009**, *10*, 453–478.
7. Morse, B.S.; Engh, C.H.; Goodrich, M.A. Uav video coverage quality maps and prioritized indexing for wilderness search and rescue. In Proceedings of the 5th ACM/IEEE international conference on Human-robot interaction, Osaka, Japan, 2–5 March 2010.
8. Doherty, P.; Rudol, P. A uav search and rescue scenario with human body detection and geolocalization. In Proceedings of the Australasian Joint Conference on Artificial Intelligence, Gold Coast, Australia, 2–6 December 2007.
9. Habib, M.K.; Baudoin, Y. Robot-assisted risky intervention, search, rescue and environmental surveillance. *Int. J. Adv. Robot. Syst.* **2010**, *7*, 1–8.
10. Tomic, T.; Schmid, K.; Lutz, P.; Domel, A.; Kassecker, M.; Mair, E.; Grixa, I.L.; Ruess, F.; Suppa, M.; Burschka, D. Toward a fully autonomous uav: Research platform for indoor and outdoor urban search and rescue. *IEEE Robot. Autom. Mag.* **2012**, *19*, 46–56. [CrossRef]
11. Waharte, S.; Trigoni, N. Supporting search and rescue operations with uavs. In Proceedings of the IEEE 2010 International Conference on Emerging Security Technologies (EST), Canterbury, UK, 6–7 September 2010.

12. Naidoo, Y.; Stopforth, R.; Bright, G. Development of an uav for search & rescue applications. In Proceedings of the IEEE AFRICON 2011, Livingstone, Zambia, 13–15 September 2011.
13. Bernard, M.; Kondak, K.; Maza, I.; Ollero, A. Autonomous transportation and deployment with aerial robots for search and rescue missions. *J. Field Robot.* **2011**, *28*, 914–931. [CrossRef]
14. Cummings, M. Designing Decision Support Systems for Revolutionary Command and Control Domains. Ph. D. Thesis, University of Virginia, Charlottesville, VA, USA, 2004.
15. Olsen, D.R., Jr.; Wood, S.B. Fan-out: Measuring human control of multiple robots. In Proceedings of the SIGCHI Conference on Human Factors in Computing Systems, Vienna, Austria, 24–29 April 2004.
16. Davis, J.W.; Keck, M.A. A two-stage template approach to person detection in thermal imagery. *WACV/MOTION* **2005**, *5*, 364–369.
17. Lee, D.J.; Zhan, P.; Thomas, A.; Schoenberger, R.B. Shape-based human detection for threat assessment. In Proceedings of the SPIE 5438, Visual Information Processing XIII, Orlando, FL, USA, 15 July 2004.
18. Mikolajczyk, K.; Schmid, C.; Zisserman, A. Human detection based on a probabilistic assembly of robust part detectors. In *European Conference on Computer Vision,* Proceedings of the 8th European Conference on Computer Vision, Prague, Czech Republic, 11–14 May 2004; Springer: Berlin/Heidelberg, Germany; pp. 69–82.
19. Rudol, P.; Doherty, P. Human body detection and geolocalization for uav search and rescue missions using color and thermal imagery. In Proceedings of the 2008 IEEE Aerospace Conference, Montana, MT, USA, 1–8 March 2008.
20. Wu, J.; Zhou, G. Real-time uav video processing for quick-response to natural disaster. In Proceedings of the 2006 IEEE International Conference on Geoscience and Remote Sensing Symposium, Denver, CO, USA, 31 July–4 August 2006.
21. Suzuki, T.; Meguro, J.; Amano, Y.; Hashizume, T.; Hirokawa, R.; Tatsumi, K.; Sato, K.; Takiguchi, J.-I. Information collecting system based on aerial images obtained by a small uav for disaster prevention. In Proceedings of the 2007 International Workshop and Conference on Photonics and Nanotechnology, Pattaya, Thailand, 16–18 December 2007.
22. Xi, C.; Guo, S. Image target identification of uav based on sift. *Proced. Eng.* **2011**, *15*, 3205–3209.
23. Li, C.; Zhang, G.; Lei, T.; Gong, A. Quick image-processing method of uav without control points data in earthquake disaster area. *Trans. Nonferrous Metals Soc. China* **2011**, *21*, s523–s528. [CrossRef]
24. United Eagle Talon Day Fatso FPV Carrier. Available online: http://www.x-uav.cn/en/content/?463.html (accessed on 20 October 2016).
25. Hardkernel Co., Ltd. Ocam: 5mp USB 3.0 Camera. Available online: http://www.hardkernel.com/main/pro-ducts/prdt_info.php?g_code=G145231889365 (accessed on 20 October 2016).
26. Chen, Y.; Hsiao, F.; Shen, J.; Hung, F.; Lin, S. Application of matlab to the vision-based navigation of UAVs. In Proceedings of the 2010 8th IEEE International Conference on Control and Automation (ICCA), Xiamen, China, 9–11 June 2010.
27. Gopro hero4 Silver. Available online: http://shop.gopro.com/APAC/cameras/hero4-silver/CHDHY-401-EU.html (accessed on 20 October 2016).
28. Hero3+ Black Edition Field of View (FOV) Information. Available online: https://gopro.com/support/articles-/hero3-field-of-view-fov-information (accessed on 20 October 2016).

sensors

MDPI

Article

An Efficient Seam Elimination Method for UAV Images Based on Wallis Dodging and Gaussian Distance Weight Enhancement

Jinyan Tian [1], Xiaojuan Li [1], Fuzhou Duan [1,*], Junqian Wang [2] and Yang Ou [1]

1 Beijing Key Laboratory for Geographic Information Systems and Environment and Resources, Capital Normal University, Beijing 100048, China; 2130901016@cnu.edu.cn (J.T.); lixiaojuan@cnu.edu.cn (X.L.); 2150902042@cnu.edu.cn (Y.O.)

2 Department of Geography and Environmental Management, University of Waterloo, Waterloo, ON N2L 2R7, Canada; j563wang@uwaterloo.ca

* Correspondence: duanfuzhou@cnu.edu.cn; Tel.: +86-10-6890-3926; Fax: +86-10-6890-3052

Academic Editors: Felipe Gonzalez Toro and Antonios Tsourdos
Received: 22 March 2016; Accepted: 2 May 2016; Published: 10 May 2016

Abstract: The rapid development of Unmanned Aerial Vehicle (UAV) remote sensing conforms to the increasing demand for the low-altitude very high resolution (VHR) image data. However, high processing speed of massive UAV data has become an indispensable prerequisite for its applications in various industry sectors. In this paper, we developed an effective and efficient seam elimination approach for UAV images based on Wallis dodging and Gaussian distance weight enhancement (WD-GDWE). The method encompasses two major steps: first, Wallis dodging was introduced to adjust the difference of brightness between the two matched images, and the parameters in the algorithm were derived in this study. Second, a Gaussian distance weight distribution method was proposed to fuse the two matched images in the overlap region based on the theory of the First Law of Geography, which can share the partial dislocation in the seam to the whole overlap region with an effect of smooth transition. This method was validated at a study site located in Hanwang (Sichuan, China) which was a seriously damaged area in the 12 May 2008 enchuan Earthquake. Then, a performance comparison between WD-GDWE and the other five classical seam elimination algorithms in the aspect of efficiency and effectiveness was conducted. Results showed that WD-GDWE is not only efficient, but also has a satisfactory effectiveness. This method is promising in advancing the applications in UAV industry especially in emergency situations.

Keywords: UAV; Wallis dodging; seam elimination; Gaussian distance weight enhancement; earthquake

1. Introduction

The development of UAVs conforms to the current increasing demand for low-altitude very high resolution (VHR) remote sensing data [1–3]. Compared with the traditional photogrammetry process, the fast reconstitution of UAV image mosaics is a precondition of its application [4,5]. However, the UAV image-processing challenges include large geometric deformity, small size, large number and uneven exposure. These challenges lead to difficulties in seam elimination when mosaicking UAV images [6,7]. The mosaic seams mainly come from two sources: (1) the color or brightness differences due to the exposure variation; and (2) the texture misplacement due to geometric deformity, projection differences caused by tall landscapes and image capture position differences [8]. These two types of seams clearly appear on the UAV remote sensing platform, therefore, the effective and efficient removal of these seams is essential for the application of UAVs.

At present, the major methods of seam elimination are the seamline detection and image fusion methods. The seamline detection method should be considered as a way of circumventing the problem of tall landscapes in the images [9], and can be attributed to two categories: the first category is seamline search by the variation of gradient degree or image texture. Davis [10] proposed the optimal seamline searching method based on Dijkstra's algorithm, which relies mainly on the calculation of adjacency matrices and distance matrices of high algorithmic complexity [11]. Yuan [12] replaced the Dijkstra algorithm with a greedy algorithm for local optimal path selection. However, the algorithm was still influenced by iterative convergence. Kerschner [13] applied the twin snake operator to automatically select the image seamline. However, the operator cannot guarantee the systematical optimization. Chon [14] eliminated seamlines by dynamic planning stitching. The computational burden of the algorithm rises exponentially with the increase of seamline length [15]. The second category is applying ancillary data to detect the seamline. Wan [16] proposed an algorithm based on the vector path ancillary data, which is only suitable for a few systems and is significantly limited by the vector data. Zuo [17] applied the greedy snake algorithm with the assistance of the DSM method to detect seamlines. The algorithm is fairly complicated and highly dependent on the ancillary data. In conclusion, all these searching seamline algorithms applied on UAV images have three limitations: (1) they require high geometric accuracy of the UAV images, but UAV remote sensing platforms are rather instable and have low parameter accuracy. The equipped camera sensors cannot meet the accuracy requirements because they are not designed for photogrammetry; (2) All of them are complicated and time-consuming. UAV images are small in size but contain large amounts of data, which requires high processing efficiency; (3) Objects in UAV images are not overlapped in a regular manner. The seamlines are difficult to detect, especially for regions with high densities of tall buildings.

In addition to the seamline detection method, image fusion can also be applied to eliminate mosaic seams [18]. Uyttendael [19] applied a feathering and interpolating function based on weighted features to reduce the color difference. However, the feathering algorithm tends to give fuzzy edges when smoothing the exposure difference, and can sometimes lead to the "ghosting" effect. Szeliski [20,21] manually selected at least four pairs of feature points, and estimated the variation of images with the function built on the variation of pixel difference of the feature points, which achieved a satisfactory layer fusion effect. However, since the estimation is based on brightness differences, it is highly sensitive to the brightness of images and can be poorly automated [22]. Su [23] proposed an image fusion method based on wavelet multi-scale decomposition. This method first applies wavelet multi-scale decomposition over the source images. Then, the wavelet weight parameters are determined and the images are reconstructed through inverse wavelet transform. The algorithm is highly complicated and it is difficult to determine wavelet parameters [24]. Zomet [25] eliminated mosaic seams by analyzing the contrast in smooth stitching areas. However, the field smoothing can lead to the appearance of "ghosting" effects [11,26]. Tian [27] developed a brightness and texture seam elimination (BTSE) method with a smooth transition effect on a one-dimensional direction in the overlap region. A "ghosting" effect tends to appear at the border when the algorithm is applied to UAV images with the large geometric deformity. In conclusion, all these image fusion methods for UAV images have two major limitations: (1) a "ghosting" effect tends to appear due to the uneven exposure and the large geometric deformity of UAV images; (2) they are fairly complicated and require long computation times, which conflicts with the fact that UAV systems require high data processing efficiency to deal with the massive amount of image data.

Therefore, the objective of this study is twofold: firstly, to adjust the difference of brightness between the two matched images with the Wallis dodging method and; secondly, to develop a new image fusion algorithm to eliminate the texture seamline based on the First Law of Geography.

2. Study Site and Data

The study site is located in Hanwang (104°09′E to 104°12′E and 31°25′N to 31°28′N) in the northwestern part of the Sichuan Basin (China) and has an overall area of 54.3 km^2. It is a city at the foot of mountains with an average elevation of 685 m above sea level and slopes of less than 5°. As an industrial city, it has a sound transportation system and a total population of 53,000, among which the non-agricultural population is 35,000 [28,29]. The major land uses of this study site are woodland, farmland, water, road, and buildings. In this task, UAV image data were acquired on 15 May 2008 after the 5.12 Wenchuan Earthquake. The flight altitude and speed of the UAV platform are 400 m and 50 km/h, respectively. The major parameters of the image sensor equipped on the UAV platform are shown in Table 1. A total of 678 images were acquired with an image resolution of 0.3 m. The average forward overlap is 70% and the side forward overlap is 40%.

Table 1. The parameters of the image sensor.

Items	Parameters
Image Sensor	Ricoh Digital
Pixel Number	3648 × 2736
Focal Distance	28 mm
CCD	1/1.75 inch
Navigation sensor	GPS
Image Format	JPEG

3. Methodology

3.1. Wallis Dodging

Image processing before image fusion contains two major steps: image matching and image dodging. Image matching aims to find corresponding points, and image dodging was used to eliminate the brightness differences between two matched images. First, in order for us to find the corresponding points between two images, an image matching method should be applied. In this study, the Scale-Invariant Feature Transform (SIFT) algorithm was used to match the two images [30,31], which consists of four stages: (1) building the scale-space; (2) keypoint localization; (3) removal of bad keypoints; and (4) keypoint description. It has been proven in many studies [32,33] that SIFT not only performs well in image rotation, scale zoom and illumination changes, but also does well in affine transformation, and noise jamming. Subsequently, the Random Sample Consensus (RANSAC) method was applied to the points matched by SIFT to remove any mismatched points [34]. Additionally, the Wallis dodging algorithm [8,35,36] was employed to adjust the difference of brightness between the two matched images before the texture seam elimination method.

The principle behind Wallis image dodging is that it can adjust the variance and mean value of the target image to the reference image's level. The Wallis filter can be defined by Equation (1):

$$I_{ij} = (I_{ij}^2 - \overline{I^2}) \times \frac{c\sigma_{I^1}}{c\sigma_{I^2} + (1-c)\sigma_{I^1}} + b\overline{I^1} + (1-b)\overline{I^2} \tag{1}$$

where I^1 is reference image, I^2 is target image, and I_{ij} is the pixel value of I^2 in i row, j column after image dodging. $\overline{I^1}$, $\overline{I^2}$ and σ_{I^1}, σ_{I^2}, are the mean and variance value of I^1 and I^2, respectively; $c\in[0.1]$ is an adjustment coefficient for variance value of the image, and $b\in[0.1]$ is an adjustment coefficient for the mean value. However, setting the two specific parameters is still a critical question in the existing research. The parameter setting method was derived in this study. First, the variance of the target image is shown in Equation (2):

$$\sigma_{I^2} = \sqrt{\sum_{i=0}^{m-1}\sum_{j=0}^{n-1} (I_{ij}^2 - \overline{I^2})^2 / m \times n} \tag{2}$$

Second, the variance and mean value of the target image was adjusted to the reference image's level. So the variance and mean value of the target image after image dodging should be roughly equal to σ_{I^1} and $\overline{I^1}$, respectively. Therefore, they can be denoted as Equation (3):

$$\sigma_{I^1} \approx \sqrt{\sum_{i=0}^{m-1}\sum_{j=0}^{n-1}(I_{ij}-\overline{I^1})^2/m \times n} \tag{3}$$

Third, both sides of the Equation (3) are multiplied by $\sigma_{I^2}/\sigma_{I^1}$:

$$\sigma_{I^2} \approx \frac{\sigma_{I^2}}{\sigma_{I^1}} \times \sqrt{\sum_{i=0}^{m-1}\sum_{j=0}^{n-1}(I_{ij}-\overline{I^1})^2/m \times n} \tag{4}$$

Then, simultaneous application of Equations (2) and (4) gives:

$$\frac{\sigma_{I^2}}{\sigma_{I^1}} \times (I_{ij} - \overline{I^1}) \approx I_{ij}^2 - \overline{I^2} \tag{5}$$

Finally, the pixel value of target image after image dodging is shown in Equation (6):

$$I_{ij} \approx \frac{\sigma_{I^1}}{\sigma_{I^2}} \times (I_{ij}^2 - \overline{I^2}) + \overline{I^1} \tag{6}$$

Comparing Equation (1) with Equation (6), it found that we will get Equation (6) when the parameters (b and c) were both set to 1 in Equation (1). Therefore, to adjust the mean and variance value of target image to reference image's level, Equation (6) with Wallis filter (b = 1, c = 1) was used for UAV image dodging.

3.2. GDWE Method

3.2.1. Theoretical Basis

The First Law of Geography proposed by Waldo Tobler in 1970 is "all attribute values on a geographic surface are related to each other, but closer values are more strongly related than are more distant ones" [37]. The law is the foundation of the fundamental concepts of spatial autocorrelation and spatial dependence [38], based on which we have developed an effective and efficient seamline elimination method (GDWE) for UAV image. The principle of GDWE is an image fusion algorithm combining relevant information from two matched UAV images into a single image in the overlapping region. As such, GDWE embraces three major steps: first, the principal point of each image was set as the optimal pixel with the minimum geometric distortion because the image sensor equipped on UAV platform is, in general, a type of non-measurement array CCD camera. Second, the weight in a certain pixel contributed by each image in the overlap region was determined by the distance between the pixel and the principal point. A two-dimensional Gaussian kernel was then employed to describe it. Third, in order to enhance the influence of distance to the weight, an exponent form adjustment coefficient was introduced and it was parameterized by a sensitive analysis method.

3.2.2. Seam Elimination

To develop the algorithm for image fusion in the overlap region of the matched UAV images, some parameters should be defined first, in which the principle points of the two matched images were O_1 and O_2; O is an arbitrary point in the overlap region; $d_1(|O - O_1|)$ and $d_2(|O - O_2|)$ are the distances between O_1, O_2 and O; The pixel values of point O in the two matched UAV images are I_{ij}^1 and I_{ij}^2. The pixel value of point O after image fusion is I_{ij}. Therefore, I_{ij} can be defined as $\omega_1 \times I_{ij}^1 + \omega_2 \times I_{ij}^2$, where w_1, w_2 are the weight contributions of the two UAV images to point O, and w_1 plus w_2 is equal to 1. Based on the theory mentioned above, a Gaussian kernel shown in Figure 1 was introduced to describe the Gaussian distance weight distribution (G_{w1}), and was defined by Equation (1):

$$G_{w_1} = a \times e^{-(|O-O_1|/|O-O_2|)^2/2\sigma^2} \tag{7}$$

where a was set to 1 because G_{w1} should be equal to 1 when d_1 is 0. In order to enhance the influence of Gaussian distance on the weight, an exponent form adjustment coefficient (λ) was introduced into Equation (2):

$$w_1 = e^{-(|O-O_1|/|O-O_2|)^{2\lambda}/2\sigma^2} \tag{8}$$

In which w_1 was set to 0.5 when d_1 equals d_2. When we apply the relationship to Equation (8), we get:

$$\sigma = \sqrt{1/(2 \times \ln 2)} \tag{9}$$

Therefore, including these terms in Equation (8) results in Equation (9), the pixel value was defined by Equation (10):

$$I_{ij} = \left(0.5^{(|O-O_1|/|O-O_2|)^{2\lambda}}\right) \times I_{ij}^1 + \left(1 - 0.5^{(|O-O_1|/|O-O_2|)^{2\lambda}}\right) \times I_{ij}^2 \tag{10}$$

Finally, we named our method Wallis dodging and Gaussian distance weight enhancement (WD-GDWE) when taking the Wallis dodging algorithm into consideration. It is shown in Equation (11):

$$I_{ij} = \left(0.5^{(|O-O_1|/|O-O_2|)^{2\lambda}}\right) \times I_{ij}^1 + \left(1 - 0.5^{(|O-O_1|/|O-O_2|)^{2\lambda}}\right) \times \left(\frac{\sigma_{I^1}}{\sigma_{I^2}} \times (I_{ij}^2 - \overline{I^2}) + \overline{I^1}\right) \tag{11}$$

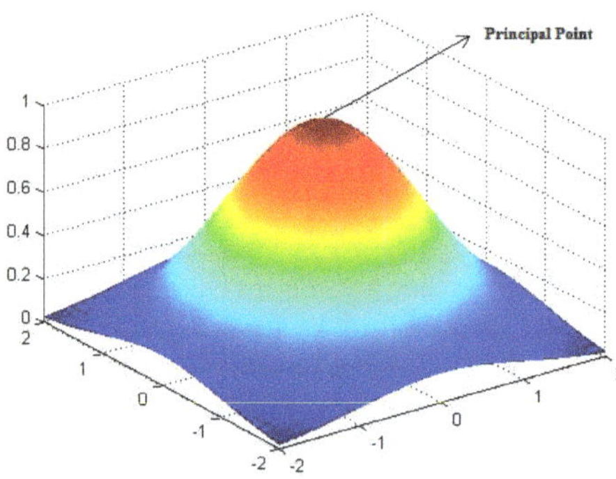

Figure 1. An example of a two-dimensional Gaussians distance weight distribution kernel.

4. Results and Discussion

4.1. Wallis Dodging

To assess the efficiency and effectiveness of WD-GDWE for seamline elimination of UAV images, the method was implemented with Visual C++ programming using 8 GB memory and an Intel Xeon 2.5 GHz CPU. The UAV images covering five different types of land use (woodland, farmland, water, road, and buildings) from the study site were tested.

Figure 2 shows the results of stacking directly *versus* stacking after Wallis dodging for two matched UAV images covering five different types of land use. From the perspective of visual effects, the results indicate that the brightness difference of two matched images has been effectively balanced by Wallis dodging, in which the left figure of each figure group in Figure 2 was stacked directly and the right figure was stacked after Wallis dodging. The root mean-square error (RMSE) values of the mean and standard deviation were calculated from the two matched UVA images in the overlap region for direct stacking and Wallis dodging, respectively. For each type of land use, at least 36 pairs of matched images were tested, and the averages of the RMSE values were recorded in Table 2. The results show that the Wallis dodging method can effectively balance the brightness differences between the two matched images, in which the RMSE of mean and stand deviation were determined to be 0.0 and less than 0.3, respectively.

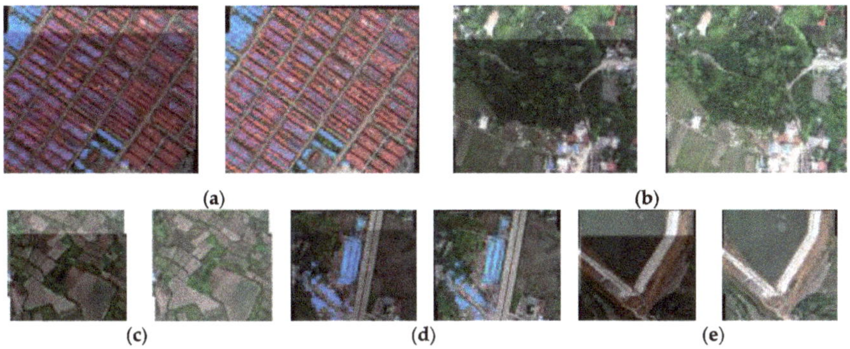

Figure 2. The results of Wallis dodging for two matched UAV images of each type land use, in which (**a**)–(**e**) correspond to buildings, woodland, farmland, road, and water, respectively. For example, in the case of (**a**), the left figure was the direct stacking result of two matched images, whereas the right figure was the stacking result of two matched images after Willis dodging.

Table 2. Average of RMSE values of mean (M) and standard deviation (SD) calculated from the matched UVA images for stacking directly and Wallis dodging, respectively, in each type of land use.

Land Use		RMSE	
		M	SD
Building	Stacking Directly	24.5	6.5
	Wallis Dodging	0.0	0.2
Woodland	Stacking Directly	23.6	6.2
	Wallis Dodging	0.0	0.1
Farmland	Stacking Directly	19.8	5.7
	Wallis Dodging	0.0	0.1
Road	Stacking Directly	17.5	3.6
	Wallis Dodging	0.0	0.1
Water	Stacking Directly	36.2	9.5
	Wallis Dodging	0.0	0.3

4.2. WD-GDWE Method

To acquire the optimal adjustment coefficient (λ) for the WD-GDWE method, a series of values from zero to five with a step size of 0.2 were set, based on which the optimal value of λ was determined when the lowest RMSE between the test images and reference images was achieved. In this study, the optimal value of λ was set to 2.6. Lastly, performance comparisons between WD-GDWE and five other classical seamline elimination algorithms were conducted in terms of efficiency and effectiveness. The specific five classical methods are: Tian's BTSE algorithm, Uyttendael's feathering algorithm, Su's Wavelet algorithm, Szeliski's algorithm, and Davis's Dijkstra algorithm, in which the first four methods are based on image fusion and the last one is based on seamline detection. Generally, the image quality assessment indicators for seamline elimination can be divided to three types [39–43]: (1) amount of information: information entropy, standard deviation, cross entropy, signal to noise ratio, and joint entropy [44,45]; (2) image quality: average gradient and wavelet energy ratio [46]; (3) spectral information reserved: RMSE, standard deviation, deviation, and spectral distortion; Taking all three types of indicators into consideration, information entropy, average gradient, and RMSE were selected to access the specific five methods of seamline elimination, respectively. In addition, processing time is also an indicator for evaluating the efficiency of the algorithm. It should be noted that orthoimages were severed as reference images of the RMSE, which was produced from the control points recorded by artificial with the help of a differential GPS.

From the perspective of visual effects, Figure 3 shows the performance comparisons of the five seamline elimination methods, in which Figure 3a is the direct stacking result, Figure 3b is the WD-GDWE method result, Figure 3c–g is the results of the other five different seamless methods, respectively.

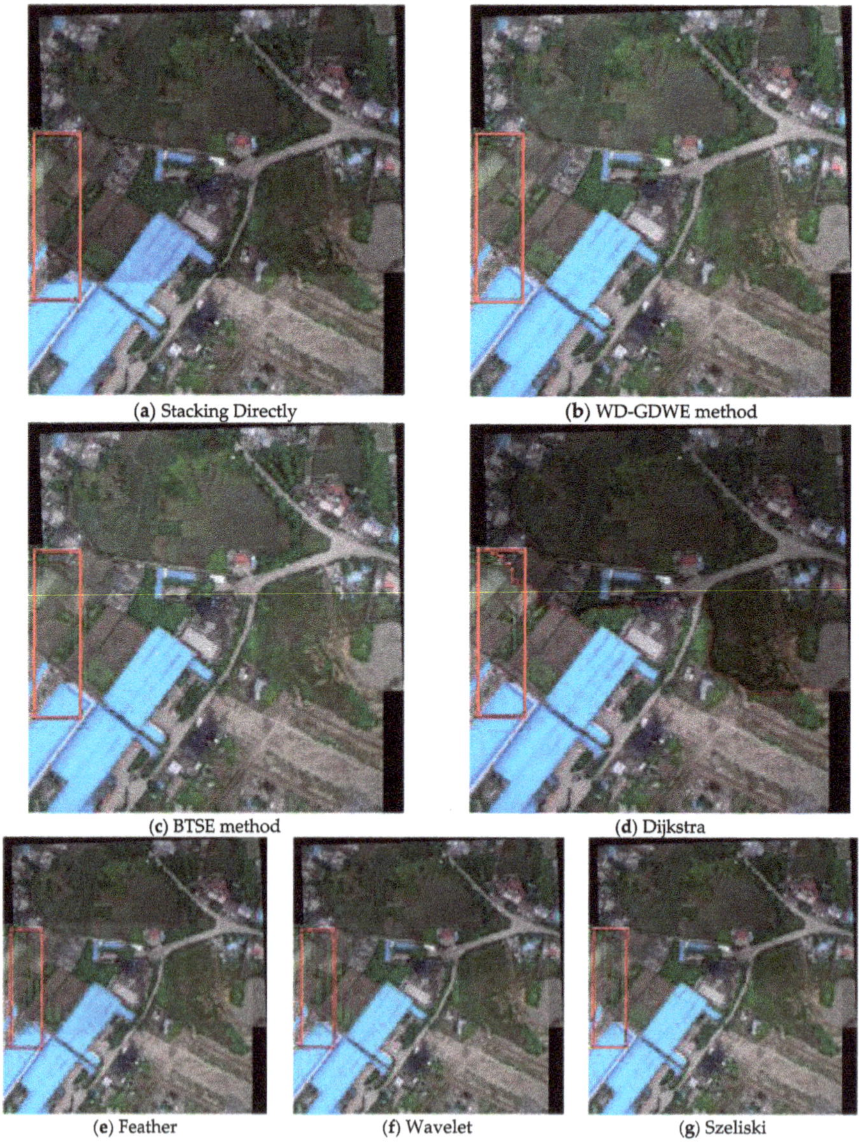

Figure 3. The performance comparison of different seam elimination algorithms.

Comparing Figure 3a,b, we find that the buildings and the roads obviously display mosaic dislocation, whereas the phenomenon has been greatly improved with the WD-GDWE method. The performance comparisons of the five seamline elimination methods shown in Figure 3c–f indicate that: (1) a "ghosting" effect tends to appear in the Feather, Wavelet, Szeliski, and BTSE algorithms; (2) the visual effects of the Dijkstra algorithm and WD-GDWE are much better than those of the other methods. From the perspective of image quality assessment indicators, the details of the performance comparisons of the six methods were shown Figure 4. Each of the four indicators is an average value calculated from lots of UAV images (at least 36 pairs) for each type of land use. Figure 4a,b show that Dijkstra method gives the most abundant amount of information and the highest definition, and the WD-GDWE method follows. The BTSE is worse than the WD-GDWE method at the border of the fusion image because it only supports smooth transitions in a one-dimensional direction in the overlap region. Considering the improvement of WD-GDWE from BTSE is not obvious in Figure 3 from the perspective of visual effects, therefore, some experimental results at the border of the fusion images with the two methods were added (Figure 5). The Wavelet and Szeliski algorithm are much worse than the BTSE method, and the Feather algorithm is the worst one. Figure 4c shows that the WD-GDWE method preserves more spectral information than the other four algorithms. Figure 4d shows that it takes a little time to run the WD-GDWE, BTSE, Szeliski, and Feather algorithms, whereas the Dijkstra and Wavelet method are time-consuming. In a word, the WD-GDWE method is not only efficient, but also has a satisfactory effectiveness.

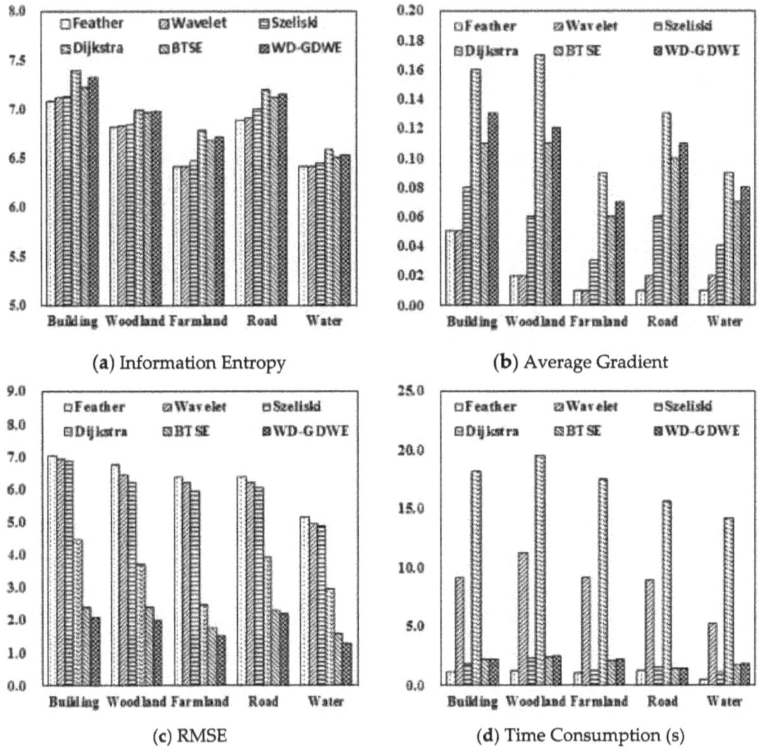

(**a**) Information Entropy

(**b**) Average Gradient

(**c**) RMSE

(**d**) Time Consumption (s)

Figure 4. Comparisons of different elimination seam algorithms. (**a**) Information entropy for describing the amount of information; (**b**) verage gradient to access the image qualities; (**c**) RMSE between the specific five methods with the orthoimages; (**d**) time consumption of the six methods.

(a) (b)

Figure 5. Both (**a**) and (**b**) are the results at the border of the fusion images, in which the left one in (**a**) or (**b**) is with the BTSE method and the right one used the WD-GDWE method.

5. Conclusions

In this study, an efficient seam elimination method for UAV images based on Wallis dodging and Gaussian distance weight enhancement was proposed. The method has successfully tested by using UAV images acquired after the 5.12 Wenchuan Earthquake. By comparison with other five classical seam elimination methods, the conclusions from this study can be summarized as follows: (1) the WD-GDWE method can effectively adjust the brightness differences between two matched images; (2) the method can successfully eliminate the texture mosaic seams which are usually caused by geometric deformity, projection differences, and image capture position differences on UAV platforms; (3) the WD-GDWE method is highly-efficient, which can meet the high processing speed requirements of massive UAV images. Time-savings are very important in advancing the applications in the UAV industry, especially in emergency situations. The results of this study can be further extended to other fields, such as aerospace remote sensing and computer vision.

Acknowledgments: We acknowledge financial support from the National Natural Science Foundation of China (No.41130744/D0107 and 41171335/D010702). The authors thank Donghai Xie for his valuable comments to the paper. We are also deeply indebted to the reviewers and editors for the thoughtful and constructive comments on improving the manuscript.

Author Contributions: Jinyan Tian carried out the analyses and wrote the manuscript. Xiaojuan Li and Fuzhou Duan were responsible for recruitment the participants, and design the experiments. Junqian Wang and Yang Ou were responsible for data processing. All the authors drafted the manuscript, and approved the final manuscript.

Conflicts of Interest: The authors declare no conflict of interest.

References

1. Mesas-Carrascosa, F.J.; Rumbao, I.C.; Berrocal, J.A. Porras AG Positional quality assessment of orthophotos obtained from sensors onboard multi-rotor UAV platforms. *Sensors* **2014**, *14*, 22394–22407. [CrossRef] [PubMed]
2. Xu, Y.; Ou, J.; He, H.; Zhang, X.; Mills, J. Mosaicking of Unmanned Aerial Vehicle Imagery in the Absence of Camera Poses. *Remote Sens.* **2016**, *8*, 204. [CrossRef]
3. Karpenko, S.; Konovalenko, I.; Miller, A.; Miller, B.; Nikolaev, D. UAV Control on the Basis of 3D Landmark Bearing-Only Observations. *Sensors* **2015**, *15*, 29802–29820. [CrossRef] [PubMed]
4. Wang, Y.C.; Liu, J.G. Evaluation methods for the autonomy of unmanned systems. *Chin. Sci. Bull.* **2012**, *57*, 3409–3418. [CrossRef]

5. Gonzalez, L.F.; Montes, G.A.; Puig, E.; Johnson, S.; Mengersen, K.; Gaston, K.J. Unmanned Aerial Vehicles (UAVs) and Artificial Intelligence Revolutionizing Wildlife Monitoring and Conservation. *Sensors* **2016**, *16*, 97. [CrossRef] [PubMed]
6. Zhou, G. Near real-time orthorectification and mosaic of small UAV video flow for time-critical event response. *IEEE Trans. Geosci. Remote Sens.* **2009**, *47*, 3. [CrossRef]
7. Wehrhan, M.; Rauneker, P.; Sommer, M. UAV-Based Estimation of Carbon Exports from Heterogeneous Soil Landscapes—A Case Study from the CarboZALF Experimental Area. *Sensors* **2016**, *16*, 255. [CrossRef] [PubMed]
8. Sun, M.W.; Zhang, J.Q. Dodging research for digital aerial images. *Int. Arch. Photogramm. Remote Sens. Spat. Inf. Sci.* **2008**, *37*, 349–353.
9. Choi, J.; Jung, H.S.; Yun, S.H. An Efficient Mosaic Algorithm Considering Seasonal Variation: Application to KOMPSAT-2 Satellite Images. *Sensors* **2015**, *15*, 5649–5665. [CrossRef] [PubMed]
10. Davis, J. Mosaics of scenes with moving objects. In Proceedings of the 1998 IEEE Computer Society Conference on Computer Vision and Pattern Recognition, Santa Barbara, CA, USA, 23–25 June 1998.
11. Philip, S.; Summa, B.; Tierny, J.; Bremer, P.T.; Pascucci, V. Distributed Seams for Gigapixel Panoramas. *IEEE Trans. Vis. Comput. Graph.* **2015**, *21*, 350–362. [CrossRef] [PubMed]
12. Yuan, X.X.; Zhong, C. An improvement of minimizing local maximum algorithm on searching Seam line on searching seam line for orthoimage mosaicking. *Acta Geod. Cartgraph. Sin.* **2012**, *41*, 199–204.
13. Kerschner, M. Seamline detection in colour orthoimage mosaicking by use of twin snakes. *ISPRS J. Photogramm. Remote Sens.* **2001**, *56*, 53–64. [CrossRef]
14. Chon, J.; Kim, H.; Lin, C.S. Seam-line determination for image mosaicking: A technique minimizing the maximum local mismatch and the global cost. *ISPRS J. Photogramm. Remote Sens.* **2010**, *65*, 86–92. [CrossRef]
15. Mills, S.; McLeod, P. Global seamline networks for orthomosaic generation via local search. *ISPRS J. Photogramm. Remote Sens.* **2013**, *75*, 101–111. [CrossRef]
16. Wan, Y.; Wang, D.; Xiao, J.; Lai, X.; Xu, J. Automatic determination of seamlines for aerial image mosaicking based on vector roads alone. *ISPRS J. Photogramm. Remote Sens.* **2013**, *76*, 1–10. [CrossRef]
17. Zuo, Z.Q.; Zhang, Z.X.; Zhang, J.Q. Seam line intelligent detection in large urban orthoimage mosaicking. *Acta Geod. Cartgraph. Sin.* **2011**, *40*, 84–89.
18. Borra-Serrano, I.; Peña, J.M.; Torres-Sánchez, J.; Mesas-Carrascosa, F.J.; López-Granados, F. Spatial Quality Evaluation of Resampled Unmanned Aerial Vehicle-Imagery for Weed Mapping. *Sensors* **2015**, *15*, 19688–19708. [CrossRef] [PubMed]
19. Uyttendaele, M.; Eden, A.; Skeliski, R. Eliminating ghosting and exposure artifacts in image mosaics. In Proceedings of the CVPR 2001 IEEE Computer Society Conference on Computer Vision and Pattern Recognition, Kauai, HI, USA, December 11–14 2001.
20. Szeliski, R. Video mosaics for virtual environments. *Comput. Graph. Appl.* **1996**, *16*, 22–30. [CrossRef]
21. Szeliski, R.; Shum, H.Y. Creating full view panoramic image mosaics and environment maps. In Proceedings of the 24th Annual Conference on Computer Graphics and Interactive Techniques, Los Angeles, CA, USA, 3–8 August 1998.
22. Szeliski, R. *Computer Vision: Algorithms and Applications*; Springer Science & Business Media: London, UK, 2010.
23. Su, M.S.; Hwang, W.L.; Cheng, K.Y. Analysis on multiresolution mosaic images. *IEEE Trans. Image Proc.* **2004**, *13*, 952–959. [CrossRef]
24. Gracias, N.; Mahoor, M.; Negahdaripour, S.; Gleason, A. Fast image blending using watersheds and graph cuts. *Image Vis. Comput.* **2009**, *27*, 597–607. [CrossRef]
25. Zomet, A.; Levin, A.; Peleg, S.; Weiss, Y. Seamless image stitching by minimizing false edges. *IEEE Trans. Image Proc.* **2006**, *15*, 969–977. [CrossRef]
26. Avidan, S.; Shamir, A. Seam carving for content-aware image resizing. *ACM Trans. Gr.* **2007**. [CrossRef]
27. Duan, F.Z.; Li, X.; Qu, X.; Tian, J.; Wang, L. UAV image seam elimination method based on Wallis and distance weight enhancement. *J. Image Graph.* **2014**, *19*, 806–813.
28. Zhang, J.; Deng, W. Multiscale Spatio-Temporal Dynamics of Economic Development in an Interprovincial Boundary Region: Junction Area of Tibetan Plateau, Hengduan Mountain, Yungui Plateau and Sichuan Basin, Southwestern China Case. *Sustainability* **2016**, *8*, 215. [CrossRef]

29. Chen, F.; Guo, H.; Ishwaran, N.; Zhou, W.; Yang, W.; Jing, L.; Cheng, F.; Zeng, H. Synthetic aperture radar (SAR) interferometry for assessing Wenchuan earthquake (2008) deforestation in the Sichuan giant panda site. *Remote Sens.* **2014**, *6*, 6283–6299. [CrossRef]
30. Lowe, D.G. Object recognition from local scale-invariant features. In Proceedings of the Seventh IEEE International Conference on Computer Vision, Kerkyra, Greek, 20–27 September 1999.
31. Lowe, D.G. Distinctive image features from scale-invariant keypoints. *Int. J. Comput. Vis.* **2004**, *60*, 91–110. [CrossRef]
32. Zhou, R.; Zhong, D.; Han, J. Fingerprint identification using SIFT-based minutia descriptors and improved all descriptor-pair matching. *Sensors* **2013**, *13*, 3142–3156. [CrossRef] [PubMed]
33. Lingua, A.; Marenchino, D.; Nex, F. Performance analysis of the SIFT operator for automatic feature extraction and matching in photogrammetric applications. *Sensors* **2009**, *9*, 3745–3766. [CrossRef] [PubMed]
34. Civera, J.; Grasa, O.G.; Davison, A.J.; Montiel, J.M.M. 1-Point RANSAC for extended Kalman filtering: Application to real-time structure from motion and visual odometry. *J. Field Robot.* **2010**, *27*, 609–631. [CrossRef]
35. Li, D.R.; Wang, M.; Pan, J. Auto-dodging processing and its application for optical remote sensing images. *Geom. Inf. Sci. Wuhan Univ.* **2006**, *6*, 183–187.
36. Pham, B.; Pringle, G. Color correction for an image sequence. *Comput. Graph. Appl.* **1995**, *15*, 38–42. [CrossRef]
37. Tobler, W.R. A computer movie simulating urban growth in the Detroit region. *Econ. Geogr.* **1970**, *46*, 234–240. [CrossRef]
38. Kemp, K. *Encyclopedia of Geographic Information Science*; SAGE: Thousand Oaks, CA, USA, 2008; pp. 146–147.
39. Efros, A.A.; Freeman, W.T. Image quilting for texture synthesis and transfer. In Proceedings of the 28th Annual Conference on COMPUTER Graphics and Interactive Techniques, Los Angeles, CA, USA, 12–17 August 2001.
40. Kwatra, V.; Schödl, A.; Essa, I.; Turk, G.; Bobick, A. Graphcut textures: image and video synthesis using graph cuts. *ACM Trans. Graph.* **2003**. [CrossRef]
41. Kuang, D.; Yan, Q.; Nie, Y.; Feng, S.; Li, J. Image seam line method based on the combination of dijkstra algorithm and morphology. *SPIE Proc.* **2015**. [CrossRef]
42. Pan, J.; Wang, M.; Li, D.; Li, J. Automatic generation of seamline network using area Voronoi diagrams with overlap. *IEEE Trans. Geosci. Remote Sens.* **2009**, *47*, 1737–1744. [CrossRef]
43. Agarwala, A.; Donteheva, M.; Agarwala, M.; Drucker, M.; Colburn, A.; Curless, B.; Salesin, D.; Cohen, M. Interactive digital photomontage. *ACM Trans. Graph.* **2004**, *23*, 294–302. [CrossRef]
44. Brown, M.; Lowe, D.G. Automatic panoramic image stitching using invariant features. *Int. J. Comput. Vis.* **2007**, *74*, 59–73. [CrossRef]
45. Summa, B.; Tierny, J.; Pascucci, V. Panorama weaving: fast and flexible seam processing. *ACM Trans. Graph. (TOG)* **2012**, *31*, 83. [CrossRef]
46. Shum, H.Y.; Szeliski, R. Systems and experiment paper: Construction of panoramic image mosaics with global and local alignment. *Int. J. Comput. Vis.* **2000**, *36*, 101–130. [CrossRef]

sensors

MDPI

Article

Diverse Planning for UAV Control and Remote Sensing

Jan Tožička * and Antonín Komenda

AI Center, Department of Computer Science, Faculty of Electrical Engineering, Czech Technical University in Prague, 166 27 Praha 6, Czech Republic; antonin.komenda@fel.cvut.cz

* Correspondence: jan.tozicka@fel.cvut.cz; Tel.: +420-224-357-657

Academic Editors: Felipe Gonzalez Toro and Antonios Tsourdos

Received: 6 October 2016; Accepted: 15 December 2016; Published: 21 December 2016

Abstract: Unmanned aerial vehicles (UAVs) are suited to various remote sensing missions, such as measuring air quality. The conventional method of UAV control is by human operators. Such an approach is limited by the ability of cooperation among the operators controlling larger fleets of UAVs in a shared area. The remedy for this is to increase autonomy of the UAVs in planning their trajectories by considering other UAVs and their plans. To provide such improvement in autonomy, we need better algorithms for generating alternative trajectory variants that the UAV coordination algorithms can utilize. In this article, we define a novel family of multi-UAV sensing problems, solving task allocation of huge number of tasks (tens of thousands) to a group of configurable UAVs with non-zero weight of equipped sensors (comprising the air quality measurement as well) together with two base-line solvers. To solve the problem efficiently, we use an algorithm for diverse trajectory generation and integrate it with a solver for the multi-UAV coordination problem. Finally, we experimentally evaluate the multi-UAV sensing problem solver. The evaluation is done on synthetic and real-world-inspired benchmarks in a multi-UAV simulator. Results show that diverse planning is a valuable method for remote sensing applications containing multiple UAVs.

Keywords: diverse planning; UAV; remote sensing

---

1. Introduction

Measuring air quality has been historically performed by ground stations. Later on, manned aircraft and satellites were used to collect necessary measurements. Unfortunately, airborne and satellite sensors are very costly which prevents their daily use. Most recently, remotely controlled unmanned aerial vehicles (UAVs) equipped with different sensors are being used to get up-to-date information with higher spatial and temporal resolution at reasonable equipment price. The use of UAVs for air quality monitoring is getting more and more attention from both the research community and industry. The most common usage of UAVs are air pollution and emission monitoring [1], climate change monitoring [2], emergency response [3], disaster monitoring (e.g., forest fires [4,5] or chemical factory explosions [6], etc.), area monitoring [7], or wildlife monitoring and protection [8,9]. In this work, we focus on autonomous UAVs which could collect required data in a coordinated manner without any human aid.

The reason to move from remotely controlled UAVs to autonomous UAVs is the fact that human operators (pilots) seem to be a bottleneck of the system when several UAVs collaborate on a single mission [10]. Each operator or a team of operators is responsible for one UAV and controls its actions. The human operators communicate among themselves and coordinate their actions in order to achieve a common goal. Such an approach has its limits in the human interactions and human control of the UAVs. Therefore, one of the main goals of research tackling UAVs is to improve management of the

UAVs such that an operator or a group of operators can control larger groups of UAVs easily. This can be achieved by two means:

- improve human–machine interface (HMI),
- increase UAV autonomy.

In this article we provide a follow-up to our previous work in [10–13] on advanced Human-Machine Interfaces (HMIs) using planning of alternatives aimed at the first approach and explore solutions tackling the second approach with the help of planning of alternative plans as well. We have already demonstrated how multiagent control algorithms can be used to control multiple UAVs [14]; therefore, the method proposed in this article also continues in this direction and provides methods of improving UAV planning capability by utilization of planning for alternatives.

An illustrative example of planning alternatives for trajectory planning is shown in Figure 1. There are two UAVs and six waypoints that need to be visited by either UAV. Planning of alternatives can propose several possible solutions to the task. Two of them are shown in the figure.

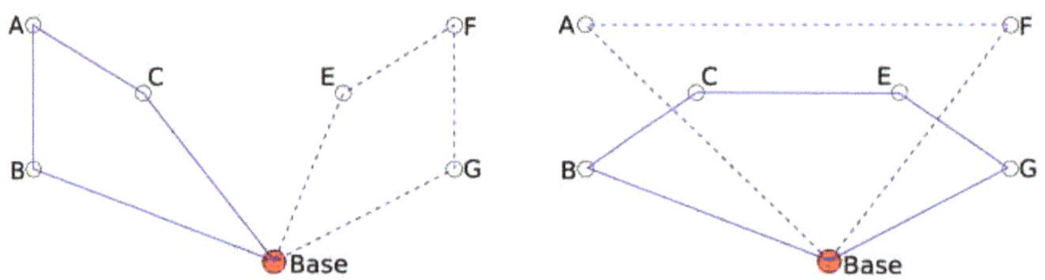

Figure 1. Example of planning alternative trajectories of two unmanned aerial vehicles (UAVs) for a task of covering a set of waypoints (A–G). Two possible solutions are shown for both UAVs (solid and dashed lines).

Unlike in the case of making alternative plans for human operators, where the utility function defining the quality of the solution is unknown (or only implicitly known only to the operator) in the case of fully autonomous UAVs, the utility function is known but the optimization problem is too complex to be solved optimally. Our proposed approach here is to use planning of alternatives to provide a diverse set of trajectories out of which final trajectories for all the UAVs are chosen. Since the set of created diverse trajectories is processed automatically, the size of the created set can be much bigger than when we want a human operator to choose. We named out approach *diverse planning*.

Although the solution is not limited to it, our task in this work is to monitor different air pollutants across a city. Even though the dense monitoring is necessary to detect sources of the pollution, it is rather an overkill for uniform continuous monitoring. Different pollutants are required to be monitored at different locations. For example, near a main road junctions, it is necessary to monitor gases produced during combustion: carbon dioxide (CO_2), methane (CH_4), and nitrous oxide (N_2O), while near schools it is necessary to monitor ultrafine particles [15], carbon monoxide (CO), and sulfur dioxide (SO_2), which negatively affect human health [16].

A UAV or a team of UAVs can be also used for monitoring of remote and inaccessible areas, as proposed, e.g., in [1,17–19]. In our case, we want to monitor a city (particularly, we are using simulation for the city of Prague), which requires low altitude flights (disallowing monitoring by conventional aircraft). There is a full range of UAVs which could be used for this task (Provided that the legal and regulation issues [20,21] are solved). These UAVs differ in their sizes, range of flights, payload and power capacities, speeds, etc. In this study we do not focus on any particular UAV and the proposed method can be easily used for any type of UAV by correctly setting few basic parameters. For our purpose we could use, for example, *Meteorological Mini-UAV* (*M2AV*) developed at the Institute of Aerospace Systems, Technical University of Braunschweig, Germany. The maximum take-off weight is 4.5 kg, including 1.5 kg of payload, with the range of 60 km at a cruising speed 20 ms^{-1}.

The presented task here is a real-world-inspired application called Multi-UAV Sensing Problem (MUSP). The goal is to gather air quality data for a large area using a group of UAVs. Different types of locations are required to be monitored by different sensors. Each UAV can be equipped with multiple sensors, but the weight of each mounted sensor negatively affects the fuel amount which can be carried and thus limits UAV flight range. As the authors in [1] mentions:

> [R]ealistically even the lightest onboard sensors would add some weight. The heavier the payload the less fuel can be added, which reduces flight duration.

This overview article implies that our algorithm is the first one considering non-zero sensors weights at such a scale. The solution of the MUSP has to specify which sensors have to be mounted on which UAV and also plan the flight trajectory which has to be shorter with each mounted sensor. The quality of the solution is measured by the total number of performed measurements.

In MUSP, we can use diverse planning to provide a set of diverse trajectories out of which the most suitable trajectories for the UAVs are chosen. This approach allows us to balance the quality of the solution and the required computational time, which is necessary for large-scale applications.

The article is structured as follows. In Section 2, we formally define the problem of multi-UAV coordination for remote sensing with two base-line solutions using the classical and greedy planning techniques. In Section 3, we present the novel algorithm DivPlan based on diverse planning techniques providing efficient solution to the defined family of remote sensing problems. In Section 4, we provide a complexity analysis for the three algorithms. Finally, we experimentally compare the proposed algorithm with the two base-line approaches in Section 5 and also evaluate the algorithm in simulation of a real-world problem.

2. Diverse Planning for Multi-UAV Coordination

The diverse planning techniques can be used directly for improved interaction with a human UAV operator, but also for algorithms planning for a team of UAVs aiming at improved autonomous behavior. Before presenting the Multi-UAV coordination planner coined DivPlan utilizing diverse planning, we present two base-line algorithms. Since the output of the algorithms are trajectories in the form of GPS coordinates it is very simple to deploy them on any UAVs supporting flight along GPS trajectories, e.g., using mixed-reality system as have been demonstrated in [14].

Firstly, we will present an pseudo-optimal planner based on translation of the problem to classical optimal planning; Secondly, we will present a greedy approach. On one hand, the pseudo-optimal algorithm provides solutions close to optima; however, at the price of high (generally intractable) computational complexity. On the other hand, the greedy approach is computationally easy (tractable); however, the solutions are often of low quality. The motivation for DivPlan was to design a middle-ground algorithm with complexity low enough for large-scale scenarios; however, with solutions of higher quality than the naive greedy approach.

The Multi-UAV Sensing Problem (MUSP) is defined as a tuple $\mathcal{M} = \langle Y, L, U, T, c, b, p \rangle$, where

- $y \in Y$ is a set of sensor types the UAVs can equip in form of particular sensors,
- $\bar{l} \in L$ is a set of target ground locations for sensing in form $\bar{l} = \langle l_x, l_y \rangle$,
- $u \in U$ is a set of (identificators of) the UAVs carrying out the mission,
- $\langle \bar{l}, y \rangle \in T$ is the set of the sensing tasks of the UAVs of sensor type y at target location $\bar{l}$ (the optimization criterion is to fulfill maximal number of these tasks),
- c is the number of sensor slots (identical for all UAVs),
- b is the maximal battery charge (identical for all UAVs), and
- p represents the battery penalty for one equipped sensor (identical for all UAVs).

The semantics of fulfilment of a task $\langle \bar{l}, y \rangle$ is following. The task has to be fulfilled by a UAV located at $\bar{l}$ with an equipped sensor of type y. Although each vehicle can be equipped by a number of sensors (maximally c), the more sensors attached, the heavier the vehicle is, therefore the smaller is its flight range. The decrease of the flight range is defined by decrease of the maximal (and initial) battery charge by equipping sensors on the vehicle. The reduced initial charge is $b - pe$ for e equipped sensors.

A solution of a MUSP problem is a mapping $\mu : u \mapsto \langle traj, eqSensors \rangle$, where for each UAV $u \in U$ a trajectory *traj* and a set of equipped sensors of types $y \in Y$ in *eqSensors* are assigned. A trajectory is an ordered sequence of locations from L (or an empty sequence) over which the UAV moves to fulfill the tasks. For the length of the trajectory (the distance between two locations $||\bar{l}_1 - \bar{l}_2||$ is computed as Euclidean distance), it must hold that the battery charge $b - pe$ is sufficient (we assume WLOG that the units of battery charge are the same units as the distance). For the number of equipped sensors, it must hold $|eqSensors| \leq c$. An optimal solution μ of a MUSP problem $\mathcal{M}$ is such that there exists no other solution μ' to $\mathcal{M}$ fulfilling more tasks $\langle \bar{l}, y \rangle \in T$ than μ. The closer is a solution to the optimum (to the maximal number of solved tasks), the higher is the quality of the solution.

After we sketch why MUSPs are hard to solve in Section 2.1, we show how to solve the discretized variant of the problem optimally in Section 2.2. The second algorithm, presented in Section 2.3, is the greedy solution.

2.1. Why Is This Task Difficult?

A MUSP combines several well-known NP-complete problems; however, to our best knowledge this particular combination has not been proposed and formally defined yet.

For one UAV, one sensor type and the case when it is possible to fulfill all tasks, the problem reduces to a Travelling Salesman Problem (TSP). The selection of sensors under the limit of the battery thus flying distance is a Knapsack Problem (KP). The combination of TSP and KP is know as Orienteering Problems (OP) [22]; however, defined over different prices of the goals, whereas we limit the total flight range in our problem. The MUSPs additionally adds the combinatorial problem of optimization for multiple UAVs.

2.2. The Pseudo-Optimal Algorithm

The (close to) optimal solution to a MUSP will be obtained by translating the problem to a classical planning problem and using top-performing optimal planner SymBA* [23] to search for a solution. Detailed description of translation MUPS into a planning problem can be found in the Appendix A. The solution is than translated to μ by means of prescription which UAV should use which sensors and how to move among the targets and which to sense. As classical planning does not directly allow modeling of continuous fluents, the proposed translation uses discretization of distances between the locations of sensor tasks and related values as battery charge. Although the discretization causes the optimal solution to the translated problem does not necessarily corresponds to optimal solution to the original MUSP the error is bounded by $|T|d$, where $|T|$ is the number of sensor tasks and d is the distance for one discrete flight "step", i.e., the discretization factor. Therefore, we denote the algorithm as pseudo-optimal.

2.3. The Greedy Algorithm

The *greedy* algorithm sequentially generates and assigns trajectories and equipped sensors to each UAV. The algorithm is listed as Algorithm 1. Each UAV is assigned a trajectory and sensors by a method `GreedyOP` listed as Algorithm 2. Sensor tasks covered by created trajectory are removed from the problem before the creation of another trajectory for the next UAV.

Algorithm 1: `GreedySolver`($\mathcal{M}$) – a greedy algorithm solving MUSPs.

```
input  : Problem M = ⟨Y, L, U, T, c, b, p⟩
output : Solution μ

foreach UAV u ∈ U do
    ⟨traj, eqSensors⟩ ⟵ GreedyOP(M, ∅);
    add u ↦ ⟨traj, eqSensors⟩ to μ;
    M ⟵ removeTasksCoveredByTrajFromProblem(M, traj);
end foreach
```

Input of the method `GreedyOP` (Algorithm 2) is a list of sensor tasks and a list of equipped sensors (this parameter contains no sensor when called from Greedy algorithm, but it will be needed later in the DivPlan algorithm). In fact, `GreedyOP` solves an Orienteering Problem (OP) together with selection of a suitable subset of sensors. Since the Orienteering Problem is currently an open problem, the proposed solution is another greedy approach. As soon as a practical solution to this problem exists, we shall replace `GreedyOP` method by a stand-alone solver.

Algorithm 2: Greedy Orienteering Problem solver (with greedy sensors selection) of one UAV.

```
Function GreedyOP (T, eqSensors)
    input  : List of sensor tasks T
    input  : Already equipped sensors eqSensors
    output : Solution ⟨traj, eqSensors⟩

    traj ⟵ (l̄_base);
    while T ≠ ∅ do
        ⟨y, l̄⟩ ⟵ closestPointToLastPointOfTrajectory(T, traj, eqSensors);
        if no suitable ⟨y, l̄⟩ found then
            break;
        end if
        remove ⟨y, l̄⟩ from M;
        add l̄ to traj minimizing;
        eqSensors ⟵ eqSensors ∪ {y});
    end while
    if |traj| > 1 then
        append l̄_base to traj;
    end if
    return ⟨traj, eqSensors⟩
end
```

`GreedyOP` firstly creates an empty trajectory containing only the location of the base $\bar{l}_{\text{base}}$ and then it sequentially adds new points to the trajectory as long as the UAV has enough battery charge to return to the base. New points are selected by the method `closestPointToLastPointOfTrajectory`(T, *traj*, *eqSensors*), which finds the closest task of T to the last point of the *traj*. All tasks requiring new sensor (not yet equipped) are penalized by p of $\mathcal{M}$ and thus tasks with available sensors are preferred. There is also a limit on maximal number of sensors equipped by a single UAV. The semantics of *adding* $\bar{l}$ *to traj minimizing* is that the new waypoint is added to the trajectory such that extension of the trajectory length is minimized.

The greedy method represents a fast algorithm with prospectively lower solution quality that is supposed to solve large MUSP instances, which cannot be solved by the pseudo-optimal algorithm.

3. Diverse Planning Based Algorithm

So far, we presented two algorithms. On one hand, a pseudo-optimal planner which can solve MUSPs nearly optimally but it can solve only very small instances. On the other hand, it is the greedy algorithm which is able to solve large problem instances but since it is a greedy method, it can produce substantially sub-optimal solutions. The main goal of the proposed planner based on diverse planning DivPlan, is to create better solutions than greedy, but in a reasonable time.

DivPlan in Algorithm 3 works in two phases. Firstly, it uses a diverse planning technique to create diverse set of possible trajectories for the UAVs together with a set of sensors it should be equipped with. Then it assigns a subset of these trajectories to individual UAVs by translation to a Constraint Optimization Problem (COP) [24].

Algorithm 3: DivPlan—A MUSP solver based on diverse planning and constraint optimization.

input :Problem $\mathcal{M} = \langle Y, L, U, T, c, b, p \rangle$
output:Solution μ

$\mu^{greedy} \longleftarrow$ `GreedySolver`$(\mathcal{M})$;
$\bar{\mu} \longleftarrow$ `CreateDiverseTrajecotries`$(\mathcal{M}) \cup \bigcup_{u \in U} \mu^{greedy}(u)$;
$\mu \longleftarrow$ `COPSolver`$(X \leftarrow U, \forall X_i \in X : D_i \leftarrow \bar{\mu}, C, f_{opt}, \mu^{greedy})$;
`// UAVs` U `as COP variables` X`,`
`//` $\bar{\mu}$ `as the COP domains` D_i `for all variables,`
`// constraints` C `forbidding selection of the same trajectory by two UAVs,`
`//` f_{opt} `maximizing the number of sensor tasks covered by the solution` μ`,`
`//` μ^{greedy} `as the initial solution, and`
`// returns` μ `assigning a value from` D_i `for each` X_i`, i.e., trajectories to UAVs`
return μ;

The strategy for creating the diverse trajectories is inspired by the experimentally most successful method for diverse planning [10], i.e., looking for good solutions of modified problems.

Method `CreateDiverseTrajecotries` (Algorithm 4) creates internally many smaller instances of the MUSP problem containing different subsets of sensor tasks and collects their solutions. The subsets of the sensor tasks are created by clustering method k-means [25], which groups tasks that are nearby and thus could be possibly covered by a single UAV. Number of clusters varies from 1 to number of sensor slots c, generating trajectories for various numbers and combinations of sensors. Then the algorithm repeatedly calls `GreedyOP` method until all tasks of the cluster are covered. This process is repeated for all subsets of sensors that can be mounted on an UAV and all the created trajectories are stored together with these required sensors.

Once a large portfolio of diverse trajectories is created, DivPlan runs a COP solver. We use `OptaPlanner` (http://www.optaplanner.org/) a popular Constraint Satisfaction and Optimization Solver. `OptaPlanner` assigns a trajectory with a set of sensors $\langle traj, eqSensors \rangle$ to each UAV optimizing the selection by f_{opt}. Such assignment solves the original MUSP problem. There is only one rule specifying the quality of the solution, i.e., how many sensor tasks are covered by this solution. The rule in form of optimization criterion follows

$$\arg\max_{\mu} |T'|, T' \subseteq T \text{ s.t. } \forall \langle y, \bar{l} \rangle \exists u : \langle y, \bar{l} \rangle \in T', \bar{l} \in traj, y \in eqSensors | \langle traj, eqSensors \rangle = \mu(u),$$

meaning the maximized number of sensor tasks $T' \subseteq T$ has to be covered by the solution μ. The rule is also listed in Algorithm algDrool in the Drools syntax (https://docs.jboss.org/drools/release/5.2.0.Final/drools-expert-docs/html/ch05.html) used by `OptaPlanner`. Note that the trajectory length is not being optimized by `OptaPlanner`, only the number of covered tasks. For practical purposes, one of the very convenient features of OptaPlanner is that it is an any-time algorithm and thus it produces better solutions as it is granted more computation time.

Algorithm 4: Creates a set of diverse trajectories.

```
Function CreateDiverseTrajecotries (M)
    input :Problem M = ⟨Y, L, U, T, c, b, p⟩
    output:A set μ̄ of ⟨traj, eqSensors⟩
    μ̄ ⟵ ∅;
    for k = 1 to c do
        foreach S : subsets of sensors of size k do
            T' ⟵ {⟨y, l̄⟩|⟨y, l̄⟩ ∈ T, y ∈ S} ;
            C ⟵ k-means(T') ;
            foreach cluster C_i ∈ C do
                while C_i ≠ ∅ do
                    ⟨traj, eqSensors⟩ ⟵ GreedyOP(C_i, S);
                    remove covered tasks by ⟨traj, eqSensors⟩ from C_i;
                    μ̄ ⟵ μ̄ ∪ ⟨traj, eqSensors⟩;
                end while
            end foreach
        end foreach
    end for
    return μ̄;
end
```

Algorithm 5: OptaPlanner rule (in the Drools syntax).

```
rule ''CoveredTasks''
    when
        $task : SensorTask()
        not UavPlan(hasSensor($task.sensor), trajectory.isCovered($task.point))
    then
        scoreHolder.addMediumConstraintMatch(kcontext, -1);
end
```

4. Complexity Analysis

A MUSP is combination of several NP-hard problems and thus it is NP-hard. In practice, that means that every algorithm solving this problem optimally needs time growing exponentially with the problem size, unless P = NP. The size of MUSP is dependent on several parameters: number of UAVs $|U|$, number of sensing tasks $|T|$, number of different sensors $|Y|$, and number of sensor slots c on a UAV. The number of locations is never more than $|L| + 1$, with +1 for the base location. The maximal battery charge b and penalty p only limit the number and size of the solutions. Let us take a closer look at how these parameters influence the computational complexity of presented algorithms.

4.1. Pseudo-Optimal Algorithm

The pseudo-optimal algorithm works in three steps, translating a MUSP to a classical planning problem, solving the translated problem by a classical planner and back translating the classical plan to the solution of the MUSP instance.

The number of planning objects used in the translation step is $n = |L| + 1 + |U| + c|U| + |Y| + \frac{b}{d}$, where d is the discretization factor. The process of grounding generates all possible parameterizations of the predicates based on the objects of the particular types. Classical planning assumes finite number of objects, therefore the grounding will be finite as well. As all the predicates are binary the asymptotic complexity of grounding of predicates will be $\mathcal{O}(n^2)$. Similarly, grounding of operators generates all possible parameterized actions. As the maximal number of predicate parameters is six for the operator equip, the asymptotic complexity of grounding of operators is $\mathcal{O}(n^6)$. Encoding of the initial state and goal conditions is $\mathcal{O}(n^2)$, because only a subset of facts is used. This gives us polynomial asymptotic complexity for the translation process $\mathcal{O}(n^6)$.

Computational complexity of classical planning (therefore also of the used SymBA* planner) grows in the worst case exponentially with the size of the input problem $|\Pi|$. The problem created during the translation is bounded by $\mathcal{O}(n^6)$. Therefore the overall complexity of the solution is exponentially dependent on the input size as follows:

$$\mathcal{O}(\exp(n^6)),$$

where the back translation process only linearly traverses the resulting plan and builds the MUSP solution μ, therefore there are no additional factors.

It is obvious that this approach is viable for smallest instances only. As we will show in the experiments only up to a dozen of monitoring tasks.

4.2. Greedy Algorithm

The greedy algorithm sequentially creates trajectories for each UAV. To create one trajectory, it repeatedly selects sensor tasks and adds the closest one to the existing trajectory. Thus, the whole computation runs in time:

$$\mathcal{O}(|U| \cdot |Y|^2) = \mathcal{O}(n^3),$$

which is polynomial in the size of the MUSP instance.

4.3. DivPlan Algorithm

DivPlan firstly creates a set $\bar{\mu}$ of diverse trajectories. Number of these trajectories can be estimated directly form Algorithm 4.

$$|\bar{\mu}| \leq \sum_{k=1}^{c} \binom{|eqSensors|}{k} \cdot \sum_{i=1}^{k} \mathsf{traj}(C_i),$$

where $\mathsf{traj}(C_i)$ is number of trajectories created from cluster C_i. In the worst case, the $\sum_{i=1}^{k} \mathsf{traj}(C_i)$ can approach the total number of sensor tasks (each trajectory covers just one location with one sensor task), but in practice this number is typically much smaller especially for large number sensor tasks. We can also limit this number by a constant t, then:

$$|\bar{\mu}| \leq \sum_{k=1}^{c} \binom{|eqSensors|}{k} \cdot k \cdot t \leq t \cdot \sum_{k=1}^{|eqSensors|} \binom{|eqSensors|}{k} \cdot k \leq \tag{1}$$

$$\leq t \cdot |eqSensors| \cdot 2^{|eqSensors|}. \tag{2}$$

Hence we bound the total number of diverse trajectories by t, the number of diverse trajectories created for one cluster of sensor tasks, and the total number of sensors, is in practice limited.

`OptaPlanner` is an anytime algorithm and thus it is difficult to evaluate its time complexity, moreover there is too many variables to theoretically estimate its performance profile (for experimental evaluations refer to the following experimental section, particularly Figure 7). Nevertheless, we can evaluate the total size of the search space as

$$|\bar{\mu}|^{|U|} \leq t^{|U|} \cdot |eqSensors|^{|U|} \cdot 2^{|eqSensors| \cdot |U|},$$

which gives us a following asymptotic bound on time complexity of DivPlan:

$$\mathcal{O}(t^{|U|} \cdot 2^{c|U|}) = \mathcal{O}(t^n \cdot \exp(n)).$$

The time complexity of DivPlan is thus exponentially dependent only on the number of UAVs and their sensor slots. Unlike the number of sensing tasks, these numbers are very limited in practice. If we consider them to be fixed parameters, we would get a polynomial complexity of DivPlan algorithm.

5. Experiments

The experimental evaluation compares the three proposed MUSP solvers. The MUSP solvers are evaluated on synthetic benchmarks and in simulated large-scale scenarios. All experiments were performed on 8 core Intel Xeon 2.5 GHz computer with 8GB RAM and Java VM heap size limited to 2 GB.

5.1. Comparison of the Multi-UAV Sensor Problem Solvers

We have evaluated more than 3000 different instances of MUSP. In these experiments we compare the three proposed algorithms. Firstly, the pseudo-optimal solver (The problem has to be discretized to be computable in reasonable time, which can cause that the solution is not always the optimal one, therefore pseudo-optimal.) (see Section 2.2 for details). The greedy algorithm represents naive fast algorithm (described in Section 2.3). And finally DivPlan shows how diverse planning together with a COP solver can provide better solution than the greedy algorithm within reasonable time (see Section 3 for details). To compare these algorithms, we designed a set of benchmark instances allowing to scale from a few sensor tasks to tens of thousands. We also demonstrate how the proposed DivPlan method works on a real-world-inspired scenario of monitoring the air pollution in the city of Prague.

The scenario for all synthetic benchmark experiments was created by random generation of a road map and random locations on each road. All locations at the same road were required to be monitored by the same sensor. The whole area was a square 1000 m × 1000 m, the range of flight is 5 km with one mounted sensor and it decreases by 1 km by each additional sensor.

The first set of benchmark tests focuses on the overall solution quality when compared to the optimal solution. These benchmarks contain 3 roads each with 1 to 7 monitoring locations, leading to 3 to 21 sensor tasks. There are 3 types of sensors and each of 2 operating UAVs can hold up maximally 2 sensors.

As expected, the results shown in Figure 2 demonstrate that the use of the pseudo-optimal solver (Section 2.2) is impractical for instances containing more than few sensor tasks. In the figure, we can also see that the discretization of the continuous space causes that the result of the optimal solver is in approx. 25 % of cases suboptimal. The DivPlan improved the quality of the greedy solution in all but the most trivial cases. The right chart shows that in average DivPlan found solutions in less 100 s while greedy algorithm required less than 1 s.

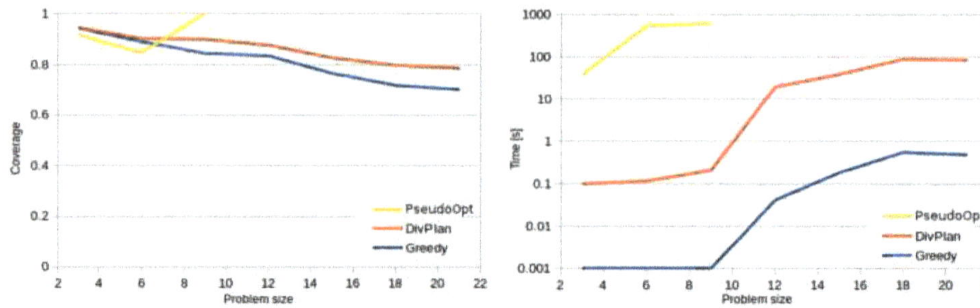

Figure 2. Time and coverage of solved sensor tasks comparison of all methods. Time of DivPlan is time needed to find final solution (DivPlan was always granted 30 min timeout, but typically the final solution has been found within few seconds). Coverage of DivPlan and greedy is always counted for all the problems while for pseudo-optimal it is only over the solved problems. For size 9, the pseudo-optimal planner solved only approximately one fourth of the problem instances. These instances were solved by DivPlan and greedy algorithms with coverage 1 too.

For larger domains with benchmark set containing 1650 problems with up to 2500 sensor tasks, 2 to 10 different sensors, 3 to 5 sensor slots and 10 to 50 UAVs, it is not feasible to find the optimal solution. Nevertheless we would like to have some estimate how good the created solution is. For this purpose we run an optimal TSP solver Concorde (http://www.math.uwaterloo.ca/tsp/concorde.html) on all covered sensor tasks by DivPlan. Its solution then corresponds to the optimal trajectory of one "omnipotent UAV" with all sensors and unlimited flight range, fulfilling the MUSP problem with the same coverage as the solution provided by DivPlan. Figure 3 shows the relative quality of the DivPlan solution (0.5 means that Concorde found trajectory of half length). We can see that even one unlimited UAV would still have to travel at least 30 % of the solution length even for the cases containing 40 to 50 UAVs. The average was computed only for problems where Concorde gave a solution within the limit of 10 min.

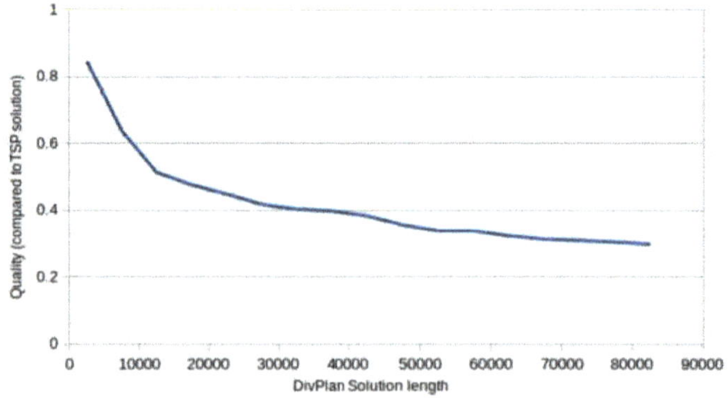

Figure 3. Comparison of DivPlan solution with the optimal solution of "omnipotent UAV" covering the same sensor tasks. We can see that the DivPlan solution was in average at most three times longer even for the longest solutions of scenarios with several dozens of UAVs.

The last set of benchmark experiments focuses on the comparison of the greedy algorithm and DivPlan on large problems. The benchmark set contained 1350 problems with up to 9000 sensor tasks, 5 to 20 different sensors, 3 to 5 sensor slots and 10 to 50 UAVs. The Figure 4 shows how many diverse trajectories have been created for different numbers of sensor tasks for UAVs with 3, 4, and 5 sensor slots. We can see that the number of created trajectories grows linearly with the number of sensor tasks beginning with approximately 3000 sensor tasks.

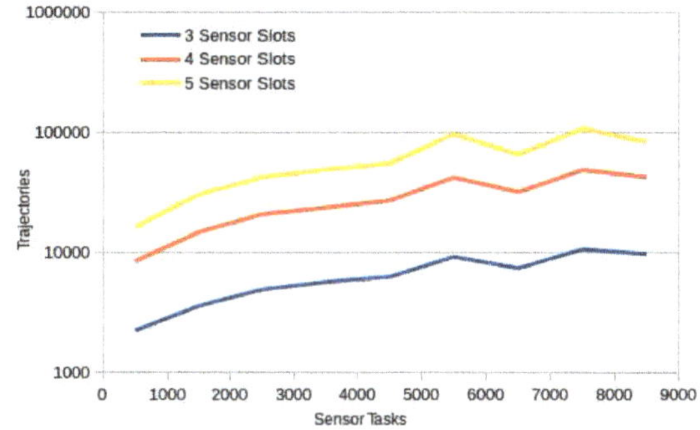

Figure 4. Number of created trajectories for different numbers of sensor tasks for UAVs with 3, 4, and 5 sensor slots. The number of trajectories is in log scale.

The last chart of this section, Figure 5, shows task coverages for different number of UAVs and different number of sensor tasks. Each line of the graph shows averages over 1060 cases of different settings (number of sensor types, number of sensor slots on UAV, number of roads, etc.). The average improvement is 33 % (11 percent points) for 10-UAVs case, 17 % (10 percent points) for 30-UAVs case and 12 % (8 percent points) for 50-UAVs case. Time limit for DivPlan has been set to 10 min and we can see that it shows a stable improvement over the greedy method for both the different numbers of sensor tasks and the different numbers of UAVs.

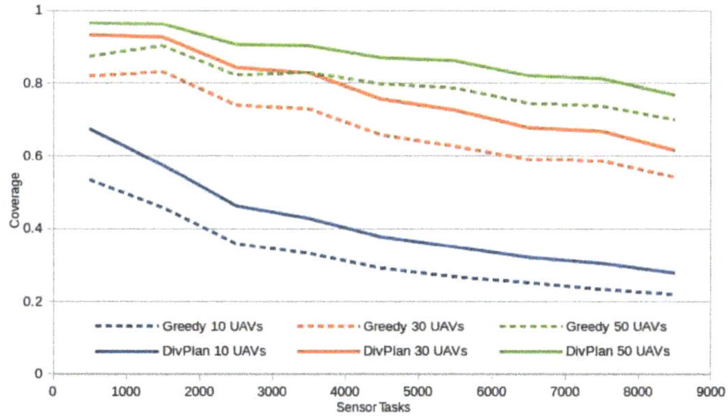

Figure 5. Task coverages for 10, 30 and 50 UAVs on different total number of tasks.

5.2. Real-World-Inspired Scenario

The motivation for the MUSP problem is monitoring of air pollution in the area 18 km × 16 km of city of Prague. There are 3506 selected locations of 6 different monitoring types, each type is requested to be monitored by 3 different sensors, which yields 10,518 sensor tasks in total. There are 20 UAVs available, each with 5 sensor slots. Since each sensor has non-zero weight, every mounted sensor decreases the UAV range of flight. Table 1 lists the used numbers of equipped sensors and related ranges of flight.

Table 1. Decrease of the UAV range of flight based on the number of equipped sensors used in the Real-World-Inspired Scenario.

Number of Equipped Sensors	Range of Flight
1	83 km
2	67 km
3	50 km
4	33 km
5	17 km

Map of monitoring locations together with the UAV plans created by DivPlan are depicted in Figure 6. The greedy solution for this problem provided a solution with sensor task coverage of 57 % in 11.5 s. DivPlan improved this solution to the coverage of 64 % within 8.1 min. The improvement of the coverage over time is shown in Figure 7.

Figure 6. Planned trajectories for 20 UAVs tasked to monitor 3506 locations in Prague. Each location is required to be monitored by 3 different sensors giving 10,518 sensor tasks in total. DivPlan reached coverage of 64%.

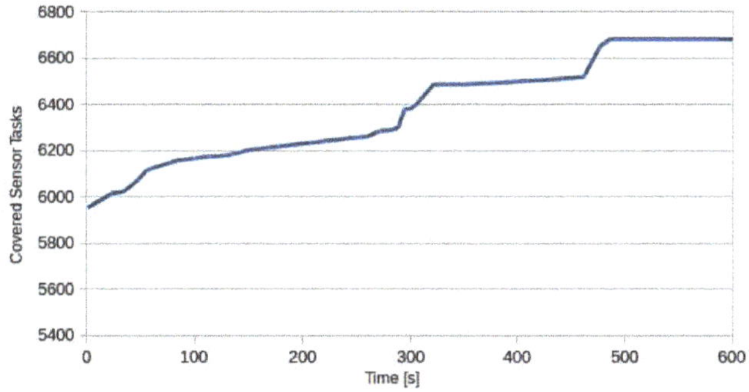

Figure 7. How the coverage improves over time. Base (time 0) is greedy solution: coverage 57 %. The time limit has been set to 10 min, but the best solution with coverage of 64 % was found after 8.1 min.

The last graph (Figure 8) compares the task coverage for different numbers of UAVs. We can see that DivPlan improvement over the greedy method was more significant for the case of UAVs with 3 sensor slots.

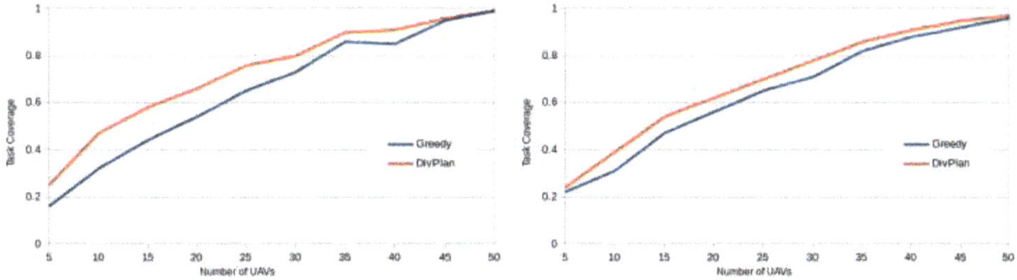

Figure 8. Task coverages for different numbers of UAVs with 3 (left) and 5 (right) sensor slots.

6. Conclusions and Future Work

To solve the problem of autonomous remote sensing with non-zero weight of sensors, we have firstly formally defined the problem and for comparison we have designed two base-line algorithms commonly used for solution of combinatorial optimization problems in the literature. The algorithms were based on two distinct paradigms: (a) translation to classical planning with appropriate discretization; and (b) a greedy approach. The base-line algorithms framed the problem from the perspective of the solution quality and efficiency metrics, respectively.

The main contribution of our work was a novel algorithm aiming at the remote sensing problem by a fleet of coordinated UAVs with practicality in mind. The DivPlan algorithm targets a middle-ground between the optimal but inefficient and low-quality but highly efficient greedy algorithms. To provide such an algorithm, we have integrated an appropriate diverse planning technique to generate the alternative trajectories and Constraint Optimization composing the final solution out of these diverse partial solutions. The solvers were both theoretically and experimentally compared. The results show that an approach based on diverse planning is a good balance between quality of the solution and planning time. Moreover, the greedy and diversity planning approaches were able to solve large problem instances, which demonstrates their good scalability.

Based on the experimental results in the simulated real-world-inspired environment and a conservative usage of waypoints as a robotic primitive, we conjecture DivPlan is a good choice for practical deployment to a Multi-UAV system, which we leave to explore in a future work.

Acknowledgments: This research was supported by the Czech Science Foundation (grant no. 15-20433Y).

Author Contributions: Jan Tožička designed and implemented the algorithms and performed the experiments; Jan Tožička also analyzed the experimental data; Antonín Komenda designed, implemented and tested the algorithm for MUSP conversion to classical planning; Antonín Komenda wrote the MUSP conversion to classical planning part and cleaned the final draft of the article.

Conflicts of Interest: The authors declare no conflict of interest. The founding sponsors had no role in the design of the study; in the collection, analyses, or interpretation of data; in the writing of the manuscript, and in the decision to publish the results.

Abbreviations

The following abbreviations are used in this manuscript:

MUSP	Multi-UAV Sensing Problem
STRIPS	Stanford Research Institute Problem Solver
TSP	Traveling Salesmen Problem
KP	Knapsack Problem
COP	Constraint Optimization Problem
UAV	Unmanned Aerial Vehicle

Appendix A. Translation of MUPS to Planning Problem

The discretized distances are interpolated by $\left\lfloor \frac{||\bar{l}_1 - \bar{l}_2||}{d} \right\rfloor$ intermediate possible positions of the UAVs. The discretized distances allow us to use also discrete number of battery charge levels $b' = \left\lfloor \frac{b}{d} \right\rfloor$. For the discretized battery penalty, we get $p' = \lfloor \frac{p}{d} \rfloor$.

A classical planning problem with costs is defined as a tuple $\Pi = \langle F, A, I, G, \mathsf{cost} \rangle$ (e.g., in [26]). A set F consists in all possible facts which describe states of the modeled world. A state $s \subseteq F$ contains only such facts that hold in the world in that particular state. A set A contains grounded actions $a = \langle \mathsf{pre}(a), \mathsf{add}(a), \mathsf{del}(a), \mathsf{cost}(a) \rangle$, where the sets of facts $\mathsf{pre}(a) \subseteq F, \mathsf{add}(a) \subseteq F, \mathsf{del}(a) \subseteq F$ represent preconditions, add effects and delete effects respectively. The cost function is defined as $\mathsf{cost} : A \to \mathbb{R}^{0+}$, where $\mathsf{cost}(a)$ represents a non-zero cost of the action a. An action can transform a state s into a new state s' when a is executed, provided that all its preconditions are satisfied s.t. $\mathsf{pre}(a) \subseteq s$ and for cost $\mathsf{cost}(a)$. The transformation function is defined as $s' = (s \setminus \mathsf{del}(a)) \cup \mathsf{add}(a)$. The set $I \subseteq F$ represents the initial state of the planning problem. The set $G \subseteq F$ represent goal conditions, such that every state s_G for that $G \subseteq s_G$ holds is a goal state. Note that the goal condition G can represent more conjunctive goal facts which all have to be satisfied. Moreover, the condition allows for satisfaction of the goal facts in various states where additional facts not present in G can but need not to hold. A solution to a classical planning problem is a sequence of successively applicable actions beginning in the initial state and ending in one of the goal states, such sequence is called a plan. A sequence of actions $\pi = (a_1, \ldots, a_m)$ is a plan to Π if for all $1 < i < m$ holds $\mathsf{pre}(a_i) \subseteq s_i$ and $s_{i+1} = (s_i \setminus \mathsf{del}(a_i)) \cup \mathsf{add}(a_i)$, where $s_1 = I$ and $G \subseteq s_m$.

The sensor tasks in a MUSP are disjunctive. Not all of them can be satisfied if the battery and equipment constraints do not allow it. The objective is defined as an optimization problem, that is we require maximizing of satisfied tasks for the particular $\mathcal{M}$. Moreover, the problem is defined over continuous variables for battery charge, distances and the equipment battery penalty. To solve such problem using a classical planing, we need to appropriately translate it, namely we need to deal with:

1. net-benefit selection of goals (selecting the maximal set of goal facts), and
2. discretization of the continuous variables (locations, moving and battery charge).

Net-benefit planning (sometimes called planning with soft goals) is a well known type of planning problem translation proposed in [27]. The discretization is based on the distance granularity parameter as explained in the previous paragraphs.

The translation will be described in form of parameterized facts, i.e., predicates and parameterized actions, i.e., operators. The initial state and the goal conditions will use parameterized representation as well.

Appendix A.1. Objects

In classical planning the parameters of predicates and operators are defined in form of typed (mathematical) object. The types used in the translated classical representation of a MUSP are:

- Pos—possible positions of UAVs and sensing targets; each location (in form of its name) $\bar{l} \in L$ is a position, each intermediate interpolation step between two locations is a position as well,
    - Base (Pos subtype)—one of the positions is marked as a base, where the UAVs can equip sensors and where their plans have to begin and end,
- Uav—objects representing the UAVs $u \in U$
- Slot—each UAV has c sensor slots, the objects of type Slot represent such slots,
- Type—types y of the sensor targets of the tasks defined as $y \in Y$ and
- Level—number b' of possible levels of the UAV battery, the objects of the type Level represent particular charge of a UAV's battery.

The types define distinct sets of typed objects, therefore we will write $x \in \mathsf{SomeType}$ to denote x is of a type $\mathsf{SomeType}$.

Appendix A.2. Predicates

The predicates define parameterized "templates" for the planning facts in F. Predicates related to the UAVs are:

- $\mathsf{at(Uav, Pos)}$—in which position a UAV is,
- $\mathsf{slotEmpty(Uav, Slot)}$—whether a slot a of a UAV is not equipped yet,
- $\mathsf{senseType(Uav, Type)}$—whether a UAV is able to sense a sensor type (by equipping appropriate sensor to one of its empty slots) and
- $\mathsf{battery(Uav, Level)}$—current level of the UAV's battery.

There are three predicates describing object adjacency:

- $\mathsf{batteryDec(Level, Level)}$—relates a battery level to its decremented value by one level (corresponds to depletion of the battery by one move action),
- $\mathsf{batteryEquipDec(Level, Level)}$—relates a battery level to its decremented value by p' (corresponds to depletion of the battery by equipping one sensor and models shorter reach by heavier UAV) and
- $\mathsf{adj(Pos, Pos)}$ - together with positions describes the movement graph (the move actions are allowed to move only between two adjacent positions).

The tasks $\langle \bar{l}, y \rangle \in T$ are described by the last predicate with equal semantics as in the MUSP:

- $\mathsf{task(Pos, Type)}$

By grounding all predicates we get the set F of all possible fact (note that not all possible facts form a state, e.g., only one fact grounding $\mathsf{battery(Uav, Level)}$ always holds for each UAV).

Example A1. *For a set of three positions* $\{\mathrm{p}_1, \mathrm{p}_2, \mathrm{p}_3\}$ *and two UAVs* $\{\mathrm{uav}_1, \mathrm{uav}_2\}$*, grounding of the predicate* $\mathsf{at(Uav, Pos)}$ *leads to 6 facts*

$$\mathsf{at}(\mathrm{uav}_1, \mathrm{p}_1), \mathsf{at}(\mathrm{uav}_1, \mathrm{p}_2), \mathsf{at}(\mathrm{uav}_1, \mathrm{p}_3),$$
$$\mathsf{at}(\mathrm{uav}_2, \mathrm{p}_1), \mathsf{at}(\mathrm{uav}_2, \mathrm{p}_2), \mathsf{at}(\mathrm{uav}_2, \mathrm{p}_3),$$

where a valid part of a state representing positions of the two UAVs is e.g., $\{\mathsf{at}(\mathrm{uav}_1, \mathrm{p}_2), \mathsf{at}(\mathrm{uav}_2, \mathrm{p}_3)\}$.

An example of a diamond-shaped movement graph with four nodes $\{\mathsf{p}_1, \mathsf{p}_2, \mathsf{p}_3, \mathsf{p}_4\}$ would be represented as:

$$\mathsf{adj}(\mathsf{p}_1,\mathsf{p}_2), \mathsf{adj}(\mathsf{p}_2,\mathsf{p}_3), \mathsf{adj}(\mathsf{p}_3,\mathsf{p}_4), \mathsf{adj}(\mathsf{p}_4,\mathsf{p}_1), \mathsf{adj}(\mathsf{p}_1,\mathsf{p}_3),$$
$$\mathsf{adj}(\mathsf{p}_2,\mathsf{p}_1), \mathsf{adj}(\mathsf{p}_3,\mathsf{p}_2), \mathsf{adj}(\mathsf{p}_4,\mathsf{p}_3), \mathsf{adj}(\mathsf{p}_1,\mathsf{p}_4), \mathsf{adj}(\mathsf{p}_3,\mathsf{p}_1).$$

Operators

The translated actions will be defined in form of parameterized operators. Let us recall the scheme of an action $a = \langle \mathsf{pre}(a), \mathsf{add}(a), \mathsf{del}(a), \mathsf{cost}(a)\rangle$. For definition of the translation operators, we will use parameters of the defined types similarly as for the predicates. The same grounding principle used from predicates to facts will be used for grounding actions from operators.

The first operator represents equipping of a sensor by a UAV:

$$\mathsf{equip}(\mathsf{Uav}, \mathsf{Base}, \mathsf{Slot}, \mathsf{Type}, \mathsf{Level}, \mathsf{Level}).$$

The grounded actions from the operator not only mark the UAV u being able to sense the type y by the fact $\mathsf{senseType}(u, y)$ (together with removing the fact $\mathsf{slotEmpty}(u, o)$, which is required in the preconditions ensuring the sensor slot o is not equipped yet). The operator also decrease the current battery level of u described by $\mathsf{battery}(u, l_i)$ to $\mathsf{battery}(u, l_{i+1})$, where the two levels are related by the p' decrement in form of the decrement relation facts $\mathsf{batteryEquipDec}(l_i, l_{i+1})$. The preconditions bind the battery levels and the position of the UAV to the base, as $p \in \mathsf{Base}$ and the preconditions contain $\mathsf{at}(u, p)$. Complete description of the actions follows:

$$\begin{aligned}
\mathsf{pre}(\mathsf{equip}(u, p, o, y, l_i, l_{i+1})) &= \mathsf{at}(u, p) \wedge \mathsf{slotEmpty}(u, o) \wedge \mathsf{battery}(u, l_i) \wedge \\
&\quad \wedge \mathsf{batteryEquipDec}(l_i, l_{i+1}), \\
\mathsf{add}(\mathsf{equip}(u, p, o, y, l_i, l_{i+1})) &= \mathsf{senseType}(u, y) \wedge \mathsf{battery}(u, l_{i+1}), \\
\mathsf{del}(\mathsf{equip}(u, p, o, y, l_i, l_{i+1})) &= \mathsf{slotEmpty}(u, o) \wedge \mathsf{battery}(u, l_i), \\
\mathsf{cost}(\mathsf{equip}(u, p, o, y, l_i, l_{i+1})) &= 1.
\end{aligned}$$

Another operator moves a UAV between two positions:

$$\mathsf{move}(\mathsf{Uav}, \mathsf{Pos}, \mathsf{Pos}, \mathsf{Level}, \mathsf{Level}).$$

The semantics is straightforward. Before applying a move action, the UAV u is in position p_i with current battery level l_i. The UAV can move only to an adjacent position p_{i+1} ensured by $\mathsf{adj}(p_i, p_{i+1})$ in the preconditions. The battery decrement is modeled similarly as in equip, but it uses $\mathsf{batteryDec}(l_i, l_{i+1})$ instead of the $\mathsf{batteryEquipDec}(l_i, l_{i+1})$ decrement relation. The UAV ends in the position p_{i+1} by adding $\mathsf{at}(u, p_{i+1})$ and deleting $\mathsf{at}(u, p_i)$. The description of the operator follows:

$$\begin{aligned}
\mathsf{pre}(\mathsf{move}(u, p_i, p_{i+1}, l_i, l_{i+1})) &= \mathsf{at}(u, p_i) \wedge \mathsf{battery}(u, l_i) \wedge \\
&\quad \wedge \mathsf{adj}(p_i, p_{i+1}) \wedge \mathsf{batteryDec}(l_i, l_{i+1}), \\
\mathsf{add}(\mathsf{move}(u, p_i, p_{i+1}, l_i, l_{i+1})) &= \mathsf{at}(u, p_{i+1}) \wedge \mathsf{battery}(u, l_{i+1}), \\
\mathsf{del}(\mathsf{move}(u, p_i, p_{i+1}, l_i, l_{i+1})) &= \mathsf{at}(u, p_i) \wedge \mathsf{battery}(u, l_i), \\
\mathsf{cost}(\mathsf{move}(u, p_i, p_{i+1}, l_i, l_{i+1})) &= 1.
\end{aligned}$$

To fulfill a sensor task $\langle \bar{l}, y\rangle \in T$, a UAV has to use the appropriate sensor at the right place. The operator modeling such situation is:

$$\mathsf{sense}(\mathsf{Uav}, \mathsf{Pos}, \mathsf{Type}).$$

The operator checks whether the UAV is in the right position $\mathsf{at}(u,p)$ and equipped with a sensor able to fulfill the task of type y by $\mathsf{senseType}(u,y)$. If so, the fulfilled $\mathsf{task}(p,y)$ is added:

$$\begin{aligned}\mathsf{pre}(\mathsf{sense}(u,p,y)) &= \mathsf{at}(u,p) \wedge \mathsf{senseType}(u,y),\\ \mathsf{add}(\mathsf{sense}(u,p,y)) &= \mathsf{task}(p,y),\\ \mathsf{del}(\mathsf{sense}(u,p,y)) &= \emptyset,\\ \mathsf{cost}(\mathsf{sense}(u,p,y)) &= 1.\end{aligned}$$

The last operator models skipping a translated task $\langle \bar{l}, y\rangle \in T$ which cannot be fulfilled by sense:

$$\mathsf{skip}(\mathsf{Pos},\mathsf{Type}).$$

The principle follows the net-benefit translation of soft goals. A soft goal is a goal which does not need to be fulfilled (in the described translation, all goals are soft as some tasks do not need to be fulfilled). The skip actions therefore model not fulfilling a task $\mathsf{task}(p,y)$. Since the problem $\mathcal{M}$ is defined as optimization over a subset of tasks, the net-benefit planning models the task selection subproblem. The classical planning problem finds a plan with minimal cost of used actions, therefore the skip actions with cost larger than any possible plan can be used by the planner only if the task cannot be fulfilled by sense. The definition of the operator follows:

$$\begin{aligned}\mathsf{pre}(\mathsf{skip}(p,y)) &= \emptyset,\\ \mathsf{add}(\mathsf{skip}(p,y)) &= \mathsf{task}(p,y),\\ \mathsf{del}(\mathsf{skip}(p,y)) &= \emptyset,\\ \mathsf{cost}(\mathsf{skip}(p,y)) &> \mathcal{O}(\exp(|\Pi|)),\end{aligned}$$

where the cost is larger than any possible solution plan of the problem, as the longest possible plan of a classical planning problem is exponential [28] in the size of the input (Practically, we use a large enough constant). By grounding all the operators with all possible parameters defined by their types, we get the set of all actions A.

Appendix A.3. Initial State and Goal Conditions

The complete sets of facts F and actions A form foundation of the translated MUSP in classical planning. The sets represent a complete transition system for the used objects originating in $\mathcal{M}$ and their interaction based on the classical representation modeling the MUSPs.

To finish the translation of $\mathcal{M}$ to Π, we need to encode the initial state I and goal conditions G. First, all UAVs $u \in U$ begin in the base position. Their battery levels are at maximum $l_{b'}$. Their sensor slots o are initially empty, therefore they cannot sense any target types yet. Formally, in the initial state holds:

$$\begin{aligned}\forall v \in \mathsf{Uav}, p \in \mathsf{Base} &: \mathsf{at}(u,p),\\ \forall v \in \mathsf{Uav}, l_{b'} \in \mathsf{Level} &: \mathsf{battery}(u,l_{b'}),\\ \forall v \in \mathsf{Uav}, \forall o \in \mathsf{Slot} &: \mathsf{slotEmpty}(u,o).\end{aligned}$$

The battery level decrement relation holds for all levels from the maximal level $l_{b'}$ to the minimal level l_0 and the equip decrement follows the linear decrement by p', such that the last decrement ends in non-negative battery level (minimally l_0):

$$\begin{aligned}\forall k \in \mathbb{N}, b' \geq k > 0 &: \mathsf{batteryDec}(l_k, l_{k-1}),\\ \forall k \in \mathbb{N}, \left\lfloor \frac{b'}{p'} \right\rfloor > k \geq 0 &: \mathsf{batteryEquipDec}(l_{b'-p'k}, l_{b'-p'(k+1)}).\end{aligned}$$

The adjacency relation is a complete graph with all positions of type Pos as vertices and $n(\bar{l}_f, \bar{l}_t) = \left\lfloor \frac{||\bar{l}_f - \bar{l}_t||}{d} \right\rfloor$ interpolated points along the edges (see Figure A1) using the distance granularity d. The formal definition follows:

$$\begin{aligned}
\forall \bar{l}_f \in \mathsf{Pos}, \forall \bar{l}_t \in \mathsf{Pos}, n(\bar{l}_f, \bar{l}_t) = 0 \quad &: \quad \mathsf{adj}(\bar{l}_f, \bar{l}_t); \\
\forall \bar{l}_f \in \mathsf{Pos}, \forall \bar{l}_t \in \mathsf{Pos}, n(\bar{l}_f, \bar{l}_t) = 1 \quad &: \quad \mathsf{adj}(\bar{l}_f, p_1), \mathsf{adj}(p_1, \bar{l}_t); \\
\forall \bar{l}_f \in \mathsf{Pos}, \forall \bar{l}_t \in \mathsf{Pos}, n(\bar{l}_f, \bar{l}_t) > 1, \quad & \\
\forall k \in \mathbb{N}, n(\bar{l}_f, \bar{l}_t) > k > 0, p_k \in \mathsf{Pos}, p_{k+1} \in \mathsf{Pos} \quad &: \quad \mathsf{adj}(p_k, p_{k+1}), \\
& \quad\;\; \mathsf{adj}(\bar{l}_f, p_1), \mathsf{adj}(p_{n(\bar{l}_f, \bar{l}_t)}, \bar{l}_t).
\end{aligned}$$

Recall the move action decreases the battery by one level. The interpolation therefore splits longer distances than d between locations by moves with an error maximally d, which correspond to decrease of one battery level.

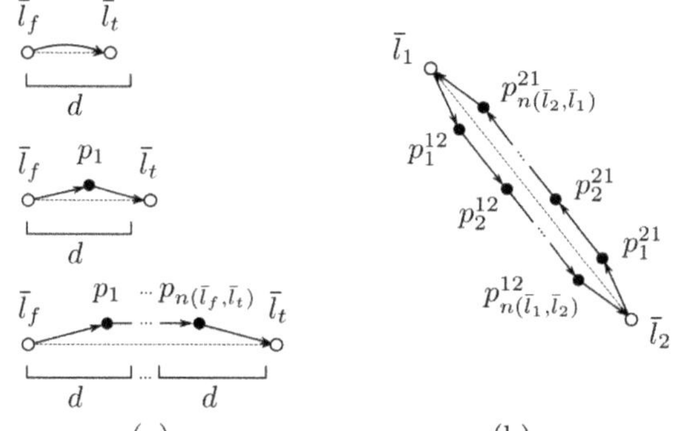

Figure A1. Interpolation of moves between two locations longer than d by intermediate positions p_i. The figure (**a**) shows additional positions $p_1, \ldots, p_{n(\bar{l}_f, \bar{l}_t)}$ between two locations $\bar{l}_f, \bar{l}_t$ based on the distance between them and the distance granularity. The solid arrows denote possible move actions. Note that the locations $\bar{l}_f, \bar{l}_t$ are described by x, y coordinates in the MUSP $\mathcal{M}$, therefore the distance between them (denoted by dashed lines) is defined. On the other hand the positions (in the translated problem), which are either location names or interpolated positions are defined only by means of the objects of type Pos; The figure (**b**) shows interpolation in both directions between two locations $\bar{l}_1, \bar{l}_2$. The superscripts distinguish direction from $\bar{l}_1$ to $\bar{l}_2$ and from $\bar{l}_2$ to $\bar{l}_1$.

The goal conditions G contain all required tasks $\langle \bar{l}, y \rangle \in T$ in the form of conjunction of facts task(Pos, Type) with the position corresponding to he location $\bar{l}$:

$$\forall \langle \bar{l}, y \rangle \in T \quad : \quad \mathsf{task}(\bar{l}, y).$$

A optimal solution to Π can be straightforwardly translated to a solution of the original MUSP $\mathcal{M}$ by mapping of the equipped sensors and moves into the trajectories.

Example A2. *Let us have an example translated problem with two UAVs* $\{\mathrm{uav}_1, \mathrm{uav}_2\}$ *each with two sensor slot* $\mathsf{slot}_1, \mathsf{slot}_2$*, initially at the* base *location, two sensor types* $\{\mathsf{a}, \mathsf{b}\}$ *and three tasks* $\{\mathsf{task}(\mathsf{left}, \mathsf{a}), \mathsf{task}(\mathsf{left}, \mathsf{b}), \mathsf{task}(\mathsf{right}, \mathsf{a})\}$*. The positions representing the locations are interconnected by the interpolated moves with different battery level costs. Moving from the* left *position to the* base *position and vice versa costs 4 battery levels and the two other moves 3 battery levels. Provided that* b' *and* p' *allows for a complete solution by one UAV (e.g.,* $b' = 12$ *and* $p' = 1$*), the resulting plan is in form:*

$$\text{equip}(\text{uav}_1, \text{base}, \text{slot}_1, \text{a}, l_{12}, l_{11}), \text{equip}(\text{uav}_1, \text{base}, \text{slot}_2, \text{b}, l_{11}, l_{10}),$$
$$\text{move}(\text{uav}_1, \text{base}, p_1^{\text{base,right}}, l_{10}, l_9), \ldots \text{move}(\text{uav}_1, p_2^{\text{base,right}}, \text{right}, l_8, l_7),$$
$$\text{sense}(\text{uav}_1, \text{right}, \text{a}),$$
$$\text{move}(\text{uav}_1, \text{right}, p_1^{\text{right,left}}, l_7, l_6), \ldots \text{move}(\text{uav}_1, p_2^{\text{right,left}}, \text{left}, l_5, l_4),$$
$$\text{sense}(\text{uav}_1, \text{left}, \text{a}), \text{sense}(\text{uav}_1, \text{left}, \text{b}),$$
$$\text{move}(\text{uav}_1, \text{left}, p_1^{\text{left,base}}, l_4, l_3), \ldots \text{move}(\text{uav}_1, p_3^{\text{left,base}}, \text{base}, l_1, l_0).$$

This example concludes the appendix on translation of MUSP solving to classical planning.

References

1. Villa, T.F.; Gonzalez, F.; Miljievic, B.; Ristovski, Z.D.; Morawska, L. An Overview of Small Unmanned Aerial Vehicles for Air Quality Measurements: Present Applications and Future Prospectives. *Sensors* **2016**, *16*, 1072.
2. Norris, G. NOAA plans fleet of 40 UAVs to monitor climate changes. *Flight Int.* **2004**, *166*, 7.
3. Boccardo, P.; Chiabrando, F.; Dutto, F.; Tonolo, F.G.; Lingua, A. UAV Deployment Exercise for Mapping Purposes: Evaluation of Emergency Response Applications. *Sensors* **2015**, *15*, 15717–15737.
4. Casbeer, D.W.; Beard, R.W.; McLain, T.W.; Li, S.M.; Mehra, R.K. Forest fire monitoring with multiple small UAVs. In Proceedings of the American Control Conference, Portland, OR, USA, 8–10 June 2005; pp. 3530–3535.
5. Merino, L.; Caballero, F.; Martínez-de Dios, J.R.; Maza, I.; Ollero, A. An Unmanned Aircraft System for Automatic Forest Fire Monitoring and Measurement. *J. Intell. Robot. Syst.* **2012**, *65*, 533–548.
6. Bruzzone, A.; Longo, F.; Massei, M.; Nicoletti, L.; Agresta, M.; di Matteo, R.; Maglione, G.L.; Murino, G.; Padovano, A. Disasters and Emergency Management in Chemical and Industrial Plants: Drones Simulation for Education and Training. In *Modelling and Simulation for Autonomous Systems*; Hodicky, J., Ed.; Springer International Publishing: Cham, Switzerland, 2016; pp. 301–308.
7. Las Fargeas, J.; Kabamba, P.; Girard, A. Cooperative Surveillance and Pursuit Using Unmanned Aerial Vehicles and Unattended Ground Sensors. *Sensors* **2015**, *15*, 1365–1388.
8. Gonzalez, L.F.; Montes, G.A.; Puig, E.; Johnson, S.; Mengersen, K.; Gaston, K.J. Unmanned Aerial Vehicles (UAVs) and Artificial Intelligence Revolutionizing Wildlife Monitoring and Conservation. *Sensors* **2016**, *16*, 97.
9. Olivares-Mendez, M.A.; Fu, C.; Ludivig, P.; Bissyandé, T.F.; Kannan, S.; Zurad, M.; Annaiyan, A.; Voos, H.; Campoy, P. Towards an Autonomous Vision-Based Unmanned Aerial System against Wildlife Poachers. *Sensors* **2015**, *15*, 31362–31391.
10. Tožička, J.; Šišlák, D.; Pěchouček, M. Diverse Planning for UAV Trajectories. In *Agents and Artificial Intelligence*; Springer: Berlin/Heidelberg, Germany, 2013; pp. 277–292.
11. Tožička, J.; Šišlák, D.; Pěchouček, M. Planning of Diverse Trajectories. In Proceedings of the 2013 5th International Conference on Agents and Artificial Intelligence (ICAART), Barcelona, Spain, 15–18 February 2013; Volume 2, pp. 120–129.
12. Tožička, J.; Šišlák, D.; Pěchouček, M. Planning of diverse trajectories for UAV control displays. In Proceedings of the International conference on Autonomous Agents and Multi-Agent Systems (AAMAS '13), Saint Paul, MN, USA, 6–10 May 2013; pp. 1231–1232.
13. Tožička, J.; Balata, J.; Mikovec, Z. Diverse trajectory planning for UAV control displays. In Proceedings of the International conference on Autonomous Agents and Multi-Agent Systems (AAMAS '13), Saint Paul, MN, USA, 6–10 May 2013; pp. 1411–1412.
14. Selecký, M.; Štolba, M.; Meiser, T.; Čáp, M.; Komenda, A.; Rollo, M.; Vokřínek, J.; Pěchouček, M. Deployment of Multi-agent Algorithms for Tactical Operations on UAV Hardware. In Proceedings of the 2013 International Conference on Autonomous Agents and Multi-agent Systems (AAMAS '13), Saint Paul, MN, USA, 6–10 May 2013; pp. 1407–1408.
15. Kumar, P.; Morawska, L.; Birmili, W.; Paasonen, P.; Hu, M.; Kulmala, M.; Harrison, R.M.; Norford, L.; Britter, R. Ultrafine particles in cities. *Environ. Int.* **2014**, *66*, 1–10.

16. Simpson, I.J.; Colman, J.J.; Swanson, A.L.; Bandy, A.R.; Thornton, D.C.; Blake, D.R.; Rowland, F.S. Aircraft measurements of dimethyl sulfide (DMS) using a whole air sampling technique. *J. Atmos. Chem.* **2001**, *39*, 191–213.
17. Chwaleba, A.; Olejnik, A.; Rapacki, T.; Tuśnio, N. Analysis of capability of air pollution monitoring from an unmanned aircraft. *Aviation* **2014**, *18*, 13–19.
18. Techy, L.; Schmale, D.G.; Woolsey, C.A. Coordinated aerobiological sampling of a plant pathogen in the lower atmosphere using two autonomous unmanned aerial vehicles. *J. Field Robot.* **2010**, *27*, 335–343.
19. Avellar, G.S.C.; Pereira, G.A.S.; Pimenta, L.C.A.; Iscold, P. Multi-UAV Routing for Area Coverage and Remote Sensing with Minimum Time. *Sensors* **2015**, *15*, 27783–27803.
20. Smith, K.W. Drone Technology: Benefits, Risks, and Legal Considerations. *Seattle J. Environ. Law* **2015**, *5*, 12.
21. Cork, L.; Clothier, R.; Gonzalez, L.F.; Walker, R. The Future of UAS: Standards, Regulations, and Operational Experiences [Workshop Report]. *IEEE Aerosp. Electron. Syst. Mag.* **2007**, *22*, 29–44.
22. Vansteenwegen, P.; Souffriau, W.; Oudheusden, D.V. The orienteering problem: A survey. *Eur. J. Oper. Res.* **2011**, *209*, 1–10.
23. Edelkamp, S.; Kissmann, P.; Torralba, A. BDDs Strike Back (in AI Planning). In Proceedings of the Twenty-Ninth AAAI Conference on Artificial Intelligenc, Austin, TX, USA, 25–29 January 2015.
24. Dechter, R. *Constraint Processing*; Morgan Kaufmann: San Francisco, CA, USA, 2003.
25. Lloyd, S. Least squares quantization in PCM. *IEEE Trans. Inf. Theory* **1982**, *28*, 129–137.
26. Nau, D.; Ghallab, M.; Traverso, P. *Automated Planning: Theory & Practice*; Morgan Kaufmann Publishers Inc.: San Francisco, CA, USA, 2004.
27. Smith, D.E. Choosing Objectives in Over-Subscription Planning. In Proceedings of the Fourteenth International Conference on Automated Planning and Scheduling (ICAPS 2004), Whistler, BC, Canada, 3–7 June 2004; pp. 393–401.
28. Bylander, T. The Computational Complexity of Propositional STRIPS Planning. *Artif. Intell.* **1994**, *69*, 165–204.

sensors

MDPI

Article

Time-of-Travel Methods for Measuring Optical Flow on Board a Micro Flying Robot

Erik Vanhoutte, Stefano Mafrica, Franck Ruffier, Reinoud J. Bootsma and Julien Serres *

Aix-Marseille Université, CNRS, ISM UMR7287, 13288 Marseille Cedex 09, France; erik.vanhoutte@univ-amu.fr (E.V.); stefano.mafrica@mpsa.com (S.M.); franck.ruffier@univ-amu.fr (F.R.); reinoud.bootsma@univ-amu.fr (R.J.B.)

* Correspondence: julien.serres@univ-amu.fr; Tel.: +33-4-9182-8365

Academic Editors: Felipe Gonzalez Toro and Antonios Tsourdos
Received: 30 December 2016; Accepted: 9 March 2017; Published: 11 March 2017

Abstract: For use in autonomous micro air vehicles, visual sensors must not only be small, lightweight and insensitive to light variations; on-board autopilots also require fast and accurate optical flow measurements over a wide range of speeds. Using an auto-adaptive bio-inspired Michaelis–Menten Auto-adaptive Pixel (M^2APix) analog silicon retina, in this article, we present comparative tests of two optical flow calculation algorithms operating under lighting conditions from 6×10^{-7} to 1.6×10^{-2} W·cm^{-2} (i.e., from 0.2 to 12,000 lux for human vision). Contrast "time of travel" between two adjacent light-sensitive pixels was determined by thresholding and by cross-correlating the two pixels' signals, with measurement frequency up to 5 kHz for the 10 local motion sensors of the M^2APix sensor. While both algorithms adequately measured optical flow between 25 °/s and 1000 °/s, thresholding gave rise to a lower precision, especially due to a larger number of outliers at higher speeds. Compared to thresholding, cross-correlation also allowed for a higher rate of optical flow output (99 Hz and 1195 Hz, respectively) but required substantially more computational resources.

Keywords: optic flow sensor; sense and avoid; VLSI retina; micro air vehicle (MAV); bionics; bio-inspired robotics; biorobotics

1. Introduction

To detect and avoid obstacles in unpredictable environments, flying insects rely heavily on optical flow (OF) [1], defined as the vector field of angular velocities of contrasted points, edges or surfaces resulting from the relative motion between the observer and the surrounding objects [2,3]. Their OF-based strategies therefore provide inspiration for the development of smart autopilots for micro-air-vehicles (MAVs) [1,4,5] and smart artificial retinas [6–12].

Insect-sized MAVs are increasingly becoming a reality [13–18] and have to be fitted in the future with sensors and flight control devices enabling them to perform all kinds of aerial maneuvers, including ground and obstacle avoidance, terrain-following and landing.

The fitted sensors should not only be non-emissive to allow an MAV to save energy resources and to be stealth in flight, but must also guarantee a high refresh rate because GPS (GPS stands for the Global Positioning System) signals are limited in both spatial (~1 m) and temporal resolution (~7 Hz). Micro cameras have been proposed for MAV applications (e.g., a PAL (PAL stands for Phase Alternating Line) -camera with 720 × 576 pixels at 25 fps [16] or a CMOS (CMOS stands for Complementary Metal-Oxide Semiconductor) camera with 752 × 480 pixels at 80 fps [19]), including visual-based simultaneous localization and mapping algorithms (SLAM).

A few cameras deliver fast and reliable OF measurement, but still remain bulky: (i) a PIX4FLOW CMOS camera with 4× binning and 188 × 120 pixels delivers only one OF vector—direction and magnitude—at 250 fps [20]; and (ii) the event-based camera called DAVIS (DAVIS stands for

Dynamic and Active-pixel VIsion Sensor) with 240 × 180 pixels delivers accurate OF measurement asynchronously [21,22]. The drawbacks of CMOS cameras moreover include their poor robustness to various lighting conditions (e.g., ~600 lux in [20]) and their high (pixel number based) demands on computational resources.

It is essential for MAV applications to use lightweight, low-power consumption and high refresh rate sensors to comply with fast dynamics and to stay reactive in unpredictable environments. Hovering requires lower refresh rates than obstacle avoidance maneuvers while flying at high speed. Inertial measurement units (IMUs) are used to stabilize an MAV in flight or simply to hover at a refresh rate up to 500 Hz. Recently, both an IMU sampled at 500 Hz and OF measurements sampled at 25 Hz were fused using an extended Kalman filter to make an MAV hover robustly [23]. The authors' idea was to use the OF direction, but not its scale to correct for inertial sensor drift in particular during changes of direction and to enhance the accuracy of a hovering positioning: this strategy helps in updating both an attitude and a positioning control outputs at 100 Hz [23]. Lately, a 40-gram pocket drone avoided obstacles on the basis of OF computed at 20 Hz [24]. The refresh rate of the visual processing appears to be the key parameter to make an MAV achieve "aggressive" maneuvers based on on-board vision (as demonstrated in [25]) with sufficient reactive abilities in unpredictable environments. More generally, the criteria for the evaluation of the potential of OF sensors for MAV applications include:

- robustness to light level variations, defined by the number of irradiance decades in which the visual sensor can operate;
- range of OF angular speeds (or magnitudes) covered, defined by the minimum and maximum values measured;
- accuracy and precision, defined by systematic errors and coefficients of variation;
- output refresh rate, defined by the instantaneous output frequency.

A recent OF sensor was based on the M^{2}APix (M^{2}APix stands for Michaelis–Menten auto-adaptive pixel) retina that can auto-adapt in a seven-decade lighting range and responds appropriately to step changes up to ±3 decades [26]. The pixels do not saturate thanks to the (normalization based) intrinsic properties of the Michaelis–Menten equation [27]. A comparison of the characteristics of auto-adaptive Michaelis–Menten and Delbrück pixels under identical lighting conditions demonstrated better performance of the Michaelis–Menten pixels in terms of dynamic sensitivity and minimum contrast detection [12].

A single local motion sensor (LMS) fitted with two auto-adaptive pixels has been demonstrated [7] to allow measuring an OF range from 50 to 350 °/s at relatively constant output refresh rate (~5–7 Hz) despite variations in lighting conditions from ~50 lux to 10,000 lux. With a larger number of LMSs, output refresh rate increased (64 Hz with 5 LMSs) in an OF range from 25 to 350 °/s (i.e., 1.1-decade) at a constant natural lighting ~1500 lux [8]. A similar OF range from 50 to 250 °/s (i.e., 0.7-decade) was measured with a semi-panoramic artificial eye, called CurvACE [9,28]. The OF range can also be adjusted by tuning the inter-pixel angle, as done for outdoor flights in [29]. An ad hoc interpolation-based "time-of-travel" algorithm relying on thresholding was embedded into a small 40 MIPS (MIPS stands for Million Instructions Per Second) dsPIC® (Microchip Technology Inc., Chandler, AZ, USA) microcontroller to allow a trade-off between OF range [25 °/s; 350 °/s], accuracy (σ_{error} < 15 °/s), the sample rate of the visual signals (200 Hz), computational resources (10.5% on average of the processing time available) and the OF refresh rate (95 Hz with 5 LMSs) [30]. With this limited OF range, the robot's operating OF was adjusted to the middle of the range: 125 °/s in [31], 160 °/s in [32], or even 200 °/s in [33] (which was by the way quite high in comparison with related works: ~60–70 °/s in [17,34–36]).

In this paper, we compare two different methods for the calculation of OF. Both methods are based on time-of-travel algorithms, relying either on thresholding or on cross-correlation of adjacent bandpass-filtered visual signals. The two algorithms are applied on the same visual signals provided by the M^{2}APix retina when operating under different combinations of the following conditions:

- in an OF range from 25 °/s to 1000 °/s;
- under irradiance conditions varying from 6×10^{-7} W·cm^{-2} to 1.6×10^{-2} W·cm^{-2};
- with sampling rates between 100 Hz and 1 kHz;
- in real flight when fitted onto a 350-gram MAV.

It will be shown that at low sampling rate ($F_e < 500$ Hz), the cross-correlation method is more precise than the thresholding method over a wide range of OF speeds. Requiring much less computational resources, the thresholding method functions adequately at a high sampling rate (F_e close to 1 kHz), with the OF range covered depending mainly on the inter pixel angle $\Delta\varphi$.

In Section 2, optics and front-end pixels of the M^{2}APix sensor will be briefly introduced with respect to previous work based on the M^{2}APix sensor. In Section 3, the two methods of OF computation based on time-of-travel algorithms will be described in detail. In Section 4, both time-of-travel algorithms applied on the same M^{2}APix sensor visual signals will be tested in an OF range from 25 °/s to 175 °/s at different levels of lighting, then in an OF range from 25 °/s to 1000 °/s at a constant lighting by using an experimental setup based on a moving pattern. In Section 5, both time-of-travel algorithms with the M^{2}APix sensor will be tested on-board a 350-gram quadrotor during flights over a flat ground textured with a natural scene. Both time-of-travel methods will be compared in terms of computational resources, accuracy, precision and output refresh rate. In Section 6, the benefits of each method will be discussed, leading to the conclusion that time-of-travel algorithms with a M^{2}APix sensor can measure a wide OF range at high refresh output frequency despite large lighting variations in real flight.

2. Optics and Front-End Pixels of the M^{2}APix Sensor

The M^{2}APix sensor is composed of two groups of light-sensitive pixels: 12 Michaelis–Menten pixels and 12 additional Delbrück pixels [12] (Figure 1a,b). Each group is composed of two rows of six pixels with rows offset by half the inter-pixel distance (Figure 1c). In this paper, we only used the 12 Michaelis–Menten pixels, sampled at the frequency F_e (from 100 Hz to 1 kHz, depending on the experimental setting) by an Overo AirSTORM computer-on-module (COM) (Gumstix, Redwood City, CA, USA) featuring a 1-GHz CPU DM3703 processor (Texas Instruments, Dallas, TX, USA) comprising an ARM Cortex-A8 architecture.

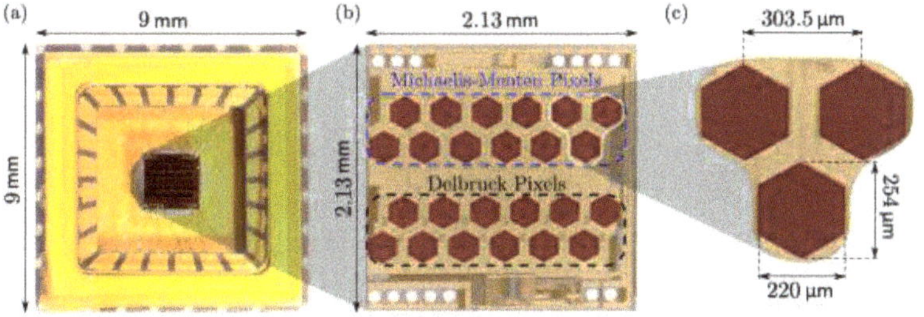

Figure 1. (**a**) The M^{2}APix sensor chip with its wire bounding; (**b**) the silicon retina composed of 12 Michaelis–Menten pixels and 12 additional Delbrück pixels; (**c**) view of three neighboring pixels. From [12].

The M^{2}APix sensor is equipped with a lens taken from a Raspberry Pi camera (focal length = 2 mm). The lens is deliberately defocused (distance between the lens and the retina = 1 mm) to provide a bio-inspired low-pass spatial filter [37]. The optical configuration can be described by the inter-pixel angle $\Delta\varphi = 4.3°$ and the full width at half height of the Gaussian sensitivity, the acceptance angle $\Delta\rho = 3.0°$, thereby respecting the $\frac{\Delta\rho}{\Delta\varphi} < 2$ rule for proper visual sampling established for diurnal insects. In this configuration, the M^{2}APix sensor has a field of view of $32.8° \times 13.4°$ in the horizontal and vertical plane, respectively.

In most animals' visual systems, the photoreceptors have been observed to auto-adapt their response to the ambient light level [27]. The M^2APix pixels are also auto-adaptive, as their responses are normalized by a signal that depends on the low-pass filtered average of all of the 12 photoreceptors' currents, according to the Michaelis–Menten equation [12,38]. The adaptation time constant of the pixels' response was set here at 0.5 s by using an external capacitor of 47 nF (see [12] for more details). This allows the amplitude of the output signals to remain within the same constant range notwithstanding variations in light level and, therefore, makes it possible for the pixels to operate in high-dynamic-range lighting conditions.

3. Optical Flow Computed by Time-Of-Travel-Based Algorithms

Time-of-travel algorithms compute the magnitude of OF (Equation (1)) by measuring the time delay Δt between the signals of two adjacent photoreceptors (here Michaelis–Menten pixels) constituting a local motion sensor (LMS):

$$OF = \frac{\Delta \varphi}{\Delta t} \tag{1}$$

In the work presented here, we used 10 LMSs (two rows of five), each computing the 1D OF in the direction parallel to the rows of pixels, as shown in Figure 2.

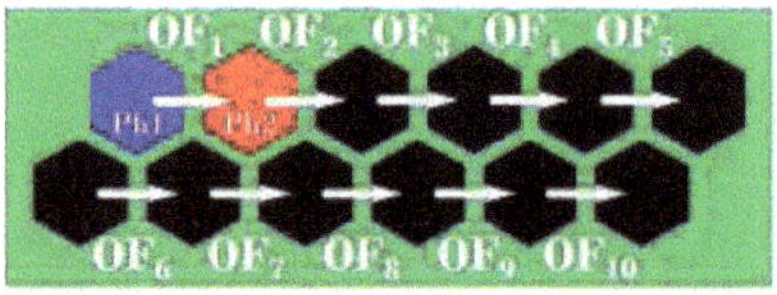

Figure 2. The 12 Michaelis–Menten pixels (Figure 1b) and the 10 corresponding local motion sensors (LMSs). LMSs are unidirectional and can only measure the OF magnitudes OF_i $i \in \{1, ..., 10\}$ in 1D. The optics geometry of the first two-pixel LMS (composed of the red and blue pixels) is described in Figure 3a.

3.1. Time-Of-Travel Based on Signal Thresholding

This time-of-travel algorithm is based on opto-electrophysiological studies of the fly's lobula plate tangential cells (H1-neuron) [39] and is described in detail in [4,7,8,29].

The algorithm is illustrated in Figure 3b and computes the "time-of-travel" Δt by means of signal thresholding. To this end, the output of each pixel is first digitally band-pass filtered with cut-off frequencies of 3 Hz and 30 Hz. Two adjacent filtered signals are then thresholded (hysteresis thresholding) in order to measure the time delay Δt between them: a counter is started when the first pixel's signal (the blue signal in Figure 3b) reaches a threshold, and it is stopped when the second pixel's signal (the red signal in Figure 3b) reaches the same threshold. Lastly, the OF magnitude is computed using Equation (1).

The key parameter in this algorithm is the threshold value. A too large threshold value will allow detection only of high-amplitude pixel signals induced by strong contrasts and is therefore likely to neglect relevant signals. A too small threshold value will allow noise to cross the threshold and induce erroneous OF values. The thresholds' values (hysteresis thresholding in Figure 3b) were therefore adjusted manually with respect to the best trade-off between the maximum refresh rate of the LMSs' output and the minimum number of outliers.

In the experiments presented here, the high threshold was set at 5% of maximum signal amplitude (half the value of the high threshold for the low threshold for the hysteresis comparator; see Figure 3b), and all time delays falling outside a specific range (Δt from 4 to 717 ms, i.e., OF from 6 to 1145 °/s) were excluded and considered as outliers. The algorithm was run at the pixels' sampling frequency F_e, which varied between 100 Hz and 1 kHz, giving a quantization time $T_e = \frac{1}{F_e}$ varying between 1 ms and 10 ms. The OF was then computed by using Equation (1), where Δt is a multiple of T_e. The OF resolution ΔOF of this method is therefore non-constant, as it is given by an inverse function of the time delay Δt.

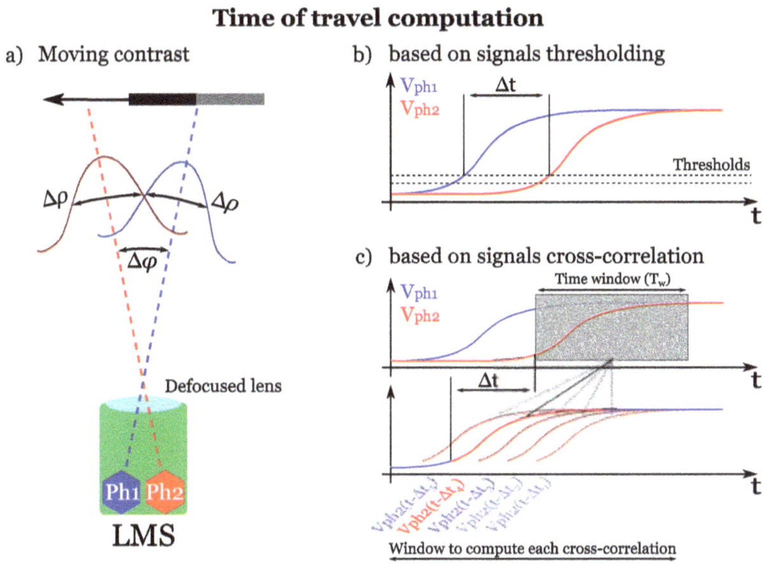

Figure 3. (**a**) Optics geometry of a LMS. A defocused lens confers an inter-pixel angle $\Delta\varphi$ between the optical axes of two adjacent pixels Ph1 (blue) and Ph2 (red) and an acceptance angle $\Delta\rho$ given by the width of the Gaussian angular sensitivity at half height. The output signals V_{ph1} and V_{ph2} coming from pixels Ph1 and Ph2, respectively, are naturally delayed by a "time-of-travel" Δt when a contrast is moving in front of them (Figure 3b,c). The OF can therefore be computed with Equation (1). (**b**) Principle of the time-of-travel algorithm based on signal thresholding: the time delay Δt is computed when both signals reach a given hysteresis threshold [4,7,8,29]. (**c**) Principle of the time-of-travel algorithm based on signals' cross-correlation: the time delay Δt is computed as the time delay giving the maximum cross-correlation coefficient ϱ between the delayed and non-delayed signal (see Figure 2 in [26] for details). Note that for simplicity, the time-of-travel computation is shown here using the raw signals V_{ph1} and V_{ph2}, but it is actually computed after band-pass filtering these signals (see [4,7,8,26,29] for details).

3.2. Time-of-Travel Based on Signals' Cross-Correlation

This algorithm was inspired by the correlation-based models [40,41] and is based on signals' cross-correlation (Figure 3c), as presented in [26]. First, the pixels' output signals are digitally filtered with the same band-pass filter used in the thresholding algorithm (see Section 3.1). Then, one of the two signals in each pair of adjacent pixels is delayed by several time delays Δt_i, and the Pearson cross-correlation coefficients ϱ_i are computed between the delayed (V_{ph2} in Figure 3c) and non-delayed signals (V_{ph1} in Figure 3c) within a fixed time window T_w. Lastly, the time delay Δt_m giving the maximum coefficient ϱ_m is obtained and used to compute the OF using Equation (1) as long as ϱ_m is greater than a given threshold ϱ_{thr}. This threshold on the cross-correlation coefficients was set at 0.99 to avoid OF measurement errors due to signals mismatching [26].

The size of the fixed time window T_w is one of the key parameters and requires being tuned as function of the signals' bandwidth in order to have reliable cross-correlation coefficients (ϱ_i). Since the signals' bandwidth is given by the band-pass filter (i.e. 3–30 Hz; see Section 3.1), T_w was fixed here at a value of $T_w = 0.14\ s$ so that $3 \leq \frac{1}{T_w} \leq 30$. The cross-correlation time window was also defined as $T_w = N \cdot T_e$, with N being the number of samples in the time window. Therefore, a sampling rate $F_e = 500$ Hz entails $N = 70$, $F_e = 250$ Hz entails $N = 35$, and so on.

To compute the cross-correlation coefficients ϱ_i, the signal time window T_w can be delayed by time steps Δt_i that are calculated from the desired OF values OF_i using the inverse function given in Equation (1). This can be done by taking time steps Δt_i between two consecutive sampling steps ($n\ T_e$ and $(n+1)\ T_e$) after linearly interpolating the sampled signals. Therefore, by choosing the desired OF values OF_i using constant steps, the computed OF will as a result be quantized with a constant resolution ΔOF, contrary to the algorithm based on signal thresholding (see Section 3.1).

4. Measuring Optical Flow with a Moving Texture

4.1. Method

In the following experiments, the M^2APix sensor was fixed perpendicularly to a moving pattern whose linear speed could be controlled (Figure 4). Lighting was provided by three sources: daylight coming through the top windows (intensity manipulated by means of blinds), fluorescent tubes attached to the ceiling (intensity manipulated by means of a dimmer) and an LED projector that could be positioned in front of the moving pattern for high luminosity conditions. The M^2APix sensor was thus confronted with a pure translational wide-range OF (from 25°/s to 1000°/s) under various lighting conditions (from 6×10^{-7} W·cm^{-2} to 1.6×10^{-2}W·cm^{-2}, i.e., from 0.2 lux to 12,000 lux for human vision).

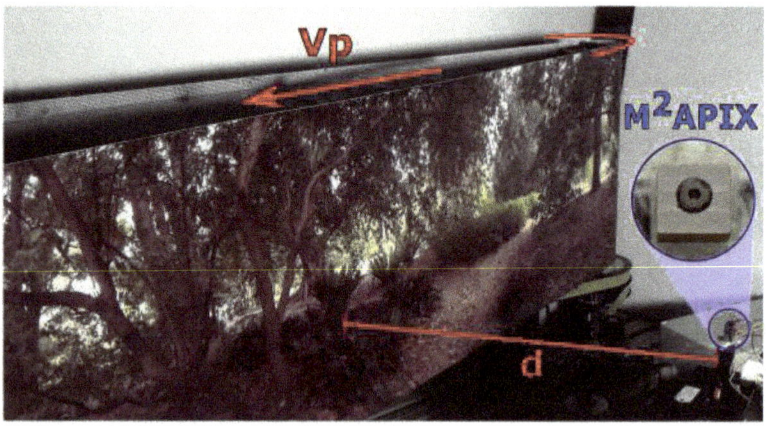

Figure 4. Experimental setup using a moving pattern and varying the artificial lighting to test algorithms with the M^2APix sensor both in various OF ranges and in various lighting conditions. The pattern speed V_p varied from 0 to 1.4 m·s^{-1}; the sensor was placed perpendicularly in front of the moving pattern at a distance d from 8 cm to 40 cm.

The moving-pattern experiments were separated in two parts: in the first part, the sensor was placed at a distance d = 40 cm from the moving pattern, and both the pattern speed and the lighting conditions were varied, whereas in the second part, only the pattern speed was varied at long constant steps, and the sensor was placed closer to the moving pattern (d = 8 cm) to obtain high OF.

During the first part, each of the two algorithms was tested at five different irradiance levels. For each irradiance level, the pattern angular speed $OF_{pattern}$ varied over 10 periods as a sinus function with an angular frequency $f = 0.2\pi\ rad/s$ and a magnitude speed V_p varying from V_{pmin} = 0.18 m·s^{-1} to V_{pmax} = 1.22 m·s^{-1} (Equation (2)). With the M^2APix sensor placed at a distance d = 40 cm (Figure 4), the moving pattern generated an OF range from 25°/s–175°/s (green lines in Figure 5b,c), according to the following equation:

$$OF_{pattern} = \frac{180}{\pi}\left(\frac{V_{pmax} - V_{pmin}}{2d} \cdot \cos(2\pi f t + \pi) + \frac{V_{pmax} + V_{pmin}}{2d}\right) \quad (2)$$

During the second part, the level of lighting was fixed at an irradiance of 7×10^{-4} W·cm^{-2}, and the two algorithms were used to compute OF in a wide OF range, from 25 °/s to 1000 °/s, by varying both the pattern speed (V_p from 0 to 1.4 m·s^{-1}) and the sensor's distance d (see Figure 4). In particular, to obtain an OF ranging from 25 °/s to 200 °/s, d was set at 40 cm, and V_p varied from 0.18 m·s^{-1} to 1.34 m·s^{-1}; whereas, for an OF ranging from 250 °/s to 1000 °/s, d was set at 8 cm, and V_p varied from 0.35 m·s^{-1} to 1.4 m·s^{-1}.

These two experiments allowed us to compare the precision and the output refresh rate of both time-of-travel algorithms when using an auto-adaptive retina, such as the M^{2}APix sensor in a wide light-level range (five-decade) and in a wide OF range (1.6-decade). However, the results should be similar with any other auto-adaptive retina implementing a similar auto-adaptation process.

The two algorithms were run at the highest operating rate/frequency that still guaranteed a proper operation for future embedded MAV applications, taking into account the constraint to maintain 40% of CPU (CPU stands for central processing unit) load-free for MAV control and navigation tasks. Therefore, the time-of-travel algorithm based on signals thresholding was run at 1000 Hz, whereas the time-of-travel algorithm based on signals cross-correlation was run at 500 Hz due to the high CPU load of this method.

4.2. Results

Figure 5 shows the OF measurements and their instantaneous refresh rates obtained when using the thresholding method (Figure 5b) and the cross-correlation method (Figure 5c) at five different light levels (five colored columns in Figure 5a–c). The OF measurements corresponding to the five light levels shown in Figure 5a were actually obtained during five distinct tests (see the oblique black line segments separating each test at 5, 10, 15 and 20 s in Figure 5b,c). Furthermore, for each light level (each column/sinusoidal period in Figure 5b,c), the OF measurements of the 10 LMSs obtained during 10 sinusoidal periods of the moving pattern (see Section 4.1) were overlaid on only one sinusoidal period in order to show all of the measurements in one figure. The refresh rates given in the lower plots in Figure 5b,c were computed by low-pass filtering the average number of OF measurements obtained at every time step over the last three and the next three steps (six centered point low-pass digital averaging filter). The OF errors were computed as the difference between the OF measurements and their ground-truth values (blue dots and green lines in Figure 5b,c, respectively). The boxplots and the distributions of these errors shown in Figure 5d were computed with 1 °/s beams in the [150 °/s; 150 °/s] range.

The number of outliers is much lower for the cross-correlation method (Figure 5c) than for the thresholding method (Figure 5b). While both error distributions (Figure 5d) have a similar standard deviation, the boxplots reveal the differences in the number of outliers.

The thresholding method running at 1 kHz (Figure 5b) is very light in terms of CPU load (2.8% ± 0.8%) and provides a good OF accuracy and precision until an irradiance of 7×10^{-6} W·cm^{-2} (blue column in Figure 5) despite more outliers than the cross-correlation method. However, the refresh rate obtained is relatively low (average of 97 Hz and maximum of 607 Hz for the full sensor). At very low irradiance, i.e., 6×10^{-7} W·cm^{-2}, both the OF precision and the refresh rate are strongly deteriorated (see the violet column in Figure 5).

In contrast, the cross-correlation method running at 500 Hz gives a much lower number of outliers (boxplots in Figure 5d) and a much higher refresh rate (average of 2326 Hz and maximum of 5000 Hz for the full sensor), but it is very heavy in terms of CPU load (52.5% ± 2.2%). Similarly to the thresholding method, at very low light levels, i.e., 6×10^{-7} W·cm^{-2} (violet column in Figure 5), the refresh rate is deteriorated, especially at very low OF values. With the cross-correlation method, the OF precision is however less deteriorated than with the thresholding method.

The cross-correlation method does not delay signals in constant steps, but with steps following an inverse function to obtain a constant OF-measurement resolution.

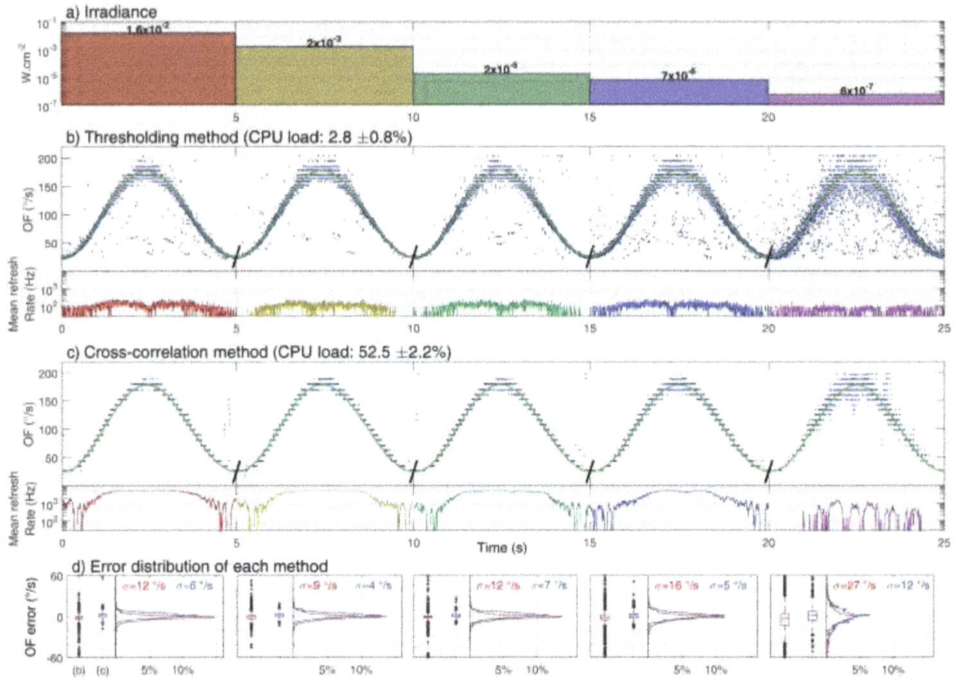

Figure 5. (**a**) The five levels of irradiance from 6×10^{-7} W·cm^{-2} to 1.6×10^{-2} W·cm^{-2} (i.e., from 0.2 lux to 12,000 lux for human vision) used during the first part of the experiments (see the experimental setup shown in Figure 4). (**b**,**c**) OF measurements obtained with the M^{2}APix sensor using (b) the thresholding method (F_e = 1 kHz) and (c) the cross-correlation method (F_e = 500 Hz) for the 10 LMSs during 10 laps of the moving pattern. Tests were run separately for each of the five lighting conditions. Each blue dot represents one OF measurement of one LMS; however, the mean output refresh rate is the average of the number of OF measurement from the 10 LMSs in 1 s. The green line represents the OF ground-truth computed using Equation (2). (**d**) Error distribution of each time-of-travel method: the thresholding method is in red and the cross-correlation method in blue.

In the second experiment, various levels of OF were tested from 25 °/s to 1000 °/s with an OF step of 50 °/s (except for the first step) at a constant irradiance level (7×10^{-4} W·cm^{-2}). For each step, the OF mean μ, standard deviation σ and coefficient of variation $\frac{\sigma}{\mu}$ were computed for both time-of-travel algorithms, as shown in Figure 6.

Inspection of Figure 6 reveals that the OF mean values μ are very close to the ground-truth values for both methods (blue circles in Figure 6). However, the standard deviation σ (blue vertical bars in Figure 6), and therefore, the coefficient of variation $C_V = \frac{\sigma}{\mu}$ (red curves in Figure 6), is higher when using the thresholding method ($C_V > 0.1$ in Figure 6a) than the cross-correlation method ($C_V < 0.1$ in Figure 6b), except for very low OF values, i.e., 25°/s, where the precision is nearly the same.

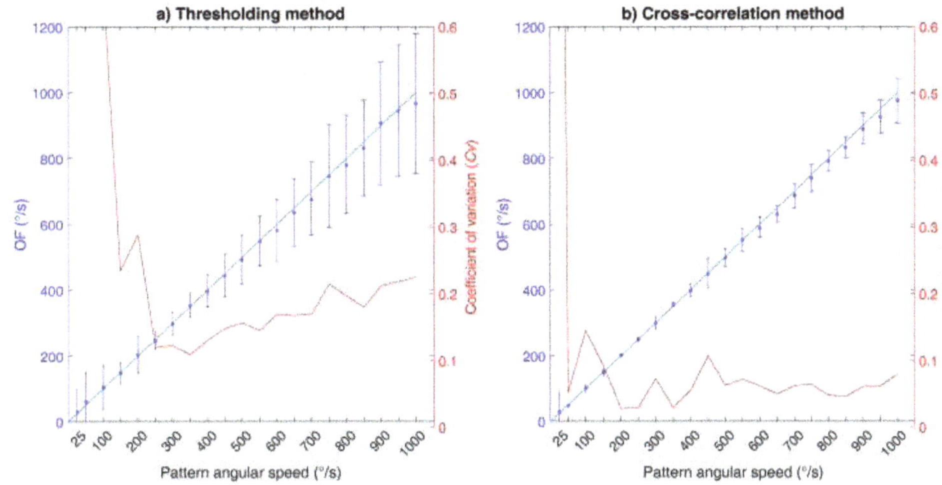

Figure 6. Accuracy of the M^2APix sensor's output when using: (**a**) the time-of-travel algorithm based on signal thresholding with F_e = 1 kHz; and (**b**) the time-of-travel algorithm based on signal cross-correlation with F_e = 500 Hz. The tests were performed at a constant level of irradiance (7×10^{-4} W·cm^{-2}), and the OF produced by the moving pattern was made to vary from 25 °/s to 1000 °/s with steps of 50 °/s (except for the first step). Each graph shows the average μ (blue circles), the standard deviation σ (blue vertical bars) and the coefficient of variation $\frac{\sigma}{\mu}$ (red lines) of the OF measurements with respect to the angular speed of the moving pattern (see Figure 4). The ground-truth values of the OF are given by the blue lines.

The increase in the coefficient of variation with the thresholding method at low angular speed could be explained by the auto-adaptive response of the M^2APix sensor that respects a time constant of 0.5 s on average on the whole sensor. Consequently, the lower the angular speed, the higher the number of outliers, as the pixels' adaptation time constant is close to the signals' dynamics, therefore generating matching errors. If one of the two signals does not reach the hysteresis threshold, no OF is measured. This does not happen with the cross-correlation method because the OF is computed by considering a time window comprising N samples (see Section 3.2).

Nevertheless, even with the cross-correlation method, the coefficient of variation increases abruptly around 25°/s. If signals coming from the same contrast on two adjacent pixels are too different due to the auto-adaptation process, the minimum cross-correlation coefficient (set at $\varrho_i = 0.99$) will not be reached, and no OF measurement will be generated. Moreover, at low speed, the signal dynamics is close to the fixed time window size set at $T_w = 0.14$ s. Different signals may therefore appear similar in this time window, thereby generating matching errors.

Toward the high angular speeds, the delay Δt between each photoreceptors' signal approaches the M^2APix sensor sampling rate F_e. The OF calculation therefore becomes less precise, especially for the thresholding method in which precision depends directly on the sampling rate F_e.

In the experiments presented in this section, the M^2APix sensor was fixed while a textured pattern was moving in front of it in a wide OF range and under controlled wide-range light levels. However, real flight conditions can generate disturbances due to vibrations or uncontrolled lighting conditions, which can change abruptly during the flight. Therefore, to test the two algorithms in real flight conditions, one M^2APix sensor was mounted underneath a 350-gram quadrotor pointing downwards to measure the OF produced by the visual motion of the ground below.

5. Measuring Optical Flow in Flight

5.1. Method

Here, we evaluate the two algorithms running on the M^2APix sensor under real flight conditions, in the presence of vibrations, movement disturbances (due to the MAV's attitude stabilization) and light variations. To this end, the M^2APix sensor was attached to the bottom of a 350-gram X4-MaG quadrotor [42], aligned to the direction of frontal movement of the MAV. In this configuration, OF measurements were only influenced by the linear speed vector's components of the quadrotor and its pitch rotation movements. Light intensity was measured with a photodiode having the same spectral sensitivity, also attached to the bottom of the X4-MaG and oriented in the same downward direction as the M^2APix sensor. To get the OF ground-truth with high accuracy, the X4-MaG flew inside the Mediterranean Flying Arena (http://www.flying-arena.eu) equipped with 17 motion-capture cameras covering a $6 \times 8 \times 6$ m volume. The X4-MaG followed a predefined 3D trajectory using the VICON™ (Oxford, UK) system outputs as feedback signals.

The trajectory pattern retained for this experiment was a half-moon shape, positioned at a constant height of 0.4 m above the textured ground (cf. Figure 7). The X4-MaG was controlled to move along the half-circle trajectory segment at a constant speed of 0.5 $m{\cdot}s^{-1}$. During the straight trajectory segment, velocity first increased progressively from 0 to 1.5 $m{\cdot}s^{-1}$ and then decreased again to 0 $m{\cdot}s^{-1}$. Between these two segments, the X4-MaG hovered in place while making a pure yaw rotation. Since the speed vector of the X4-MaG is tangential to its trajectory, the half-moon pattern confronted the M^2APix sensor with three distinct types of movement: translation, rotation and a combination of both.

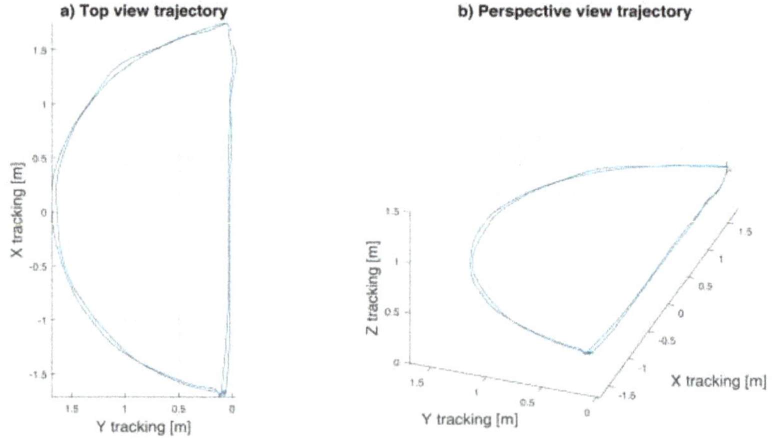

Figure 7. MAV trajectory: (**a**) top view; (**b**) perspective view. The two laps were performed at a flight height of 0.4 m.

Lighting was provided by three sources: daylight coming through the top windows, fluorescent tubes attached to the ceiling and infra-red light emitted by the motion capture system. Manipulation of the window blinds not only modulated the global intensity of the entering daylight, but also gave rise to high-range variations in lighting on the ground texture (see Figure 8). The intensity of the light emitted by the fluorescent tubes could be adjusted manually by means of a dimmer. Because it was necessary for the proper functioning of the tracking, the 9×10^{-5} $W{\cdot}\ cm^{-2}$ irradiance provided by the motion capture system could not be altered.

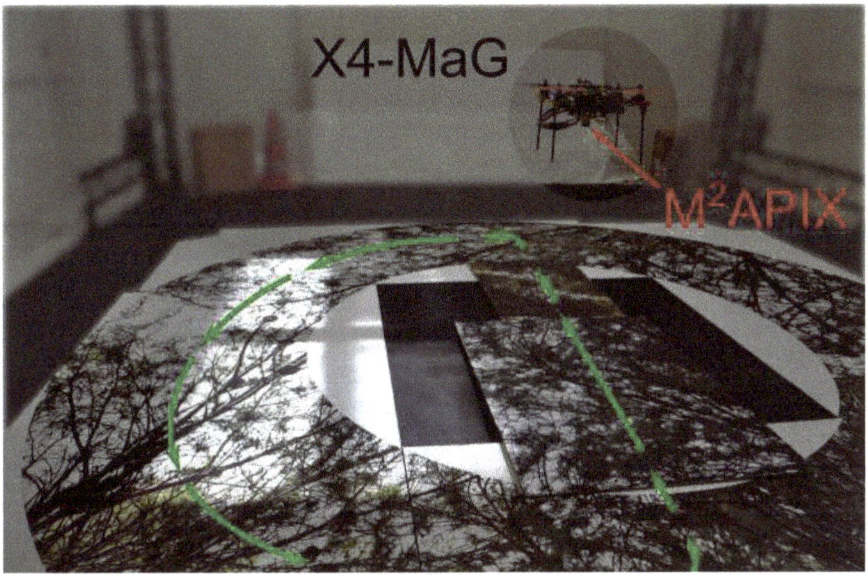

Figure 8. X4-MaG drone hovering at 1.2 m above the natural texture in the flight arena. The half-moon shaped trajectory is represented in green. The M^{2}APix sensor is embedded under the quadrotor looking downwards and measuring the OF coming from the ground below. See Video S1 to watch the experiment.

5.2. In-Flight Results

Two flight laps of the pattern were recorded, during which the light was arbitrarily modulated. During the first lap, the blinds were left open, and daylight projected very luminous bands on the textured ground (Figure 8) that the drone encountered between the third and eighth seconds of the experiment (Figure 9a). As a result, irradiance during this first lap varied between 10^{-3} W·cm^{-2} and 10^{-1} W·cm^{-2}. The blinds were then fully closed (at 10 s in Figure 9). Manual modulation of the fluorescent tubes induced irradiance variations ranging from 9×10^{-5} W·cm^{-2} to 2×10^{-3} W·cm^{-2}.

In Figure 9b1,c1, white zones correspond to periods of forward movement of the drone, whereas gray zones correspond to hovering periods with 90° yaw rotation, orienting the drone in preparation of the next trajectory segment. Half-circle segments are identified by larger white zones between 0 and 10 s and between 18 and 28 s, and straight segments are identified by smaller white zones between 13 and 16 s and between 31 and 34 s. Thus, one large white zone and one small white zone represent one lap. For a trajectory height of 0.4 m and a speed range between 0.5 and 1.5 m·s^{-1} without pitch rotations, the OF theoretically varies from 72 to 215 °/s.

The OF oscillations observed during the half-circle part are mainly due to the drone's attitude stabilization system, with the resulting pitch rotations inducing changes in OF.

Inspection of Figure 9b revealed that the thresholding method provided a virtually uninterrupted stream of OF measurements. Even in the presence of forward drone movement, measurements however showed quite some variability, as captured by the overall (excluding gray areas) standard deviation (σ) of 43 °/s. During forward movement (white zones), the OF-output refresh rate reached an average of 99 Hz for the 10 LMSs. The cross-correlation method (see Figure 9c) occasionally gave rise to measurement gaps, especially during the high acceleration phases (see Figure 9c at 13 s). However, compared to the thresholding method, measurement variability was lower with $\sigma = 16$ °/s, and the OF-output refresh rate was much higher, reaching an average of 1195 Hz for the full M^{2}APix sensor (10 LMSs).

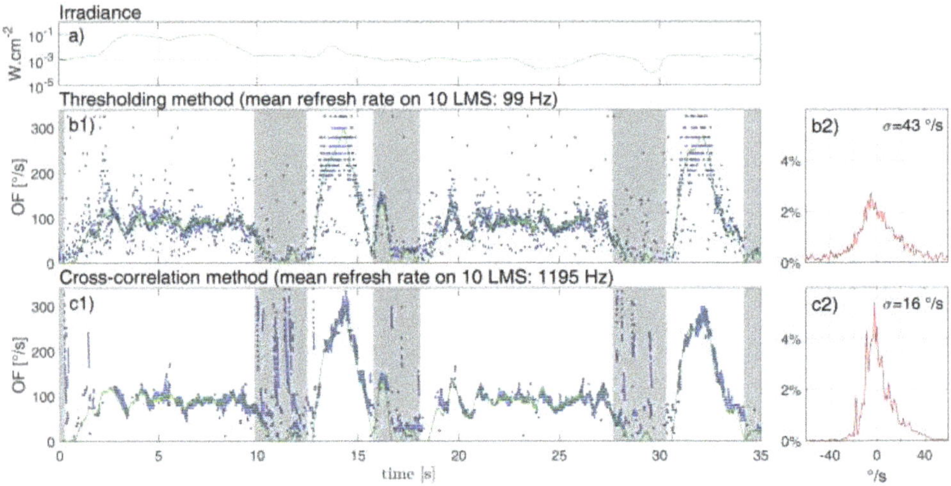

Figure 9. (**a**) Light intensity in W·cm^{-2} measured with a photodiode oriented in the same direction as the M^{2}APix sensor embedded under the X4-MaG; (**b**,**c**) In the left panels (**b1**,**c1**), blue points represent single-LMS OF measurements; the green line is the ground-truth provided by high-precision Vicon tracking. Right panels (**b2**,**c2**) present the OF error distribution, with 1°/s beams in the [300°/s; 300°/s] range, with respect to the ground-truth during forward drone movement (white zones in left panels). The excluded (gray) zones correspond to periods of pure yaw rotation during the MAV trajectory. See Supplemental Movie 1 to watch the experiment.

The foregoing results bring out the differential capabilities of the two methods in different situations. Although, overall, the cross-correlation method provides better results in terms of precision, robustness, working range and output refresh rate, its use in MAV applications may be compromised by a crippling CPU load. Indeed, as demonstrated in Section 4 , for a single M^{2}APix sensor, the cross-correlation method, operating at F_e = 500 Hz, gave rise to an average CPU load of 52.5%, whereas with the thresholding method operating at F_e = 1000 Hz, the CPU load was always lower than 3%. We therefore explored OF-output under reduced working frequencies, allowing lessening of the CPU load.

5.3. *Offline Results at a Low Sampling Rate*

The recorded flight data of the online calculations presented in Figure 9 were down-sampled before running the two algorithms again; the results of this offline OF computation are presented in the same format in Figure 10 with F_e = 250 Hz and in Figure 11 with F_e = 100 Hz.

Comparison of Figures 9 to 11 (see also Table 1) revealed that for the thresholding method, a reduction of the sampling rate resulted in a stronger OF quantization (especially visible with higher OF values), accompanied by a decrease in OF-output refresh rate (from 99 Hz at F_e = 1000 Hz to 36 Hz at F_e = 100 Hz) and a widening of the error distribution. For the cross-correlation method, a reduction in sampling rate did not affect the error distribution to any noticeable extent, whereas the OF-output refresh rate decreased proportionally with the sampling rate (from 1195 Hz at F_e = 500 Hz to 264 Hz at F_e = 100 Hz).

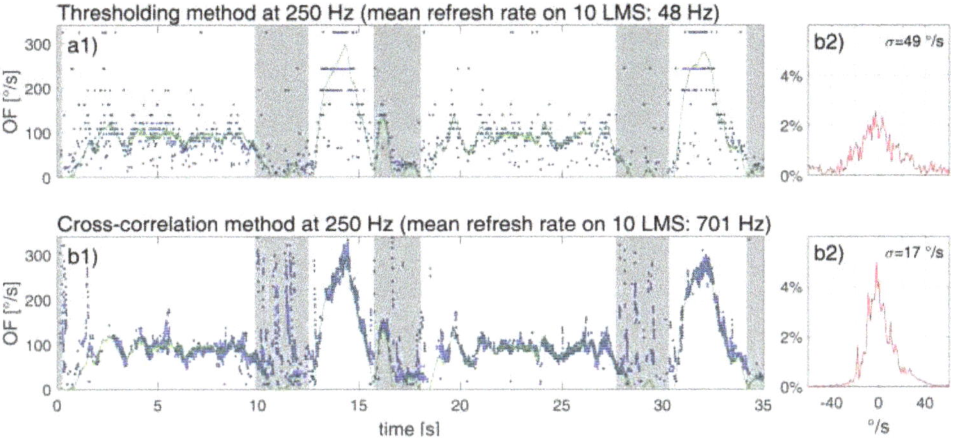

Figure 10. OF computed with (**a**) the thresholding method and (**b**) the cross-correlation method after down-sampling the data underlying Figure 9 to F_e = 250 Hz.

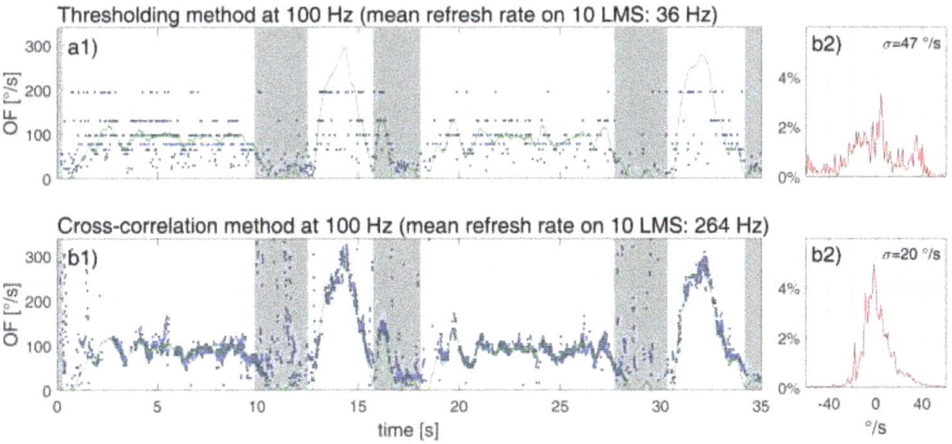

Figure 11. OF computed with (**a**) the thresholding method and (**b**) the cross-correlation method after down-sampling the data underlying Figure 9 to F_e = 100 Hz.

Table 1. Comparison between the results obtained with the two time-of-travel algorithms.

	Thresholding Method				**Cross-Correlation Method**			
Sampling Rate (F_e)	**1 kHz**	**500 Hz**	**250 Hz**	**100 Hz**	**1 kHz**	**500 Hz**	**250 Hz**	**100 Hz**
CPU Load (%)	2.2	1.1 *	< 1 *	< 1 *	overload	52.5	26.3 *	10.5 *
Precision σ (°/s)	43	44	49	47	-	16	17	20
Refresh rate (Hz/10 LMSs)	99	51	48	36	-	1195	701	264

* Theoretical value because corresponding tests were made offline.

6. Discussion and Conclusions

Experiments were performed to test two time-of-travel algorithms with an auto-adaptive bio-inspired silicon retina, called M^{2}APix, intended for MAV applications. The M^{2}APix sensor is a prototype sensor [12] composed of only 12 auto-adaptive pixels forming 10 LMSs. The auto-adaptive pixels allowed us to evaluate each time-of-travel algorithm largely independently of lighting conditions. In the field of robotics, working in a seven-decade range of irradiance is a considerable advantage because robots may then be led to work in environments subject to strong lighting variations, such as obstacle forests, urban canyons or inside buildings.

In the experiments underlying Figure 5, we controlled the lighting in a five-decade range of irradiance. The lower limit was reached at around 10^{-6} W·cm^{-2} irradiance, when the M^{2}APix sensor was no longer able to deliver visual signals.

As can be seen in Figure 6, our results revealed that both time-of-travel algorithms are accurate at all OF speeds tested. However, the precision of the cross-correlation method ($C_V < 0.1$ in Figure 6b) is considerably better than the thresholding method ($C_V > 0.1$ in Figure 6a) over a wide OF range. At the lowest OF (25°/s), both methods strongly lack precision, probably due to the auto-adaption: working at a 0.5-s time constant, the auto-adaptive process may affect OF measurement at low speed when a time delay Δt approaches 0.5 s. It is in fact difficult to measure both slow and fast OFs with a constant inter-pixel angle $\Delta\varphi$; with $\Delta\varphi = 1.5°$, the thresholding allowed measuring an OF range of [1.5°/s; 25°/s] in real flight conditions [29]. Modulating $\Delta\varphi$ as a function of the OF required appears to be the only way to measure the OF in the three-decade range (e.g., [1°/s; 1000°/s]) with a time-of-travel algorithm.

Experiments with a moving pattern and in real flight showed that even under varying lighting conditions, both time-of-travel algorithms are quite accurate with respect to ground-truth OF. The cross-correlation method however appears to be more precise with a higher output refresh rate than the thresholding method. For MAV applications with fully-embedded computational resources, the cross-correlation method requires 18-times more computational resources than the thresholding method. The cross-correlation was therefore tested offline at a low sampling rate (250 Hz in Figure 10b1, 100 Hz in Figure 11b1) and found to work well despite some brief measurement gaps when OF accelerated beyond 260°/s^2 (e.g., Figure 10b1). The thresholding method, on the other hand, only worked adequately at a high sampling rate (1 kHz in Figure 9b1); results were poor at a low sampling rate (250 Hz in Figure 10a1 or 100 Hz in Figure 11a1). For future work, it appears worthwhile to explore combinations of both methods by merging their OF measurements (e.g., a cross-correlation method running at 250 Hz with a thresholding method working in parallel and running at 1 kHz), as this would allow profiting from the benefits of each method without too strongly loading the CPU.

Results presented in Table 1 will allow future research to choose between the time-of-travel algorithms for measuring the OF as a function of embedded computational resources. If an algorithm using little computational resources is required (e.g., for computing dozens or hundreds of LMSs, while using one and the same embedded target), the thresholding method may be preferred. By adding a median filter eliminating matching errors [8], the thresholding method moreover becomes more precise. Its precision can also be improved when running at 200 Hz $\leq F_e <$ 1 kHz using a linear interpolation applied to the photoreceptor signals [30]. While the cross-correlation method is inherently both accurate and precise and can provide a high output refresh rate (that is, about proportional to the sampling rate F_e), it requires much more computational resources (also approximately proportional to the sampling rate F_e), which will limit the number of LMSs that can be embedded into the same target. The CPU load of the cross-correlation method may be reduced by using on-board digital signal processing (DSP).

Finally, from a robotics point of view, our 350-gram X4-MaG quadrotor was equipped with an IMU working at 250 Hz to stabilize attitude and with a downward OF sensor, composed of 10 LMSs running at a refresh output frequency close to the IMU sampling rate. A high refresh rate for both inertial and visual sensors will be a perquisite in the near future to endow MAVs with strong reactive abilities in unpredictable environments. Moreover, this study has demonstrated that a time-of-travel algorithm coupled with a M^{2}APix sensor can accurately measure the OF in a wide range [25°/s; 1000°/s], allowing MAVs to fly close to surrounding obstacles at high speed in any lighting conditions.

Supplementary Materials: The following are available online at http://www.mdpi.com/1424-8220/17/3/571/s1, Video S1: a video of the experiment with the micro flying robot.

Acknowledgments: We thank Julien Diperi for the mechanical design of the flying robot and the test bench and Marc Boyron for his involvement in the overall electronic design of the flying robot. This work was supported by the French Direction Générale de l'Armement (DGA), CNRS, Aix-Marseille University, the Provence-Alpes- Côte d'Azur region and the French National Research Agency for Research (ANR) with the Equipex/Robotex project.

Author Contributions: E.V., J.S. and F.R. conceived of the experiments. E.V. and S.M. implemented the algorithms. E.V. conducted the experiments. All authors actively participated in writing the manuscript.

Conflicts of Interest: The authors declare no conflict of interest.

Abbreviations

The following abbreviations are used in this manuscript:

OF	Optical flow
LMS	Local motion sensor
M^2APix	Michaelis–Menten auto-adaptive pixel

References

1. Srinivasan, M.V. Honeybees as a model for the study of visually guided flight, navigation, and biologically inspired robotics. *Physiol. Rev.* **2011**, *91*, 413–460.
2. Gibson, J.J. The perception of the visual world. *J. Philos.* **1951**, *48*, 788.
3. Nakayama, K.; Loomis, J. Optical velocity patterns, velocity-sensitive neurons, and space perception: A hypothesis. *Perception* **1974**, *3*, 63–80.
4. Franceschini, N.; Ruffier, F.; Serres, J.; Viollet, S. *Optic Flow Based Visual Guidance: From Flying Insects to Miniature Aerial Vehicles*; INTECH Open Access Publisher: Rijeka, Croatia, 2009.
5. Serres, J.; Ruffier, F. Optic Flow-Based Robotics. In *Wiley Encyclopedia of Electrical and Electronics Engineering*; John Wiley & Sons, Inc.: Hoboken, NJ, USA, 2016.
6. Moeckel, R.; Liu, S.C. Motion detection chips for robotic platforms. In *Flying Insects and Robots*; Springer: Berlin, Germany, 2009; pp. 101–114.
7. Expert, F.; Viollet, S.; Ruffier, F. A mouse sensor and a 2-pixel motion sensor exposed to continuous illuminance changes. In Proceedings of the IEEE Sensors, Limerick, Ireland, 28–31 October 2011; pp. 974–977.
8. Roubieu, F.L.; Expert, F.; Sabiron, G.; Ruffier, F. Two-Directional 1-g Visual Motion Sensor Inspired by the Fly's Eye. *IEEE Sens. J.* **2013**, *13*, 1025–1035.
9. Floreano, D.; Pericet-Camara, R.; Viollet, S.; Ruffier, F.; Brückner, A.; Leitel, R.; Buss, W.; Menouni, M.; Expert, F.; Juston, R.; et al. Miniature curved artificial compound eyes. *Proc. Natl. Acad. Sci. USA* **2013**, *110*, 9267–9272.
10. Song, Y.M.; Xie, Y.; Malyarchuk, V.; Xiao, J.; Jung, I.; Choi, K.J.; Liu, Z.; Park, H.; Lu, C.; Kim, R.H.; et al. Digital cameras with designs inspired by the arthropod eye. *Nature* **2013**, *497*, 95–99.
11. Duhamel, P.E.J.; Perez-Arancibia, C.O.; Barrows, G.L.; Wood, R.J. Biologically inspired optical-flow sensing for altitude control of flapping-wing microrobots. *IEEE/ASME Trans. Mechatron.* **2013**, *18*, 556–568.
12. Mafrica, S.; Godiot, S.; Menouni, M.; Boyron, M.; Expert, F.; Juston, R.; Marchand, N.; Ruffier, F.; Viollet, S. A bio-inspired analog silicon retina with Michaelis-Menten auto-adaptive pixels sensitive to small and large changes in light. *Opt. Express* **2015**, *23*, 5614.
13. Duhamel, P.E.J.; Pérez-Arancibia, N.O.; Barrows, G.L.; Wood, R.J. Altitude feedback control of a flapping-wing microrobot using an on-board biologically inspired optical flow sensor. In Proceedings of the 2012 IEEE International Conference on Robotics and Automation (ICRA), Saint Paul, MN, USA, 14–18 May 2012; pp. 4228–4235.
14. Kushleyev, A.; Mellinger, D.; Powers, C.; Kumar, V. Towards a swarm of agile micro quadrotors. *Auton. Robots* **2013**, *35*, 287–300.
15. Ma, K.Y.; Chirarattananon, P.; Fuller, S.B.; Wood, R.J. Controlled flight of a biologically inspired, insect-scale robot. *Science* **2013**, *340*, 603–607.

16. Dunkley, O.; Engel, J.; Sturm, J.; Cremers, D. Visual-inertial navigation for a camera-equipped 25g nano-quadrotor. In Proceedings of the IROS2014 Aerial Open Source Robotics Workshop, Chicago, IL, USA, 14–18 September 2014.
17. Moore, R.J.; Dantu, K.; Barrows, G.L.; Nagpal, R. Autonomous MAV guidance with a lightweight omnidirectional vision sensor. In Proceedings of the 2014 IEEE International Conference on Robotics and Automation (ICRA), Hong Kong, China, 31 May–7 June 2014; pp. 3856–3861.
18. Floreano, D.; Wood, R.J. Science, technology and the future of small autonomous drones. *Nature* **2015**, *521*, 460–466.
19. Liu, C.; Prior, S.D.; Teacy, W.L.; Warner, M. Computationally efficient visual–inertial sensor fusion for Global Positioning System–denied navigation on a small quadrotor. *Adv. Mech. Eng.* **2016**, *8*, doi:10.1177/1687814016640996.
20. Honegger, D.; Meier, L.; Tanskanen, P.; Pollefeys, M. An open source and open hardware embedded metric optical flow cmos camera for indoor and outdoor applications. In Proceedings of the 2013 IEEE International Conference on Robotics and Automation (ICRA), Karlsruhe, Germany, 6–10 May 2013; pp. 1736–1741.
21. Brandli, C.; Berner, R.; Yang, M.; Liu, S.C.; Delbruck, T. A 240 × 180 130 dB 3 µs latency global shutter spatiotemporal vision sensor. *IEEE J. Solid-State Circuits* **2014**, *49*, 2333–2341.
22. Rueckauer, B.; Delbruck, T. Evaluation of event-based algorithms for optical flow with ground-truth from inertial measurement sensor. *Front. Neurosci.* **2016**, *10*, 176.
23. Briod, A.; Zufferey, J.C.; Floreano, D. A method for ego-motion estimation in micro-hovering platforms flying in very cluttered environments. *Auton. Robots* **2016**, *40*, 789–803.
24. McGuire, K.; de Croon, G.; De Wagter, C.; Tuyls, K.; Kappen, H. Efficient Optical flow and Stereo Vision for Velocity Estimation and Obstacle Avoidance on an Autonomous Pocket Drone. *arXiv* **2016**, arXiv:1612.06702.
25. Falanga, D.; Mueggler, E.; Faessler, M.; Scaramuzza, D. Aggressive Quadrotor Flight through Narrow Gaps with Onboard Sensing and Computing. *arXiv* **2016**, arXiv:1612.00291.
26. Mafrica, S.; Servel, A.; Ruffier, F. Minimalistic optic flow sensors applied to indoor and outdoor visual guidance and odometry on a car-like robot. *Bioinspir. Biomim.* **2016**, *11*, 066007.
27. Normann, R.A.; Perlman, I. The effects of background illumination on the photoresponses of red and green cones. *J. Physiol.* **1979**, *286*, 491.
28. Viollet, S.; Godiot, S.; Leitel, R.; Buss, W.; Breugnon, P.; Menouni, M.; Juston, R.; Expert, F.; Colonnier, F.; L'Eplattenier, G.; et al. Hardware architecture and cutting-edge assembly process of a tiny curved compound eye. *Sensors* **2014**, *14*, 21702–21721.
29. Sabiron, G.; Chavent, P.; Raharijaona, T.; Fabiani, P.; Ruffier, F. Low-speed optic-flow sensor onboard an unmanned helicopter flying outside over fields. In Proceedings of the IEEE International Conference on Robotics and Automation, Karlsruhe, Germany, 6–10 May 2013; pp. 1742–1749.
30. Expert, F.; Roubieu, F.L.; Ruffier, F. Interpolation based "time of travel" scheme in a Visual Motion Sensor using a small 2D retina. In Proceedings of the IEEE Sensors, Taipei, Taiwan, 28–31 October 2012; pp. 1–4.
31. Roubieu, F.L.; Serres, J.R.; Colonnier, F.; Franceschini, N.; Viollet, S.; Ruffier, F. A biomimetic vision-based hovercraft accounts for bees' complex behaviour in various corridors. *Bioinspir. Biomim.* **2014**, *9*, 036003.
32. Roubieu, F.L.; Serres, J.; Franceschini, N.; Ruffier, F.; Viollet, S. A fully-autonomous hovercraft inspired by bees: Wall following and speed control in straight and tapered corridors. In Proceedings of the 2012 IEEE International Conference on Robotics and Biomimetics, ROBIO, Guangzhou, China, 11–14 December 2012; pp. 1311–1318.
33. Expert, F.; Ruffier, F. Flying over uneven moving terrain based on optic-flow cues without any need for reference frames or accelerometers. *Bioinspir. Biomim.* **2015**, *10*, 026003.
34. Zufferey, J.C.; Floreano, D. Fly-inspired visual steering of an ultralight indoor aircraft. *IEEE Trans. Robot.* **2006**, *22*, 137–146.
35. Zufferey, J.C.; Klaptocz, A.; Beyeler, A.; Nicoud, J.D.; Floreano, D. A 10-gram vision-based flying robot. *Adv. Robot.* **2007**, *21*, 1671–1684.
36. Beyeler, A.; Zufferey, J.C.; Floreano, D. Vision-based control of near-obstacle flight. *Auton. Robots* **2009**, *27*, 201–219.
37. Land, M.F. Visual Acuity in Insects. *Annu. Rev. Entomol.* **1997**, *42*, 147–177.
38. Michaelis, L.; Menten, M.L. Die kinetik der invertinwirkung. *Biochem. z* **1913**, *49*, 352.

39. Franceschini, N.; Riehle, A.; Le Nestour, A. Directionally Selective Motion Detection by Insect Neurons. In *Facets of Vision*; Springer: Berlin, Germany, 1989.
40. Hassenstein, B.; Reichardt, W. Systemtheoretische analyse der zeit-, reihenfolgen-und vorzeichenauswertung bei der bewegungsperzeption des rüsselkäfers chlorophanus. *Z. Naturforsch. B* **1956**, *11*, 513–524.
41. Albus, J.S.; Hong, T.H. Motion, depth, and image flow. In Proceedings of the 1990 IEEE International Conference on Robotics and Automation, Cincinnati, OH, USA, 13–18 May 1990; pp. 1161–1170.
42. Manecy, A.; Marchand, N.; Ruffier, F.; Viollet, S. X4-MaG: A Low-Cost Open-Source Micro-Quadrotor and Its Linux-Based Controller. *Int. J. Micro Air Veh.* **2015**, *7*, 89–110.

sensors

MDPI

Article

Differential GNSS and Vision-Based Tracking to Improve Navigation Performance in Cooperative Multi-UAV Systems

Amedeo Rodi Vetrella *, Giancarmine Fasano, Domenico Accardo and Antonio Moccia

Department of Industrial Engineering, University of Naples Federico II, Piazzale Tecchio 80, Naples 80125, Italy; g.fasano@unina.it (G.F.); domenico.accardo@unina.it (D.A.); antonio.moccia@unina.it (A.M.)

* Correspondence: amedeorodi.vetrella@unina.it; Tel.: +39-81-768-3365

Academic Editors: Felipe Gonzalez Toro and Antonios Tsourdos

Received: 7 November 2016; Accepted: 14 December 2016; Published: 17 December 2016

Abstract: Autonomous navigation of micro-UAVs is typically based on the integration of low cost Global Navigation Satellite System (GNSS) receivers and Micro-Electro-Mechanical Systems (MEMS)-based inertial and magnetic sensors to stabilize and control the flight. The resulting navigation performance in terms of position and attitude accuracy may not suffice for other mission needs, such as the ones relevant to fine sensor pointing. In this framework, this paper presents a cooperative UAV navigation algorithm that allows a chief vehicle, equipped with inertial and magnetic sensors, a Global Positioning System (GPS) receiver, and a vision system, to improve its navigation performance (in real time or in the post processing phase) exploiting formation flying deputy vehicles equipped with GPS receivers. The focus is set on outdoor environments and the key concept is to exploit differential GPS among vehicles and vision-based tracking (DGPS/Vision) to build a virtual additional navigation sensor whose information is then integrated in a sensor fusion algorithm based on an Extended Kalman Filter. The developed concept and processing architecture are described, with a focus on DGPS/Vision attitude determination algorithm. Performance assessment is carried out on the basis of both numerical simulations and flight tests. In the latter ones, navigation estimates derived from the DGPS/Vision approach are compared with those provided by the onboard autopilot system of a customized quadrotor. The analysis shows the potential of the developed approach, mainly deriving from the possibility to exploit magnetic- and inertial-independent accurate attitude information.

Keywords: cooperative navigation; unmanned aerial vehicles; multi-UAV Systems; differential GNSS; vision-based tracking; vision-based navigation; TRIAD method; sensor fusion; flight tests

---

1. Introduction

In the last few years, miniaturization of flight control systems and payloads, and the availability of computationally affordable algorithms for autonomous guidance, navigation and control (GNC), have contributed to an increasing diffusion of micro-unmanned aircraft systems (micro-UAS). Besides military applications, micro-UAS can play a key role in several civil scenarios, and the attention of international top level companies and research centers has been focused on the adoption of these systems for commercial purposes [1,2] and on the paradigms for a safe and profitable access of micro-unmanned aerial vehicles (micro-UAVs) to civil airspace [3,4].

Micro-UAV navigation is typically based on the integration of low cost GNSS receivers and commercial grade Micro-Electro-Mechanical Systems (MEMS)-based inertial and magnetic sensors. An extensive review of techniques based on the integration of low cost Inertial Measurement Units (IMUs) and GNSS can be found in [5]. However, these navigation systems, only provide position accuracies of approximately 5–10 m and attitude accuracies of approximately 1°–5°, which are good enough to realize automated waypoint following, but are insufficient for most of the remote sensing or surveying applications in which fine sensor pointing is required [6,7]. Furthermore, collaborative sensing and data fusion frameworks are based on data registration as a fundamental pre-requisite, which may be directly correlated with navigation accuracy. There exist different approaches to improve UAV navigation performance.

A direct solution is to utilize high performance navigation systems. As an example, high accuracy aerial mapping systems, besides adopting dual frequency GPS receivers for accurate positioning with respect to fixed ground stations, usually exploit tactical grade IMU and/or dual GPS antenna architectures explicitly aimed at improving heading accuracy. The main disadvantages of this approach are in terms of cost and challenges related to installing dual antenna configurations on-board small UAVs. In fact, in [6] a heading accuracy of the order of 0.2°–0.5° is attained by installing the two antennas with a baseline of 1 m.

Some authors have instead followed an approach based on developing upgraded algorithmic solutions to enhance navigation performance for given low accuracy MEMS sensors. As an example in de Marina et al. [8], the Three-Axis Determination (TRIAD) algorithm [9] is used to measure the Direct Cosine Matrix (DCM) of a fixed wing aircraft, where the two reference vectors are the Earth's magnetic field vector and gravity in North East Down (NED) coordinates, while the two observation vectors are obtained by means of magnetometers and accelerometers in the Body Reference Frame (BRF). This method is not applicable in all flight conditions, especially in presence of an external magnetic field that corrupts the magnetic measurements or when accelerations acting on the aircraft do not allow a precise identification of the gravity observation unit vector. The authors assume an attitude accuracy requirement of 1.0° on pitch and roll and 4.0° on heading, as required by industry for a fixed wing aircraft [10]. In Valenti et al. [11], attitude is obtained from the observation of the gravity and magnetic fields, where the degrading effects of magnetic disturbances on pitch and roll are mitigated separating the problem of finding the tilt and the heading quaternion with an improvement in attitude estimation. However the heading angle uncertainty is still of the order of 10°. In both cases the algorithmic improvements cannot overcome technological limitations of consumer grade IMUs.

Another approach is to integrate electro-optical sensors to detect and track natural or manmade features. Some of these vision-based techniques require the a-priori knowledge of ground control points of known appearance [12,13] in order to find homologous pairs between an on-board geotagged database and the images taken by the flying viewing system. Others estimate the egomotion of the vehicle relying only on the motion of features in consecutive images [14–17]. Moreover, several Simultaneous Localization and Mapping (SLAM) techniques have been developed [18,19] in which the vehicle build a map of the environments while simultaneously determining its location. Indeed, SLAM techniques are usually considered to limit the drift induced by inertial sensors when flying in GPS-denied and unknown environments, more than to improve navigation accuracy under nominal GPS coverage.

Furthermore, vision-aided SLAM approaches present limits such as the necessity to detect and track natural or manmade features in a sequence of overlapping images which require a static and textured scene in good illumination conditions. This is not the case when UAS are flying over areas covered by snow, wood, sand or water e.g., day and night Search and Rescue (S&R) or natural hazards missions. Open issues to be solved remain to render the approach generally operational such as accumulated drift over time, computational complexity and data association.

The aforementioned methods exploit only one micro-UAS. However, due to single micro-UAS limits in terms of reliability, coverage and performance, multi-UAV systems have encountered increasing interest in the unmanned systems community [20–23] both for military and civil applications.

Within this framework, most of the research on cooperative navigation techniques focuses on GPS-challenging or denied environments. In Merino et al. [24] navigation in GPS-denied areas is performed acquiring overlapping images of the scene from different UAS in which at least one of them has GPS coverage. Once blob features [25] are matched among those images, it is possible to recover UAS's relative positions and consequently the absolute position of each vehicle. A similar approach has been followed by Indelman et al. [26] and Melnyk et al. [27] in which overlapping views are processed in order to evaluate relative positions in swarms. Heredia et al. [28] continued the work presented in [24] addressing the open issue of reliability, developing a Fault Detection and Identification (FDI) technique.

These Cooperative Localization (CL) techniques, in addition to the vision-based approaches drawbacks mentioned above, are affected by the need of acquiring multiple images from different platforms with an overlap that ranges from 50% up to 80% which limits the vehicles speed and requires an assigned distance between the platforms depending on the flight height and the Field of View (FOV).

This paper presents a new approach to improve the absolute navigation performance of a formation of UAVs flying in outdoor environments under nominal GPS coverage, with respect to the one achievable by integrating low cost IMUs, GNSS and magnetometers. The developed concept is to use the required formation of UAVs to build virtual navigation sensors that provide additional measurements, which are based on DGPS among flying vehicles and visual information and are not affected by magnetic and inertial disturbances. The architecture exploits cooperative GPS, as in multi-antenna attitude estimation architectures [29–31]. In particular, while the latter ones exploit carrier phase processing and short baselines (known by calibration) between antennas rigidly mounted on the vehicle, the new approach described in this paper is to exploit differential GPS using antennas embarked on different vehicles, where the exact geometry among them is unknown, but the line of sight between antennas can be estimated by vision sensors.

The main innovative points are: the cooperative nature of the UAV formation is exploited to obtain drift-free navigation information; for the first time, UAV absolute attitude is estimated combining DGPS among flying vehicles and vision-based information. Indeed, the exploitation of DGPS among flying vehicles to derive attitude information is a novel concept itself. Compared with traditional navigation systems, the main advantages of our method are:

- The possibility to attain high accuracy navigation performance (e.g., sub-degree attitude measurement accuracy) without requiring high cost avionics technologies, known ground features, or textured ground surfaces.
- Reduced computational complexity.
- Independency of the navigation information from magnetic disturbances and inertial errors, also allowing better estimation of biases for these sensors [32].
- Absence of error drifts in time.

The main disadvantage is the need of keeping vehicles within the camera(s) FOV in a multi-UAV scenario. However, considering typical performance limitations of micro-UAVs, a multi-vehicle architecture could be adopted, regardless of navigation needs, in order to improve coverage and reliability. Then, the proposed approach does not require close formation control and precise relative navigation, which makes its implementation much easier. Finally, considering the cost of micro-UAV systems with consumer grade avionics, having several UAVs can be more cost effective than equipping a single vehicle with high performance navigation hardware.

The objectives of this work are as follows:

- To present a cooperative navigation architecture that is able to ensure improved navigation performance in outdoor environment, with a major focus on attitude estimation based on differential GPS and vision-based tracking. This is done in Sections 2 and 3;
- To evaluate the achievable attitude accuracy in a numerical error analysis, pointing out the effects of DGPS/Vision measurement uncertainties and formation geometry. This is presented in Section 4;

- To compare, in flight tests, cooperative navigation output with traditional single vehicle-based data fusion implementations, thus highlighting main performance advantages of the proposed approach Sections 5 and 6. In particular, in Section 5 a validation strategy is presented in which a ground control point is used to evaluate DGPS/Vision attitude accuracy. Experimental results are reported in Section 6.

2. Cooperative Navigation Architecture

As stated above, cooperation is here exploited to improve the absolute navigation performance of formation flying UAVs in outdoor environments. This is done thanks to an architecture that integrates differential GPS and relative sensing by vision (defined as "DGPS/Vision" in what follows) within a customized sensor fusion algorithm.

Considering a formation of at least two UAVs flying cooperatively, the objective is to improve the absolute navigation performance for a "chief" vehicle, equipped with inertial and magnetic sensors, a GPS receiver, and a vision system, thanks to "deputy" vehicles equipped with GPS antenna/receivers and flying in formation with the chief. On the other hand, if GPS observables are exchanged among all the vehicles, and if each vehicle is able to track at least another UAV by one or more onboard cameras, each vehicle can exploit cooperation to improve its absolute navigation performance (i.e., each vehicle can be a chief exploiting information from other deputies). The proposed cooperative navigation technique can be used either in real time or in post processing phase. In the former case, proper communication links have to be foreseen among vehicles.

In the following we assume that:

- GPS is available for all vehicles that comprise the formation.
- Each vehicle attains the same absolute positioning accuracy.

On the basis of these assumptions, the DGPS/Vision method, described in this paper, focuses on improving the UAV attitude accuracy, leaving to the sensor fusion algorithm the position and velocity improvement, as shown in [33].

The overall cooperative navigation architecture is shown in Figure 1 where input data include: GPS measurements from the chief and the deputies; images (taken by the chief) of deputies within the FOV; inertial/magnetic data provided by the chief onboard IMU. Three processing steps are then involved:

- The vision-based tracking algorithm that allows extracting chief-to-deputies unit vectors in the BRF.
- The Differential-GPS (DGPS) block which returns chief to deputies baselines in a stabilized NED reference frame.
- A multi-sensor fusion algorithm based on an Extended Kalman Filter (EKF), which can be used to combine different information sources obtaining a more accurate and reliable navigation solution.

As it will be made clearer in the next section, depending on the processing scheme, DGPS and vision-based information can be directly integrated in the EKF, or it can be used to provide an attitude estimate that is then integrated as an additional measurement within the state estimation filter. The latter solution is the one considered in this work.

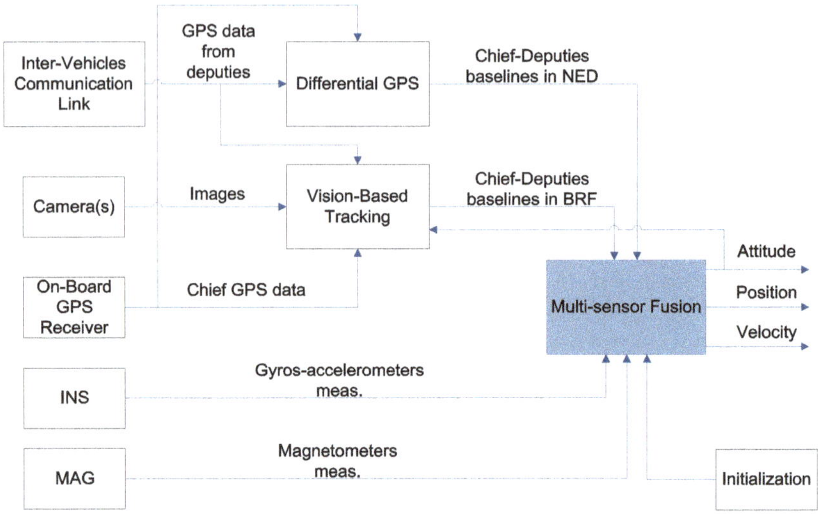

Figure 1. Logical architecture.

3. DGPS/Vision Attitude Determination Method

Vision-based tracking provides chief-deputies unit vectors in the Camera Reference Frame (CRF), which are then converted into the BRF either by only using a constant rotation matrix (accurately estimated off-line) in the case of strapdown installation, or by exploiting gimbal rotation angles in the case of gimbaled installation.

The available GPS data, and the estimated chief attitude, can be used to cue the vision-based tracking system and individuate search windows within acquired images, improving target detection reliability and significantly reducing processing time. As an example, deputies can be tracked in the video sequences by adopting template matching approaches based on computing and maximizing the Normalized Cross Correlation (NCC) [34]. This provides an estimate of deputy centroid in pixel coordinates, which are then converted into line-of-sight (LOS) information by the intrinsic camera model.

Vision-based tracking performance for given chief/deputy platforms basically depends on the range to deputies, on environmental conditions (impacting deputy appearance, contrast and background homogeneity), and on camera(s) parameters, such as quantum efficiency and instantaneous field of view (IFOV). Moreover, camera FOV limits the maximum angular separation between deputies that can be exploited.

In order to increase the detection range performance for a given sensor, the IFOV can be reduced increasing optics focal length, and thus reducing the overall FOV and the possibility to detect widely separated deputies. The trade-off between coverage and detection range can be tackled by installing higher resolution sensors and/or multiple camera systems.

As concerns DGPS, it can be carried out in different ways such as carrier phase differential and code-based differential processing [35]. Dual frequency carrier phase DGPS provides the most accurate relative positioning solution (cm-level error) adopting relatively expensive onboard equipments and also paying the cost of a significant computational weight [35]. Dual frequency GPS receivers are uncommon on micro UAVs, with some exceptions [36]. Indeed, a lower accuracy can be obtained even with single frequency carrier phase differential processing, provided that the integer ambiguity is solved.

The solution adopted in this work, is code-based DGPS, which requires hardware that is affordable for commercial micro UAVs, less observables to be exchanged between different vehicles (basically, only pseudoranges from common satellites in view) and much lighter processing.

A basic estimate of code-based DGPS relative positioning accuracy can be obtained multiplying typical Diluition of Precision (DOP) values by the average User Equivalent Range Error (UERE) accuracy in DGPS operating scenarios. Indeed, this is a conservative approach since pseudorange measurements correlation is increased in differential architectures. This computation leads to a typical 1-sigma accuracy of the order of 0.99 m (horizontal) and 1.86 m (vertical) [37–39].

While this uncertainty would correspond to a very rough angular accuracy, in the case of short baselines among rigidly mounted antennas on a single aerial platform [29–31], in our case a fine angular accuracy can still be attained by increasing the baselines between the chief and the deputies, thus converting position uncertainties into relatively small angular errors.

It is also interesting to underline that within our scenarios, it is likely that the different GPS receivers use the same satellites for position fix, which leads to the possibility of a significant cancellation of common errors even by adopting position-based DGPS, i.e., by simply calculating the difference between the various position fixes.

The unit vectors (camera and DGPS) obtained by using the DGPS/Vision method, can be used to provide navigation information in different ways, such as:

- Directly integrating line of sight measurements within an EKF (which works for any number of deputies);
- TRIAD [40] (which works for two deputies);
- QUEST [41] (which works for two and more deputies).

The second and the third approach provide a straightforward attitude estimate. In this work, the focus is set on demonstrating the potential of DGPS/Vision attitude determination, thus we assume a formation with two deputies and TRIAD-based processing.

In particular, once the attitude is estimated by TRIAD, the estimate can be included as an additional measurement in a classical EKF-based aided navigation algorithm [37], which works on the basis of a prediction-correction scheme (Figure 2). The reader is referred to [42] for a detailed explanation of the navigation filter, while this paper focuses on attitude estimation aspects.

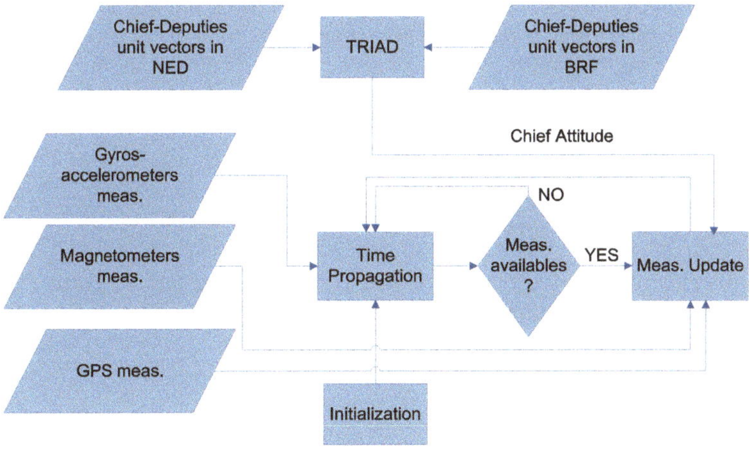

Figure 2. Multi-sensor fusion.

The main advantage in using two deputies lies in the possibility to obtain direct inertial- and magnetic-independent attitude information, with a reduced computational load. Also, having two deputies improves reconfiguration capabilities of the distributed navigation sensor, as two lines-of-sight can be used to the chief advantage. The main disadvantage is the need of keeping both deputies within the FOV of chief camera(s). However, the possible challenges related to this point depend on the specifications of the adopted vision sensor(s) and the consequent trade-offs between angular accuracy and coverage.

The TRIAD algorithm [40,43,44] is an analytical method to determine the rotation matrix between two reference frames in a straightforward manner. In particular, given two nonparallel reference unit vectors $\widehat{V}_1, \widehat{V}_2$ in a primary reference frame and two corresponding observation unit vectors $\widehat{W}_1, \widehat{W}_2$ represented with respect to a secondary reference frame, TRIAD starts defining two orthonormal triads of vectors $\left\{ \widehat{r}_1 \quad \widehat{r}_2 \quad \widehat{r}_3 \right\}$ and $\left\{ \widehat{o}_1 \quad \widehat{o}_2 \quad \widehat{o}_3 \right\}$ given by:

$$\widehat{r}_1 = \widehat{V}_1,\ \widehat{r}_2 = \frac{\widehat{V}_1 \times \widehat{V}_2}{\left|\widehat{V}_1 \times \widehat{V}_2\right|},\ \widehat{r}_3 = \widehat{r}_1 \times \widehat{r}_2 \tag{1}$$

$$\widehat{o}_1 = \widehat{W}_1,\ \widehat{o}_2 = \frac{\widehat{W}_1 \times \widehat{W}_2}{\left|\widehat{W}_1 \times \widehat{W}_2\right|},\ \widehat{o}_3 = \widehat{o}_1 \times \widehat{o}_2 \tag{2}$$

and determines the unique orthogonal matrix R which converts from the primary to the secondary reference frame as follows:

$$R = M_{obs} M_{ref}^T \tag{3}$$

where $M_{ref} = \left\{ \widehat{r}_1 \quad \widehat{r}_2 \quad \widehat{r}_3 \right\}$ and $M_{obs} = \left\{ \widehat{o}_1 \quad \widehat{o}_2 \quad \widehat{o}_3 \right\}$ are 3×3 matrices.

As shown in Figure 2, the two vector pairs needed by the DGPS/Vision method to compute the attitude matrix are the chief-to-deputies BRF and NED (DGPS) unit vectors which are computed as explained in the following sections.

The two unit vectors in BRF are obtained starting from the pixel coordinates (u_i, v_i) of the two deputies $(i = 1, 2)$ within images acquired by the Pelican camera(s), which can be extracted by proper vision-based techniques. The normalized pixel coordinates u_i^n, v_i^n of the two deputies are then obtained by applying the intrinsic camera model [45,46] which takes into account the focal length, the principal point coordinates, the radial and tangential distortion coefficients, and the skew coefficient. Consequently Azimuth and Elevation and the unit vectors in CRF are then computed as follows:

$$Az_{CRF_i} = \tan^{-1}(u_i^n) \tag{4}$$

$$El_{CRF_i} = \tan^{-1}(-v_i^n \cos(Az_i)) \tag{5}$$

$$\hat{r}_i^{CRF} = \begin{bmatrix} \hat{r}_{i,1} \\ \hat{r}_{i,2} \\ \hat{r}_{i,3} \end{bmatrix} = \begin{bmatrix} \cos(El_{CRF_i})\cos(Az_{CRF_i}) \\ \cos(El_{CRF_i})\sin(Az_{CRF_i}) \\ -\sin(El_{CRF_i}) \end{bmatrix} \tag{6}$$

where $\hat{r}_i^{CRF}$ is the unit vector of components $(\hat{r}_{i,1}, \hat{r}_{i,2}, \hat{r}_{i,3})$ in CRF. The unit vectors in BRF, to be used as reference vectors within the TRIAD algorithm, are given by:

$$\hat{r}_i^{BRF} = R_{CRF \to BRF} \hat{r}_i^{CRF} \tag{7}$$

where $R_{CRF \to BRF}$ is the constant rotation matrix from CRF to BRF and $\hat{r}_i^{BRF}$ is the unit vector in BRF.

The other two vectors needed to apply the TRIAD algorithm are the chief-to-deputies NED unit vectors. In this work, these vectors are obtained by adopting a Double Difference (DD) code-based DGPS solution [35], which offers significant advantages due to the cancellation of receiver and satellite clock biases, as well as most of the ionospheric and tropospheric propagation delays. To this end, it is assumed that the chief and the two deputies are in view of the same n satellites and consequently their pseudorange measurements are available which allow to calculate single and DD observables.

In particular, considering the chief and the i-th deputy, and two GPS satellites, one of which named pivot, DD observables are obtained as follows

$$PR_{ci}^{pk} = (PR_i^k - PR_i^p) - (PR_c^k - PR_c^p) \tag{8}$$

where the superscript p refers to the pivot GPS satellite, which is chosen to be the one with the highest elevation, k refers to the generic satellite ($k = 1, \ldots, n-1$), the subscripts c and i represent the chief and the i-th deputy vehicle GPS receiver respectively, PR_i^k stands for the pseudorange estimated by the i-th receiver with respect to the k-th satellite, and a similar interpretation holds for the other estimated pseudoranges.

The DD observation model is a non linear function of the baseline $\Delta \underline{r}_i^{ECEF}$ between the chief and i-th deputy in the Earth Centered Earth Fixed (ECEF) reference frame as shown in the following equation:

$$\begin{gathered} PR_{ci}^{pk} = \rho_{ci}^{pk} + \nu_{ci}^{pk} = \\ \|\underline{R}_k - (\underline{r}_c + \Delta \underline{r}_i^{ECEF})\| - \|\underline{R}_k - \underline{r}_c\| - \|\underline{R}_p - (\underline{r}_c + \Delta \underline{r}_i^{ECEF})\| + \|\underline{R}_p - \underline{r}_c\| + \nu_{ci}^{pk} \end{gathered} \tag{9}$$

where ρ_{ci}^{pk} represents the DD between the true pseudoranges, $\underline{r}_c$ is the chief ECEF position, $\underline{R}_p$ and $\underline{R}_k$ are the pivot and k-th satellites ECEF positions and ν_{ci}^{pk} are the non-common mode pseudorange errors. The problem of finding $\Delta \underline{r}_i^{ECEF}$ is solved applying a recursive least square estimation method based on the linearization of the DD observation model [35].

The i-th baseline $\Delta \underline{r}_i^{NED}$ in the NED reference frame, with origin in the chief center of mass (usually defined as "navigation frame" [37]), is then given by:

$$\Delta \underline{r}_i^{NED} = R_{ECEF \to NED} \Delta \underline{r}_i^{ECEF} \tag{10}$$

where $R_{ECEF \to NED}$ is the rotation matrix from the ECEF to the navigation frame that depends on the chief longitude λ and geodetic latitude μ.

$$R_{ECEF \to NED} = \begin{bmatrix} -\sin\mu\cos\lambda & -\sin\mu\sin\lambda & \cos\mu \\ -\sin\lambda & \cos\lambda & 0 \\ -\cos\mu\cos\lambda & -\cos\mu\sin\lambda & -\sin\mu \end{bmatrix} \tag{11}$$

Of course, the accuracy of the resulting attitude estimate depends on several factors such as DGPS and vision-based tracking errors, formation geometry, and chief vehicle attitude. These aspects are analyzed in the following section.

For the case of n deputies ($i = 1, \ldots, n$), n unit vectors can be computed in BRF and NED by applying Equations (4)–(7) and (8)–(11), respectively. In this case, the optimal solution is given by the QUEST algorithm [41] which minimizes a quadratic cost function involving an arbitrary number of vector measurements made in BRF and NED.

4. Error Analysis

In order to analyze the performance of the DGPS/Vision sensor, a numerical approach has been followed. In particular, for a given chief attitude and formation geometry, DGPS and optical measurements are simulated by random extractions, and the resulting attitude measurement error is analyzed with statistical tools.

Indeed, an analytical solution exists which allows estimation of TRIAD attitude error covariance matrix [40] as a function of formation geometry and line-of-sight uncertainties. The basic assumptions underlying this derivation are that errors must to first order lie in the plane perpendicular to the respective unit vector, and have an axially symmetric distribution about it.

The second assumption clearly does not hold in the architecture considered in this paper, due to the difference in GPS performance between the horizontal plane and the vertical direction [37–39]. Worst case or averaging approaches in using TRIAD covariance matrix equation are clearly sub-optimal and can produce over or under-conservative results.

Regarding the numerical simulation architecture, the usual 321 sequence of Euler angles (heading, pitch, and roll), and the case of chief null attitude angles are assumed. In general, each deputy is instantaneously located at given azimuth and elevation angles (ε_i and μ_i) with respect to the body reference frame of the chief, and at a different range (L_i). In order to underline the main effects on measurement accuracy, formation geometry can be conveniently described in terms of azimuth center ε_c and azimuth separation $\Delta\varepsilon$ between the deputies:

$$\varepsilon_c = \frac{\varepsilon_1 + \varepsilon_2}{2} \quad (12)$$

$$\Delta\varepsilon = \varepsilon_1 - \varepsilon_2 \quad (13)$$

Formation geometry is depicted in Figure 3.

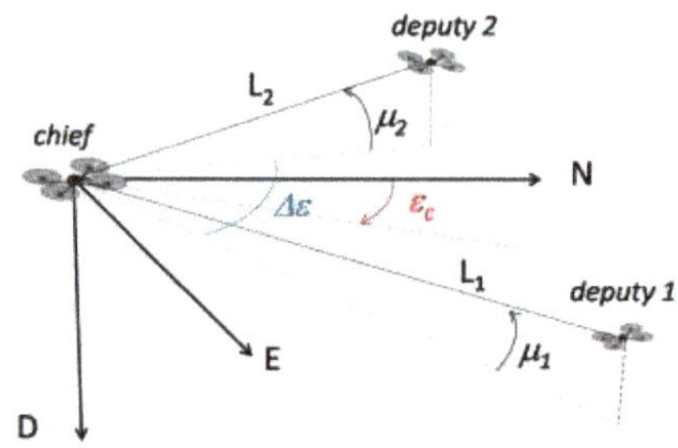

Figure 3. Formation geometry parameters.

EO-based azimuth and elevation errors are simulated as zero-mean gaussian noises with standard deviation equal to 0.05°, which is consistent with typical IFOV values of sensors commonly found onboard micro-UAVs. For the sake of simplicity, the case of equal range and elevation is considered for the two deputies. Then, three cases are analyzed:

- Case 1: horizontal geometry with azimuth center at 0°;
- Case 2: "tilted" geometry with azimuth center at 0°;
- Case 3: horizontal formation with azimuth center at 45°.

For each case, a variable angular separation between deputies is considered. The three cases are summarized in Table 1.

Table 1. Considered formation geometries.

	$\mu_1 = \mu_2$ (°)	ε_c (°)	$\Delta\varepsilon$ (°)	$L_1 = L_2$ (m)
Case 1	0	0	[10, 90]	[50, 500]
Case 2	45	0	[10, 90]	[50, 500]
Case 3	0	45	[10, 90]	[50, 500]

Results based on 200 numerical simulations are shown in Figure 4 for the three considered cases. Figure 4 offers several points of discussion.

The above diagrams show as expected that increasing the baseline improves the attitude estimation accuracy, with an error dependency on 1/L. This is mainly related to how angular differential GPS uncertainty decreases by increasing the baseline between the antennas. As already stated above, this advantage must be traded-off against the performance of vision-based tracking, given the decreasing dimensions in pixels of the deputies. As an example, an instantaneous field of view of 0.05° corresponds at a distance of 100 m to a geometric resolution of 9 cm. Consequently, a longer baseline requires an improvement of the geometric resolution of the vision sensors which can be attained by decreasing the FOV or installing a multi-camera system on the chief. Furthermore, differences among attitude angles are due to the different contributions of horizontal and vertical DGPS uncertainties.

As regards the relation between formation geometry and attitude accuracy, it is clear that in all cases the impact of line of sight uncertainties on attitude determination errors mainly depends on the angle between unit vectors and the considered chief body axes.

In Case 1 (Figure 4a), heading error does not depend on $\Delta\varepsilon$, since in all cases unit vectors are normal to the third axis of the BRF. In the considered formation geometry the heading uncertainty basically depends only on the horizontal GPS error and thus exhibits the finest accuracy: the uncertainty is always well below 1°, and fast approaches 0.1° for increasing baselines. This is a very useful result coming out from the proposed DGPS/Vision sensor to be pointed out considering typical high uncertainties in estimating magnetic heading on board small and micro UAVs and the consequent interest in high cost compact navigation systems (e.g., based on high cost IMUs and/or dual antenna GNSS) [6,7].

As regards the other angles (Figure 4b,c), for increasing angular separation pitch accuracy decreases while the roll accuracy increases. These effects are due to the varying angles between unit vectors and the coordinate axes.

In particular, effects on pitch are limited for the considered range of angular separations, while a much larger effect is present for roll. These effects would further increase if the angular separation approached 180°. The considered geometry, especially for small angular separation, is optimal for pitch estimation. However, the final pitch accuracy does not reach the heading one due to the fact that it (only) depends on the (larger) GPS vertical error.

Case 2 (Figure 4) highlights the effects of non-horizontal formation geometry. Compared with Case 1, it is clear how heading accuracy decreases due to the non-optimal observation geometry (smaller angles between unit vectors and third body axis). Instead, pitch and roll estimates are positively influenced by the non-null elevation angle. In particular, pitch estimate takes advantage from depending not only on the vertical GPS error, but also on the smaller horizontal component. Instead, benefits for roll mainly derive from the increased angles among unit vectors and the first body axis. Deputy separation does not influence significantly pitch accuracy, while its increase is beneficial for both heading and roll angles. In this case, among the three angles, pitch has the highest accuracy.

As regards Case 3 (Figure 4), this asymmetric geometry does not impact significantly heading accuracy, which is very similar to Case 1 and thus very weakly influenced by deputy separation. Instead, both pitch and roll errors are positively impacted by increasing separation, due to the fact that one of the unit vectors tends to be normal to the first or second body axis, as $\Delta\varepsilon$ approaches 90°. In fact, the latter geometry represents an optimal compromise in terms of accuracy for all the three angles, which represents a well known result exploited in designing single vehicle multi-antenna GPS configurations [29–31]. For example, at 100 m baseline we have an error standard deviation of 0.4° for heading, and of 1° for roll and pitch.

In summary, these budgets show that even the relatively coarse code-based DGPS processing is very promising in terms of attitude estimation, when combined with vision-based sensing. In particular, horizontal geometries provide a very fine heading accuracy even at relatively short baselines, while asymmetries can be useful to find a balance between roll and pitch errors. In practical cases, the choice of formation geometries will depend on the requirements for final navigation performance (sensor fusion output) and on eventual constraints deriving from the flight environment. Also, formation reconfiguration strategies can be envisaged to keep required levels of attitude estimation performance.

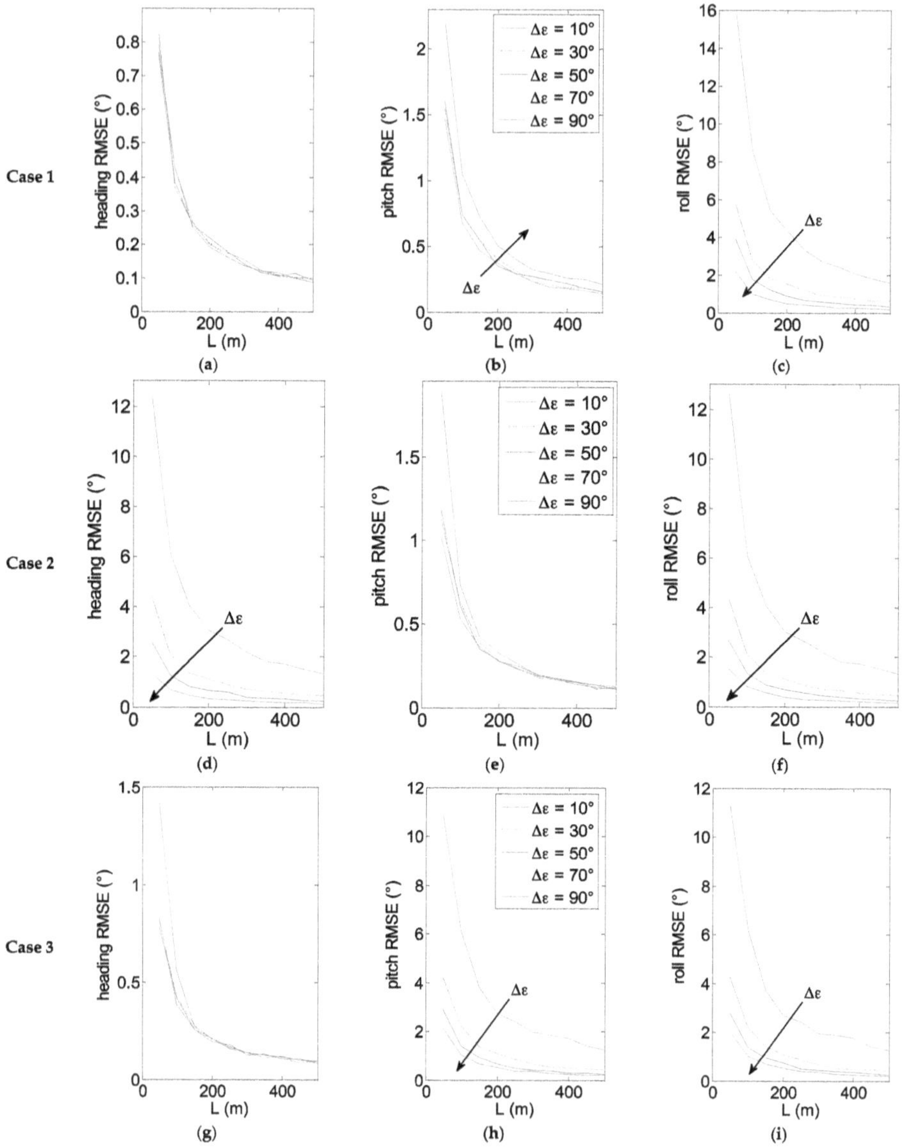

Figure 4. Attitude root mean square errors (RMSE) as a function of baseline L, for different values of angular separation. Case 1 (**a–c**) heading pitch and roll RMSE for $\mu = \varepsilon_c = 0°$; Case 2 (**d–f**) heading pitch and roll RMSE for $\mu = 45°$ and $\varepsilon_c = 0°$; Case 3 (**g–i**) heading pitch and roll RMSE for $\mu = 0°$ and $\varepsilon_c = 45°$.

5. Testing and Validation Strategy

In order to evaluate the performance of the novel DGPS/Vision algorithm, experimental tests are described where the chief vehicle is a customized quadrotor, and two ground-based GPS antennas/receivers are used as surrogate deputies.

5.1. Experimental Setup

Tests have been conducted by using two surrogate deputies, consisting in two ground stations equipped with an AV59 antenna (Trimble™, Sunnyvale, CA, USA) and BD960 receiver (Trimble™, Sunnyvale, CA, USA, Figure 5) observed by a Pelican quadrotor (Figure 6) (Ascending Technologies™, Krailling, Germany) [47] that plays the role of chief.

In particular, the Pelican quadrotor, besides being equipped with a controller, a set of onboard sensors (Table 2) and an onboard computer (AscTec™ Mastermind, Ascending Technologies™, Krailling, Germany), has been customized with a miniaturized electro optical sensor (BlueFox™ MLC200wC, Matrix Vision, Brescia, Italy) and an additional uBlox™ GPS Receiver (LEA 6T, uBlox™, Thalwil, Switzerland) which provides raw measurements that are used in the DGPS processing.

Figure 5. Ground antennas/receivers used as deputy vehicles.

Figure 6. Chief vehicle (Customized Ascending Technologies™ Pelican).

Table 2. Pelican Navigation Sensors.

Component	Model
Gyroscopes	Analog Devices™ ADXRS610
Accelerometer	Memsic™ R9500
Barometer	NXP™ MPXA6115A
Compass	Honeywell™ HMC5843
PS Receiver	uBlox™ LEA-6S

The additional GPS receiver and the optical sensor have been connected to the Mastermind computer via a USB link, while other raw and calibrated/processed sensor data are read using the UART connection between Mastermind and autopilot. The acquisition software that runs on the Mastermind has been coded in C++ and gathers all the necessary data with an accurate time-tag based on GPS time and the CPU clock. In particular, IMU data are acquired with the aid of the Asctec Communication Interface (ACI) at a frequency of 100 Hz while images and GPS raw data are gathered simultaneously at a frequency of 1 Hz.

This setup has been used to validate the presented DGPS/Vision attitude determination approach with particular emphasis on estimation of heading angle.

5.2. Pointing/Attitude Accuracy Evaluation Strategy

Given the sub-degree attitude determination accuracy obtained by the DGPS/Vision method for the baselines experimented during flight tests, in particular for the heading angle, it is nontrivial to find a reference measurement that provides a ground truth of better accuracy level. In fact, this level of accuracy can be reached installing a tactical grade IMU on board the Pelican, or a dual antenna navigation system [6,7], or using very accurately geoferenced ground control points. Besides the cost, dual antenna navigation systems have a significant limit related to the necessity to install the two antennas with a sufficiently large baseline (1 m or more), which is hard to obtain onboard the Pelican.

In this work, the DGPS/Vision attitude accuracy has been evaluated by identifying on an open source 1:1000 georeferenced map [48] (planar error of the order of 25 to 50 cm) the position of a Ground Control Point (GCP) which is visible in the acquired images.

The idea is to compute azimuth and elevation in the navigation frame of an identifiable GCP, to be used as reference measurements to evaluate the pointing accuracy.

The logical scheme of the pointing accuracy analysis is shown in Figure 7 where the main processing steps are highlighted. As mentioned above, the focus is on heading performance, as a consequence only the processing strategy and results concerning azimuth in NED will be further analyzed showing that azimuth accuracy represents a good benchmark to evaluate and compare heading performance.

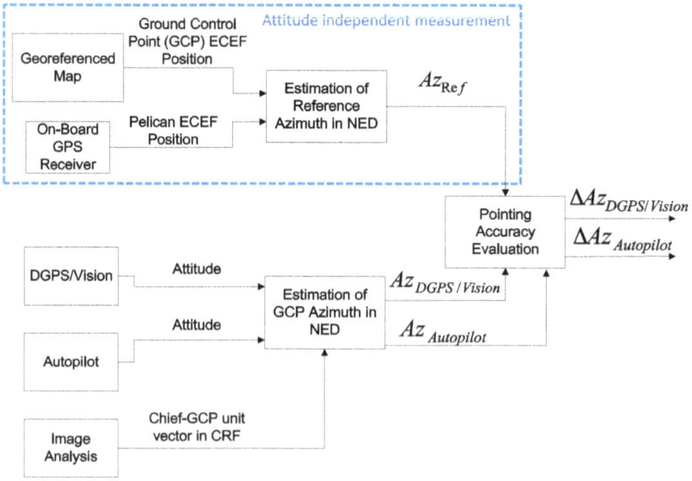

Figure 7. Pointing accuracy logical scheme.

The (attitude independent) reference measurement is obtained starting from the ECEF relative position vector $\Delta \underline{r}^{ECEF}$ between the Pelican and the GCP:

$$\Delta \underline{r}^{ECEF} = \begin{bmatrix} x_{GCP} - x_{Pelican} \\ y_{GCP} - y_{Pelican} \\ z_{GCP} - z_{Pelican} \end{bmatrix} = \begin{bmatrix} \Delta x \\ \Delta y \\ \Delta z \end{bmatrix} \tag{14}$$

where x_{GCP}, y_{GCP} and z_{GCP} are the ECEF coordinates of the GCP (provided by the map) and $x_{Pelican}$, $y_{Pelican}$ and $z_{Pelican}$ are the Pelican ECEF coordinates given by the on-board GPS receiver. $\Delta \underline{r}^{ECEF}$ is then converted in the NED reference frame as follows

$$\Delta \underline{r}^{NED} = R_{ECEF \to NED} \Delta \underline{r}^{ECEF} \tag{15}$$

Once $\Delta \underline{r}^{NED}$ is obtained, the reference Azimuth Az_{Ref} in NED is given by

$$Az_{\mathrm{Ref}} = \tan^{-1}\left(\frac{\Delta y^{NED}}{\Delta x^{NED}}\right) + k\pi, \begin{cases} k = 0 \text{ if } \Delta \mathrm{x}^{\mathrm{NED}} > 0 \\ k = 1 \text{ if } \Delta \mathrm{x}^{\mathrm{NED}} < 0 \end{cases} \tag{16}$$

where Δx^{NED} and Δy^{NED} are the NED-referenced relative position vector components.

Az_{Ref} is used as a reference to evaluate the accuracy of the DGPS/Vision and Autopilot heading. To this end, the GCP (Figure 8) is identified on each image to compute the Pelican-to-GCP Line of Sight (LOS) in the CRF $\hat{r}^{CRF}$ and then in BRF $\hat{r}^{BRF}$ according to Equation (7). $\hat{r}^{BRF}$ is then transformed in two unit vectors, one, $\hat{r}^{NED}_{DGPS/Vision}$, applying the attitude matrix $R^{DGPS/Vision}_{BRF \to NED}$ computed using the DGPS/Vision method, and the other, $\hat{r}^{NED}_{Autopilot}$, applying the attitude matrix $R^{Autopilot}_{BRF \to NED}$ computed by the Autopilot:

$$\hat{r}^{NED}_{DGPS/Vision} = R^{DGPS/Vision}_{BRF \to NED} \hat{r}^{BRF}; \ \hat{r}^{NED}_{Autopilot} = R^{Autopilot}_{BRF \to NED} \hat{r}^{BRF} \tag{17}$$

Consequently, the two azimuth errors are

$$\Delta Az_{DGPS/Vision} = Az_{DGPS/Vision} - Az_{\mathrm{Ref}}, \ \Delta Az_{Autopilot} = Az_{Autopilot} - Az_{\mathrm{Ref}} \tag{18}$$

where

$$Az_{DGPS/Vision} = \tan^{-1}\left(\frac{\hat{r}^{NED}_{DGPS/Vision,2}}{\hat{r}^{NED}_{DGPS/Vision,1}}\right), \ Az_{Autopilot} = \tan^{-1}\left(\frac{\hat{r}^{NED}_{Autopilot,2}}{\hat{r}^{NED}_{Autopilot,1}}\right), \tag{19}$$

are the estimates of the GCP azimuth in NED ($Az_{Autopilot}$, $Az_{DGPS/Vision}$) and [$\hat{r}^{NED}_{DGPS/Vision,1}$, $\hat{r}^{NED}_{DGPS/Vision,2}$] and [$\hat{r}^{NED}_{Autopilot,1}$, $\hat{r}^{NED}_{Autopilot,2}$] are the NED-referenced unit vector components related to DGPS/Vision and autopilot, respectively.

Two important factors for effective application of the proposed accuracy evaluation strategy regard the uncertainty of the reference azimuth measurement given by Equation (16), and the relation between the uncertainties on attitude measurements and the azimuth angles computed in Equation (19).

Figure 8. Example of flight image showing the observation geometry.

As regards the first point, the georeferenced map has a planar sub-metric accuracy, while the error on Pelican positioning depends on horizontal accuracy of standalone GPS. Due to the large distance from the GCP (about 600 m), the linear uncertainty of the baseline is converted into a relatively small angular error. For the sake of concreteness, if one assumes 6 m of horizontal relative positioning error, the worst case uncertainty on the reference azimuth measurement (i.e., error vector normal to the Pelican-GCP line of sight) is given by:

$$\tan^{-1}(\frac{6}{600}) \approx 0.01 \text{ rad} = 0.57^\circ \quad (20)$$

As concerns the relation between azimuth and heading, it is intuitive that for small roll and pitch angles (as it indeed happens in the considered flight tests), azimuth accuracy depends primarily on heading measurement performance, with very little effect produced by the other errors. This can be demonstrated analytically by deriving a first order error budget.

In particular, given Equation (19) it is possible, both for DGPS/Vision and Autopilot, to express the azimuth uncertainty as

$$\sigma_{Az}^2 \cong \left(\frac{\partial Az}{\partial \psi}\right)^2 \sigma_\psi^2 + \left(\frac{\partial Az}{\partial \vartheta}\right)^2 \sigma_\vartheta^2 + \left(\frac{\partial Az}{\partial \varphi}\right)^2 \sigma_\varphi^2 + \left(\frac{\partial Az}{\partial Az_{CRF}}\right)^2 \sigma_{Az_{CRF}}^2 + \left(\frac{\partial Az}{\partial El_{CRF}}\right)^2 \sigma_{El_{CRF}}^2 \quad (21)$$

where ψ, ϑ and φ are heading, pitch, and roll, respectively, while Az_{CRF} and El_{CRF} are the GCP azimuth and elevation angles computed in the camera reference frame. The squared derivatives can be computed analytically starting from Equation (19). They measure the sensitivity of azimuth uncertainty on the input errors, and depend themselves on ψ, ϑ, φ, Az_{CRF}, and El_{CRF}.

For the sake of concreteness in view of the analyzed experiment, assuming CRF coincident with BRF, null attitude angles, $Az_{CRF} = -20^\circ$, and $El_{CRF} = 5^\circ$, one gets

$$\left(\frac{\partial Az}{\partial \psi}\right)^2 \cong 1, \left(\frac{\partial Az}{\partial \vartheta}\right)^2 \cong 9 \cdot 10^{-4}, \left(\frac{\partial Az}{\partial \varphi}\right)^2 \cong 7 \cdot 10^{-3}, \left(\frac{\partial Az}{\partial Az_{CRF}}\right)^2 \cong 1, \left(\frac{\partial Az}{\partial El_{CRF}}\right)^2 \cong 0 \quad (22)$$

which shows that the main contributions to azimuth pointing error are the uncertainties on ψ and Az_{CRF}.

Even degree-level errors on roll and pitch have a limited contribution, since they are strongly attenuated. On the other hand, since uncertainties in Az_{CRF} and El_{CRF} are related to the camera IFOV and thus of the order of 0.05° in the considered case, the final uncertainty on azimuth pointing is given by a small amplification of the heading one.

As an example, assuming the uncertainties obtained in case 1 of the error analysis approach $\sigma_\psi = 0.35^\circ, \sigma_\vartheta = 0.7^\circ, \sigma_\varphi = 3.5^\circ$, and considering $\sigma_{Az_{CRF}} = 0.05^\circ, \sigma_{El_{CRF}} = 0.05^\circ$, one gets $\sigma_{Az} \cong 0.46^\circ$. This shows that azimuth accuracy evaluation represents a good benchmark to evaluate and compare heading estimation performance.

5.3. Flight Tests

Experimental tests have been carried out in an outdoor area that allows baselines among chief and deputies of the order of 100 m. As noted above, this is necessary to reduce the DGPS angular error and thus to improve attitude determination uncertainty. Tests have been designed to compare DGPS/Vision estimates with heading measurements based on the onboard magnetometers and the output of the real time data fusion algorithm running on Pelican autopilot, based on a filter which combines accelerometers, gyroscopes and magnetometers. In particular, among the flights that have been conducted, two tests have been chosen as representative of attitude dynamics and varying formation geometry, as follows:

- Test 1: an almost constant horizontal formation geometry (Figure 9) has been kept, with a baseline between the two deputies of the order of 40 m, and the chief at a distance slightly larger than 100 m from the two ground antennas (Figure 10). A number of attitude maneuvers has been commanded, including four 360° heading rotations and 1 Hz heading oscillation with about 30° amplitude. This test has been selected to point out the different levels of robustness of onboard fusion and DGPS/Vision with respect to flight dynamics history.

- Test 2: the chief vehicle has been commanded to fly along a path of about 200 m (Figure 11), thus generating a significant change of formation geometry in NED coordinates. The two ground antennas have been positioned in order to provide a baseline of about 100 m with respect to the chief vehicle at the starting and end point of the flight path (Figure 12). This test has been selected to point out the effects on the onboard filter and on DGPS/Vision estimates of both flight dynamics history and magnetic effects.

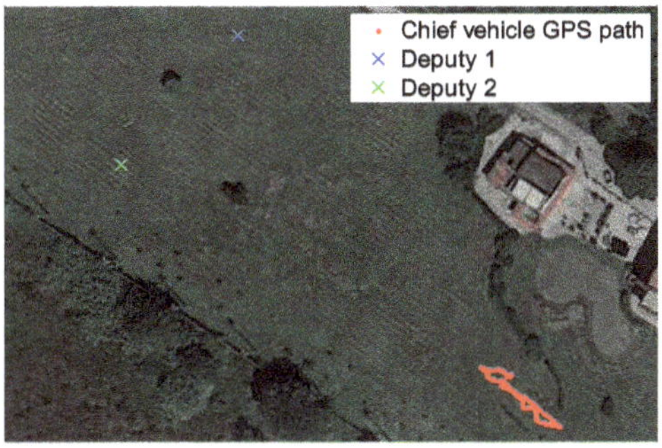

Figure 9. Formation geometry Test 1.

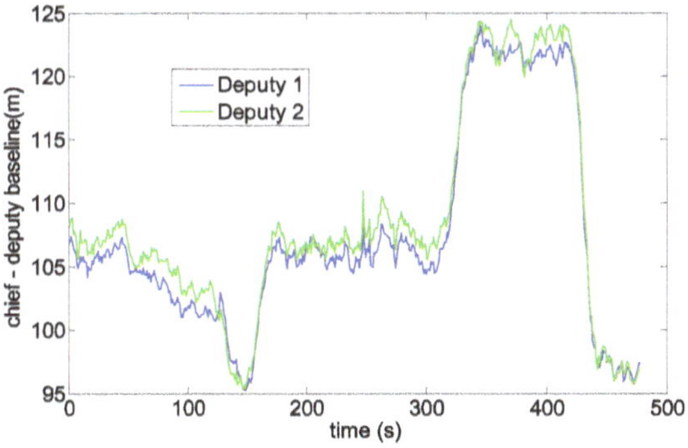

Figure 10. Chief-deputies baselines Test 1.

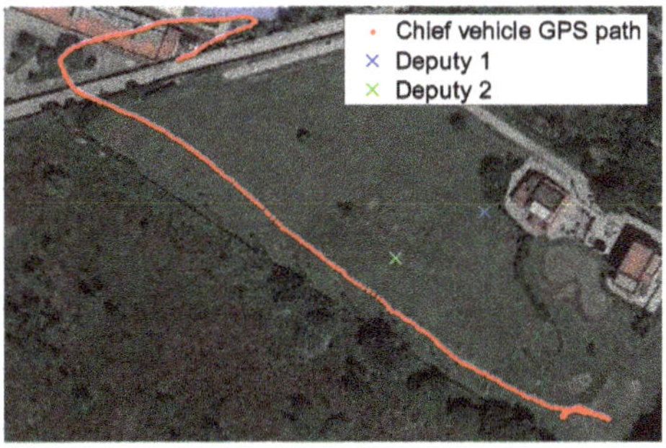

Figure 11. Formation geometry Test 2.

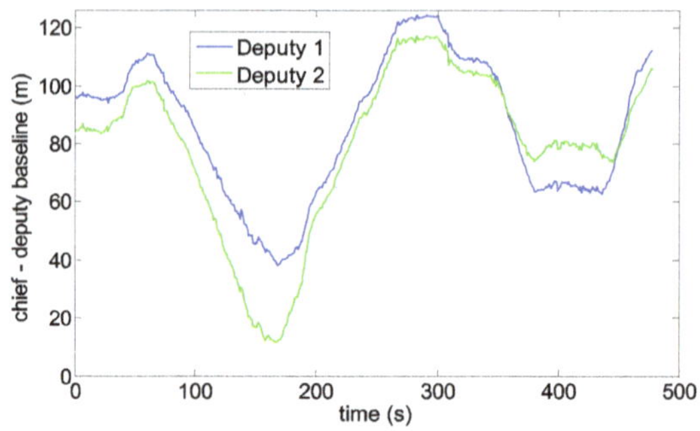

Figure 12. Chief-deputies baselines Test 2.

6. Experimental Results

During the experimental tests, the (forward-looking) camera embarked on the chief vehicle has acquired images of the two deputy antennas while measurements from GPS receivers and other onboard sensors have also been gathered. In particular, attitude measurements obtained from the filter running on the autopilot have been stored, while attitude estimates based on differential GPS and vision have been obtained by off-line processing.

Stability and noise properties of the DGPS solution can be verified by analyzing the baseline estimated between static ground antennas. This is shown in Figure 13 that regards test 1. It can be seen that the estimated baseline exhibits sub-meter oscillations, which is consistent with the error analysis presented in Section 4.

Experimental data have been acquired in both static and dynamic conditions. Results from a static test are described in Subsection 6.1, while Subsections 6.2–6.5 report results from the previously presented flight tests.

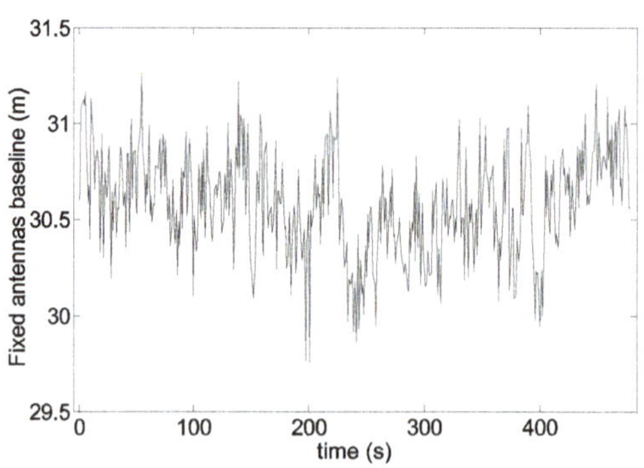

Figure 13. Fixed antennas baseline (DGPS) as a function of time.

6.1. Static Test

Static data acquisitions have been conducted mainly to verify the precision properties of the DGPS/Vision solution. During this test the chief vehicle has been positioned and held stationary on the ground in an almost horizontal formation geometry keeping the two ground antennas in the camera FOV at a distance slightly larger than 100 m.

Static results have confirmed the drift-free behavior of DGPS/Vision measurements, while also showing a standard deviation consistent with formation geometries considered in Case 1 in the numerical simulation error budget (0.7° pitch, 3.5° roll and 0.35° heading). Indeed, the experimented attitude measurement noise has been a little smaller than numerical predictions, in fact, the heading estimated by DGPS/Vision (Figure 14) has a standard deviation of about 0.23°.

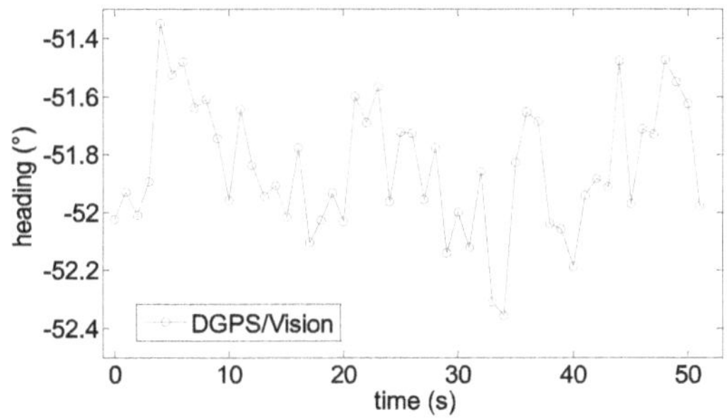

Figure 14. Heading angle as estimated by the DGPS/Vision as a function of time.

6.2. Heading—Test 1

Test 1 allows comparing DGPS/Vision and magnetometers/autopilot attitude estimates from the point of view of sensitivity to attitude dynamics.

Figure 15 shows the heading angle during the flight where DGPS/Vision attitude estimates are available for most of the considered time interval. Some DGPS/Vision isolated losses are produced by the impossibility to detect both antennas within images, due to the observation geometry, the limited FOV of the camera and the maneuvers executed during the tests.

In the following, to get a clearer insight into the DGPS/Vision performance and robustness, it is worthwhile to focus the attention on three flight segments, as indicated in Figure 15. In fact, while DGPS/Vision measurements are independent from magnetic and inertial information, the Pelican data fusion shows significant limits particularly at the end of the heading rotation maneuvers. This seems due to different weights which probably the Pelican data fusion algorithm applies to the input data coming from the gyroscopes, the accelerometers and the magnetometer.

In Figure 16 the low frequency profile of the heading angle during the first time interval does not introduce any significant disturbance and consequently the pelican data fusion algorithm follows quite well the heading given by the MEMS magnetometer.

The first flight segment lies before the heading rotation maneuvers (Figure 15). Within this time interval, the difference of about 5.9° (Table 3) between DGPS/Vision and autopilot heading is almost constant, mainly due to magnetic biases. In addition, Figure 16 shows a good consistency between magnetometer-based and autopilot estimates (Table 3).

The situation changes completely after the 360° heading rotations (Figure 17). While the offset between DGPS/Vision and magnetometer-based heading is very similar to the first flight segment, a significant drift of about 15° (Table 3) is generated with respect to the autopilot solution, which is mostly due to the coarse gyroscopes accuracy and the consequent data fusion limits in tracking the actual vehicle dynamics. The difference between DGPS/Vision and autopilot heading achieves a maximum value of about 20°, while a significant offset is also generated between magnetometer-based heading and autopilot heading (about 22°).

As shown in Table 3, only after several tens of seconds, and further maneuvers (third flight segment, Figure 18), the autopilot solution recover the offset with respect to DGPS/Vision and magnetometer-based heading, thus achieving a final performance level that resembles the one experimented in the first flight segment.

During all the flight test, the difference, of about 7° (Table 3), between DGPS/Vision and magnetometer-based heading does not show significant variations.

In summary, compared with classical attitude determination techniques, this test confirms the potential of DGPS/Vision to provide small noise measurements which are completely independent from attitude dynamics history.

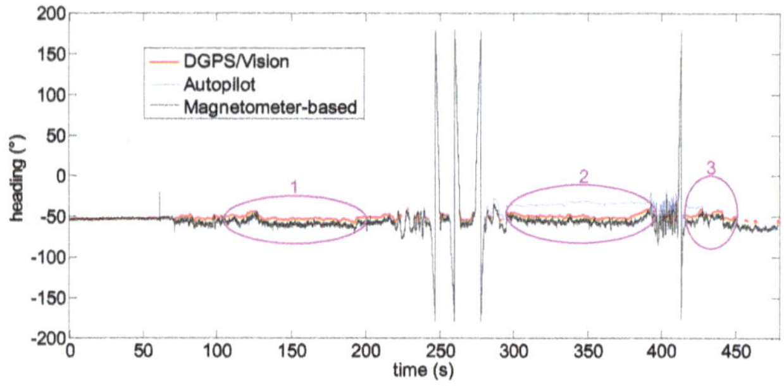

Figure 15. Heading angle as a function of the flight time (Test 1).

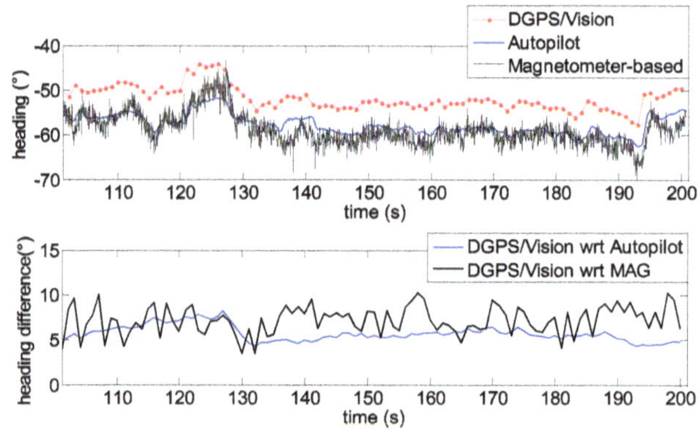

Figure 16. Heading angle as a function of time (**Top**), differences of DGPS/Vision with respect to autopilot and magnetometers (**Bottom**) during the first flight segment (Test 1).

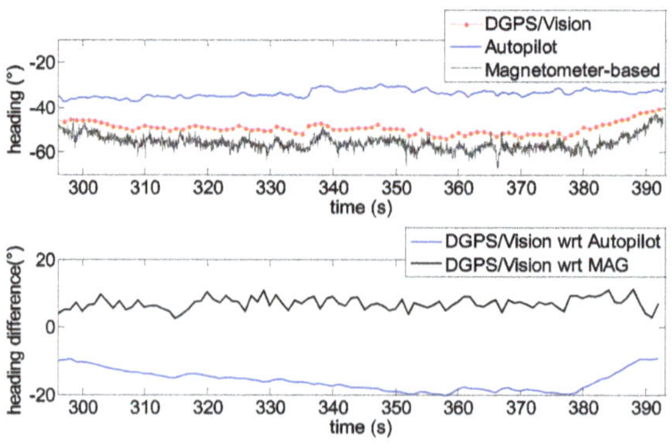

Figure 17. Heading angle as a function of time (**Top**), differences of DGPS/Vision with respect to autopilot and magnetometers (**Bottom**) during the second flight segment (Test 1).

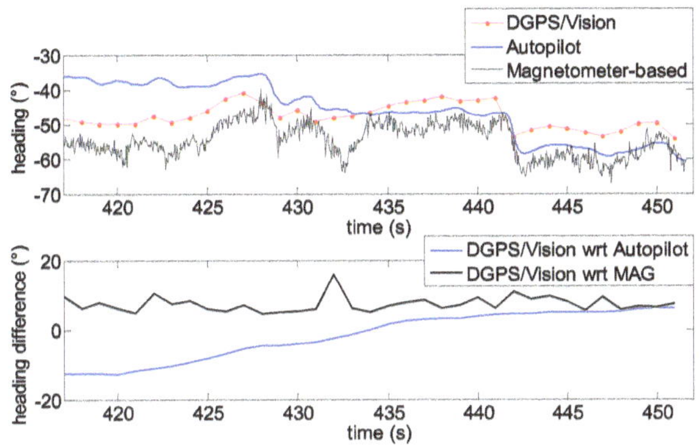

Figure 18. Heading angle as a function of time (**Top**), differences of DGPS/Vision with respect to autopilot and magnetometers (**Bottom**) during the third flight segment (Test 1).

Table 3. Heading comparison (mean values in degrees).

Time Intervals	DGPS/Vision	Magnetometer	Autopilot Data Fusion	Difference between DGPS/Vision and	
				Magnetometer	Autopilot
1	−51.5	−58.8	−57.4	7.3	5.9
2	−48.8	−55.8	−33.3	7	−15.5
3	−47.5	−54.7	−45.9	7.2	−1.6

6.3. Heading—Test 2

During the second test, the Pelican has been commanded to fly along a relatively long path, covering the whole test field and generating a heading change of almost 180°. Due to the necessity to keep the antennas within camera FOV, chief motion actually combines a forward-lateral-backward translation and a very slow heading rotation. The changing aircraft orientation within the Earth magnetic field allows pointing out DGPS/Vision potential with respect to both inertial and magnetic effects.

Heading as a function of time is depicted in Figure 19, also in this case two flight segments are focused to point out DGPS/Vision and autopilot/magnetometer-based performance. Both flight segments are characterized by a small velocity (order of 1 m/s) and an almost constant small heading rate, of the order of 0.8°/s. Moreover, during these flight segments the baseline with deputies has been kept large enough to achieve sub-degree DGPS/Vision heading uncertainty. In these conditions, it is particularly challenging for the onboard data fusion algorithm to track heading dynamics. In fact, the first flight segment (Figure 20) shows an increasing drift of the difference between DGPS/Vision and autopilot, reaching a mean value of 14.8° (see Table 4), with the latter being almost insensitive to heading variations. On the other hand, if one neglects high frequency noise of magnetic estimates, the difference between DGPS/Vision and magnetometer-based estimates tends to be an almost constant offset of about 7.1° (Table 4). Given the similar aircraft heading, the offset is of the same order of the one experimented during test 1 (Table 3).

From a qualitative point of view, the second flight segment (Figure 21) shows a similar behavior of DGPS/Vision and autopilot estimates, with an increasing drift of the difference that reaches a mean value of about 13.3° (Table 4), since data fusion output is relatively insensitive to very slow rotations. At a quantitative level the estimated difference is impacted by the fact that data fusion output depends on attitude dynamics history and is increasingly affected by bias instability of inertial sensors.

Table 4. Heading comparison (mean values in degrees).

Time Intervals	DGPS/Vision	Magnetometer	Autopilot Data Fusion	Difference between DGPS/Vision and	
				Magnetometer	Autopilot
1	−31	−38.1	−45.4	7.1	14.8
2	114.8	123.1	101.5	−8.3	13.3

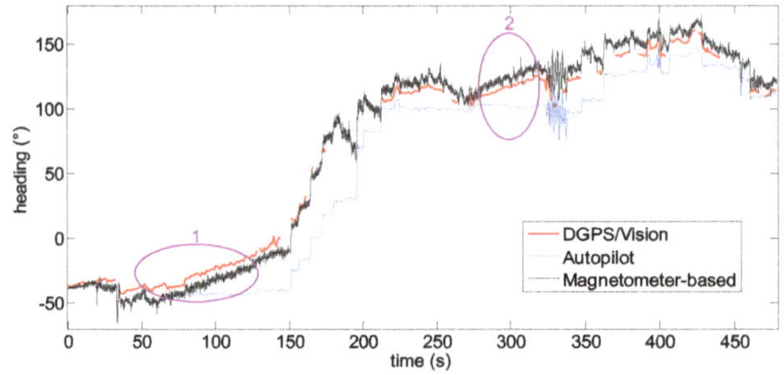

Figure 19. Heading as a function of time (Test 2).

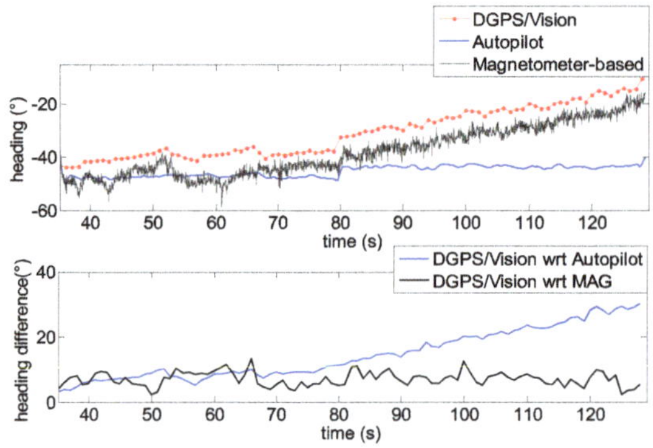

Figure 20. Heading angle as a function of time (**Top**), differences of DGPS/Vision with respect to autopilot and magnetometers (**Bottom**) during the first flight segment (Test 2).

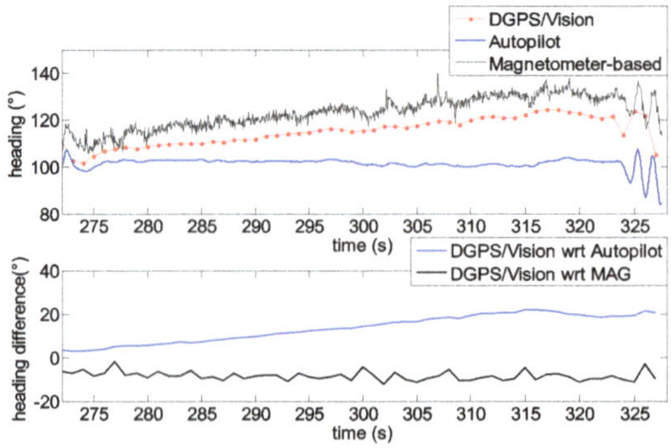

Figure 21. Heading angle as a function of time (**Top**), differences of DGPS/Vision with respect to autopilot and magnetometers (**Bottom**) during the second flight segment (Test 2).

If one compares DGPS/Vision and magnetometer-based solutions (Table 4), as in the first flight segment an almost constant offset is obtained (removing high frequency noise of magnetic measurements). However, the estimated bias of about $-8.3°$ is almost opposite with respect to the first flight segments in which the bias is about $7.1°$. This confirms that the origin of this difference lies in the effect of uncompensated magnetometer bias and onboard magnetic fields, which are constant in the body reference frame and thus generate an effect on heading estimation error which strongly depends on quadrotor orientation within the Earth's magnetic field.

6.4. Pitch and Roll

Although formation geometries and experimental tests have been mainly designed to optimize and analyze in detail DGPS/Vision heading angle estimation performance, for the sake of completeness it is also useful to compare pitch and roll angles as estimated by DGPS/Vision and onboard data fusion algorithm. This is done in Figures 22 and 23 which are relevant to Test 1. In both cases, estimated differences fall within the accuracy levels predicted in DGPS/Vision error budgets. Due to the baseline and the formation geometry, pitch and roll accuracy levels are of the order of $0.7°$ and $3.5°$ respectively, while autopilot estimates take great advantage from gravity contribution and accelerometers measurements. However, both pitch and roll diagrams allow appreciating the good consistency between different measurements.

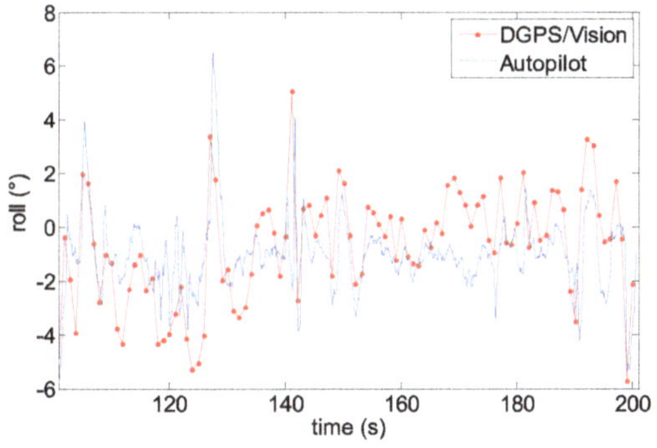

Figure 22. Roll angle as a function of time (Test 1).

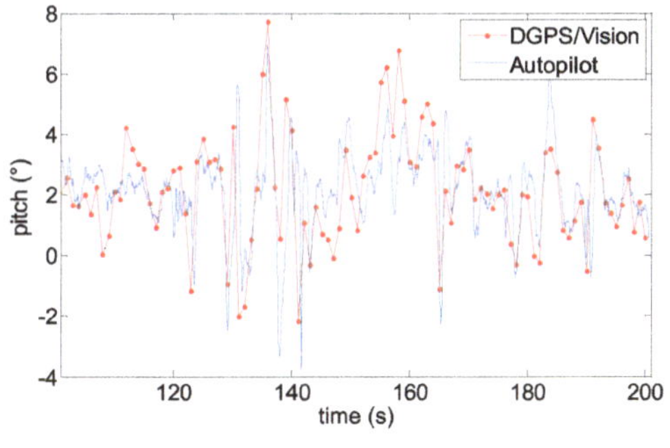

Figure 23. Pitch angle as a function of time (Test 1).

6.5. Pointing/Attitude Accuracy Results

Time frames 1 and 2 of Test 1 have been chosen as an example to show the accuracy achievable by using the DGPS/Vision approach, based on the strategy described in Subsection 5.2. Figures 24 and 25 show the computed errors in the two time frames, both for DGPS/Vision and the Autopilot. In particular, considering the time frame 1 of Test 1 (Figure 24), the DGPS/Vision azimuth error $\Delta Az_{DGPS/Vision}$ has a mean of about 0.09° with a standard deviation of 0.3° while the autopilot azimuth error $\Delta Az_{Autopilot}$ is about 4.4° (see Table 5). Considering the time frame 2 of Test 1, (Figure 25) after the three 360° heading rotations, the $\Delta Az_{DGPS/Vision}$ mean remains of the order of 0.06° while $\Delta Az_{Autopilot}$ mean increases to about 17.6° (Table 5). These results show that, unlike the Autopilot, the DGPS/Vision errors do not show a clear dependence on flight dynamics history, and fall within the uncertainty of the reference azimuth measurements. This is consistent with [11], which shows that, depending on flight dynamics and history, the Pelican autopilot heading error can increase up to about 20°.

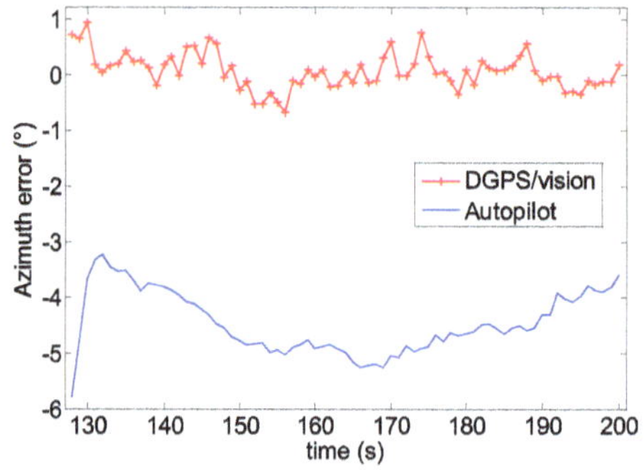

Figure 24. Azimuth error (°) as a function of time (Test 1-time frame 1).

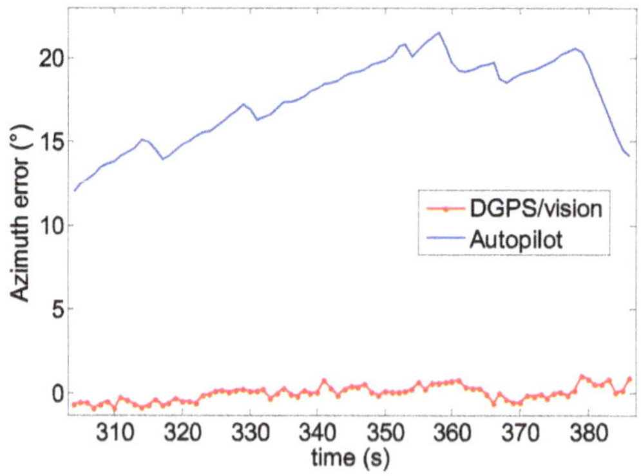

Figure 25. Azimuth error (°) as a function of time (Test 1-time frame 2).

Table 5. Azimuth error (°) comparison ($\Delta Az_{DGPS/Vision}$, $\Delta Az_{Autopilot}$) Test 1.

Time Intervals	$\Delta Az_{DGPS/Vision}$ Mean	$\Delta Az_{DGPS/Vision}$ Std.	$\Delta Az_{Autopilot}$ Mean	$\Delta Az_{Autopilot}$ Std.
1	0.09	0.3	−4.4	0.5
2	0.06	0.46	17.6	2.5

7. Conclusions

This paper presented an algorithm developed to improve UAV navigation performance in outdoor environments by exploiting cooperation among UAVs, differential GNSS and relative sensing by vision. In particular, the focus was set on attitude determination based on TRIAD algorithm.

Both numerical simulations and flight results showed the potential of sub-degree angular accuracy. In particular, proper formation geometries, and even relatively small baselines, allow achieving a heading uncertainty that can approach 0.1°, which represents a very important result taking into account typical performance levels of IMUs onboard small UAVs. Furthermore, the dependency of attitude estimation performance on formation geometry can be exploited to the navigation advantage if proper cooperative guidance laws are used to reconfigure the UAV formation as needed.

Flight experiments showed that the main factor enabling highly accurate attitude estimates is the information independence from both inertial and magnetic measurements. On the one hand, DGPS/Vision estimates are not influenced by flight history and changing inertial sensors biases, thus being insensitive to error accumulation phenomena. On the other hand, they are not affected by magnetic phenomena which are difficult to counteract in single vehicle applications since resulting errors depend on vehicle orientation.

Acknowledgments: This research was carried out in the frame of Programme STAR, financially supported by UniNA and Compagnia di San Paolo.

Author Contributions: We underline that this work is the result of a tight collaboration between the authors in which everyone contributed to conceive the proposed method and to design the experiments. In particular, A.R. Vetrella integrated the hardware, developed real time and post processing software, and wrote the draft version of the paper. A.R. Vetrella and G. Fasano conducted the flight tests. G. Fasano, D. Accardo and A. Moccia supervised the work contributing to software development and data analysis and interpretation, and revised the paper.

Conflicts of Interest: The authors declare no conflict of interest.

References

1. El-Sheimy, N.; Wright, B. Real-time Airborne Mapping System for Forest Fire Fighting (F3) System. *PERS* **2004**, *70*, 381–383.
2. Valavanis, K.; Vachtsevanos, G.J. *Handbook of Unmanned Aerial Vehicles*; Springer: Dordrecht, The Netherlands, 2015.
3. Brooks, M. Welcome to the personal drone revolution. *New Sci.* **2012**, *216*, 42–45. [CrossRef]
4. Chan, W.L.; Hsiao, F.B. Implementation of the Rauch-Tung-Striebel smoother for sensor compatibility correction of a fixed-wing unmanned air vehicle. *Sensors* **2011**, *11*, 3738–3764. [CrossRef] [PubMed]
5. Hasan, A.M.; Samsudin, K.; Ramli, A.R.; Azmir, R.S.; Ismaeel, S.A. A review of navigation systems (integration and algorithms). *Aust. J. Basic Appl. Sci.* **2009**, *3*, 943–959.
6. Eling, C.; Wieland, M.; Hess, C.; Klingbeil, L.; Kuhlmann, H. Development and Evaluation of UAV based Mapping Systems for Remote Sensing and Surveying Applications. In *International Archives of the Photogrammetry, Remote Sensing and Spatial Information Sciences*; Copernicus GmbH: Toronto, ON, Canada, 2015.
7. Hirokawa, R.; Ebinuma, T. A low-cost tightly coupled GPS/INS for small UAVs augmented with multiple GPS antennas. *J. Inst. Navig.* **2009**, *56*, 35–44. [CrossRef]
8. De Marina, H.G.; Pereda, F.J.; Giron-Sierra, J.M.; Espinosa, F. UAV attitude estimation using unscented kalman filter and TRIAD. *IEEE Trans. Ind. Electron.* **2012**, *59*, 4465–4474. [CrossRef]
9. Black, H.D. A passive system for determining the attitude of a satellite. *AIAA J.* **1964**, *2*, 1350–1351. [CrossRef]
10. Whitmore, S.A.; Fife, M.; Brasher, L. *Development Closed-Loop Strap down Attitude System for an Ultralight Altitude Flight Experiment*; NASA: Washington, DC, USA, 1997.
11. Valenti, R.G.; Dryanovski, I.; Xiao, J. Keeping a good attitude: A quaternion-based orientation filter for IMUs and MARGs. *Sensors* **2015**, *15*, 19302–19330. [CrossRef] [PubMed]
12. Majdik, A.L.; Albers-Schoenberg, Y.; Scaramuzza, D. Air-ground matching: Appearance-based GPS-denied urban localization of micro aerial vehicles. *J. Field Robot.* **2015**, *32*, 1015–1039. [CrossRef]
13. Wu, A.; Johnson, E.; Kaess, M.; Dellaert, F.; Chowdhary, G. Autonomous flight in GPS-denied environments using monocular vision and inertial sensors. *J. Aerosp. Comput. Inf. Commun.* **2010**, *10*, 172–186.
14. Kaiser, M.K.; Gans, N.R.; Dixon, W.E. Vision-Based Estimation for Guidance, Navigation, and Control of an Aerial Vehicle. *IEEE Trans. Aerosp. Electron.* **2010**, *46*, 1064–1077. [CrossRef]
15. Amidi, O.; Kanade, T.; Fujita, K. A visual odometer for autonomous helicopter flight. *Robot. Autom. Syst.* **1999**, *28*, 185–193. [CrossRef]

16. Andert, F.; Bathge, F.; Frehse, S.; Dittrich, J. Vision-Based Navigation and Exploration Strategies for Unmanned Helicopters in Disaster Scenarios. In Proceedings of the AHS Internetional Specialists Meeting on Unmanned Rotorcraft, Scottsdale, AZ, USA, 22–24 January 2013.
17. Milella, A.; Siegwart, R. Stereo-Based Ego-Motion Estimation Using Pixel Tracking and Iterative Closest Point. In Proceedings of the IEEE International Conference on Computer Vision Systems, New York, NY, USA, 4–7 January 2006.
18. Durrant-Whyte, H.; Bailey, T. Simultaneous localization and mapping: Part I. *IEEE Robot. Autom. Mag.* **2006**, *13*, 99–110. [CrossRef]
19. Durrant-Whyte, H.; Bailey, T. Simultaneous localization and mapping: Part II. *IEEE Robot. Autom. Mag.* **2006**, *13*. [CrossRef]
20. Daniel, K.; Dusza, B.; Lewandowski, A.; Wietfeld, C. AirShield: A system-of-systems MUAV remote sensing architecture for disaster response. In Proceedings of the 3rd Annual IEEE International Systems Conference (SysCon), Vancouver, BC, Canada, 23–26 March 2009.
21. Teacy, W.T.L.; Nie, J.; McClean, S.; Parr, G.; Hailes, S.; Julier, S.; Trigoni, N.; Cameron, S. Collaborative sensing by unmanned aerial vehicles. In Proceedings of the 3rd International Workshop on Agent Technology for Sensor Networks, Budapest, Hungary, 12 May 2009.
22. Hauert, S.; Leven, S.; Zufferey, J.-C.; Floreano, D. Communication-based Swarming for Flying Robots. In Proceedings of the IEEE International Conference on Robotics and Automation, Workshop on Network Science and Systems Issues in Multi-Robot Autonomy, Anchorage, AK, USA, 3–7 May 2010.
23. Bürkle, A.; Segor, F.; Kollmann, M. Towards Autonomous Micro UAV Swarms. *J. Intell. Robot. Syst.* **2011**, *61*, 339–353. [CrossRef]
24. Merino, L.; Wiklund, J.; Caballero, F.; Moe, A.; De Dios, J.R.M.; Forssen, P.-E.; Nordberg, K.; Ollero, A. Vision-based Multi-UAV position estimation. *IEEE Robot. Autom. Mag.* **2006**, *13*, 53–62. [CrossRef]
25. Forssen, P.E.; Moe, A. View matching with blob features. In Proceedings of the 2nd Canadian Conference on Computer and Robot Vision, Victoria, BC, Canada, 9–11 May 2005; pp. 228–235.
26. Indelman, V.; Gurfil, P.; Rivlin, E.; Rotstein, H. Graph-Based distributed cooperative navigation for a general multi-robot measurement model. *Int. J. Robot. Res.* **2012**, *31*, 1057–1080. [CrossRef]
27. Melnyk, I.V.; Hesch, J.A.; Roumeliotis, S.I. Cooperative Vision-aided Inertial Navigation Using Overlapping Views. In Proceedings of the 2012 IEEE International Conference on Robotics Automation, Saint Paul, MN, USA, 14–18 May 2012.
28. Heredia, G.; Caballero, F.; Maza, I.; Merino, L.; Viguria, A.; Ollero, A. Multi-unmanned aerial vehicle (UAV) cooperative fault detection employing differential global positioning (DGPS), inertial and vision sensors. *Sensors* **2009**, *9*, 7566–7579. [CrossRef] [PubMed]
29. Park, C.; Teunissen, P.J.G. A new carrier phase ambiguity estimation for GNSS attitude determination systems. In Proceedings of the International, GPS/GNSS Symposium, Tokyo, Japan, 15–18 November 2003; pp. 283–290.
30. Giorgi, G.; Teunissen, P.J.G.; Verhagen, S.; Buist, P.J. Testing a new multivariate GNSS carrier phase attitude determination method for remote sensing platforms. *Adv. Space Res.* **2010**, *46*, 118–129. [CrossRef]
31. Renga, A.; Fasano, G.; Accardo, D.; Grassi, M.; Tancredi, U.; Rufino, G.; Simonetti, A. Navigation facility for high accuracy offline trajectory and attitude estimation in airborne applications. *Int. J. Navig. Obs.* **2013**, *2013*, 397686. [CrossRef]
32. Alonso, R.; Shuster, M.D. TWOSTEP: A fast robust algorithm for attitude-independent magnetometer-bias determination. *J. Astronaut. Sci.* **2002**, *50*, 433–451.
33. Vetrella, A.R.; Fasano, G.; Renga, A.; Accardo, D. Cooperative UAV Navigation Based on Distributed Multi-Antenna GNSS, Vision, and MEMS Sensors. In Proceedings of the International Conference on Unmanned Aircraft Systems, Denver, CO, USA, 9–12 June 2015.
34. Gonzalez, R.C.; Woods, R.E. *Digital Image Processing*, 3rd ed.; Prentice Hall (Pearson International Edition): Upper Saddle River, NJ, USA, 2008.
35. Kaplan, E.D.; Leva, J.L.; Milbert, D.; Pavloff, M.S. Fundamentals of Satellite Navigation. In *Understanding GPS—Principles and Applications*, 2nd ed.; Kaplan, E.D., Hegarty, C.J., Eds.; Artech House: Boston, MA, USA, 2006.
36. V-Map. Available online: http://v-map.net/ (accessed on 6 November 2016).
37. Farrell, J.A. *Aided Navigation: GPS with High Rate Sensors*; McGraw-Hill: New York, NY, USA, 2008.

38. *GPS Standard Positioning Service Performance Standard*, 4th ed.; Office of the Secretary of Defense: New York, NY, USA, 2008.
39. Cosentino, R.J.; Diggle, D.W.; de Haag, M.U.; Hegarty, C.J.; Milbert, D.; Nagle, J. Differential GPS. In *Understanding GPS—Principles and Applications*, 2nd ed.; Kaplan, E.D., Hegarty, C.J., Eds.; Artech House: Boston, MA, USA, 2006.
40. Shuster, M.D.; Oh, S.D. Three-axis attitude determination from vector observations. *AIAA J. Guid. Control* **1981**, *4*, 70–77. [CrossRef]
41. Wertz, J.R. *Spacecraft Attitude Determination and Control*; D. Reidel Publishing Company: Boston, MA, USA, 1978.
42. Vetrella, A.R.; Fasano, G.; Accardo, D. *Vision-Aided Cooperative Navigation for UAV Swarms*; AIAA Infotech@ Aerospace: San Diego, CA, USA, 2016.
43. Markley, F.L. Attitude determination using vector observations: A fast optimal matrix algorithm. *J. Astronaut. Sci.* **1993**, *41*, 261–280.
44. Cheng, Y.; Shuster, M.D. QUEST and the anti-quest good and evil attitude estimation. *J. Astronaut. Sci.* **2005**, *53*, 337–351.
45. Camera Calibration Toolbox for Matlab. Available online: http://www.vision.caltech.edu/bouguetj/calib_doc/#start (accessed on 2 March 2016).
46. Heikkilla, J.; Silvén, O. A Four-step Camera Calibration Procedure with Implicit Image Correction. In Proceedings of the 1997 Computer Society Conference on Computer Vision and Pattern Recognition, San Juan, Puerto Rico, 17–19 June 1997.
47. Ascending Technologies. Available online: http://www.asctec.de/en/uav-uas-drone-products/asctec-pelican/ (accessed on 2 March 2016).
48. Rilievo Aerofotogrammetrico. 1:1000 del Comune di Napoli. Available online: http://www.comune.napoli.it/flex/cm/pages/ServeBLOB.php/L/IT/IDPagina/26177 (accessed on 20 September 2016).

sensors

MDPI

Article

An Overview of Small Unmanned Aerial Vehicles for Air Quality Measurements: Present Applications and Future Prospectives

Tommaso Francesco Villa [1], Felipe Gonzalez [2], Branka Miljievic [1], Zoran D. Ristovski [1] and Lidia Morawska [1,*]

[1] International Laboratory for Air Quality and Health (ILAQH), Queensland University of Technology (QUT), 2 George St, Brisbane QLD 4000, Australia; tf.villa@hdr.qut.edu.au (T.F.V.); b.miljevic@qut.edu.au (B.M.); z.ristovski@qut.edu.au (Z.D.R.)

[2] Australian Research Centre for Aerospace Automation (ARCAA), Queensland University of Technology (QUT), 2 George St, Brisbane QLD 4000, Australia; felipe.gonzalez@qut.edu.au

* Correspondence: l.morawska@qut.edu.au; Tel.: +61-7-3138-2616

Academic Editor: Assefa M. Melesse
Received: 21 April 2016; Accepted: 5 July 2016; Published: 12 July 2016

Abstract: Assessment of air quality has been traditionally conducted by ground based monitoring, and more recently by manned aircrafts and satellites. However, performing fast, comprehensive data collection near pollution sources is not always feasible due to the complexity of sites, moving sources or physical barriers. Small Unmanned Aerial Vehicles (UAVs) equipped with different sensors have been introduced for in-situ air quality monitoring, as they can offer new approaches and research opportunities in air pollution and emission monitoring, as well as for studying atmospheric trends, such as climate change, while ensuring urban and industrial air safety. The aims of this review were to: (1) compile information on the use of UAVs for air quality studies; and (2) assess their benefits and range of applications. An extensive literature review was conducted using three bibliographic databases (Scopus, Web of Knowledge, Google Scholar) and a total of 60 papers was found. This relatively small number of papers implies that the field is still in its early stages of development. We concluded that, while the potential of UAVs for air quality research has been established, several challenges still need to be addressed, including: the flight endurance, payload capacity, sensor dimensions/accuracy, and sensitivity. However, the challenges are not simply technological, in fact, policy and regulations, which differ between countries, represent the greatest challenge to facilitating the wider use of UAVs in atmospheric research.

Keywords: air quality; UAVs; sensors; atmosphere; pollution; aerosols

1. Introduction

The composition of ambient air changes continuously, due to both natural and anthropogenic emissions which, when released into the atmosphere as aerosols or gaseous pollutants, affect air quality and human health [1]. The association between adverse health outcomes and poor air quality has been clearly demonstrated [2,3], and ambient air pollution has been recognized as the ninth largest health risk factor globally [4]. However, atmospheric pollution also reduces agriculture yields, visibility, sunlight at ground level and snowfall, and increases atmospheric heating as well [5,6]. These impacts highlight the need for continuous air quality assessment.

Detailed information on the characteristics of aerosol distribution and gaseous pollutant concentrations is needed when quantifying their effects on human health and the environment [7–11]. However, spatial and temporal resolution of data from ground, manned aircraft [7,12–20] and satellite measurements is relatively low and often inadequate for local and regional applications. In addition, satellite and airborne sensors can be prohibitively costly, restricting the use of these platforms to sporadic tests rather than routine analysis. Furthermore, taking measurements close to pollutant sources may not always be possible and it could be too dangerous or risky for manned aircraft to fly close to the ground [21–24]. Together, these reasons promote the use of small, lightweight UAVs for a range of applications, including atmospheric measurements.

Small, lightweight UAVs can provide more accurate information on aerosol distribution throughout the atmospheric column, which is needed to better understand air quality and composition in specific atmospheric layers [25–27]. UAVs cover large areas and can monitor remote, dangerous or difficult to access locations, increasing operational flexibility and resolution over land-based methods [28,29]. Since the application of UAV is relatively new, the aims of this review were to: (1) compile information on the use of UAVs for air quality studies; and (2) assess their major benefits and range of applications.

The review is organized as follows: the 'Materials and Methods' section explains how multiple bibliographic databases were used to search for published literature. The 'Results and Discussion' section describes the key aspects of UAV use which are critical for ambient air pollution monitoring [30], as well as an analysis of current UAV applications in air quality. The latter is divided into three keys areas of investigation: atmospheric composition, pollution and climate change; earth surface, interior and atmospheric phenomena; and prevention, patrolling and intervention. Finally, the current challenges, including possible solutions and future applications of UAVs, are presented.

2. Materials and Methods

This review used three different bibliographic databases, Web of Knowledge, Scopus and Google Scholar. Scopus provided extensive coverage of the topic area, but was limited to articles published after 1995. Google Scholar was useful for retrieving even the most obscure information, although its use was marred by inadequate or out-of-date citation information [31]. Different keywords were used and original peer-reviewed research articles and literature reviews were included in the search. In total, 61 search terms (See Table S1 in Supplementary Materials) and different combinations of these were used, including a combination of at least three individual terms. Data was compiled from published journal articles, conference proceedings, books and grey literature, namely technical reports published by government agencies, academic institutions, trade publications and information gathered from manufacturer websites, which are not typically subjected to peer review and may contain biased data. Although some cited reports came directly or indirectly from industries with a financial interest in promoting the use of UAVs, the data were cross-checked to ensure validity of the conclusions. The literature review ended in March 2016.

3. Results and Discussion

3.1. Air Pollutants Which Need to Be Monitored

3.1.1. Atmospheric Composition

The composition of the atmosphere is strongly related to emission processes which release a wide variety of aerosols and gases, but are also connected to the Earth's geography, meteorology and rotation which affect the movement of air masses [32,33]. The emission sources are both natural, such as vegetation, deserts, volcanoes and wild fires, and anthropogenic, such as power plants, domestic heating, transportation and industrial production. However, anthropogenic sources generally have a higher impact on health and the environment [34]. The main particulate matter (PM) and volatile organic compound (VOC) sources in urban environments are vehicle emissions, domestic heating and industrial processes.

Combustion processes release four principal greenhouse gasses (GHGs) into the atmosphere: carbon dioxide (CO_2), methane (CH_4), nitrous oxide (N_2O) and halocarbons (gases with fluorine, chlorine and bromine) [35]. In addition, nitrogen monoxide (NO), sulphur dioxide (SO_2), VOCs, PM, black carbon (BC) and organic carbon (OC) are also emitted [36,37].

3.1.2. Health and Environmental Impact

Combustion-generated particles are the main contributors of PM in the urban atmosphere and they are strongly related to a number of adverse health effects [38]. Previous studies have identified that long-term exposure to combustion-related fine PM might be associated with an increased risk of cardiopulmonary and lung cancer mortality [39]. The PM emitted by anthropogenic combustion sources contributes a small fraction in terms of mass, but a very large number of particles, predominantly in the ultrafine range (Ultrafine particles (UFPs) < 100 nm) [40]. UFPs have a significant impact on human health and the environment [2,41]. They can penetrate and deposit deep inside the lungs [2,41,42] and because their lifetime is longer than particles with a larger diameter, they can be transported further from their sources, due to slow removal from the atmosphere [43]. In addition to their effects on human health, monitoring combustion-generated PM (including BC) is critical due to its capacity to adsorb sunlight, as well as its subsequent release of heat [44–46]. Assessing atmospheric VOC concentrations is also important, not only due to their direct effects on health, but also because they can act as precursors and become photo-chemically oxidized (atmospheric aging), leading to the formation of secondary organic aerosols (SOA), which have also been linked to poor health outcomes [47].

3.1.3. Air Quality Measurements

Data collection in relation to gaseous pollutants, PM [48–53] and VOCs from traffic, power plants, urban and industrial air quality is extensive and includes ground level sampling using both on-line and off-line techniques [54–56], as well as manned aircrafts [57]. However, direct measurements at the source are not always feasible due to the complexity of sites, moving sources or physical barriers, such as direct measurements of shipping emissions or biomass burning.

It is critical to characterize in-situ air pollutant properties, in terms of origin, concentration mixing state, size, chemo-physical composition and reactivity, for both air quality and climate change research, as well as for policy development and the regulation of combustion source emissions. Therefore, UAVs may be a viable option for such in-situ air quality data collection [25].

3.2. UAV Types and Requirements for Outdoor Air Composition Studies

UAVs are operationally more versatile and visible compared to land-based approaches or other aerial methods, such as manned aircraft and satellites. Conducting atmospheric measurements in remote locations is one situation where the use of small, lightweight UAVs is of particular benefit [58,59]. In fact, the reduced size, weight and power needs of these flying robots, along with the reduced cost of the platforms and instrumentation, make them highly suitable for these operations.

Unmanned aircrafts encompass a wide range of different platforms which, due to their physical size and power, differ in terms of their capability and simplicity of operation. These factors impact the payload carrying capacity, speed, altitude and range of flight, which determines the different applications that can be performed by each type of UAV. Figure 1 shows examples of fixed and rotary wing UAVs. Several platform classifications have already been proposed, however, the nomenclature adopted for civil and scientific use has generally followed the existing military descriptions of size, flight endurance and capabilities [60–62].

(**a**) CyberEye II (**b**) Silvertone Flamingo (**c**) SenseFly Swinglet

Figure 1. *Cont.*

(**d**) AscTec Pelican (**e**) DJI F550 (**f**) DJI S800

Figure 1. Example of a small fixed wing (**a**) CyberEye II [63]; (**b**) Silvertone Flamingo [64]; (**c**) SenseFly Swinglet [65] and rotary wing (**d**) AscTec Pelican [66]; (**e**) DJI F550 [67]; (**f**) DJI S800 [68] unmanned aircraft (All the UAVs shown are part of the fleet of the Australian Research Centre for Aerospace Automation).

Even though small UAVs are subject to significant payload restrictions compared to larger (manned) aircrafts, they have a distinct advantage over their manned counterparts in terms of relatively low platform cost, capability to perform autonomous flight operations from take-off to landing, and to fly closer to the ground with no risk for a manned crew [69]. Pre-programmed flight plans can be issued and automatically controlled on-board, which means that small UAVs can be flown with greater accuracy and less workload than aircraft with human operators. Some platforms even have the ability to work in environments without GPS signals and/or to follow local linear infrastructure without the use of a GPS [70], and therefore, could provide efficient and accurate monitoring inside buildings, forests or canyons [71].

3.2.1. Performance and Capability of UAVs

UAV performance and capability are closely linked to aircraft size, and therefore, small, low-cost aircraft will inherently have payload, speed, power and endurance limitations. They will have a limited ability to carry on-board sensors/equipment and potentially short flight times. Airframe dimensions and shape, for example, can make the mounting of sensing equipment difficult, and power may have to be shared with the sensors, depending on the propulsion system. As such, large amounts of energy may be required, which will reduce both flight time and the spatial diversity of collected data. Low speed operation, governed by low stall speed characteristics, is often possible with such platforms, allowing for the spatially dense data collection often required for localized, site-specific inspections [72–74].

Larger aircraft have less stringent payload limitations and can accommodate an increased number and diversity of on-board sensors and equipment. A de-coupled propulsion and payload power system is typical, which further increases the potential for long-range, high endurance applications, such as large scale regional inspection. UAVs have been categorized into five groups by the U.S. Department of Defense (DoD), as shown in Table 1 [58].

To date, UAVs have been equipped with small size, lightweight visible-spectrum cameras or, in some cases, near-infrared cameras for conducting air quality measurements [60]. Chwaleba et al. [75] reviewed optical sensors that can be carried on-board a UAV for air pollution monitoring, ranking Light Detecting and Ranging (LIDAR) sensors as the best optical device to be used as a payload for air quality monitoring. Depending on the sensors used, multiple data sets may be collected with a high spatial and temporal resolution [60,76,77]. However, with more complexity and capability comes more maintenance, and additional specialist skills may be required. Larger platforms are costly and require a significant financial investment. Perhaps, the most important consideration is the safety of using such platforms in commercial applications, since they have the potential to cause considerable damage (to humans and property) and as such, fall under stricter operating guidelines than smaller UAVs [78,79].

Table 1. UAV categorization used by the U.S. Department of Defense [58].

UAV Category	Max Takeoff Weight (Gross)	Normal Operation Altitude (ft)	Airspeed
Group 1	<20 pounds (9.07 kg)	<1200 (365.76 m) above ground level (AGL)	<100 knots (<185.20 km/h)
Group 2	21–55 pounds (9.53–24.95 kg)	<3500 (1066.8 m) AGL	<250 knots (<463.00 km/h)
Group 3	<1320 pounds (<598.74 kg)	<18,000 (5486.4 m) mean sea level (MSL)	Any airspeed
Group 4	>1320 pounds		
Group 5		>18,000 MSL	

Note: if an UAV has one characteristic of the next higher level, it is classified as being part of that group.

3.2.2. UAVs as Platform for Air Quality Research

Aircraft capability in the context of air quality monitoring is a critical aspect that needs to be considered, based on the purpose of the research. Fixed wing aircraft can typically cover a greater area over a given time interval and provide flexibility in terms of sensor mounting points. As they are unable to hover and have minimum operating height requirements, high spatial diversity can be achieved at the cost of decreased spatial resolution. In some cases, they can be inherently stable, allowing some forgiveness when failure occurs, such that payload recovery is likely. They may also require short runways (30–200 m), with some clearance for take-off and landing, or the use of a capture and release device [80–83]. They can be designed for compact and easy transport and deployment, allowing for operation in remote locations with minimal infrastructure requirements [84].

Rotary wing aircraft, such as helicopters and multirotor (quadrotors, hexacopters or octocopters), generally have a lower operating speed, but allow discontinuous trajectories, such as hovering, for close proximity inspection. Typically, increased spatial resolution can be achieved at the cost of decreased spatial diversity. Recent advances in control have made these inherently unstable platforms more reliable and easier to operate, reducing the risk of payload damage and accidents in general. They do not require specialized equipment or a runway for launch and recovery, and, depending on their size, can be easier to transport. Sensor placement has to be well matched to the platform, with the sensor intake typically located away from the propeller or rotor downwash effect.

Other aircraft types, such as parasails and blimps (balloons), can also be utilized [85]. They allow for larger payloads and slower operating speeds, compared to fixed wing aircraft. They may also have long operating times, but can, in some cases, be difficult to control and with less maneuverability [86]. This is mainly due to their high susceptibility to ambient weather conditions [87].

Unmanned aircraft systems can also be used in standalone operations involving a single platform, or more advanced systems utilizing multiple aircraft. In each case, a ground station is usually required for remote piloting and mission command. Multiple UAVs can be flown in a swarm, or coordinated to fly separately but with complementary trajectories for a given application [88]. This requires advanced centralized or decentralized control and guidance algorithms, but has the potential to increase the quality and quantity of data collected with a reduced operator workload. Currently, the use of multiple UAV systems has been demonstrated for a range of related applications [89].

Unmanned aircraft can also be viewed as useful tools for plume monitoring control and management within the disaster management framework [90,91]. They have the potential to provide high resolution spatial and temporal data sampling over large areas or in site-specific locations. They can also be used on a local, district or field level, depending on the type of data being collected and the platform characteristics. These data sets may then be coupled directly with other measurements, depending on the number of auxiliary sensors and aircraft/agents used. Cost per unit, including platform and auxiliary equipment, such as a ground station, may be significant (500–1,000,000 AUD), but in the appropriate circumstances, subsequent operation may be less expensive than manned aircraft operations or satellite based approaches.

3.2.3. Regulations and Advisories

Importantly, unmanned aircraft cannot be deployed without restrictions. Under current aviation safety operating regulations, restrictions are placed on their use in commercial, research and private applications [92,93]. For example, in most countries, it is a requirement that UAVs be controlled by a certified operator. This adds an additional safety measure, but bears a cost and dictates who can legally conduct UAV operations [94]. This, in turn, has a direct effect on the frequency, quality and type of atmospheric research-related applications that can be conducted.

Considering that unmanned aircraft are flying robots operating in the same national airspace as manned aircraft, they too need to comply with governing aviation safety regulations. Aviation regulatory bodies, such as the Federal Aviation Authority of the U.S. (FAA) and Civil Aviation Safety Australia (CASA) define the rules and restrictions that govern who can access the national airspace and under what conditions, with the aim of protecting the general public and other airspace users by ensuring that safety standards are met. Some rules may only apply to a particular type of aircraft, while some may apply to all aircraft operating under certain weather conditions or flight types. Globally, a complete set of operating regulations for UAVs are yet to be defined [95,96].

3.3. *UAVs for Ambient and Air Quality Composition Measurements*

UAVs have been used by researchers and commercial organizations to sense atmospheric gases and aerosols [97], and have been shown to be capable of reaching remote areas and survey large regions [98,99]. The following sub-sections highlight the contribution of UAVs to the air quality research domain.

3.3.1. Study of Atmospheric Composition, Pollution and Climate Change

The application of UAVs to measure atmospheric composition, pollution and climate change includes taking in-situ samples above, below and within the atmospheric boundary layer (ABL), as reported by a number of studies reviewed below.

Wind Vector Measurements

In order to measure wind vector, a UAV can perform a spiral flight trajectory. A Pitot tube mounted on the nose of a fixed-wing UAV could be used to measure and calculate the horizontal wind. However, no airflow angles are available, and so the wind vector can only be calculated by performing special maneuvers, giving a horizontal resolution of about 300 m [100].

A meteorological Mini-UAV "M^2AV" (5 kg takeoff weight, 1 kg payload, with more than 50 km flight range) capable of measuring T, H and wind vector data was developed by Spiess et al. [101,102]. The data collected by the UAV (temperature), were shown to be in good agreement, with a maximum difference of less than 0.5 K [100].

Van den Kroonenberg et al. [103] used the same UAV, the M^2 AV, to develop a UAV system to collect wind data. The system had a five-hole probe (5HP), a GPS receiver and inertial measurement unit (IMU), meaning that an inertial sub-range of locally isotropic turbulence can be measured up to 40 Hz (or 0.55 m at 22 m$\cdot$ s^{-1} airspeed). During weak wind (3–4 m$\cdot$ s^{-1}), the M^2 AV data agreed with sonic detection and ranging (SODAR) and meteorological tower data to within 1 m$\cdot$ s^{-1}.

However, UAV measurements were accompanied by large standard deviations of up to 0.4 m$\cdot$ s^{-1} when measuring stronger wind (6–7 m$\cdot$ s^{-1}). The M^2 AV measured higher mean wind speeds compared to nearby SODAR profiles, but agreed well with tower measurements.

The results of the work of Van den Kroonenberg et al. [103] suggest that to accurately measure the wind vector, any UAV would need a 5HP, GPS receiver and IMU.

Martin et al. [104] flew the "M^{2}AV" UAV during a campaign over Lindenberg, Germany, to measure morning and evening ABL transitions. With a climbing rate of 3 m/s and equipped with fast reading sensors (response frequency of 30 Hz), the "M^{2}AV" was able to provide a 10 cm vertical resolution for temperature, humidity, wind direction and speed. The collected data was in very good agreement (not larger than 1 K) with ground-based sensing systems.

A UAV called small unmanned meteorological observer (SUMO), was flown in Iceland during several campaigns to measure temperature, humidity, wind speed and direction up to 3500 m above ground level [88,105]. The SUMO operated successfully under polar conditions, reaching 1500 m in altitude, at a ground temperature of −20°C and a wind speed of 15 m/s. The results of these measurements were used to compare and evaluate the results of an Advanced Research Weather Forecasting (AR-WRF) model. The observed and modelled data were in good agreement; however, in some cases the model did not reproduce small inversions measured by the SUMO, which reported two layers at heights of 200–500 m and 1000–1300 m that were dryer than expected. In fact, the model overestimated humidity by around 25%. The collection of wind data was only possible during autonomous flight mode, which for safety reasons, could only be activated above 200 m, and therefore, resulted in no wind data being collected below that threshold. The need to operate under clear sky or thin cloud conditions was another limitation, since the SUMO stabilization system, which is based on infrared sensors, requires a radiation temperature difference of around 8 K between the ground and the sky [105]. Reuder et al. [88] worked on the optimization of the SUMO system to include: a new autopilot to enhance in-cloud flying capacity, a faster temperature sensor to reduce the measuring time from 5 to 1 s and by adopting the five-hole probe (5HP) system. The 5HP was the same approach suggested by Van den Kroonenberg et al. [103], which enabled the SUMO to determine 3D turbulent flow vector with a temporal resolution of 100 Hz.

The work of Reuder et al. [88] showed: (1) that was possible to integrate sensors onboard a UAV for air quality measurements (optimized within 650 flight missions since the first attempt in 2009 [106]); but also (2) that temperature, humidity and wind data, collected during a field campaign (Lannemezan, Franced) were significant for a case study on anabatic flow from the lowlands towards the foothills of the Pyrenees.

Atmospheric Aerosols Data Collection

The versatility of UAVs has been shown in several missions focused on the investigation of atmospheric aerosol properties, in particular light-adsorption and light-scattering properties. The application of a stacked configuration of three UAVs was successfully adopted during the Maldives Autonomous UAV Campaign (2006) to simultaneously measure aerosol-cloud-radiation parameters within the same column [99]. The aim was to clarify the nature of discrepancies between modelled and observed data under clear sky conditions. The UAVs flew with a horizontal separation of 10 m and a delay of less than 10 s to avoid ambiguities that arise from spatial and temporal variability in aerosols when passing over the same geographic point (or clouds). Results showed that the model was within an experimental error of 15% compared to the data collected when flying between 500 m and 3000 m. Both Ramana et al. [98] and Ramanathan et al. [99,107,108] stated that there was no need to invoke anomalous or excess absorption or unknown physics in clear skies.

The work of Ramanthan et al. [108] indicates useful information for sensor integration, flight path design and real flight of a UAV swarm. In addition to this, the work is significant in terms of data collected, overcoming the challenge of direct measurement of the solar heating caused by BC. This was possible thanks to the use of the three UAVs flown in a stacked formation over the same environment at different altitudes to measure flux divergences (heating rates) for an extensive period of time. The study also demonstrated that atmospheric brown clouds with a visible absorption optical depth as low as 0.02 are sufficient to enhance solar heating of the lower atmosphere by 50%. Corrigan et al. [25] used two twin fixed wing UAVs in a stack formation to monitor total particle mass concentration, particle size distributions, aerosol absorption and BC concentrations within the mixed layer over the Indian Ocean. Each platform had a weight of only 27 kg and 5 kg payload. A cruising speed of 60 knots (111 k/h) enables the UAVs to fly for up to 5 h, giving a nominal range of 550 km. The difference between data collected by airborne and ground instruments, and also aircraft-to-aircraft measurements, was <10%. However, measurements of aerosol optical depth taken by the UAV differed by up to 20% with those taken by a columnar AERONET sun-photometer. In this case, the data collected by the UAV allowed Corrigan et al. [25] to observe a large aerosol plume above the mixed layer, with a peak concentration near 2000 m. This result is in agreement with previous studies which disproved the common assumption that the mixed layer has a uniform concentration of constituents which decrease exponentially once above this layer.

The study by Bates et al. [109] aimed to generate a vertical profile of atmospheric BC concentrations, using a MANTA UAV in an 18-flight (38 flight hours) campaign. The payload was customized for online particle number concentration and aerosol light absorption (at three wavelengths) measurements, as well as particle collection using an 8-filter sampler (for off-line analysis). The flight plan for the UAV was to climb to 2700 m, descend to the altitude of maximum aerosol concentration, and then conduct sampling at that altitude. BC concentration varied from below the detection limit (0.04 $\mu g \cdot m^{-3}$) to 0.51 $\mu g \cdot m^{-3}$. Using the UAV, Bates et al. [109] were able to measure BC transport and distribution above the ABL, which was previously made with manned aircraft, being impossible to measure from a ground station.

Altstädter et al. [110] developed the ALADINA (Application of Light-weight Aircraft for Detecting In-situ Aerosol) system to investigate the 3D distribution of UFPs within the ABL. The UAV, equipped with two condensation particle counters (CPCs) and an optical particle counter (OPC), to give a total payload weight of 2.8 kg, was able to measure aerosols distribution from the ground up to 1000 m. The authors reported that the concentrations measured by the ALADINA were consistent with those obtained using a scanning mobility particle sizer (SMPS) system and an aerodynamic particle sizer (APS) at ground-level, however, the percentage of agreement between ground and airborne measurements was not presented. The work of Altstädter et al. [110] demonstrated the possibility of integrating two CPCs onboard a UAV to measure UPFs. However, the instruments measured the number of particles in the range of 11 nm and 2 μm and not only the UFP (<100 nm) fraction as stated in the title of the work. The research focused and provided details on integration of the sensors rather than validation of the data collected with the UAV system.

Harrison et al. [111] proposed the use of a remote-controlled aircraft to investigate the horizontal, vertical and temporal variability of PM within the first 150 m of the atmosphere. They aimed to prove that UAV-based systems could be the next generation validation methodology for satellites. Harrison et al. [111] used a modified 3 m fixed wing model aircraft (already successful in another study [112]) to perform four different flights: the first three followed an oval pattern with the altitude increasing after each loop, starting at 30 m and finally reaching up to 140 m. The UAV sensor package system was an aerosol spectrometer with the intake probe mounted in a cowl to allow clean, undisturbed air sample collection (payload specification not given). The mean $PM_{2.5}$ concentration for three flights with varying altitude was 36.3 $\mu g/m^3$, and the highest concentration was recorded below a 10 m altitude. Results showed an overall vertical variation with a standard deviation of only 3.6 $\mu g/m^3$. The $PM_{2.5}$ concentration did not change significantly across the day, with mean concentrations for the first three flights being 35.1, 37.2, and 36.8 μg/m^3. A lower concentration of 23.5 μg/m^3 was recorded during the constant altitude flight. Harrison et al. [111] believe that this reduction in concentration is significant compared to the concentration variations in any of the first three flights, therefore, that the constant altitude does not account for the change. For this reason, more data is needed to explain the cause of the change.

The flight at near constant altitude of 60 m was conducted to characterize variation on the satellite data subpixel scale; data was divided into a 120 m × 65 m grid. The mean $PM_{2.5}$ concentration across the grid cells was within the standard deviation of any grid cell, with a variation from 20.5 to 24.9 μg/m^3. This meant that the PM concentration could be accurately characterized as the data pixel area equivalent to the flight path area.

Other researchers have used multi-rotors to overcome the limitations of a ground-based station when measuring near-surface gradients of trace gases and PM concentrations. Brady et al. [113] used a quad-copter based system to study the vertical and horizontal concentration of CO_2 and PM at a high spatial resolution (1 m) within the atmospheric mixed layer (0–100 m). A 3D Robotics Iris+ was used as a platform and the system made by integrating two commercial sensors onboard: (1) a dual channel optical sensor (MetOne 80080) to measure PM with an atmospheric diameter between 0.5 and 1 μm (channel one), and greater than 1 μm (channel two); (2) a NDIR CO_2 sensor (CO_2Meter-30) to detect CO_2 between 0 and 10^3 ppm. The payload weight was 510 g, limiting the flight time to 5 min, which was enough to observe sea-spray aerosol generation in the surf zone (a high-intensive production zone for sea spray aerosol due to the abundant wave breaking), showing high precision in both vertical (±0.5 m) and horizontal positions (±1 m). The UAV system was flown horizontally at 5, 10, 15 and 25 m from the beach to the surf zone to characterize the aerosol plume, providing both vertical and horizontal profiles. The profile, obtained from a total of 13 vertical samplings, demonstrated a maximum aerosol plume height of 40 m above the surf zone, and also a turnover in the 0.5–1 μm particle concentrations at an altitude of 70 m. These horizontal and vertical aerosol profiles enabled researchers to calculate ambient humidity emission rates for small and large particles in the surf zone. Future work aims to make the system lighter to increase flight endurance. Overall, the UAV system provided an efficient sampling platform to measure the vertical and horizontal profiles of sea spray aerosol generated within the ABL.

Mölders et al. [114] theoretically examined the capability of UAVs to provide spatial distribution of mean pollution concentrations using the 2009 Crazy Mountain complex fire in Alaska as a case study. The Evaluated Weather Research and Forecasting model in line with chemistry (WRF/Chem) data was used to represent ABL conditions. A virtual Scan Eagle capable of 20 h flight time (assuming zero-weight payload), was flown at different altitudes, speeds and following different path to collected data from the WRF/Chem. UAV measurements of CO correspond to the ground measurements within a factor of two at 1000 m, demonstrating that the UAV system can provide good 20 h mean distributions of CO at 1000 m for 60 km × 60 km. However, it is necessary to derive separate mean distributions for daytime and nighttime when considering pollutants involved in photochemical reaction chains (SO_2, NO). In fact, results showed that the diurnal cycle of SO_2 and NO concentrations led to overestimation compared to the ground measurements. Moreover, due to the relatively short period of darkness at high latitudes in late summer, a UAV swarm would be required for nighttime data collection. Each UAV would sample flying different pattern for 20 h.

Aviation advisory would benefit from the possibility of UAV to sample around the top of the ABL, which would give information on plume's dispersion. This is particularly important when satellite imagery cannot provide such information due to cloud cover in the mid and upper troposphere.

Sampling strategies for meteorological and chemical quantities data collection are different. Air quality forecasts and the virtual sampling technique could help in effective, optimized flight planning, and data collecting. However, the authors assumed zero payload, while realistically even the lightest onboard sensors would add some weight. The heavier the payload the less fuel can be added, which reduces flight duration.

UAV deployment for air quality advisories requires long flight durations to cover large areas such as in the downwind of wildfires. Mölders et al. [114] asserted that instruments light and small enough to fit a UAV system capable of sampling at high frequency still have to be developed.

Greenhouse Gases and Other Gaseous Pollutants Measurements

Berman and Fladeland used a SIERRA UAV fitted with a custom-made GHG analyzer to conduct highly accurate (1 Hz) measurements of CO_2, CH_4, and water vapor concentrations at low altitudes ($\geqslant$10 m) in Svalbard, Norway [115,116]. These SIERRA results are consistent with those measured by a Zeppelin station at 475 m above sea level (the percentage of agreement between readings at the different stations was not reported).

Malaver et al. [117–120] explored the possibility of flying a solar UAV as part of solar powered wireless sensor network system (WNS), to monitor GHGs continuously, using solar energy to solve the power consumption issue.

Rossi et al. [121,122] worked on the UAV and sensor energy consumption issue and successfully developed a new VOC and gas sensing system. The device is based on a fully automated metal-oxide (MOX) sensor, capable of running and storing data for 30 min. The researchers mounted the device underneath a hexacopter and performed two experiments. The first, hovering above an opened solvent (isopropyl alcohol) bottle, and the second, flying above the University refectory chimney, demonstrated that the sensing performance is not affected by the air and capability of the system to spatially describe VOC concentration.

Both the studies by Malaver et al. [117–119] and by Rossi et al. [121,122] were focused on sensors' integration and illustrated the possibility of integrating low cost, low power consumption sensors to allow longer sampling times, for air quality applications rather than quantification of the measured gases.

Watai et al. [97] demonstrated that UAVs are a suitable and cost-effective method for measuring spatial and temporal variations of atmospheric CO_2 in and above the ABL. The researchers integrated, calibrated and tested a 3.5 kg CO_2 sensitive payload, with a 20 s response time and a precision of $\pm$0.26 ppm. With a maximum flight time of 1.5 h, the UAV was able to rise to 2000 m (main observation area) and then descend in a spiral to about 650 m, before recovering and landing at the starting point. After 15 flight tests, Watai et al. [97] stated that the system was capable of conducting precise, high-frequency measurements, in order to determine the temporal trend of CO_2 which tended to change between 200 and 400 m and 400–600 m layers. However, CO_2 data were not collected at ground stations, and therefore direct comparisons were not possible.

Illingwoth et al. [123] confirmed both the cost efficiency of UAVs and the limitations of ground-based instruments to monitor the variability of target gases over a localized area as well as to provide important information in relation to the rapid characterization of micrometeorology and chemical composition. Illingwoth et al. [123] mounted an ozonesonde (manufactured by Science Pump Corporation., Camden, NJ, USA) on-board an inexpensive UAV (Skywalker Technology, Wuhan, China) to measure variations in ozone concentration on an urban scale. Flying close by Manchester, UK, they captured a peak concentration of approximately 39 ppm, which was associated with a short-term shift in wind direction. On the other hand, data collected by two nearby ground stations did not show this variance, but rather a constant O_3 concentration of about 19 ppm and 26 ppm, respectively.

Lawrence et al. [124] developed a low-cost (400 USD airframe, 300 USD sensors) UAV system to address the emerging need for fine-scale measurements of atmospheric variables throughout the troposphere and lower stratosphere. The UAV system, DataHawk (0.7 kg, 1.0 m wingspan), was capable of in situ temperature, humidity, wind vector and turbulence sensing at high spatial resolution (1 m over a horizontal scale of 1 km), over altitudes ranging from a few meters up to at least 9 km. The work showed not only the possibility to integrate low-cost sensor to have accurate data (temperature, humidity and wind speed accuracy: 2 °C, 2%, 0.2 m$\cdot$s^{-1}), but also the ability of collecting data up to the top of the ABL.

Overall, UAVs can make significant contributions in terms of atmospheric data collection within the ABL, as well as in more complex environments such as mountainous areas [105]. Knowledge in relation to aerosol vertical profiles within the atmospheric column can also be improved when using UAVs, particularly where limited infrastructure is available, such as in remote or hostile areas. Furthermore, obtaining lower atmosphere profiles, up to about 3 km above ground, every 30 min, provides sufficient temporal resolution to investigate the relevant chemo-physical processes in the atmosphere. UAV systems could validate satellite data, and measure horizontal, vertical and temporal variability of gaseous and aerosol pollutants in the lower atmosphere, however, the intended application(s) of the systems have to inform its design. These design factors include the type of monitoring required (local, regional, etc.), the targeted activity required (VOC sampling, gases or data collection), operational frequency, safety, cost and long-term flexibility for that application and therefore the capabilities and limitations of the UAV platform.

Despite the cost and flexibility advantages, small UAVs are still limited by their short flight endurance, low payload capacity and network integration. Sensor technology is another limitation for the use of lightweight UAVs when fast sampling (1 s) and high resolution (ppb) are required. It is critical for researchers to select or design a new UAV system for air quality by taking into account how to validate system measurements, given that the sensors used in them may not have been originally developed to be mounted onboard a UAV. The use of small UAVs under a mass of 7 kg is favorable to maximizing their productivity. Using such class of UAVs, single flight duration and daily sampling are enhanced. The issue is that most commercially available, reasonably priced fixed wing systems are not designed to facilitate these operations. Set-up and breakdown times are excessive and individual components are not designed for heavy use [100].

These limitations need to be overcome, along with current aviation flight restrictions, in order to facilitate the extensive use of UAVs to assess air quality.

3.3.2. Measurement of Surface, Interior and Atmospheric Phenomena

Volcano Emissions Data Collection

The limitations of using manned aircraft to conduct in-situ observation of volcanic plumes, which are transient and difficult-to-access, can now be overcome using UAVs.

In 2002, the first fixed wing UAV, an Aerosonde, carrying a miniaturized SO_2 sensor [125] was freely flown for volcanic gas sampling [59]. Although the technical report did not show the specifications of the sensor or precise details of the analysis, researchers reported issues regarding the cost deployment, together with regulatory difficulties. At about that time (2000–2001), the Yamaha Corporation flew its latest unmanned helicopter to conduct surveillance imaging at the Unzen Volcano and at Mt. Usu, Japan [126], promoting the use of UAVs for the surveillance and investigation of natural disasters, such as earthquakes and volcanic eruptions.

Several years later, Astuti et al. [127] developed a novel "VOLCAN UAV system" for the measurement of volcanic plume chemical composition. This included the design of a new custom-made platform, an electric engine (required by volcanologists, in order to avoid instrument reading contamination, in the case of a gas engine), about 30 min of flying autonomy with a 3 km operational range (the distance between the plume and the base camp was roughly 2000 m), 5 kg payload capacity, a 40 km/h minimum cruise speed (to collect enough in-situ samples) and a maximum cruise altitude of 4000 m. A flight simulator (X-Plane), with a suitable plane model and interfaced with the real electronic boards, was adopted to simulate the UAV dynamics, in order to allow for easy tuning of control parameters [128,129]. However, based on time and money constraints, the researches decided to use a low cost petrol-powered model airplane (80 cm^3, 2-stroke engine) instead, because the VOLCAN UAV still needed a proper launch system and real world tests before it could be used for the required measurements. The work of Astuti et al. [127] showed the possibility of integrating a gas sensor system onboard a UAV. However, the experiment focused on integration rather than quantification of trace gas species.

Longo et al. [130] went a step further and used data collected by robotic vehicles in the air and from the ground to design a new mixed Terrestrial Aerial Robotic PLAtform for Volcanic and Industrial Surveillance (TARPLAVIS). The TARPLAVIS system was comprised of several modules: a base station with a tele-control system, a six-wheeled Unmanned Ground Vehicle (UGV; ROBOVOLC model) equipped with different cameras for orientation and scenario screening (IR camera to thermally map the environment), a climbing robot ALICIA, and the VOLCAN UAV. Although the authors concluded that the potential and capability of this new multi-platform approach was tested for volcanic and industry surveillance, there are currently no data available in this regard. The aim of the study of Longo et al. [130] was to integrate different technologies in order to build a platform to collect data for industrial or volcanic emergencies. It was a success from an engineering point of view; however, it did not address the challenges of the application of the technology to real world capability as well as of the accurate sensor selection and validation.

In 2004, Saggiani et al. [131] flew a small fixed wing UAV (3.5 m of wingspan, 4 m long, max take-off weight 30 kg, 2 kg payload) around the Stromboli volcano (Italy). Three main issues were highlighted: difficulty finding a suitable area close to the volcano to use as a runway to take-off and landing (min 100 m); the inability to conduct a fully autonomous flight due to the limited UAV operational range (2 km max); and the limited payload capacity, which highlighted the need for a larger platform with better flight stability during bad weather. Based on the above considerations, Saggiani et al. [132] went on to test an INGV Raven UAV, which had a 6 kg payload (including an IR camera, a micro-interferometer (for O_3 and H_2O analysis) and a SO_2 sensor), a flight range of 50 km and a flight time of 4.5 h. Although the researchers declared the mission a success, the data were fragmented, the landing system needed improving, and a ground control station was required to allow mission control from remote sites [133]. The use of UAVs for volcanology studies continued with the work of Amici et al. [134], who used a rotary wing UAV carrying an IR camera to thermally map a mud volcano (lower south-east flank of Mt Etna, Italy). Amici et al. [134] stated that the data from IR cameras at the ground level and those from the UAV were in good agreement, with the temperature of a cold pool detected during a flight found to be 34 °C by the UAV and 34.7 °C by the ground cameras.

Patterson et al. [135] flew an autonomous Silver Fox (lightweight UAV with a 9 kg take-off weight, max speed 110 km/h, service ceiling up to 4850 m) at an active volcano crater (Mount St Helens, Washington USA), delivering real time data to a remote location using visual and IR video sensors. The Silver Fox was also equipped with electro-chemical gas sensors (weight of less than 1 kg) capable of detecting seven different gases (authors did not provide further details). While this showed the potential for UAVs to provide useful information, even when using a relatively low-cost aerial platform, it was determined that both the sensitivity (1–100 ppm) and response time (30–60 s) of the sensors were insufficient in terms of range and reading speed to detect volcanic plume gases. The researchers highlighted the need for miniaturized sensors, as well as an optimized engine and airframe, to allow flights within acid and ash-bearing volcanic plumes, as well as close collaboration with scientists to improve the accuracy of data collection, interpretation and hazard analysis.

The potential of rotary wing UAVs for use in volcanology studies was confirmed by McGonigle et al. [136], who used an off-the-shelf platform (for less than 2000 USD) equipped with two different payloads below 2 kg: a UV spectrometer for SO_2 monitoring and a multi-gas system for in plume analysis. The rotary-wing UAVs were flown less than 200 m downwind of the crater rim at Vulcano, Italy, following an eight traverses path underneath the plume to determine the SO_2 flux. A reading error of 25% was estimated (±15% uncertainty from the spectral SO_2 concentration retrievals, ±20% in plume transport speed measurement). After the first test, McGonigle et al. [136] used the second payload to measure gas concentrations by hovering the UAV inside the plume for 22 min, 10–100 m downwind of the crater rim. CO_2 and SO_2 were measured and the resultant CO_2/SO_2 ratio of 30 ± 5 was found to be in good agreement with the ratio of 35 found by Aiuppa et al. [137], who previously tested the performance of a portable gas analyzer at the same crater.

The study of McGonigle et al. [136] showed an innovative approach in addition to remote sensing, measuring CO_2 and SO_2 with a UAV from a volcanic plumes. The focus was not only on the sensor integration, but on the data analysis which showed the possibility of measuring gas concentrations in the plume downwind of the volcano. Collecting such data is normally very difficult due to SO_2 cross-sensitivities of many gases and in particular O_3. The study highlighted and detailed the possibility of using a combination of multiple sensors onboard a UAV, as an independent method for in-situ gas data collection.

It is critical to calibrate sensing systems and validate the data they produce, before mounting them onboard the UAV. Normally manned aircraft are used for this purpose, however, Pieri et al. [138] stated this approach in volcanology studies is an undue risk for the crew, even in dilute plumes (~1000–2000 $\mu g/m^3$). The authors claimed that UAV observations proximal to the eruption site can provide the direct data collection necessary for model inputs and are more reliable compared to those currently collected from ground-based and satellite stations. The research suggested the use of three different UAVs: the Dragon Eye μUAV, the Wing Vector 100 and the SIERRA, and also highlighted the need for civil aviation authorities to recognize the unique benefits of UAVs for conducting measurements over volcanoes or within eruption plumes that are too dangerous for manned aerial vehicles.

Diaz et al. [139] worked to provide geochemical data using mass spectrometry techniques to investigate volcanic activity. The ULISSES is a small (10 kg weight and 21,000 cm^3 volume), low powered (24 V) mass spectrometer (MS), developed in 2008 by Diaz et al. [140] in collaboration with NASA [141]. Continuing with their miniaturization studies for spectrometers, Diaz et al. [142] recently developed a MS with the world's smallest turbo-molecular pump, to be mounted on-board a UAV for in-situ volcanic plume analysis. Two platforms were considered, the NASA Sierra UAV and the Wing 500 (3.3 m wingspan, 30 min of endurance and 4.5 kg payload), to carry the MS and other sensors for temperature, humidity, pressure, SO_2, H_2S and CO_2. The MS was laboratory calibrated using H_2, He, N_2, O_2, and CO_2 gases at different concentrations (0 ppmv, 100 ppmv, 1000 ppmv and 10,000 ppmv). The UAV-MS system was tested in three locations: (1) Miravelles volcano, Costa Rica, where it found a CO_2 concentration close to 75%, while H_2S was 168 ppmv and SO_2 was 24 ppmv; (2) La Bocca Grande fumarole, Italy, where the CO_2 concentration was about 86%, the H_2S concentration was 66 ppmv and the SO_2 concentration was 13 ppmv; and (3) La Bocca Nuova fumarole, Italy, recording a CO_2 concentration of 76%, and 68 ppmv and 11 ppmv for H_2S and SO_2, respectively.

Typhoons Data Collection

Atmospheric phenomena, such as typhoons, are another situation where UAVs have successfully been used to collect data to provide an independent check for ground-based measurements. In a study by Lin et al. [143], the Aerosonde was flown up to a height of 3000 m for a 10 h-long reconnaissance inside the eye of Typhoon Longwang. The measured wind speed was about 35 m/s, which increased to over 50 m/s for a period of more than 20 min and then dropped to less than 10 m/s, indicating that the Aerosonde was located inside the eye. The temperature measurements showed a more complicated structure when the Aerosonde flew inward, penetrating through the rain band and the eyewall along flight leg 1. At the convective region outside the eyewall (80–160 km radius), the temperature was higher at the 140 km radius, where the specific humidity was slightly lower. The temperature at a radius of 40–70 km (eyewall region) was slightly higher than that at the 80–160-km radius (the convective region outside the eyewall). The results pointed that the Aerosonde data, collected inside the typhoon, can serve as an independent check for Doppler radar wind retrieval.

Arctic Environment Data Collection

UAVs have also been shown to be effective for conducting environment and sea surface temperature studies in the Arctic [144,145]. Curry et al. [28] customized and tested the Aerosonde UAV in an industrial freezer (at temperatures down to less than −20 °C) to safely fly in the Arctic. The avionics were properly insulated, a fuel-injection engine was used to avoid carburetor icing and a servo-system was adopted to force ice breaking over the leading edge of the air-foil. Although icing and flight restrictions remain critical challenges, the Aerosonde demonstrated its capability in Barrow (Alaska), where three different Aerosondes provided continuous sampling for 48 h along a 30 km^2 box. Curry et al. [28] reported that approximately 20 Aerosonde profiles flown at Point Barrow coincided with National Weather Service (NWS) soundings. The greatest variability occurred between 300 and 900 m, where the NWS sounding was 1–2 °C cooler than that obtained by the Aerosonde, probably due to a combination of different sonde instruments used by Aerosonde and NWS, and slight differences in locations of the soundings that could had given different results in a coastal environment. The aim of the study by Curry et al. [28] was to customize and integrate sensors onboard an Aerosonde to prove the UAV's capability to fly in an hostile environment such as the Arctic. It was a success from an engineering point of view, however, the specifications of the payload used and a quantification of the error between the two measurement approaches were not provided.

Research into air pollution and atmospheric composition has already benefited from the use of UAVs, due to their flexibility and ability to access environments that are risky for manned platforms, such as volcanic plumes, typhoons and thunderstorms [146].

When sampling from a volcano plume, a rotary wing UAV would be particularly beneficial, as it can hover inside the plume [136], however, covering longer distances and higher altitudes to sense different atmospheric layers may be more suitable for a fixed wing UAV [133]. Physical, technological and regulatory limits as well as the best environments where UAVs can do the most, have no yet been identified. Defining environments where small UAVs can or cannot operate is still an open question (as with long coverage missions), but the integration into national airspace, i.e., regular, routine operations above 400 feet, emerges as the main issue. It is imperative for civil aviation authorities to acknowledge the potential of UAVs in order to facilitate their wide-spread use in atmospheric research and related scientific research in volcanology, typhoon study and meteorology.

3.3.3. Measurements for Prevention, Patrolling and Intervention

UAVs can also be used for regular patrols around industrial areas [147] and metal ore sites. This for the purposes at air quality measurement extends the monitoring range beyond ground based fixed locations to collect/transmit data helping for faster decision making [148]. UAVs can also be used as a tool for the prevention and early diagnosis of environmental disasters, such as monitoring nuclear radiation levels in order to identify radiation leakage [149,150].

Alvarado et al. [151] developed both a small fixed wing and a multi-rotor UAV as part of a system to collect data after blasting at open-pit mines. The Teklite (commercially available) was selected for its portability, ease of integration of sensors, successful flight testing, light weight and low (<100 feet) target flight altitude. A custom made quad-copter of 2.5 kg (including batteries) was used to record data 35 m below ground level. The sensing payload, consisting of a SHARP GP2Y10 sensor to monitor $PM_{2.5}$ and PM_{10}, was calibrated in a lab, and demonstrated a high correlation (correlation coefficient R2 greater than 0.9) with an industry grade dust monitoring device. Talcum powder was used as the PM source to validate the system, with results showing the system to be capable of collecting significant data. Alvarado et al. [151] pointed out, however, that individual calibration equations are needed if the payload has different sensors, and that a different optical sensor is desirable to measure concentration with a precision higher than 1 mg/m^3.

The research of Alvarado et al. [151] showed that integration of air quality sensor and an autopilot onboard a UAV is feasible. The UAV system could help to characterize airborne particulates in time and space, however, such system needs to be tested for real world application. Finally, an analysis of the data collected is required, to feed atmospheric modeling software and for flight path-planning algorithms.

Pollanen at al. [152] tested a new UAV payload below 0.5 kg, using a small gamma-ray spectrometer to detect alpha-particle emitting radionuclides in the air at the level of 0.3 Bq/m^3, assuming 0.5 h sampling and 1 h counting times. Pollanen at al. [152] used a low-active ^{137}Cs source (normally used for detector energy calibration) and autonomously flew the UAV over the unshielded sources several times within a time frame of approximately 50 s. The peaks were not well distinguished when flying above 100 m, since detection limits for the unshielded ^{137}Cs point source on the ground were approximately 1 GBq or larger, depending on the flying altitude. Significant data collection was possible at 75 m, demonstrating the capability of the Mini-UAV for aerial radiation surveillance, searching for lost or stolen unshielded point sources, and mapping radioactive fallout.

Hausamann et al. [91] proposed the use of UAVs to monitor oil and gas transmission pipelines, and investigated the use of two different UAVs for two different scenarios: the first using a small light-weight UAV for low altitude and high resolution data collection; and the second using a medium-size, medium-weight UAV, capable of carrying a heavier payload and with longer endurance. Although the results are yet to be published, Hausamann et al. [91] highlighted the need for a complete mission to determine the total operational capability of the system, including: the UAV platform, sensors, data processing and alarm detection. They also outlined the need for certification and operation standards, for the safe and efficient operation of UAVs. The work of Hausamann et al. [91] presented a comprehensive analysis of both the current technology useful for pipeline monitoring and the selection of two UAV platforms that could be used for such an application. However, questions of the real world capability as well as the integration and use of a UAV system for this application into national airspace still need to be addressed.

When fitted with pathogen traps and programmed for multi-location flights, UAVs can also be used for routine spatial and temporal data collection to facilitate early detection and prevent the spread of pathogens in the air [153], even in remote locations [154–156]. A multi-channel fluorimeter for real time fluorescent airborne particle detection was successfully integrated and operated on-board a UAV [157,158]. Similarly, Roldán et al. [159] developed a Mini-UAV to monitor atmospheric parameters while flying through greenhouses. Using a lightweight quadcopter (400 mm wheelbase, 125 mm square center), Roldán et al. [159] mapped CO_2, T, H and luminosity profiles inside greenhouses. One of the challenges was to decide the most effective mounting point for the sensors on-board the UAV. For this reason, a computational flow dynamic (CFD) study was performed to confirm the findings of two previous quadrotor aerodynamic studies [160,161], which found maximum air speed at the rotor perimeters and minimum air speed at the center of the UAV and outside it, for both single rotor and quad rotor operation. Therefore, only two possible sensor mounting points were feasible: at the center of the UAV and some distance away from it. In order to confirm the results of the CFD study, Roldán et al. [159] attached the UAV to a cardan joint to maintain its precise location and, using an anemometer, they measured the air speed in a 24 point grid above and below the platforms, and found results consistent with the CFD simulations. A CO_2 source was then placed on the ground and used to validate the system. The UAV flew at a height of 0.5 m, starting at 5 m away from the source. Results showed negligible relative errors for T (3.71%), H (1.65%) and CO_2 (3.84%). Although the significance of the results was well explained throughout the paper, the same cannot be said for the sensor mounting point (in the UAV center), the height of the cardan joint, which CO_2 source was used, and if any consideration was given to a possible ground effect. The results of a real-world test within a greenhouse (106 m × 47 m, 3 m height) showed an increase in temperature from the first measurement (25.3 °C) to the last one (29.6 °C), due to the transition from nighttime to daytime temperatures (from 9:00 a.m. to 9:22 a.m.) and the overall warming of the greenhouse. The humidity declined from 43% to 33%, while the CO_2 concentration showed spatial variation, likely due to different ventilation efficiency in some areas of the greenhouse. Roldán et al. [159] stated that unlike UGVs, UAVs can collect data at nearly any point in the three dimensional space of a greenhouse, which facilitates activities such as local climate control and crop monitoring. This study showed the possibility of integrating cheap and light gas sensors onboard a small UAV. However, the study was focused on sensor integration and testing the system, rather than on quantification of the concentrations of the gases. Also, routine real world work might need longer flight times (the UAV used had less than 12 min endurance) and more than only 100 g payload capability.

UAVs as a Tool to Monitor Local Gas Emissions

Neumann et al. [90,162] examined the use of a gas-sensitive micro-UAV (200 g max payload, total take-off weight of 1.3 kg for 30 min flight time) to identify plume origins at ground level, as well as gas source localization and prevention. Neumann et al. [90,162] considered three different plume tracking algorithms: surge-cast, zig-zag and pseudo-gradient algorithms, and they performed a total of 5600 simulation experiments. The surge-cast algorithm was a combination of upwind surge and cross-wind casting. In this case, the UAV moved directly upwind into the plume and continued to travel for a distance, d, until it was no longer inside the plume. The UAV then flew cross-wind for a set distance (d cast), first on one side and then on the other, in an attempt to reacquire the plume. To maximize the chances of hitting the plume in the first cross-wind movement, the robot measured the wind direction to estimate from which side it left the plume. If the robot did not reacquire the plume by casting, the run was considered unsuccessful [163]. With the zig-zag algorithm, the UAV moved upwind at an angle, α, across the plume and when it sensed a concentration below a given threshold, the UAV was assumed to have reached the edge of the plume. It then re-measured the wind direction and continued moving upwind at an angle, α, with respect to the upwind direction. The UAV repeated this procedure moving in a zig-zag fashion within the plume, stopping when it reached the source. Finally, the pseudo-gradient algorithm included a new measuring strategy to deal with the strong disturbances induced by the rotors of a rotary-wing UAV, since measuring a local concentration gradient with spatially separated sensors as part of the gradient algorithm was not feasible in this case. The pseudo gradient algorithm also considered wind information in order to overcome the limitations of older gradient ascent methods that are unable to determine whether they are following a plume towards or away from the source. The simulation experiment results showed that the zigzag algorithm, with $\alpha = 15°$ (where α is the angle used by the UAV to move upwind) and $\alpha = 30°$, as well as the surge-cast algorithm, were the most effective algorithms, but also the least robust. Higher robustness was shown by the pseudo-gradient-based algorithm and the zigzag algorithm with $\alpha = 60°$. Subsequent experiments to determine the capability of the UAVs for real-world plume tracking were conducted using a CH_4 bottle placed in the corner of a 20×16 m^2 area. A fan was used to blow the gas towards the UAV, which was placed 1.5 m away. The plume tracking experiments were stopped just after the UAV passed the source, but the algorithm was able to locate the gas source with a success rate of 83% $\pm$ 3%. Data collection behind the gas source would have also been advantageous to obtain a more accurate and reliable gas source location estimate. The study showed that it was difficult to locate gas sources in a dynamic environment where changes in wind speed and direction, together with high turbulence, were present.

Bennetts et al. [164] demonstrated the weakness of algorithms that directly mimic insect behavior, however, they proposed their use in source identification measurements where the plume is at a higher level, such as industrial chimneys, or where the airflow is more stable.

When using UAVs, and in particular rotary-wing platforms, the location of air sampler or air sensor intakes is crucial for accurate sampling [159].

Neumann et al. [165] described the results of experiments aiming to identify the best location for the gas sensor intake on-board a quad rotor UAV. Three different approaches were tested: active, using additional fans (12 V axial fan diameter 24 mm $\times$ 30 mm with an airflow of 5.4 m^3/h) connected to a tube going beyond the prop diameter; semi-active, using the airflow generated by the rotors; and passive, without any auxiliary device directing the airflow. Significant differences were found between the different gas sampling approaches, however, none of them were capable of measuring the reference gas concentration (0.5% by volume of CO_2). The highest measured concentration was found using the active approach, which was 78% of the reference concentration. The average measured concentration was 0.33 ± 0.02 (active) % by volume, which was 66% of the reference concentration. Although the active gas transport approach was most effective at reducing the propeller dilution effect, the additional weight (76 g) of the long carbon fiber tube, the position of the tube inlet (which strongly dictates the measurement results) and the enlarged drone size (which makes it more susceptible to wind conditions) were found to be significant drawbacks. Gerhardt et al. [166] investigated the possibility of using a VOC sampler onboard a quadcopter, excluding the pump built into the device, and using instead the flow created by propellers to convey the air to the device. A steady flow rate above 150 mL/min was needed to remove the pump and the experimental results showed that sufficient flow rates are generated at the hover RPM (more than 80,000 times the minimum required for a reliable measurement).

Nathan et al. [112] investigated the possibility of using a 3 m fixed wing model aircraft as an airborne-based platform to quantify local emission sources to the atmosphere. The sensing payload consisted in a custom made laser-based open path methane sensor with a precision of 0.1 ppmv at 1 Hz, a mass of 3.1 kg and power consumption of 25 W. The UAV was an AMR Payload Master 100, modified to a dual-motor, each with 7.5 horsepower and 50 cm^3 equivalent electric. The methane sensor was calibrated in a lab and compared to a LICOR LI-7700 before and after the flights. An uncertainty of airborne CH_4 mixing ratio was estimated to be ±10%, normally insufficient for atmospheric measurements, but the nature of the mission means that the methane concentration above the background greatly exceeded this level of uncertainty. Data was collected from 22 flights around a natural compressor station to monitor CH_4 leakage. The measured gas concentrations showed a high variability, from 0 to 73 (±40) g CH_4 s^{-1}, for hours to days. To understand if the reason was due to real changes in the emission rates, atmospheric variability or flight sampling issues, Nathan et al. [112] analyzed pairs of flights close in time and with similar weather conditions. The results highlighted an uncertainty due to flight sampling and atmospheric conditions based upon reproducibility at ±43% g· CH_4 s^{-1}. During the campaign, two additional and independent research groups measured the methane concentrations on the ground. Both of these independent measurements are consistent, but lower than the mean UAV measured rate of 14 (±8) g of CH_4. This is likely due to the impossibility of measuring the lofted emissions by the ground stations.

Statistically, the interpolated methane downwind concentration was the main reason for the flux error analysis. In order to quantify this error, an interpolated map of areas where the flight did not sample is needed. The variance of the Kriged results at each grid point was assumed to be representative of the total uncertainty of the methane mixing ratio, and the standard deviation of the Kriged results gave a relative error of 46%. Nathan et al. [112] chose a 19 × 19 Kriging grid resolution to be consistent with the typical scales of individual plumes (10–30 m). The error due to the sensitivity of this grid resolution resulted in a 4% difference after the comparison with two other grids: 10 × 10 and 40 × 40, respectively. Gaussian-linear, Gaussian-cosine, linear and exponential were compared to the exponential-bessel variogram and no significant (5%) difference was observed. Based on this analysis, the overall statistical uncertainty of the emission rate was ±55% and for the median emission rate of 7.4 g CH_4 s^{-1} is 4.1 g CH_4 s^{-1}, comparable to those estimated from the field derivate estimates (4.2 g CH_4 s^{-1}). The temporal distribution of fugitive CH_4 emission rates from compressor stations needs further study, but the study highlighted that UAVs can be a valuable approach to quantify emission from point sources.

The UK Environment agency recently issued a report based on Allen et al. [167], who outlined the development of UAV systems equipped for GHG sampling with a focus on CH_4 emission measurements from landfill sites. In this work a rotary (DJI F550) and a fixed wing (Bormatec Explorer aircraft) UAVs; operated successfully for 10 days. Allen et al. [167] ascertained that there are no high precision (<40 ppb at 1 Hz) CH_4 instruments suitable for a small UAVs on the market. However, the results of the uncertainty analysis showed that measurement uncertainty is a very small component of the total uncertainty (at ~3% of the total flux). This would suggest that it may be possible to yield satisfactory flux uncertainties (<10%) with less precise CH_4 instrumentation. An alternative could be to obtain CH_4 measurements from a tethered rotary platform which involves vertical sampling (profiling) from a series of locations along a downwind transect. Suitable instruments for use with a rotary system and a sampling line are currently available (LGR-UGGA [168]). Using portable equipment (i.e., CH_4 monitor in back-packs are commercially available) would overcome the long sampling time limitation required to move the system to different locations. Using a tethered UAS platform the size of the methane instrument would not be a constraint.

UAV systems have been demonstrated to be an appropriate platform for in-situ inspection systems to sample closer to the source. However, bias can be introduced in a number of ways, including sampling a region with significantly few pollutants and assuming a uniform distribution.

Quantifying emissions on a UAV is inherently challenging. A Gaussian single plume model is not ideal to create estimates of airborne data collection when there is more than one central plume. In fact, all basic Gaussian plumes would assume that constant and continuous emissions create a steady-state system. Therefore, the largest and most problematic assumption for these in-flight data sets would be to have one central plume. Although the point source approximation might be appropriate for most point source studies, the emissions from other locations around the investigated source could significantly be affected by the single-plume-model approach. To overcome this issue, researchers have investigated other approaches and the Kriging interpolation seems to be a better methodology to analyze UAV-acquired measurements from a point source [112]. Future studies need to optimize a more refined model that considers either a segmented Gaussian plume or a Gaussian puff. Currently, mechanical design limitations affect UAV applications. This is because the majority of commercially available UAVs are not designed to facilitate operations like quantification of CH_4 emissions from compressor stations. These constraints can be addressed through UAV design, construction, and careful flight and sampling design [114]. Research should be directed towards effective mission/path planning, and validation targeted at ensuring unbiased sampling. Since most operations take place on controlled access sites, developing procedures that allow UAV use across the entire site during normal operational hours is required. Finally, payload and endurance limits, system network integration and aerospace legislation will affect UAV applications.

4. Current Challenges, Possible Solutions and Future Applications of UAVs

Air quality research is moving towards the more wide-spread use of UAVs to: measure gases at different altitudes; compare ground based data; and autonomously track plumes emitted by combustion sources, revealing the origin of pollutants [169]. New miniaturized instruments have been, and are still being developed, such as ultra-portable (10 kg weight and 21,000 cm^3 volume) and low powered (24V) mass spectrometer (MS), for the study and visualization of gaseous volcanic emissions [140]. Similarly, research into UAV navigation is empowering more efficient bio-inspired algorithms [170] to enable aircraft to autonomously locate, and fly into or out of, an existing plume [171]. UAVs can complement existing wireless stationary networks (WSN) to improve the accuracy of data assimilation [90,172] by providing data comparable to that obtained using the stationary network, but with more than 30 sensors [173]. Table S2 in the Supplementary Materials summarizes the on-board UAV technologies which have been used for air quality data collection to date.

A recent literature review [174] suggested the use of UAVs to measure VOCs emitted from healthy plants and those under environmental stress. While this might enable researchers to detect plant stress and disease in its early stages, difficulties arise in terms of sampling methodologies and the chemical species to be sampled. Therefore, ground-based robots would be better suited for such an application, however, they could work in co-ordination with UAVs, to sample the broader area using optical sensors and relay information regarding site-specific points of interest. Alternatively, the sensor could be suspended from the UAV and a slow hover-like flight path would allow for the sensor to move through the field and obtain reliable measurements.

An analysis of the literature presented in this paper has highlighted potential applications for the use of small lightweight UAVs in the atmospheric research domain. However, there are a number of challenges that may prevent the science community from moving towards a widespread use of UAVs for air quality measurements [88,115,116]. The most significant of them relates to aviation policy and regulations that cover using unmanned aircraft to investigate atmospheric composition and phenomena in the ABL within national airspace systems [175]. In addition, limitations such as flight time, sensor integration and limited autonomy are driving research into the use of small UAVs. Table 2 shows the overall benefits and limitations and other challenges to the use of these small robotic platforms for air quality research.

Another potential research area is plume tracking using UAVs, but a system capable of such a task under real world conditions, including changing wind speed and direction, has not yet been developed. To date, experimental tests have shown that it is possible to track plumes where the airflow is more stable, like those emitted by industrial chimneys in a higher atmospheric layer, or on wide open landfill sites, geo-dynamically active regions or waste disposal sites [90].

Table 2. Overall benefits and limitations for the use of small lightweight UAVs in the atmospheric research domain.

Benefits	**Limitations**
Cost—small lightweight platforms are less expensive vs. manned aircraft, ground based instruments and satellites	Endurance limitations—flight time is still one the greatest limitations
Flexibility—wide range of UAV applications for atmospheric research	Payload capacity
Time—deploying a small UAV platform saves time vs. large manned platforms as well as ground stations	Sensor availability—limited choice for professional sensors suitable for mounting on-board a small lightweight UAV
Safety—there is no risk for crew when flying UAVs in dangerous situations such as close to the ground	Sensors limitations—smaller sensors may have less sensitivity, selectivity
Repeatability—ground station allows following the same programmed flight path every time	Aerospace regulation—a complete set of UAV operating regulations has not yet been globally defined
Routine flights—data collection for routine flights can be tedious/stressful for humans	Civil aviation authority recognition—UAVs has unique benefit for air quality research, over volcanoes or within eruption plumes
Dirty environments—UAVs can fly in dangerous environments such as when contaminated by radiological, biological, chemical hazards or volcanic plume	System network integration
Easy to deploy—small UAVs do not need airport runways, with fixed-wing UAVs able to take-off in less than 10–30 m while rotary-wing UAVs do not need runways	Autonomous plume tracking—although few algorithms for autonomous plume tracking have been developed and successfully tested in simulation environments, the real world feasibility still needs to be proved
Data collection—small UAVs can take measurements at any point in three dimensional space	
Challenges	
Overall deployment cost—to allow more dangerous missions where there is the risk of losing the UAV	
Public Perception	
High resolution 4D (time & space) atmospheric data sets—with the use of UAV swarms performing simultaneous operations. These goals will assist in providing adequate flight time for various integrated and time series data collection missions, real-time data links and imaging, and also to develop autonomous methods for sample collection	
Comprehensive in situ chemo-physical characterization of combustion source emissions, (including, UFPs, VOCs and above mentioned compounds)	

Small instrument technology is developing quickly, with promising methods for both gas measurements, including tunable diode lasers, cascade quantum lasers and fiber chemical sensing [176], and PM measurements [177]. For example, a pulsed differential adsorption LIDAR (DIAL) system has been proposed as a potential UAV payload for atmospheric CH_4 detection [178].

Future instruments for GHG analysis should also be equipped with on-board calibration cylinders for real time and in-flight instrument calibration [115], and autopilot improvements will enable more precise flight paths, as well as optimized sampling procedures [179].

There are no current high precision CH_4 instruments (defined in the feasibility study as <40 parts per billion at 1 Hz) suitable for a small fixed wing UAS platform. Future methane instruments may also be suitable for rotary UAS platforms. The flight control and flight management software for multi-rotors is more mature than the equivalent software for fixed wing UAVs. The reason comes from both the requirement for sophisticated software to control multi-rotors and the current dominance of photography in the market for small multi-rotors UAVs. The result is that, for some applications, multi-rotors are the preferred solution; however, their lower endurance would preclude them from use on larger sites [100,167].

Established ground-based methods for air quality sampling, such as the sorbent tube or solid phase micro extraction (SPME) sampling techniques for VOCs, could be re-designed for use on-board UAVs, as well as integration with real-time display and data logging. Algorithm developments using multiple UAVs for plume tracking is possible where one UAV is flying upwind while the other tracks the plume downwind, in order to predict the path of the plume and provide early warning for areas of contamination. Genetics algorithms could also be used to obtain faster and more efficient reactions.

5. Conclusions

UAVs can offer high resolution spatiotemporal sampling, which is not possible or feasible with manned aircraft. Some UAV systems are more flexible than others, with the ability to carry multiple sensors and operate in different flight modes (hybrid rotary/fixed wing designs).

The future of UAVs for use in air quality applications is promising, thanks to the capability and flexibility of these robotic platforms. At the same time, new technologies in fields such as chemistry, physics and information technology are also developing very fast, resulting in smaller and lighter devices, with higher sensitivity and the ability to work remotely.

Questions still need to be addressed regarding the miniaturization of sensors, which seems to be the main issue when working with lightweight UAVs. In fact, the diverse range of UAV payload capacity is primarily made by the difference between rotary and fixed wing UAVs. The key limiting factors in new sensor development include: power, mass, and size, because these are intimately connected to the type of platform (rotary instead of fixed wings), engine (electrically or gas powered) and the type of mission (speed, longer distances with low altitude, rather than flight stability and low speed for a higher spatial resolution). Strict civil aviation regulations mean UAVs are commonly deployed for weather forecasting, atmospheric monitoring or as a tool for volcanology research.

Supplementary Materials: The following are available online at http://www.mdpi.com/1424-8220/16/7/1072/s1, Table S1: Search terms, Table S2: Summary of the lightweight UAV onboard technology for air quality research.

Acknowledgments: The authors like to thank John Paul Cunningham, Queensland University of Technology (QUT), for the collaboration and helpful discussion during the preparation of the manuscript. The authors also like to thank Tobias Schripp, Fraunhofer WKI, for the valuable comments on the manuscript.

Author Contributions: Tommaso Francesco Villa conducted the literature search following the research plan developed with the other authors and drafted the entire manuscript. Felipe Gonzalez contributed to the sections: 3.2; 3.2.1; 3.2.2 and 3.2.3; and reviewed the entire manuscript. Branka Miljievic contributed to the literature search and reviewed the entire manuscript. Zoran D. Ristovski contributed to the literature search and reviewed the entire manuscript. Lidia Morawska directed the research project, contributed to the revision of the entire manuscript and to the discussion of UAV systems for air quality studies.

Conflicts of Interest: The authors declare no conflict of interest.

Abbreviations

The following abbreviations are used in this manuscript:

UAV	Unmanned Aerial Vehicles
UGV	Unmanned Ground Vehicles
GHGs	Greenhouse Gasses
PM	Particulate Matter
VOC	Volatile Organic Compound
UFP	Ultrafine Particle
BC	Black Carbon
OC	Organic Carbon
FAA	Federal Aviation Authority of the U.S.
CASA	Civil Aviation Safety Australia
WSN	Wireless System Network
NASA	National Aeronautics and Space Administration
MOX	Metal Oxide
NDIR	Non-dispersive Infrared
SPME	Solide Phase Microextraction

References

1. Monks, P.S.; Granier, C.; Fuzzi, S.; Stohl, A.; Williams, M.L.; Akimoto, H.; Amann, M.; Baklanov, A.; Baltensperger, U.; Bey, I.; et al. Atmospheric composition change—Global and regional air quality. *Atmos. Environ.* **2009**, *43*, 5268–5350. [CrossRef]
2. Dockery, D.W.; Pope, C.A.; Xu, X.; Spengler, J.D.; Ware, J.H.; Fay, M.E.; Ferris, B.G.; Speizer, F.E. An association between air pollution and mortality in six U.S. Cities. *N. Engl. J. Med.* **1993**, *329*, 1753–1759. [CrossRef] [PubMed]
3. Schwartz, J.; Laden, F.; Zanobetti, A. The concentration-response relation between $PM_{2.5}$ and daily deaths. *Environ. Health Perspect.* **2002**, *110*, 1025–1029. [CrossRef] [PubMed]
4. Lim, S.S.; Vos, T.; Flaxman, A.D.; Danaei, G.; Shibuya, K.; Adair-Rohani, H.; AlMazroa, M.A.; Amann, M.; Anderson, H.R.; Andrews, K.G.; et al. A comparative risk assessment of burden of disease and injury attributable to 67 risk factors and risk factor clusters in 21 regions, 1990–2010: A systematic analysis for the global burden of disease study 2010. *Lancet* **2012**, *380*, 2224–2260. [CrossRef]
5. Hansen, J.; Nazarenko, L. Soot climate forcing via snow and ice albedos. *Proc. Natl. Acad. Sci. USA* **2004**, *101*, 423–428. [CrossRef] [PubMed]
6. Jacobson, M.Z.; Kaufman, Y.J. Wind reduction by aerosol particles. *Geophys. Res. Lett.* **2006**, *33*. [CrossRef]
7. Ramanathan, V.; Crutzen, P.J.; Lelieveld, J.; Mitra, A.P.; Althausen, D.; Anderson, J.; Andreae, M.O.; Cantrell, W.; Cass, G.R.; Chung, C.E.; et al. Indian ocean experiment: An integrated analysis of the climate forcing and effects of the great Indo-Asian haze. *J. Geophys. Res. Atmos.* **2001**, *106*, 28371–28398. [CrossRef]
8. Kaufman, Y.J.; Tanre, D.; Boucher, O. A satellite view of aerosols in the climate system. *Nature* **2002**, *419*, 215–223. [CrossRef] [PubMed]
9. Podgorny, I.A.; Ramanathan, V. A modeling study of the direct effect of aerosols over the tropical Indian ocean. *J. Geophys. Res. Atmos.* **2001**, *106*, 24097–24105. [CrossRef]
10. Menon, S.; Hansen, J.; Nazarenko, L.; Luo, Y.F. Climate effects of black carbon aerosols in China and India. *Science* **2002**, *297*, 2250–2253. [CrossRef] [PubMed]
11. Lelieveld, J.; Berresheim, H.; Borrmann, S.; Crutzen, P.J.; Dentener, F.J.; Fischer, H.; Feichter, J.; Flatau, P.J.; Heland, J.; Holzinger, R.; et al. Global air pollution crossroads over the mediterranean. *Science* **2002**, *298*, 794–799. [CrossRef] [PubMed]
12. Seinfeld, J.H.; Pandis, S.N. *Atmospheric Chemistry and Physics: From Air Pollution to Climate Change*, 2nd ed.; Wiley: Hoboken, NJ, USA, 2012.
13. Jacob, D.J.; Crawford, J.H.; Maring, H.; Clarke, A.D.; Dibb, J.E.; Emmons, L.K.; Ferrare, R.A.; Hostetler, C.A.; Russell, P.B.; Singh, H.B.; et al. The arctic research of the composition of the troposphere from aircraft and

satellites (arctas) mission: Design, execution, and first results. *Atmos. Chem. Phys.* **2010**, *10*, 5191–5212. [CrossRef]

14. Cho, J.Y.N.; Newell, R.E.; Bui, T.P.; Browell, E.V.; Fenn, M.A.; Mahoney, M.J.; Gregory, G.L.; Sachse, G.W.; Vay, S.A.; Kucsera, T.L.; et al. Observations of convective and dynamical instabilities in tropopause folds and their contribution to stratosphere-troposphere exchange. *J. Geophys. Res. Atmos.* **1999**, *104*, 21549–21568. [CrossRef]
15. Toon, O.B.; Starr, D.O.; Jensen, E.J.; Newman, P.A.; Platnick, S.; Schoeberl, M.R.; Wennberg, P.O.; Wofsy, S.C.; Kurylo, M.J.; Maring, H.; et al. Planning, implementation, and first results of the tropical composition, cloud and climate coupling experiment (tc4). *J. Geophys. Res. Atmos.* **2010**, *115*, D00J04. [CrossRef]
16. Pearman, G.I.; Beardsmore, D.J.; Obrien, R.C. *The CSIRO (Australia) Atmospheric Carbon Dioxide Monitoring Program: Ten Years of Aircraft Data*; Commonwealth Scientific and Industrial Research Organization: Melbourne, Australia, 1983; p. 115.
17. Simpson, I.J.; Colman, J.J.; Swanson, A.L.; Bandy, A.R.; Thornton, D.C.; Blake, D.R.; Rowland, F.S. Aircraft measurements of dimethyl sulfide (DMS) using a whole air sampling technique. *J. Atmos. Chem.* **2001**, *39*, 191–213. [CrossRef]
18. Buhr, M.; Sueper, D.; Trainer, M.; Goldan, P.; Kuster, B.; Fehsenfeld, F.; Kok, G.; Shillawski, R.; Schanot, A. Trace gas and aerosol measurements using aircraft data from the north atlantic regional experiment (nare 1993). *J. Geophys. Res. Atmos.* **1996**, *101*, 29013–29027. [CrossRef]
19. Brenninkmeijer, C.A.M.; Crutzen, P.J.; Fischer, H.; Gusten, H.; Hans, W.; Heinrich, G.; Heintzenberg, J.; Hermann, M.; Immelmann, T.; Kersting, D.; et al. Caribic-civil aircraft for global measurement of trace gases and aerosols in the tropopause region. *J. Atmos. Ocean. Technol.* **1999**, *16*, 1373–1383. [CrossRef]
20. Karion, A.; Sweeney, C.; Wolter, S.; Newberger, T.; Chen, H.; Andrews, A.; Kofler, J.; Neff, D.; Tans, P. Long-term greenhouse gas measurements from aircraft. *Atmos. Meas. Tech.* **2013**, *6*, 511–526. [CrossRef]
21. Wich, S.; Koh, L.P. Conservation drones. *GIM Int.* **2012**, *26*, 29–33.
22. Martin, R.V. Satellite remote sensing of surface air quality. *Atmos. Environ.* **2008**, *42*, 7823–7843. [CrossRef]
23. Wespes, C.; Emmons, L.; Edwards, D.P.; Hannigan, J.; Hurtmans, D.; Saunois, M.; Coheur, P.F.; Clerbaux, C.; Coffey, M.T.; Batchelor, R.L.; et al. Analysis of ozone and nitric acid in spring and summer arctic pollution using aircraft, ground-based, satellite observations and Mozart-4 model: Source attribution and partitioning. *Atmos. Chem. Phys.* **2012**, *12*, 237–259. [CrossRef]
24. Miller, D.J.; Sun, K.; Zondlo, M.A.; Kanter, D.; Dubovik, O.; Welton, E.J.; Winker, D.M.; Ginoux, P. Assessing boreal forest fire smoke aerosol impacts on U.S. Air quality: A case study using multiple data sets. *J. Geophys. Res. Atmos.* **2011**, *116*. [CrossRef]
25. Corrigan, C.E.; Roberts, G.C.; Ramana, M.V.; Kim, D.; Ramanathan, V. Capturing vertical profiles of aerosols and black carbon over the Indian ocean using autonomous unmanned aerial vehicles. *Atmos. Chem. Phys.* **2008**, *8*, 737–747. [CrossRef]
26. Dubovik, O.; Smirnov, A.; Holben, B.N.; King, M.D.; Kaufman, Y.J.; Eck, T.F.; Slutsker, I. Accuracy assessments of aerosol optical properties retrieved from aerosol robotic network (AERONET) sun and sky radiance measurements. *J. Geophys. Res. Atmos.* **2000**, *105*, 9791–9806. [CrossRef]
27. Levy, R.C.; Remer, L.A.; Kaufman, Y.J. Effects of neglecting polarization on the modis aerosol retrieval over land. *IEEE Trans. Geosci. Remote Sens.* **2004**, *42*, 2576–2583. [CrossRef]
28. Curry, J.; Maslanik, J.; Holland, G.; Pinto, J. Applications of aerosondes in the arctic. *Bull. Am. Meteorol. Soc.* **2004**, *85*, 1855–1861. [CrossRef]
29. Soddell, J.R.; McGuffie, K.; Holland, G.J. Intercomparison of atmospheric soundings from the aerosonde and radiosonde. *J. Appl. Meteorol.* **2004**, *43*, 1260–1269. [CrossRef]
30. Pope, C.A.; Dockery, D.W. Health effects of fine particulate air pollution: Lines that connect. *J. Air Waste Manag. Assoc.* **2006**, *56*, 709–742. [CrossRef] [PubMed]
31. Falagas, M.E.; Pitsouni, E.I.; Malietzis, G.A.; Pappas, G. Comparison of pubmed, scopus, web of science, and google scholar: Strengths and weaknesses. *FASEB J.* **2008**, *22*, 338–342. [CrossRef] [PubMed]
32. Solomon, P.A.; Sioutas, C. Continuous and semicontinuous monitoring techniques for particulate matter mass and chemical components: A synthesis of findings from epa's particulate matter supersites program and related studies. *J. Air Waste Manag. Assoc.* **2008**, *58*, 164–195. [PubMed]

33. Goldberg, M.S.; Burnett, R.T.; Bailar, J.C.; Brook, J.; Bonvalot, Y.; Tamblyn, R.; Singh, R.; Valois, M.F.; Vincent, R. The association between daily mortality and ambient air particle pollution in Montreal, Quebec 2. Cause-specific mortality. *Environ. Res.* **2001**, *86*, 26–36. [CrossRef] [PubMed]
34. Lamarque, J.F.; Bond, T.C.; Eyring, V.; Granier, C.; Heil, A.; Klimont, Z.; Lee, D.; Liousse, C.; Mieville, A.; Owen, B.; et al. Historical (1850–2000) gridded anthropogenic and biomass burning emissions of reactive gases and aerosols: Methodology and application. *Atmos. Chem. Phys.* **2010**, *10*, 7017–7039. [CrossRef]
35. Forster, P.; Ramaswamy, V. Changes in atmospheric constituents and in radiative forcing. In *Climate Change 2007, The Physical Science Basis*; Cambridge University Press: Cambridge, UK, 2007; pp. 129–234.
36. Fowler, D.; Amann, M.; Anderson, R.; Ashmore, M.; Cox, P.; Depledge, M.; Derwent, D.; Grennfelt, P.; Hewitt, N.; Hov, O.; et al. *Ground-Level Ozone in the 21st Century: Future Trends, Impacts and Policy Implications*; The Royal Society: London, UK, 2008.
37. Warnatz, J.; Maas, U.; Dibble, R.W. *Combustion: Physical and Chemical Fundamentals, Modeling and Simulation, Experiments, Pollutant Formation*; Springer: Berlin, Germany, 2006.
38. Donaldson, K.; Tran, L.; Jimenez, L.A.; Duffin, R.; Newby, D.E.; Mills, N.; MacNee, W.; Stone, V. Combustion-derived nanoparticles: A review of their toxicology following inhalation exposure. *Part. Fibre Toxicol.* **2005**, *2*, 10. [CrossRef] [PubMed]
39. Pope, C.A., III; Burnett, R.T.; Thun, M.J.; Calle, E.E.; Krewski, D.; Ito, K.; Thurston, G.D. Lung cancer, cardiopulmonary mortality, and long-term exposure to fine particulate air pollution. *JAMA* **2002**, *287*, 1132–1141.
40. Gramotnev, G.; Ristovski, Z. Experimental investigation of ultra-fine particle size distribution near a busy road. *Atmos. Environ.* **2004**, *38*, 1767–1776.
41. Anderson, K.R.; Avol, E.L.; Edwards, S.A.; Shamoo, D.A.; Peng, R.C.; Linn, W.S.; Hackney, J.D. Controlled exposures of volunteers to respirable carbon and sulfuric-acid aerosols. *J. Air Waste Manag. Assoc.* **1992**, *42*, 770–776.
42. Nemmar, A.; Hoet, P.M.; Vanquickenborne, B.; Dinsdale, D.; Thomeer, M.; Hoylaerts, M.; Vanbilloen, H.; Mortelmans, L.; Nemery, B. Passage of inhaled particles into the blood circulation in humans. *Circulation* **2002**, *105*, 411–414. [PubMed]
43. Niemi, J.V.; Tervahattu, H.; Vehkamäki, H.; Kulmala, M.; Koskentalo, T.; Sillanpää, M.; Rantamäki, M. Characterization and source identification of a fine particle episode in Finland. *Atmos. Environ.* **2004**, *38*, 5003–5012.
44. Mikhailov, E.F.; Vlasenko, S.S.; Podgorny, I.A.; Ramanathan, V.; Corrigan, C.E. Optical properties of soot-water drop agglomerates: An experimental study. *J. Geophys. Res. Atmos.* **2006**, *111*. [CrossRef]
45. Helmig, D.; Bottenheim, J.; Galbally, I.E.; Lewis, A.; Milton, M.J.T.; Penkett, S.; Plass-Duelmer, C.; Reimann, S.; Tans, P.; Thiel, S. Volatile organic compounds in the global atmosphere. *Eos Trans. Am. Geophys. Union* **2009**, *90*, 513–514. [CrossRef]
46. Hewitt, N. *Preface. Reactive Hydrocarbons in the Atmosphere*; Hewitt, C.N., Ed.; Academic Press: San Diego, CA, USA, 1999; pp. xi–xii.
47. Ashworth, K.; Folberth, G.; Hewitt, C.N.; Wild, O. Impacts of near-future cultivation of biofuel feedstocks on atmospheric composition and local air quality. *Atmos. Chem. Phys.* **2012**, *12*, 919–939. [CrossRef]
48. Hitchins, J.; Morawska, L.; Wolff, R.; Gilbert, D. Concentrations of submicrometre particles from vehicle emissions near a major road. *Atmos. Environ.* **2000**, *34*, 51–59. [CrossRef]
49. Gramotnev, G.; Brown, R.; Ristovski, Z.; Hitchins, J.; Morawska, L. Determination of average emission factors for vehicles on a busy road. *Atmos. Environ.* **2003**, *37*, 465–474. [CrossRef]
50. Zhu, Y.; Fung, D.C.; Kennedy, N.; Hinds, W.C.; Eiguren-Fernandez, A. Measurements of ultrafine particles and other vehicular pollutants inside a mobile exposure system on los angeles freeways. *J. Air Waste Manag. Assoc.* **2008**, *58*, 424–434. [PubMed]
51. Nikolova, I.; Janssen, S.; Vrancken, K.; Vos, P.; Mishra, V.; Berghmans, P. Size resolved ultrafine particles emission model—A continues size distribution approach. *Sci. Total Environ.* **2011**, *409*, 3492–3499. [CrossRef] [PubMed]
52. Hudda, N.; Cheung, K.; Moore, K.F.; Sioutas, C. Inter-community variability in total particle number concentrations in the eastern los angeles air basin. *Atmos. Chem. Phys.* **2010**, *10*, 11385–11399. [CrossRef]

53. Moore, K.; Krudysz, M.; Pakbin, P.; Hudda, N.; Sioutas, C. Intra-community variability in total particle number concentrations in the san pedro harbor area (Los Angeles, California). *Aerosol Sci. Technol.* **2009**, *43*, 587–603. [CrossRef]
54. Ras, M.R.; Borrull, F.; Marcé, R.M. Sampling and preconcentration techniques for determination of volatile organic compounds in air samples. *TrAC Trends Anal. Chem.* **2009**, *28*, 347–361. [CrossRef]
55. Ras, M.; Marcé, R.; Borrull, F. Volatile organic compounds in air at urban and industrial areas in the tarragona region by thermal desorption and gas chromatography–mass spectrometry. *Environ. Monit. Assess.* **2010**, *161*, 389–402. [CrossRef] [PubMed]
56. Bouvier-Brown, N.C.; Goldstein, A.H.; Gilman, J.B.; Kuster, W.C.; de Gouw, J.A. In-situ ambient quantification of monoterpenes, sesquiterpenes, and related oxygenated compounds during bearpex 2007: Implications for gas- and particle-phase chemistry. *Atmos. Chem. Phys.* **2009**, *9*, 5505–5518. [CrossRef]
57. Toscano, P.; Gioli, B.; Dugheri, S.; Salvini, A.; Matese, A.; Bonacchi, A.; Zaldei, A.; Cupelli, V.; Miglietta, F. Locating industrial voc sources with aircraft observations. *Environ. Pollut.* **2011**, *159*, 1174–1182. [CrossRef] [PubMed]
58. Barnhart, R.K. *Introduction to Unmanned Aircraft Systems*; CRC Press: Boca Raton, FL, USA, 2012.
59. Holland, G.; McGeer, T.; Youngren, H. Autonomous aerosondes for economical atmospheric soundings anywhere on the globe. *Bull. Am. Meteorol. Soc.* **1992**, *73*, 1987–1998. [CrossRef]
60. Watts, A.C.; Ambrosia, V.G.; Hinkley, E.A. Unmanned aircraft systems in remote sensing and scientific research: Classification and considerations of use. *Remote Sens.* **2012**, *4*, 1671–1692. [CrossRef]
61. Colomina, I.; Molina, P. Unmanned aerial systems for photogrammetry and remote sensing: A review. *ISPRS Photogramm. Remote Sens.* **2014**, *92*, 79–97. [CrossRef]
62. Fahlstrom, P.G.; Gleason, T.J. Classes and missions of UAVs. In *Introduction to UAV Systems*, 4th ed.; John Wiley & Sons, Ltd.: Hoboken, NJ, USA, 2012; pp. 17–31.
63. Technology, Cyber. 2016. Available online: http://www.cybertechuav.com.au/ (accessed on 7 July 2016).
64. Silvertone. 2013. Available online: http://www.silvertone.com.au/content/flamingo-uav-overview (accessed on 7 July 2016).
65. Sensefly. 2015. Available online: https://www.sensefly.com/drones/overview.html (accessed on 7 July 2016).
66. Asctec. 2016. Available online: http://www.asctec.de/en/uav-uas-drones-rpas-roav/asctec-pelican/ (accessed on 7 July 2016).
67. DJI. 2016. Available online: http://www.dji.com/product/flame-wheel-arf (accessed on 7 July 2016).
68. DJI. DJI S800-evo. 2014. Available online: http://www.dji.com/product/spreading-wings-s800-evo (accessed on 7 July 2016).
69. Gatewing. Available online: http://www.gatewing.com (accessed on 8 October 2014).
70. (ARCAA), Australian Research Centre for Aerospace Automation. Enhanced Flight Assist System (eFAS) for automated Aerial Survey of Powerline Networks. Available online: http://www.arcaa.net/research/enhanced-flight-assist-system-efas-for-automated-aerial-survey-of-powerline-networks/ (accessed on 8 October 2014).
71. Bachrach, A.; He, R.; Roy, N. Autonomous flight in unknown indoor environments. *Int. J. Micro Air Veh.* **2009**, *1*, 217–228. [CrossRef]
72. Fahlstrom, P.G.; Gleason, T.J. Mission planning and control. In *Introduction to UAV Systems*, 4th ed.; John Wiley & Sons, Ltd.: Hoboken, NJ, USA, 2012; p. 99.
73. Fahlstrom, P.G.; Gleason, T.J. Mission planning and control station. In *Introduction to UAV Systems*, 4th ed.; John Wiley & Sons, Ltd.: Hoboken, NJ, USA, 2012; pp. 101–118.
74. Fahlstrom, P.G.; Gleason, T.J. Air vehicle and payload control. In *Introduction to UAV Systems*, 4th ed.; John Wiley & Sons, Ltd.: Hoboken, NJ, USA, 2012; pp. 119–130.
75. Chwaleba, A.; Olejnik, A.; Rapacki, T.; Tuśnio, N. Analysis of capability of air pollution monitoring from an unmanned aircraft. *Aviation* **2014**, *18*, 13–19. [CrossRef]
76. Antonio, P.; Grimaccia, F.; Mussetta, M. Architecture and methods for innovative heterogeneous wireless sensor network applications. *Remote Sens.* **2012**, *4*, 1146–1161. [CrossRef]
77. Skoglar, P.; Orguner, U.; Törnqvist, D.; Gustafsson, F. Road target search and tracking with gimballed vision sensor on an unmanned aerial vehicle. *Remote Sens.* **2012**, *4*, 2076–2111. [CrossRef]
78. Novaković, Z.; Medar, N. Analysis of a UAV bungee cord launching device. *Sci. Tech. Rev.* **2013**, *63*, 41–47.
79. Drury, J.L.; Scott, S.D. Awareness in unmanned aerial vehicle operations. *Int. C2 J.* **2008**, *2*, 1–10.

80. Wyllie, T. Parachute recovery for UAV systems. *Aircr. Eng. Aerosp. Technol.* **2001**, *73*, 542–551. [CrossRef]
81. Woolley, C.C.A.; Beggs, K.W.; Bakewell, R.A.; Axford, R.D.J.; Wainwright, J. Launch System. U.S. Patent 8,584,985, 19 November 2013.
82. Su, Z.-Z.; Zhang, B.; Guo, W.; Liu, X.-C.; Qu, W.-B. A review of electromagnetic launch technology used in UAV. *J. Gun Launch Control* **2011**, *1*, 023.
83. Fahlstrom, P.G.; Gleason, T.J. Launch systems. In *Introduction to UAV Systems*, 4th ed.; John Wiley & Sons, Ltd.: Hoboken, NJ, USA, 2012; pp. 249–260.
84. Ozdemir, U.; Aktas, Y.; Vuruskan, A.; Dereli, Y.; Tarhan, A.; Demirbag, K.; Erdem, A.; Kalaycioglu, G.; Ozkol, I.; Inalhan, G. Design of a commercial hybrid vtol UAV system. *J. Intell. Robot. Syst.* **2014**, *74*, 371–393. [CrossRef]
85. Everaerts, J. The use of unmanned aerial vehicles (UAVs) for remote sensing and mapping. *Int. Arch. Photogramm. Remote Sens. Spat. Inf. Sci.* **2008**, *37*, 1187–1192.
86. Jensen, T.; Apan, A.; Young, F.; Zeller, L. Detecting the attributes of a wheat crop using digital imagery acquired from a low-altitude platform. *Comput. Electron. Agric.* **2007**, *59*, 66–77. [CrossRef]
87. Inoue, Y.; Morinaga, S.; Tomita, A. A blimp-based remote sensing system for low-altitude monitoring of plant variables: A preliminary experiment for agricultural and ecological applications. *Int. J. Remote Sens.* **2000**, *21*, 379–385. [CrossRef]
88. Reuder, J.; Jonassen, M.; Ólafsson, H. The small unmanned meteorological observer sumo: Recent developments and applications of a micro-uas for atmospheric boundary layer research. *Acta Geophys.* **2012**, *60*, 1454–1473. [CrossRef]
89. Techy, L.; Schmale, D.G., III; Woolsey, C.A. Coordinated aerobiological sampling of a plant pathogen in the lower atmosphere using two autonomous unmanned aerial vehicles. *J. Field Robot.* **2010**, *27*, 335–343.
90. Neumann, P.P.; Bennetts, V.H.; Lilienthal, A.J.; Bartholmai, M.; Schiller, J.H. Gas source localization with a micro-drone using bio-inspired and particle filter-based algorithms. *Adv. Robot.* **2013**, *27*, 725–738. [CrossRef]
91. Hausamann, D.; Zirnig, W.; Schreier, G.; Strobl, P. Monitoring of gas pipelines—A civil UAV application. *Aircr. Eng. Aerosp. Technol.* **2005**, *77*, 352–360. [CrossRef]
92. Melnyk, R.; Schrage, D.; Volovoi, V.; Jimenez, H. Sense and avoid requirements for unmanned aircraft systems using a target level of safety approach. *Risk Anal.* **2014**, *34*, 1894–1906. [CrossRef] [PubMed]
93. Smith, K.W. Drone technology: Benefits, risks, and legal considerations. *Seattle J. Environ. Law* **2015**, *5*, 12.
94. Valavanis, K.; Vachtsevanos, G.J. *Handbook of Unmanned Aerial Vehicles*; Springer: Berlin, Germany, 2011.
95. Cork, L.; Clothier, R.; Gonzalez, L.F.; Walker, R. The future of UAS: Standards, regulations, and operational experiences. *IEEE Aerosp. Electron. Syst. Mag.* **2007**, *22*, 29–45.
96. Clothier, R.A. Overview of Australian civil UAS regulations and supporting research. In Proceedings of the Technical Cooperation Panel Meeting AER (Aerospace Group), Technical Panel 6 "UAV Systems and Operations", Fishermans Bend, Australia, 28 March 2012.
97. Watai, T.; Machida, T.; Ishizaki, N.; Inoue, G. A lightweight observation system for atmospheric carbon dioxide concentration using a small unmanned aerial vehicle. *J. Atmos. Ocean. Technol.* **2006**, *23*, 700–710. [CrossRef]
98. Ramana, M.V.; Ramanathan, V.; Kim, D.; Roberts, G.C.; Corrigan, C.E. Albedo, atmospheric solar absorption and heating rate measurements with stacked UAVs. *Q. J. R. Meteorol. Soc.* **2007**, *133*, 1913–1931. [CrossRef]
99. Ramanathan, V. *Maldives AUAV Campaign (MAC): Observing Aerosol-Cloud-Radiation-Climate Interactions Simultaneously from three Stacked Autonomous Unmanned Aerial Vehicles (AUAVs)*; Report of the Field Campaign Held from March; National Science Foundation: Arlington, VA, USA, 2006; Volume 5.
100. Allen, G.; Hollingsworth, P.; Illingworth, S.; Kabbabe, K.; Perciva, C. *Feasibility of Aerial Measurements of Methane Emissions from Landfills*; Environmental Agency: Rotherham, UK, 2014.
101. Spiess, T.; Bange, J.; Buschmann, M.; Voersmann, P. First application of the meteorological mini-UAV 'm(2)av'. *Meteorol. Z.* **2007**, *16*, 159–169. [CrossRef] [PubMed]
102. Buschmann, M.; Bange, J.; Vörsmann, P. 6.7 MMAV—A Miniature Unmanned Aerial Vehicle (Mini-UAV) for Meteorological Purposes. Available online: http://citeseerx.ist.psu.edu/viewdoc/summary?doi=10.1.1.134.3104. (accessed on 7 July 2016).
103. Kroonenberg, A.V.D.; Martin, T.; Buschmann, M.; Bange, J.; Vörsmann, P. Measuring the wind vector using the autonomous mini aerial vehicle m2av. *J. Atmos. Ocean. Technol.* **2008**, *25*, 1969–1982. [CrossRef]

104. Martin, S.; Bange, J.; Beyrich, F. Meteorological profiling of the lower troposphere using the research UAV "m2av carolo". *Atmos. Meas. Tech.* **2011**, *4*, 705–716. [CrossRef]
105. Mayer, S.; Sandvik, A.; Jonassen, M.; Reuder, J. Atmospheric profiling with the UAS sumo: A new perspective for the evaluation of fine-scale atmospheric models. *Meteorol. Atmos. Phys.* **2012**, *116*, 15–26. [CrossRef]
106. Reuder, J.; Brisset, P.; Jonassen, M.; Müller, M.; Mayer, S. The small unmanned meteorological observer sumo: A new tool for atmospheric boundary layer research. *Meteorol. Z.* **2009**, *18*, 141–147. [CrossRef]
107. Ramanathan, V.; Roberts, G.; Corrigan, C.; Ramana, M.; Nguyen, H. *Aerosol, Cloud, and Radiometric Measurements with Small Autonomous Unmanned Aerial Vehicles*; AGU Fall Meeting Abstracts; American Geophysical Uninion: Washington, DC, USA, 2005.
108. Ramanathan, V.; Ramana, M.V.; Roberts, G.; Kim, D.; Corrigan, C.; Chung, C.; Winker, D. Warming trends in Asia amplified by brown cloud solar absorption. *Nature* **2007**, *448*, 575–578. [CrossRef] [PubMed]
109. Bates, T.S.; Quinn, P.K.; Johnson, J.E.; Corless, A.; Brechtel, F.J.; Stalin, S.E.; Meinig, C.; Burkhart, J.F. Measurements of atmospheric aerosol vertical distributions above svalbard, norway, using unmanned aerial systems (UAS). *Atmos. Meas. Tech.* **2013**, *6*, 2115–2120. [CrossRef]
110. Altstädter, B.; Platis, A.; Wehner, B.; Scholtz, A.; Lampert, A.; Wildmann, N.; Hermann, M.; Käthner, R.; Bange, J.; Baars, H. Aladina—An unmanned research aircraft for observing vertical and horizontal distributions of ultrafine particles within the atmospheric boundary layer. *Atmos. Meas. Tech. Discussions* **2014**, *7*, 12283–12322. [CrossRef]
111. Harrison, W.A.; Lary, D.J.; Nathan, B.J.; Moore, A.G. Using remote control aerial vehicles to study variability of airborne particulates. *Air Soil Water Res.* **2015**, *8*, 43–51. [CrossRef]
112. Nathan, B.J.; Golston, L.M.; O'Brien, A.S.; Ross, K.; Harrison, W.A.; Tao, L.; Lary, D.J.; Johnson, D.R.; Covington, A.N.; Clark, N.N.; et al. Near-field characterization of methane emission variability from a compressor station using a model aircraft. *Environ. Sci. Technol.* **2015**, *49*, 7896–7903. [CrossRef] [PubMed]
113. Brady, J.M.; Stokes, M.D.; Bonnardel, J.; Bertram, T.H. Characterization of a quadrotor unmanned aircraft system for aerosol-particle-concentration measurements. *Environ. Sci. Technol.* **2016**, *50*, 1376–1383. [CrossRef] [PubMed]
114. Mölders, N.; Butwin, M.K.; Madden, J.M.; Tran, H.N.; Sassen, K.; Kramm, G. Theoretical investigations on mapping mean distributions of particulate matter, inert, reactive, and secondary pollutants from wildfires by unmanned air vehicles (UAVs). *Open J. Air Pollut.* **2015**, *4*, 149. [CrossRef]
115. Berman, E.S.F.; Fladeland, M.L.J.; Kolyer, R.; Gupta, M. Greenhouse gas analyzer for measurements of carbon dioxide, methane, and water vapor aboard an unmanned aerial vehicle. *Sens. Actuators B Chem.* **2012**, *169*, 128–135. [CrossRef]
116. Fladeland, M.; Sumich, M.; Lobitz, B.; Kolyer, R.; Herlth, D.; Berthold, R.; McKinnon, D.; Monforton, L.; Brass, J.; Bland, G. The nasa sierra science demonstration programme and the role of small-medium unmanned aircraft for earth science investigations. *Geocarto Int.* **2011**, *26*, 157–163. [CrossRef]
117. Malaver, A.; Gonzalez, F.; Depari, A.; Corke, P.; Motta, N. Towards the development of a gas sensor system for monitoring pollutant gases in the low troposphere using small unmanned aerial vehicles. In Proceedings of Workshop on Robotics for Environmental Monitoring, Sydney, Australia, 9–13 July 2012.
118. Malaver, A.; Motta, N.; Corke, P.; Gonzalez, F. Development and integration of a solar powered unmanned aerial vehicle and a wireless sensor network to monitor greenhouse gases. *Sensors* **2015**, *15*, 4072–4096. [CrossRef] [PubMed]
119. Malaver, A.J.R.; Gonzalez, L.F.; Motta, N.; Villa, T.F. Design and flight testing of an integrated solar powered UAV and WSN for remote gas sensing. In Proceedings of the IEEE Aerospace Conference 2015, Big Sky, MT, USA, 7–14 March 2015.
120. Malaver Rojas, J.A.; Motta, N.; Peter, C.; John, B.; Alessandro, D. Development of a gas nanosensor node powered by solar cells. In Proceedings of the Solar2011, the 49th AuSES Annual Conference, Australian Technology Park, Sidney, Austalia, 30 November–2 December 2011.
121. Rossi, M.; Brunelli, D.; Adami, A.; Lorenzelli, L.; Menna, F.; Remondino, F. Gas-drone: Portable gas sensing system on UAVs for gas leakage localization. In Proceedings of the 2014 IEEE SENSORS, Valencia, Spain, 2–5 November 2014; pp. 1431–1434.
122. Gallego, V.; Rossi, M.; Brunelli, D. Unmanned aerial gas leakage localization and mapping using microdrones. In Proceedings of the 2015 IEEE Sensors Applications Symposium (SAS), Zadar, Croatia, 13–15 April 2015; pp. 1–6.

123. Illingworth, S.; Allen, G.; Percival, C.; Hollingsworth, P.; Gallagher, M.; Ricketts, H.; Hayes, H.; Ładosz, P.; Crawley, D.; Roberts, G. Measurement of boundary layer ozone concentrations on-board a skywalker unmanned aerial vehicle. *Atmos. Sci. Lett.* **2014**, *15*, 252–258. [CrossRef]
124. Lawrence, D.A.; Balsley, B.B. High-resolution atmospheric sensing of multiple atmospheric variables using the datahawk small airborne measurement system. *J. Atmos. Ocean. Technol.* **2013**, *30*, 2352–2366. [CrossRef]
125. Pieri, D. In-Situ Observations of Volcanic Plumes for Applications and Research. NASA Technical Report 20060022664. 2005. Available online: http://ntrs.nasa.gov/archive/nasa/casi.ntrs.nasa.gov/20060022664_2006156292.pdf (accessed on 23 June 2016).
126. Sato, A. *Civil UAV Applications in Japan and Related Safety & Certification*; Yamaha Motor CO., LTD.: Shizuoka, Japan, 2003.
127. Astuti, G.; Giudice, G.; Longo, D.; Melita, C.D.; Muscato, G.; Orlando, A. An overview of the "volcan project": An uas for exploration of volcanic environments. *J. Intell. Robot. Syst.* **2009**, *54*, 471–494. [CrossRef]
128. Astuti, G.; Longo, D.; Melita, C.; Muscato, G.; Orlando, A. Hil tuning of UAV for exploration of risky environments. *Int. J. Adv. Robot. Syst.* **2008**, *5*, 419–424. [CrossRef]
129. Astuti, G.; Caltabiano, D.; Giudice, G.; Longo, D.; Melita, D.; Muscato, G.; Orlando, A. "Hardware in the loop" tuning for a volcanic gas sampling UAV. In *Advances in Unmanned Aerial Vehicles*; Valavanis, K., Ed.; Springer Netherlands: Rotterdam, The Netherlands, 2007; Volume 33, pp. 473–493.
130. Longo, D.; Melita, D.; Muscato, G.; Sessa, S. A mixed terrestrial aerial robotic platform for volcanic and industrial surveillance. In Proceedings of the 2007 IEEE International Workshop on Safety, Security and Rescue Robotics, Rome, Italy, 27–29 September 2007; pp. 1–6.
131. Saggiani, G.; Amici, S. *UAV Systems Volcano Monitoring: First Test on Stromboli on October 2004*; Geophysical Research Abstracts; European Geoscience Union: Munich, Germany, 2006; p. 02901.
132. Saggiani, G.M.; Persiani, F.; Ceruti, A.; Tortora, P.; Troiani, E.; Giulietti, F.; Amici, S.; Buongiorno, M.F.; Bentini, G.G.; Bianconi, M.; et al. UAV system development for the monitoring and study volcanic and natural hazard events. In Proceedings of the 2007 Annual Conference Remote Sensing and Photogrammetry Society, Newcastle, UK, 11–14 September 2007.
133. Saggiani, G.P.F.; Ceruti, A.; Tortora, P.; Troiani, E.; Giuletti, F.; Amici, S.; Buongiorno, M.; Distefano, G.; Bentini, G. *A UAV System for Observing Volcanoes and Natural Hazards*; American Geophysical Union: Washington, DC, USA, 2007; p. 5.
134. Amici, S.; Turci, M.; Giammanco, S.; Spampinato, L.; Giulietti, F. UAV thermal infrared remote sensing of an italian mud volcano. *Adv. Remote Sens.* **2013**, *2013*, 41248. [CrossRef]
135. Patterson, M.; Mulligan, A.; Douglas, J.; Robinson, J.; Wardell, L.; Pallister, J. Volcano surveillance by acr silver fox. *Infotech. Aerosp.* **2005**, 26–29. [CrossRef]
136. McGonigle, A.J.S.; Aiuppa, A.; Giudice, G.; Tamburello, G.; Hodson, A.J.; Gurrieri, S. Unmanned aerial vehicle measurements of volcanic carbon dioxide fluxes. *Geophys. Res. Lett.* **2008**, *35*. [CrossRef]
137. Aiuppa, A.; Federico, C.; Giudice, G.; Gurrieri, S. Chemical mapping of a fumarolic field: La fossa crater, vulcano island (Aeolian Islands, Italy). *Geophys. Res. Lett.* **2005**, *32*, L13309. [CrossRef]
138. Pieri, D.; Diaz, J.A.; Bland, G.; Fladeland, M.; Madrigal, Y.; Corrales, E.; Alegria, O.; Alan, A.; Realmuto, V.; Miles, T.; et al. In situ observations and sampling of volcanic emissions with NASA and UCR unmanned aircraft, including a case study at Turrialba Volcano, Costa Rica. *Geol. Soc. Spec. Publ.* **2013**, *380*, 321–352. [CrossRef]
139. Diaz, J.; Giese, C.; Gentry, W.R. Sub-miniature exb sector-field mass spectrometer. *J. Am. Soc. Mass Spectrom.* **2001**, *12*, 619–632. [CrossRef]
140. Andres Diaz, J.; Pieri, D.; Arkin, C.R.; Gore, E.; Griffin, T.P.; Fladeland, M.; Bland, G.; Soto, C.; Madrigal, Y.; Castillo, D.; et al. Utilization of in situ airborne ms-based instrumentation for the study of gaseous emissions at active volcanoes. *Int. J. Mass Spectrometry* **2010**, *295*, 105–112. [CrossRef]
141. Diaz, J.A.; Corrales, E.; Madrigal, Y.; Pieri, D.; Bland, G.; Miles, T.; Fladeland, M. Volcano monitoring with small unmanned aerial systems. In Proceedings of the AIAA Infotech at Aerospace Conference and Exhibit 2012, Garden Grove, CA, USA, 19–21 June 2012.
142. Diaz, J.; Pieri, D.; Wright, K.; Sorensen, P.; Kline-Shoder, R.; Arkin, C.; Fladeland, M.; Bland, G.; Buongiorno, M.; Ramirez, C.; et al. Unmanned aerial mass spectrometer systems for in-situ volcanic plume analysis. *J. Am. Soc. Mass Spectrom.* **2015**, *26*, 292–304. [CrossRef] [PubMed]

143. Lin, P.-H.; Lee, C.-S. The eyewall-penetration reconnaissance observation of typhoon longwang (2005) with unmanned aerial vehicle, aerosonde. *J. Atmos. Ocean. Technol.* **2008**, *25*, 15–25. [CrossRef]
144. Darack, E. UAVs: The new frontier for weather research and prediction. *Weatherwise* **2012**, *65*, 20–27. [CrossRef]
145. Wong, K.; Bill, C. UAVs over Australia-market and capabilities. In Proceedings of the 13th Bristol International Conference on RPVs/UAVs, Bristol, UK, 30 March–1 April 1998.
146. Austin, R. *Unmanned Aircraft Systems: UAVs Design, Development and Deployment*; American Institute of Aeronautics and Astronautics: Reston, VA, USA; Chichester, UK, 2010.
147. Chao-Chung, P.; Chao-Yung, H. Integration of an unmanned vehicle and its application to real-time gas detection and monitoring. In Proceedings of the 2015 IEEE International Conference on Consumer Electronics—Taiwan (ICCE-TW), Taipei, Taiwan, 6–8 June 2015; pp. 320–321.
148. Danilov, A.; Smirnov, Y.; Petrova, T.; Pashkevich, M. Using drones of preconstruction monitoring conducting in mining enterprise. *Int. J. Ecol. Dev.* **2015**, *30*, 36–42.
149. Han, J.; Xu, Y.; Di, L.; Chen, Y. Low-cost multi-UAV technologies for contour mapping of nuclear radiation field. *J. Intell. Robot. Syst.* **2013**, *70*, 401–410. [CrossRef]
150. Behnke, D.; Bok, P.B.; Wietfeld, C. UAV-based connectivity maintenance for borderline detection. In Proceedings of the 2013 IEEE 77th Vehicular Technology Conference (VTC Spring), Dresden, Germany, 2–5 June 2013; pp. 1–6.
151. Alvarado, M.; Gonzalez, F.; Fletcher, A.; Doshi, A. Towards the development of a low cost airborne sensing system to monitor dust particles after blasting at open-pit mine sites. *Sensors* **2015**, *15*, 19667. [CrossRef] [PubMed]
152. Pollanen, R.; Toivonen, H.; Perajarvi, K.; Karhunen, T.; Smolander, P.; Ilander, T.; Rintala, K.; Katajainen, T.; Niemela, J.; Juusela, M.; et al. Performance of an air sampler and a gamma-ray detector in a small unmanned aerial vehicle. *J. Radioanal. Nucl. Chem.* **2009**, *282*, 433–437. [CrossRef]
153. Gottwald, T.R.; Tedders, W.L. A spore and pollen trap for use on aerial remotely piloted vehicles. *Phytopathology* **1985**, *75*, 801–807. [CrossRef]
154. Schmale, D.G.; Dingus, B.R.; Reinholtz, C. Development and application of an autonomous unmanned aerial vehicle for precise aerobiological sampling above agricultural fields. *J. Field Robot.* **2008**, *25*, 133–147. [CrossRef]
155. Lin, B.; Ross, S.D.; Prussin, A.J., II; Schmale, D.G., III. Seasonal associations and atmospheric transport distances of fungi in the genus fusarium collected with unmanned aerial vehicles and ground-based sampling devices. *Atmos. Environ.* **2014**, *94*, 385–391. [CrossRef]
156. Gonzalez, F.; Castro, M.P.G.; Narayan, P.; Walker, R.; Zeller, L. Development of an autonomous unmanned aerial system to collect time-stamped samples from the atmosphere and localize potential pathogen sources. *J. Field Robot.* **2011**, *28*, 961–976. [CrossRef]
157. Anderson, G.P.; King, K.D.; Cuttino, D.S.; Whelan, J.P.; Ligler, F.S.; MacKrell, J.F.; Bovais, C.S.; Indyke, D.K.; Foch, R.J. Biological agent detection with the use of an airborne biosensor. *Field Anal. Chem. Technol.* **1999**, *3*, 307–314. [CrossRef]
158. Ligler, F.S.; Anderson, G.P.; Davidson, P.T.; Foch, R.J.; Ives, J.T.; King, K.D.; Page, G.; Stenger, D.A.; Whelan, J.P. Remote sensing using an airborne biosensor. *Environ. Sci. Technol.* **1998**, *32*, 2461–2466. [CrossRef]
159. Roldán, J.; Joossen, G.; Sanz, D.; del Cerro, J.; Barrientos, A. Mini-UAV based sensory system for measuring environmental variables in greenhouses. *Sensors* **2015**, *15*, 3334–3350. [CrossRef] [PubMed]
160. Poyi, G.T.; Wu, M.H.; Bousbaine, A.; Wiggins, B. Validation of a quad-rotor helicopter matlab/simulink and solidworks models. In Proceedings of the IET Conference on Control and Automation 2013: Uniting Problems and Solutions, Birmingham, UK, 4–5 June 2013; pp. 1–6.
161. Aleksandrov, D.; Penkov, I. Optimal gap distance between rotors of mini quadrotor helicopter. In Proceedings of the 8th DAAAM Baltic conference. Tallinn, Estonia, 19–21 April 2012.
162. Neumann, P.P.; Asadi, S.; Lilienthal, A.J.; Bartholmai, M.; Schiller, J.H. Autonomous gas-sensitive microdrone: Wind vector estimation and gas distribution mapping. *IEEE Robot. Autom. Mag.* **2012**, *19*, 50–61. [CrossRef]
163. Lochmatter, T.; Martinoli, A. Theoretical analysis of three bio-inspired plume tracking algorithms. In Proceedings of the ICRA '09. IEEE International Conference on Robotics and Automation, Kobe, Japan, 12–17 May2009; pp. 2661–2668.

164. Bennetts, V.H.; Lilienthal, A.J.; Neumann, P.P.; Trincavelli, M. Mobile robots for localizing gas emission sources on landfill sites: Is bio-inspiration the way to go? *Front. Neuroeng.* **2011**, *4*, 20.
165. Neumann, P.P. Gas Source Localization and Gas Distribution Mapping with a Micro-Drone. Ph.D. Thesis, Freie Universität Berlin, Berlin, Germany, 2013.
166. Gerhardt, N.; Clothier, R.; Wild, G.; Mohamed, A.; Petersen, P.; Watkins, S. Analysis of inlet flow structures for the integration of a remote gas sensor on a multi-rotor unmanned aircraft system. In Proceedings of the ACUS 2014: Fourth Australasian Unmanned Systems Conference, Melbourne, Australia, 15–16 December 2014; pp. 1–6.
167. Allen, M.J.P.G.; Hollingsworth, P.; Mead, I.; Kabbabe, K.; Roberts, G.; Percival, C. Measuring Landfill Methane Emissions Using Unmanned Aerial Systems: Field Trial and Operational Guidance. Environment Agency: Horizon House, Deanery Road, Bristol, BS1 5AH, 2016; Available online: www.gov.uk/government/organisations/environmentagency (accessed on 7 July 2016).
168. Research, L.G. Ultraportable Greenhouse Gas Analyzer (CH_4, CO_2, H_2O). Available online: http://www.lgrinc.com/analyzers/ultraportable-greenhouse-gas-analyzer/ (accessed on 7 July 2016).
169. Koppmann, R. *Volatile Organic Compounds in the Atmosphere*; John Wiley & Sons: Hoboken, NJ, USA, 2008.
170. Neumann, P.P.; Bennetts, V.H.; Lilienthal, A.J.; Bartholmai, M. From insects to micro air vehicles—A comparison of reactive plume tracking strategies. In *Intelligent Autonomous Systems 13*; Springer: Berlin, Germany, 2016; pp. 1533–1548.
171. Porter, M.J., III; Vasquez, J.R. Bio-inspired, odor-based navigation—Art. No. 62280v. In *Modeling and Simulation for Military Applications*; Schum, K., Sisti, A.F., Eds.; SAGE Publishing: London, UK, 2006; Volume 6228, pp. V2280–V2280.
172. Evangelatos, O.; Rolim, J. An airborne wireless sensor system for near-real time air pollution monitoring. *Sens. Transducers* **2015**, *189*, 12–21.
173. Smidl, V.; Hofman, R. Tracking of atmospheric release of pollution using unmanned aerial vehicles. *Atmos. Environ.* **2013**, *67*, 425–436. [CrossRef]
174. Sankaran, S.; Mishra, A.; Ehsani, R.; Davis, C. A review of advanced techniques for detecting plant diseases. *Comput. Electron. Agric.* **2010**, *72*, 1–13. [CrossRef]
175. Houston, A.L.; Argrow, B.; Elston, J.; Lahowetz, J.; Frew, E.W.; Kennedy, P.C. The collaborative colorado-nebraska unmanned aircraft system experiment. *Bull. Am. Meteorol. Soc.* **2012**, *93*, 39–54. [CrossRef]
176. Holland, G.J.; Webster, P.J.; Curry, J.A.; Tyrell, G.; Gauntlett, D.; Brett, G.; Becker, J.; Hoag, R.; Vaglienti, W. The aerosonde robotic aircraft: A new paradigm for environmental observations. *Bull. Am. Meteorol. Soc.* **2001**, *82*, 889–901. [CrossRef]
177. Morawska, L.; Afshari, A.; Bae, G.N.; Buonanno, G.; Chao, C.Y.H.; Hänninen, O.; Hofmann, W.; Isaxon, C.; Jayaratne, E.R.; Pasanen, P.; et al. Indoor aerosols: From personal exposure to risk assessment. *Indoor Air* **2013**, *23*, 462–487. [CrossRef] [PubMed]
178. Refaat, T.F.; Ismail, S.; Nehrir, A.R.; Hair, J.W.; Crawford, J.H.; Leifer, I.; Shuman, T. Performance evaluation of a 1.6-μm methane dial system from ground, aircraft and UAV platforms. *Opt. Express* **2013**, *21*, 30415–30432. [CrossRef] [PubMed]
179. Ippolito, C.; Fladeland, M.; Yoo Hsiu, Y. Applications of payload directed flight. In Proceedings of the 2009 IEEE Aerospace Conference, Big Sky, MT, USA, 7–14 March 2009; pp. 1–15.

sensors

MDPI

Article

Development and Validation of a UAV Based System for Air Pollution Measurements

Tommaso Francesco Villa [1], Farhad Salimi [2], Kye Morton [3], Lidia Morawska [1,*] and Felipe Gonzalez [3]

1. International Laboratory for Air Quality and Health (ILAQH), Queensland University of Technology (QUT), 2 George St, Brisbane QLD 4000, Australia; tf.villa@hdr.qut.edu.au
2. Menzies Institute for Medical Research, University of Tasmania, Hobart, Tasmania 7000, Australia; farhad.salimi@utas.edu.au
3. Australian Research Centre for Aerospace Automation (ARCAA), Queensland University of Technology (QUT), 2 George St, Brisbane QLD 4000, Australia; kye.morton@hdr.qut.edu.au (K.M.); felipe.gonzalez@qut.edu.au (F.G.)

* Correspondence: l.morawska@qut.edu.au; Tel.: +61-7-3138-2616

Academic Editor: Hans Tømmervik
Received: 23 November 2016; Accepted: 19 December 2016; Published: 21 December 2016

Abstract: Air quality data collection near pollution sources is difficult, particularly when sites are complex, have physical barriers, or are themselves moving. Small Unmanned Aerial Vehicles (UAVs) offer new approaches to air pollution and atmospheric studies. However, there are a number of critical design decisions which need to be made to enable representative data collection, in particular the location of the air sampler or air sensor intake. The aim of this research was to establish the best mounting point for four gas sensors and a Particle Number Concentration (PNC) monitor, onboard a hexacopter, so to develop a UAV system capable of measuring point source emissions. The research included two different tests: (1) evaluate the air flow behavior of a hexacopter, its downwash and upwash effect, by measuring air speed along three axes to determine the location where the sensors should be mounted; (2) evaluate the use of gas sensors for CO_2, CO, NO_2 and NO, and the PNC monitor (DISCmini) to assess the efficiency and performance of the UAV based system by measuring emissions from a diesel engine. The air speed behavior map produced by test 1 shows the best mounting point for the sensors to be alongside the UAV. This position is less affected by the propeller downwash effect. Test 2 results demonstrated that the UAV propellers cause a dispersion effect shown by the decrease of gas and PN concentration measured in real time. A Linear Regression model was used to estimate how the sensor position, relative to the UAV center, affects pollutant concentration measurements when the propellers are turned on. This research establishes guidelines on how to develop a UAV system to measure point source emissions. Such research should be undertaken before any UAV system is developed for real world data collection.

Keywords: UAV remote gas sensing; downwash effect; air quality; hexacopter; optical sensor; air pollution; particle number concentration monitor

1. Introduction

Unmanned Aerial Vehicles (UAVs), carrying onboard sensors, can be used to directly measure shipping emissions, emissions from industrial stacks or ground vehicles when it is too difficult or dangerous to use both manned aircrafts [1] and ground level stations [2]. However, accurate sampling of small plumes emitted by combustion sources such as trucks, petrol locomotives, ships and dredgers, industrial and even domestic chimneys demands appropriate location of the air sensor intakes onboard the UAV. Therefore, the use of UAVs for air pollution measurement, particularly at slow speeds or stationary flights, can only be effective if the location point of the air sensor intake is optimized, such that sampling of gaseous and particulate matter pollutants is done before the propellers of the UAV mix or disperse the plume. The air is deflected by the action of the UAV propellers in motion, when producing lift, and this is called the downwash and upwash effect. It is the transport of gas to the gas sensors that is the critical process, and accurate sampling relies on assessing the contribution of rotor disturbance to this process.

Only two studies were found [3,4] to investigate the sensor mounting point onboard a multirotor UAV to measure air pollution. The first study designed three different experiments to identify the best location of a CO_2 gas sensor intake onboard a quadrotor and how to bring the air to the sensor [3]. The three experiments included: an active system using an additional fan, a semi-active system using the airflow generated by the rotors, and a passive system, without any auxiliary device directing the airflow. Significant differences between the three different gas transport solutions were found and no system was capable of measuring the reference gas concentration (0.5% by volume of CO_2). Overall, the authors [3] established that the position of the tube inlet strongly affected the gas measurement results and wind resistance caused by the enlarged drone.

The second study [4] investigated the airflow of a small quadcopter using a Computational Flow Dynamic (CFD) simulation and found that the maximum air speed was at the rotor perimeters and that the minimum was at the center of the UAV. The researchers proved this by attaching the UAV to a cardan joint to measure air speed above and below the platforms. Above the UAV center was the best mounting point. The paper failed, however, to provide information about the height of the cardan joint, ground effect, or mounting point along aside the UAV, which was a quadcopter (Parrot AR drone 2.0) for hobby use with both a limited payload (<100 g) and a flight endurance of less than 12 min with a zero-weight payload.

Optimizing the sensor mounting location can improve how the pollutant samples emitted by a point source are represented. This research aimed to: (1) determine the best location for mounting a gas and PNC sensing payload intake onboard a hexacopter by measuring air speed along its three axes; (2) to validate the UAV based system capable of measuring CO_2, CO, NO_2 and NO gases and PNCs by measuring emissions from a diesel engine.

2. Unmanned Airborne Plume Assessing System

2.1. Overview

Hexacopter UAVs provide a larger payload capacity (>2 kgs payload), and more in-flight stability and maneuverability compared to quadrotors, making them suitable for UAV and air quality studies where the capability to carry different sensors and maintain an in-flight fixed position are needed [5]. However, a comprehensive study on sensor location on such platforms has not been done yet.

The system developed and evaluated in this research consists of a multi-rotor UAV, four gas sensors for CO, CO_2, NO and NO_2, a DISCmini (Portable PNC monitor, Testo AG, Lenzkirsch, Germany), temperature and humidity sensors as well as a real-time visualization interface. All sensors are integrated with an Arduino MEGA 2560 microcontroller board.

2.2. System Architecture

Figure 1 shows an overview of the system, the gas sensing payload and the PNC monitor, while the UAV system architecture is shown in Figure S1 in the Supplementary Material.

The system presented in Figure 1 comprises: (a) the UAV system components including DJI S800, DISCmini and gas sensors, Arduino board, telemetry, and RC receiver, and (b) Ground Control Components including RC transmitter for the pilot, and GSO (ground station + telemetry link).

Figure S1 (Supplementary Material) shows UAV system hardware and software, including the ground control station, with the connection types between the different UAV components highlighted.

The UAV pilot or the UAV ground station operator (GSO) can communicate with the UAV wirelessly, using a radio controller (RC) transmitter or a computer, respectively. The purple arrows show pilot and GSO inputs.

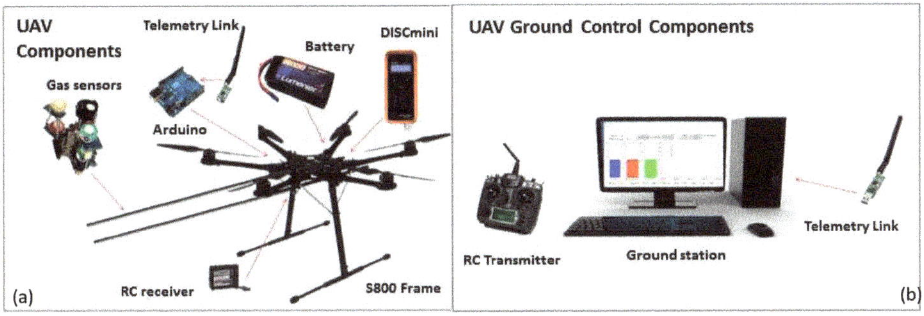

Figure 1. (**a**) Unmanned aerial vehicle (UAV) system components including gas sensors, DISCmini, Arduino MEGA 2560, battery and RC receiver; (**b**) UAV Ground Control Components for both manual (using RC transmitter) and autonomous (using PC) operations.

2.3. UAV

The UAV used in this research is a modified hexacopter S800 EVO manufactured by DJI (Shenzhen, China) [6]. The frame measures 800 mm in width and 320 mm in height and is made of composite materials. The built-in damping system allows the multi-rotor to be assembled without additional frames and dampers. The UAV weighs 3.7 kg with a maximum take-off weight of 8 kg. The S800 is designed to operate under the 20 kg all up weight (AUW) class of UAS (Unmanned Air System) as defined by the Civil Aviation Safety Australia (CASA) to reduce operation costs, and avoid being subject to the tighter CASA regulations for larger UAVs [7]. Different countries may have different aviation regulations which should be considered in the design of a UAV system. For example, in the US, the Federal Aviation Administration (FAA) considers small UAVs those with a weight less than 55 lb (25 kg) [8].

The UAV uses a 16,000 mAh LiPo 6 cell battery, which provides a hover time of approximately 20 min with no additional payload. However, the flight time with the payload used in this study has been calculated to be about 12–13 min. The hovering motor power consumption is 800 W operating with a minimum take-off weight. Motors run in conjunction with 38 cm $\times$ 13 cm propellers of 13 g each.

2.4. Sensor Selection

Gas sensors are classified according to their operational principles with the most common being thermal, mass, electrochemical, potentiometric, amperometric, conductometric, and optical sensors [9,10]. PM sensors differ in terms of monitoring PM in the range of PM_{10} (mass concentration of particles with aerodynamic diameter <10 μm) and $PM_{2.5}$ (<2.5 μm) and the ultrafine fraction of PM (<0.1 μm). Many "built in" devices, with a sensor incorporated as an integral part of the device itself, already exist for PM_{10} and $PM_{2.5}$. However, limited options are available for PNC and size distribution monitoring devices and those existing include: the Nanotracer Monitor from Philips [11] and the Mini Diffusion Size Classifier (DISCmini) developed by the University of Applied Sciences, Windisch, Switzerland [12,13]. Both devices operate on similar principles, and the later was selected for this research.

The DISCmini is a portable instrument of relatively small dimensions (180 mm$\times$ 90 mm$\times$ 40 mm), low weight (640 gr, 780 gr including probe provided by the supplier) and long battery duration (8 h operative). It is used to measure the number of particles with diameters ranging between 10 and 500 nm, with a time resolution of 1 s (Supplementary Material, Table S1) and an experimentally tested response time of 7.3 s [14]. The DISCmini can measure a concentration range from about 10^3 to over

10^6 p/cm^3, based on the electrical charging of the aerosols. The sensor works by generating positive air ions in a corona discharge subsequently mixed with the aerosol. Measurement accuracy depends on the shape of the particle size distribution and number particle concentration; it is usually around 10%–15% compared to a reference condensation particle counter (CPC). Unlike the CPC or other larger instruments, the DISCmini does not need a liquid to grow and count particles and therefore works in any position and does not require a liquid refill.

The gas sensing payload includes three Alphasense gas sensors (Alphasense, B4 type, Great Notley, Essex, UK) and one SprintIR CO_2 sensor. The Alphasense sensors are electrochemical cells that operate in the amperometric mode and generate a current that is linearly proportional to the fractional volume of the measured gas. They are used to measure CO, NO and NO_2 [15]. The SprintIR CO_2 for CO_2 concentration measurements is based on Non-Dispersive Infra-Red (NDIR) technology [16].

2.5. Payload Design and Telemetry

The UAV system architecture uses a radio modem to transmit real-time data including information on the three-dimensional location of the UAV and payload parameters to the ground station. The Arduino MEGA 2560 microcontroller (Arduino, Ivrea, Italy) transmits the data and was chosen over other devices such as the Raspberry Pi B+ microprocessor which might have both better hardware (e.g., memory) and software (e.g., operating system). However, such devices need more power and are slower. The Arduino is easier to use for its connectivity and programming, better speed for receiving and transmitting data, and lower power consumption. The Arduino can power all four gas sensors simultaneously.

2.6. Integration

The DiSCmini [17] can be easily integrated on the UAV as a small and lightweight monitor. However, careful positioning of the sensor to avoid possible issues with the aircraft center of gravity is needed. The custom made gas sensor payload (shown in Figure 2) includes the four gas sensors. Further details on each sensor are provided in Table S2 in the Supplementary Materials. The integration process resulted in a system that is capable of measuring PNCs and five gases simultaneously.

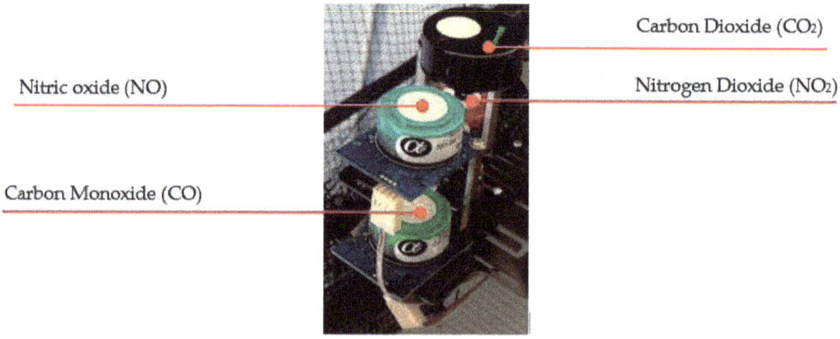

Figure 2. Queensland University of Technology (QUT) payload sensor system capable of measuring CO_2, CO, NO_2, NO.

2.7. Gas Visualization Interface

The UAV system includes a ground control interface to visualize and store the data sent from the on-board sensors, in real-time. Figure S2 in the Supplementary Material presents a screenshot of this interface.

The interface summarizes all data received from the gas sensors in a graph displayed on the right-hand side, while pressure, humidity and temperature measurements are displayed on the left-hand side. The instantaneous measurements are displayed at the bottom of the screen. The Ground Control Interface allows the user to store data using control buttons on top of the screen.

3. Experimental Design

This research conducted two different tests to assess the optimal sensor intake location and the effectiveness of the sensors. For test 1, a professional standard anemometer (model 9565-P, TSI, Shoreview, MN, USA) was used to measure the air speed caused by the UAV propellers. The anemometer had 3% air velocity accuracy, 0–50 m/s measuring range and an accuracy of $\pm$1.5% of reading and a resolution of 0.01 m/s (Figure 3) to create an airspeed map, to quantify the downwash and upwash effect of the S800 hexacopter. The portable anemometer was activated manually to store a reading, taken at 20 s intervals. Test 1 was designed to be conducted indoors to avoid any wind or external input that could have modified the UAV wash effect behavior.

As in Test 1, Test 2 was also conducted indoors and validated the functionality of the onboard sensors and their mounting location by measuring gas and particles of a plume emitted by a diesel engine as a source. Test 2 was further divided into three parts to record measurements inside, below and above the plume.

3.1. Test 1

To solve the sensor mounting point issue so a boom was attached to the frame of the UAV to avoid turbulence and the air mixing effect of the propellers. The boom worked as both the mounting point for the gas sensors which work passively, and as sampling port for the DISCmini. The boom could extend alongside the UAV, above or even below it, using a cardan joint to fold it when the hexacopter lands. The air speed behavior was mapped along the UAV axes to determine which of the three available options was the most effective.

The experiment was conducted at the Australian Research Centre for Aerospace Automation (ARCAA) Indoor Flying Lab located in Brisbane, Queensland. The position of the anemometer was recorded by the VICON (Vicon Motion System Ltd., Oxford, UK) position mapping system an indoor tracking and localization system, used at ARCAA [18] to provide the position and orientation of an object inside the flight area [18]. The system is accurate to a sub millimeter level and provides 100 position readings per second. VICON provides the timestamp of each reading with an average position of 20 s at each interval to associate the air speed reading for each time interval.

An electrical forklift was used to fix the UAV, and provide an adjustable height mechanism (Figure 4a,b). The UAV was firmly attached to the forklift using zip ties, without blocking the space underneath the UAV so as to not affect propeller airflow (Figure S3 Supplementary Materials).

Test 1 was divided into four experiments where measurements were taken along the Y axis, for both the negative, Experiment 1: $-$Y, and positive, Experiment 2: +Y, and along the X axis, including the horizontal, Experiment 3, and the vertical, Experiment 4, directions of the air flow. The UAV hexacopter has a symmetrical shape, and so it is assumed to have the same air speed profile along the Z and X axis.

Figure 4a illustrates the set-up for Test 1 equipment and Figure 5b shows the direction of measurements taken every 100 mm. The maximum distance from the propellers where the sensors could be mounted was limited to 1200 mm, to maintain the UAV balance without the need of additional counterbalance weight at the other end. The total weight of the carbon tube that holds the sensors and the sensors themselves, is counterbalanced (offset) by the weight of the battery which is attached towards the back, underneath the UAV frame.

Infrared detection markers were attached to the anemometer probe to record the anemometer position using the VICON system. Four infrared markers were placed on the UAV to record its position and as a reference (Figures 3 and 4a). During each Test 1 experiment, one researcher was in charge of monitoring the marker positions, while another moved the anemometer along the three UAV axes, to ensure the measurements were taken correctly in terms of position and time. For both

Experiments 1 and 2, the anemometer probe was moved from the reference, considered to be the center of the UAV at 0–1500 mm. During Experiments 3 and 4, the anemometer probe was moved along the X axis, from the closest point to the propellers, at 600 mm from the UAV center, to 1900 mm. During Test 1, the forklift to which UAV was attached was placed at a nominal height 'h' of three meters (h = 3000 mm).

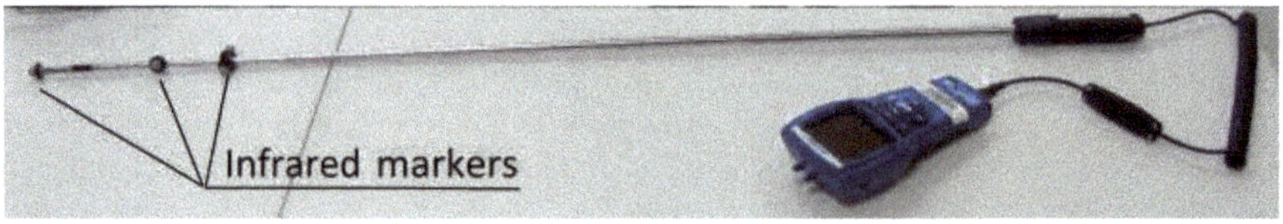

Figure 3. The Portable anemometer TSI, model 9565-P [19] with infrared targets to track and record the anemometer probe positon during Test 1 with the VICON system.

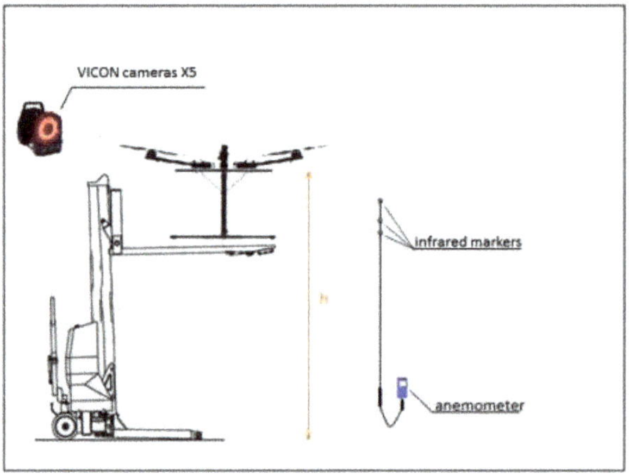

(a)

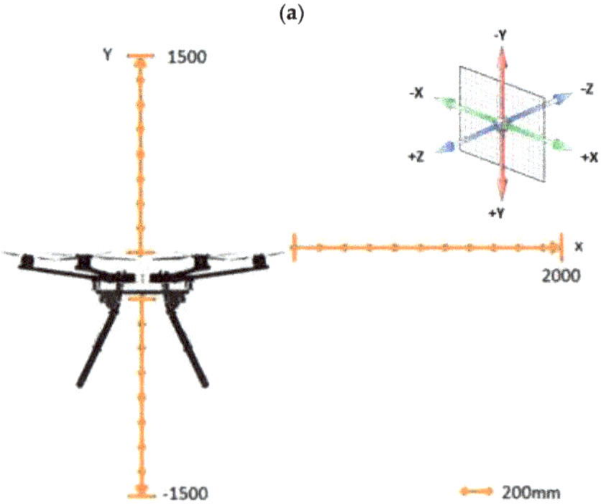

(b)

Figure 4. (**a**) Test 1 schematic representation; (**b**) Direction the anemometer probe was moved during the different experiments of Test 1.

3.2. Test 2

The overall aim of Test 2 was to quantify the propeller downwash effect on the sensor readings and to validate the UAV system in terms of its capability to collect data of gaseous and particle pollutants emitted by a pollution source. Test 2 compared the sensor measurements above, inside and below a plume emitted by the source. The sensors' electrical signals were converted to concentrations using the manufacturer's calibration values and equations. During Test 2, ground measurements of CO_2 were also taken, while temperature and humidity were recorded by the onboard sensors. The two hangar doors were open in Test 2 to create an air current, and ensure that the plume from the combustion source was going in one direction. The UAV system was placed downwind to be in the plume.

The aim of Test 2 was to answer the following questions:

(1) How does the status of the propeller (on/off) affect the measured gas and particle number concentrations?
(2) How does the position of the sensors (below, above or inside) affect the measured concentrations?
(3) How does the distance from the UAV center affect the measured gas and particle number concentrations?

The set-up for Test 2 is shown in Figure 5. The pollutant source was an Isuzu D-Max 4 $\times$ 4 Crew Cab Chassis (3.0 liter turbo diesel, in-line 4-cylinder, DOHC, 16-valve, with intercooled turbo charger, 380 nm, 130 kW, automatic). A 4000 mm heat-resistant aluminum flexible air ducting outlet tube was attached to the car exhaust. The pipe end was attached to the top of a ladder at a height of 2500 mm (Figure S4 Supplementary Materials). The indoor flight lab at ARCAA is a hangar with two specular doors. For this test, the D-Max was parked downwind with the engine running at the minimum RPM, to have a constant emission and to create a stable plume blowing towards the UAV system. As in Test 1, the UAV was fixed to the forklift in order have a precise and stable UAV system location. Based on the results of Test 1, two carbon fiber booms of 1000 mm each were attached to the UAV frame, along the X axis. The gas sensors were mounted on one boom, while the DISCmini nozzle was mounted on the other. The DISCmini was attached to the UAV frame. Two different positions along the carbon fiber booms were tested for the sensors and DISCmini nozzle at 700 and 1100 mm from the UAV center (distance 'd1' in Figure 5). These two positions were chosen because they offer the best compromise between minimum air speed and flight stability. Using a boom with the sensors mounted at a maximum distance of 1100 mm allows the UAV to fly, as the overall weight (sensors plus boom) is offset by the battery which is attached directly to the opposite side of the UAV frame. Figure S5 in the Supplementary Materials shows the sensors mounted onboard the UAV.

In Test 2, the UAV was placed in three different positions: 700 mm above the pipe exhaust, in front of the exhaust, and 700 mm below the exhaust. The distance of 700 mm was chosen as the position where the sensor recorded the same values of gases and PNCs of the background, confirming a reading outside the pollutant plume. The sensors were also moved along the boom in two positions to test how much the downwash effect affects the sensor readings. The distance between the sensor intakes and the plume source (exhaust pipe) was set to 700 mm and maintained for both sensor positions, respectively (d2 in Figure 5). The forklift was moved toward the source when the sensors were moved from 1100 mm (first mounting position for the sensors and DISCmini nozzle) to 700 mm along the boom. Test 2 comprised four different experiments which compared the sensor readings when the UAV system was in three different positions related to a plume from a pollutant source, and the sensor readings at two different mounting points. Experiment 1 was conducted with the propellers turned off, with the purpose of measuring the gases and PNC with no interference due to the propeller operation. During Experiments 2, 3 and 4, the propellers were turned on, the gas sensor distance from the UAV center varied (d1 in Figure 5) and measurements taken inside the pollutant plume at 2500 mm from the ground, outside the plume at 3200 mm (above the plume), and at 1800 mm from the ground (below the plume), respectively. The summary of these experiments is presented in Table 1.

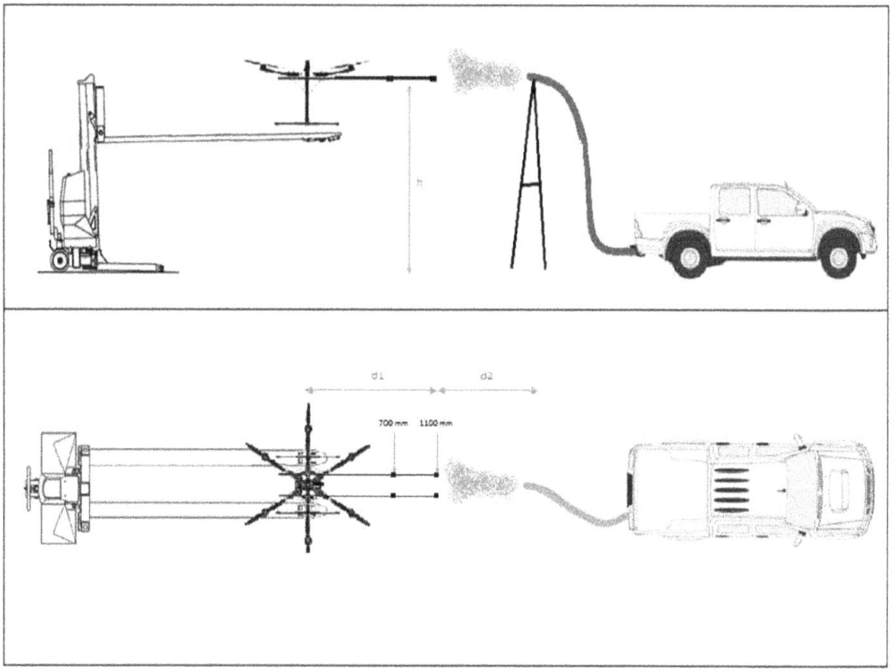

Figure 5. Test 2 schematic representation.

Table 1. Summary of Test 2 Experiments 1–4.

	Propeller Status	UAV Position, UAV Height from the Ground (mm)	Sensor Intake Position; Distance from UAV Center (mm)	Sensor Position; Distance from Exhaust (mm)
Experiment 1	Off	Inside the plume, 2500	-	700
Experiment 2a	On	Inside the plume, 2500	1100	700
Experiment 2b	On	Inside the plume, 2500	700	700
Experiment 3a	On	Above the plume, 3200	1100	700
Experiment 3b	On	Above the plume, 3200	700	700
Experiment 4a	On	Below the plume, 1800	1100	700
Experiment 4b	On	Below the plume, 1800	700	700

4. Results

4.1. Test 1

Results of each of the three experiments, Experiment 1 (+Y), Experiment 2 (−Y), Experiment 3 and 4 (X), as part of Test 1, are shown in Figure 6a–d, respectively.

Figure 6a shows the air speed (m/s) measurements taken above the UAV during Test 1 Experiment 1. The data trend indicates that air speed decreases with distance from UAV to propeller tip, but plateaus at approximately 900 mm with an average speed of 0.9 m/s. Figure 6b shows the results of Experiment 2 for air speed readings below the UAV. The anemometer recorded an increase in air speed to 600 mm, starting from the center of the UAV and moving down. From that point, the airspeed decreased until 1200 mm and apart from a few fluctuations, stabilized with an average of 6.5 m/s.

Figure 6c shows the result of Experiment 3 (horizontal direction of air flow), for air speed readings along the X axis. The hexacopter propellers do affect air flow, as expected, however air speed dropp quickly and become insignificant at distances over 700 mm. Figure 6d shows the result of Experiment 4 (vertical direction of air flow), for air speed readings along the same X axis, which demonstrated that the air speed decreased steadily at distances over 700 mm.

The zero reference in both Figure 6c,d is just beyond the tip of the propeller. These results suggest that the propeller effect is more pronounced in the region closer to the propeller, and its effect lessens as distance increases.

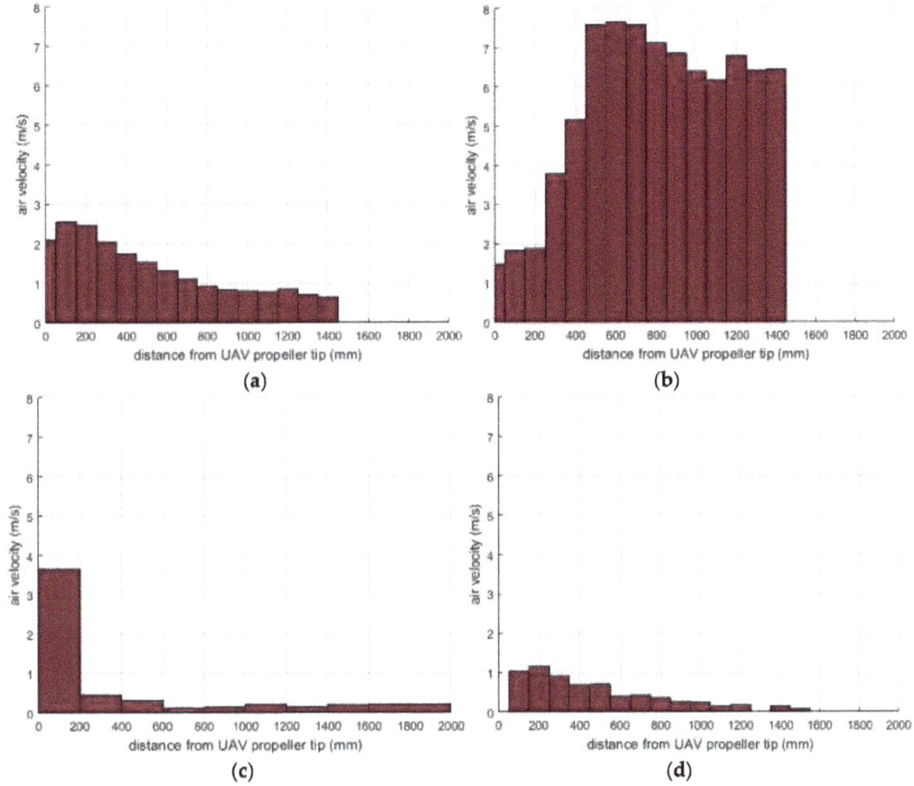

Figure 6. (**a**) Air speed along the Y axis of the UAV system, considering the positive direction component (+Y, above the UAV system); (**b**) Air speed along the Y axis of the UAV system, considering the negative direction of the values; (**c**) Air speed along the X axis of the UAV system—horizontal direction of air flow; (**d**) Air speed along the X axis of the UAV system—vertical direction of air flow.

The air speed data collected in all four experiments are presented in Figure 7 as an air speed map. Arrows of different size and color represent air speed. Arrows increase in size with an increase in air speed, while color moves from blue to red to indicate speed increment.

Air speed measured above and below the UAV, along the Y axis (Experiment 1: +Y, Experiment 2: −Y) is represented with arrows pointing down. For example, air speed recorded during Experiment 2: −Y, at 500 mm underneath the UAV, was 5 m/s and increased to 7.5 m/s at 700 mm. Air speed measured along the X axis (Experiment 3: horizontal, Experiment 4: vertical) is represented with arrows pointing to the right of the UAV and down. For example, Experiment 3 showed that the air speed just beyond the propellers was about 4 m/s, and at 700 mm the air speed was less than 1 m/s.

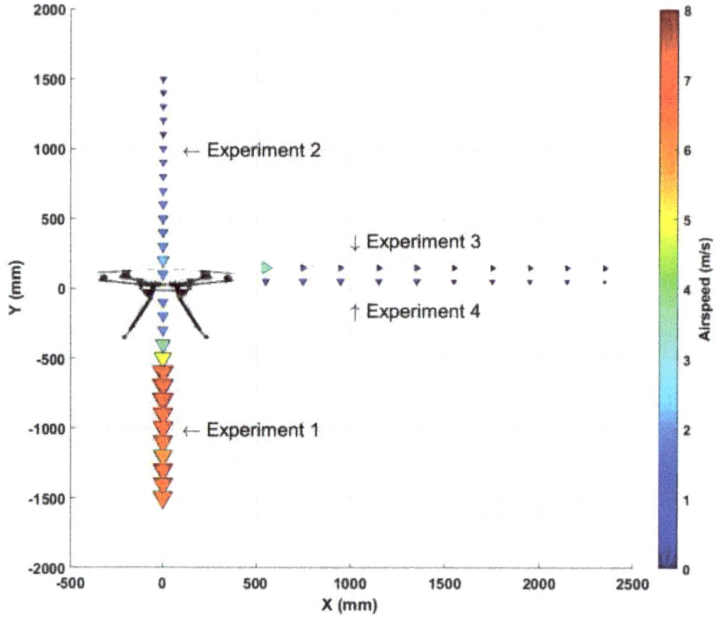

Figure 7. UAV system air velocity map.

4.2. Test 2

The distribution of Test 2 results is represented using violin plots (Figure 8a–c) where the width of each violin plot represents the density of data as measured for a particular concentration value. Violin plots are more informative than a plain box plot which only show a summary of mean/median and interquartile ranges for example [20]. Concentrations of measured pollutants were higher when the propellers were off (Figure 8a) and the UAV was inside the plume (Figure 8b).

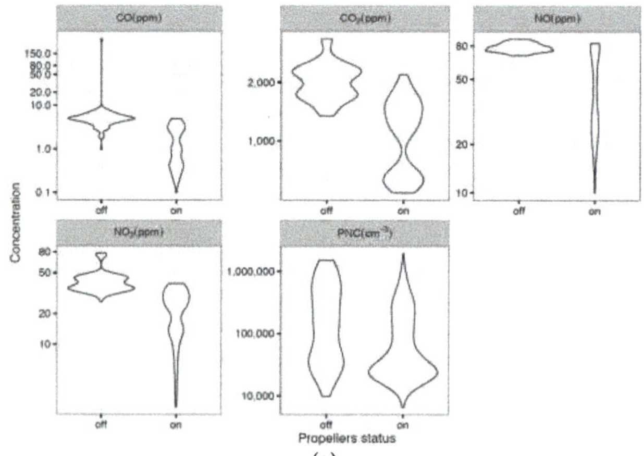

(a)

Figure 8. *Cont.*

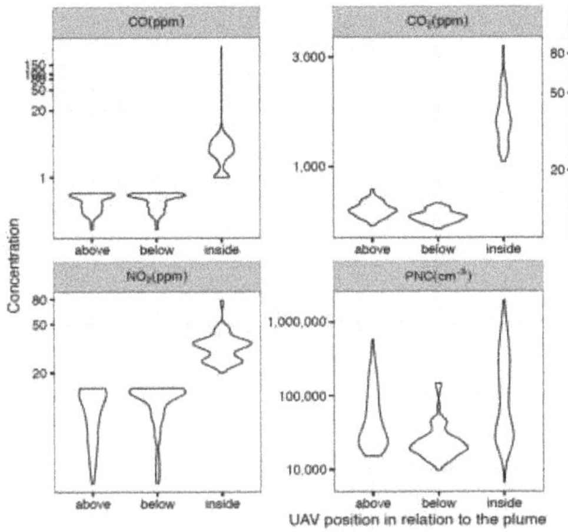

(**b**)

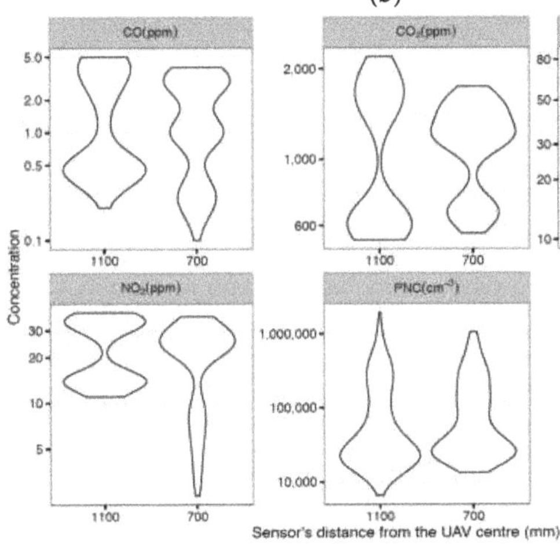

(**c**)

Figure 8. (**a**) Measured gas (ppm) and PNC (pt/cm^3) concentrations with propellers turned off and on; (**b**) Measured gas (ppm) and PNC (pt/cm^3) concentrations at three different positions relative to plume (inside, below and above); (**c**) Measured gas (ppm) and PNC (pt/cm^3) concentrations measured at 1100 mm and 700 mm from the UAV center.

Pollutant distribution under different conditions is presented as violin plots (Figure 8a–c) to show that measured concentrations were higher when the propeller was off (Figure 8a) and the UAV inside the plume (Figure 8b).

The Linear Regression model was developed from data collected when propellers were on, to predict the simultaneous effect of distance and position on the measured concentrations. The explanatory variables used are, distance from the UAV (in meters), and position (above, below, or inside (reference point).

The Linear Regression model was used to estimate the sensor position effects and their distance from the UAV center on measurements results. The model was structured as follows Equation (1):

$$y_i = \beta_0 + \beta_1 p_i + \beta_2 d_i \tag{1}$$

where: y is the measured concentration, i is the measurement number, p is a categorical variable indicating the position of the sensor (inside, above, or below), β_0 is the intercept, d is a continuous variable indicating the distance of the sensor from the center of the UAV in mm, β_1 and β_2 are the estimates. The Linear Regression model results identify the average changes in concentration with 95% confidence intervals. The inside-plume position was set as the reference position, in order to compare the effect of the positions below and above it.

Humidity and temperature measurements taken during the four experiments of Test 2 are presented in Figure S6a–g (Supplementary Material document). These measurements were recorded in real-time and show an increase in humidity when the UAV was inside the plume compared with the above and below positions. Temperature measurements were stable during all experiments with no significant variation between different UAV positions. Ground CO_2 measurements were taken as a reference during Test 2, with results shown in Table 2.

Table 2. Test 2 ground measurements summary.

Experiment	1	2a	2b	3a	3b	4a	4b
CO_2 (ppm)	558	548	546	580	524	554	576

5. Discussion

During Test 1, there was no air circulation within the room from air conditioning/filtering. Airspeed measurements in the room showed air currents were insignificant (<0.1 m/s) and stable before and after testing, with measurements taken during testing showing that the downwash caused by the UAV is mostly diffused within 3 m from the testing location. The effect of the airflow caused by the UAV on the measured parameters is expected to be negligible.

The UAV system air velocity map, obtained from Test 1 (Figure 7), showed that the best mounting point is along the X axis with a distance beyond the propellers between 1000 and 1200 mm. Figure 9a indicates that the concentrations of measured gases and PNC decrease significantly when the propellers are turned on. The 95% confidence interval (Figure 9a) represents the range where there is a 95% probability to establish the actual value of the pollutant concentrations measured. If the 95% confidence interval includes zero, then the estimate is not statistically significant, as was the case for the change of distance for NO, and PNC. However, the decrease of PNC is so small that it lies within the level of instrumental uncertainty. The decrease in measured concentration when the propellers are on is simply due to the dispersion of the plume.

The effect of position (below or above) compared to inside the plume is negative, which means that the concentration of both the gases and PN is lower when the UAV is above or below the plume compared to when the UAV is inside the plume. Figure 9b and the Linear Regression model results (Equation (1)) indicate that the measured concentration of gases and PNC changes per 1 meter increase with sensor distance for all pollutants, except PNC and NO. Furthermore, the measured concentrations decrease significantly for all pollutants when the sensor is located above or below, compared with inside the plume (Figure 9b). Test 2 results demonstrate that UAV propellers cause a dispersion effect, as shown by a decrease of gas concentration as read by the sensors when mounted closer to the UAV center. In fact, moving the sensors farther from the UAV center reduces the dispersion effect. The difference in value is evident by comparing the results in Figure 9a,b where their significance is visualized by confidence intervals. During Test 2, it was found that while the response time of the sensors was somewhat decreased, it was still an acceptable response, as the sensors are designed to work passively. The main reason for not using active sampling, however, is because the additional weight added to the end of the boom began to offset the amount of thrust available to counterbalance the additional weight created by using such a set-up. It was found that the alphasense sensors had an average response time in the order of 3–6 s, with the added airflow caused by the propeller wash actually helping with the response time by cycling the air around the entire system. Test 2 also demonstrated the capability of the UAV system to collect and transmit data with the position of the UAV system inside and outside a pollutant plume. This finding is important when the aim is to stay inside the plume. The UAV system should be flown in a way that the sensors are inside the plume and as far from the UAV center as possible. This positioning is confirmed by an increase in relative humidity while the temperature is stable due to the quick cooling of the outdoor air when it mixes with the plume.

The ISUZU D-Max was the pollution source used to validate the UAV system presented in this research. Real world applications include measurement of small plumes emitted by a cargo or cruise ship.

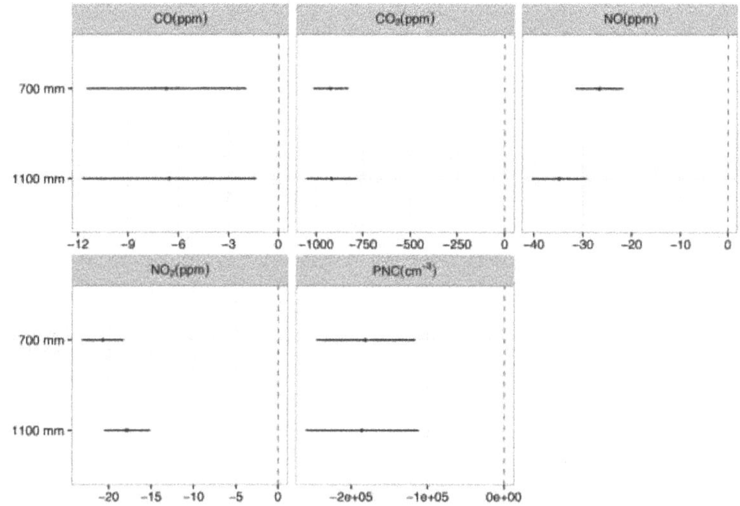

(a)

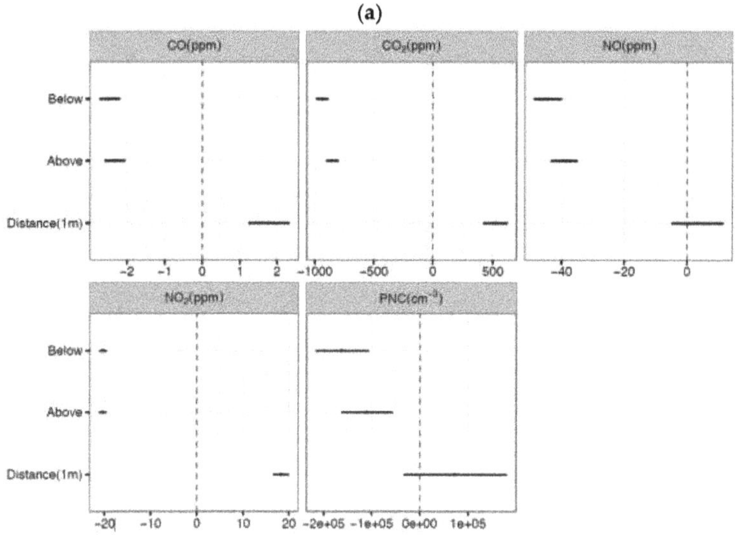

(b)

Figure 9. (**a**) Change in concentrations and 95% confidence intervals when propellers are on compared to off.; (**b**) Change in concentrations and 95% confidence intervals (per 1 m increase in sensor distance and position of sensor compared to inside the plume).

6. Conclusions

Monitoring air quality using small sensors onboard a UAV is very complicated, not only for constraints such as power consumption, weight and propeller effect, but also because the choice of sensors depends on the pollution source being measured.

Real-test data collection must consider the nature of the mission to evaluate the best compromise between feasibility to fly and accuracy of data. That is, a small plume emitted by a stationary (industrial stacks) or a moving (ships, in-land vehicles) source. The UAV system presented in this research has been designed for the purpose of combustion source emission measurements. These types of missions normally comprise low speed and linear flight paths, such as taking off, flying to and hovering in the sampling position, and landing, which means the UAV system should be flown in a way where the sensors are inside the plume and as far from the UAV center as feasible.

An alternative is required when there is strong wind or where the flight path involves fast, frequent and steep changes of direction, such a chasing a fast moving source. This action could damage the payload, which makes positioning of the sensor closer to the UAV system center preferable, even though the error of the reading is higher. This research presents for the first time a comprehensive analysis of what needs to be done to optimize the development of UAV systems so they are more accurate in real time applications. The Linear Regression model is an example of a model that could be used for further studies and a similar approach could be used by other researchers who use different UAV platforms where the focus could be not only on the integration of sensors onboard UAV, but also on quantitative data collection. It is feasible to integrate air quality sensors onboard a UAV system to measure gaseous and particle pollutants in time and space. Further work will focus on real-world applications and the analysis of near-real time data to help atmospheric modelling software development and flight path planning algorithms.

Supplementary Materials: The Supplementary Materials are available online at http://www.mdpi.com/1424-8220/16/12/2202/s1.

Acknowledgments: The authors would like to acknowledge the ARCAA for the use of indoor and outdoor flying facilities, Steven Bulmer for his technical support, Barbara Lissa De Oliveira Lara for her help during the tests and in particular the excellent support of the ARCAA Operations Team (Dirk Lessner, Dean Gilligan, Gavin Broadbent and Dmitry Bratanov) who operated the Unmanned Aerial Vehicle (S800). The authors would also like to acknowledge the help of the QUT Academic Language and Learning Service, in particular Karyn Gonano and Christian Long for the language editing of this manuscript.

Author Contributions: Tommaso Francesco Villa conducted the entire research following the plan developed with the other authors and drafted the entire manuscript. Farhad Salimi helped analyzing the data, developed the Linear Regression model and contributed to the writing of the manuscript. Kye Morton helped in the development of the gas sensing payload, the integration onboard the UAV and during all the indoor tests. Lidia Morawska directed the research project, contributed to the revision of the entire manuscript. Felipe Gonzalez contributed to the writing and to the revision the entire manuscript.

Conflicts of Interest: The authors declare no conflict of interest. The founding sponsors had no role in the design of the study; in the collection, analyses, or interpretation of data; in the writing of the manuscript, and in the decision to publish the results.

Abbreviations

The following abbreviations are used in this manuscript:

UAV	Unmanned aerial vehicle
UAS	Unmanned aerial system
PM	Particulate matter
VOC	Volatile organic compounds
UFP	Ultrafine particles
FAA	Federal Aviation Authority of the U.S.
CASA	Civil Aviation Safety Australia
ARCAA	Australian Research Center for Aerospace Automation
PNC	Particle Number Concentration

References

1. Toscano, P.; Gioli, B.; Dugheri, S.; Salvini, A.; Matese, A.; Bonacchi, A.; Zaldei, A.; Cupelli, V.; Miglietta, F. Locating industrial VOC sources with aircraft observations. *Environ. Pollut.* **2011**, *159*, 1174–1182. [CrossRef] [PubMed]
2. Maria, R.; Marcé, R.M.; Borrull, F. Volatile organic compounds in air at urban and industrial areas in the Tarragona region by thermal desorption and gas chromatography-mass spectrometry. *Environ. Monit. Assess.* **2010**, *161*, 389–402.
3. Neumann, P.P. *Gas Source Localization and Gas Distribution Mapping with a Micro-Drone*; Freie Universität Berlin, Fachbereich Mathematik und Informatik: Berlin, Germany, 2013.
4. Juan, R.; Joossen, G.; Sanz, D.; del Cerro, J.; Barrientos, A. Mini-UAV based sensory system for measuring environmental variables in greenhouses. *Sensors* **2015**, *15*, 3334–3350.
5. Imam, A.S.; Bicker, R. Design and construction of a small-scale rotorcraft UAV system. *Int. J. Eng. Sci. Innov. Technol. (IJESIT)* **2014**, *3*, 96–109.
6. DJI. DJI S800-evo. 2014. Available online: http://www.dji.com/product/spreading-wings-s800-evo (accessed on 14 November 2016).
7. NPRM 1309OS—Remotely Piloted Aircraft Systems. Available online: https://www.casa.gov.au/standard-page/nprm-1309os-remotely-piloted-aircraft-systems?WCMS%3ASTANDARD%3A%3Apc=PC_102028 (accessed on 13 December 2016).
8. U.S. Department of Transportation, Federal Aviation Administration. The NEW Small UAS Rule (Part 107), Including All Pilot and Operating Rules, Is in Effect as of 12:01 a.m. Available online: https://www.faa.gov/uas/ (accessed on 29 August 2016).
9. Jiří, J. *Principles of Chemical Sensors*; Springer: Atlanta, GA, USA, 2009.
10. Xiao, L.; Cheng, S.; Liu, H.; Hu, S.; Zhang, D.; Ning, H. A survey on gas sensing technology. *Sensors* **2012**, *12*, 9635–9665.
11. Johan, M.; Voetz, M.; Kiesling, H.-J. Monitor for detecting and assessing exposure to airborne nanoparticles. *J. Nanopart. Res.* **2010**, *12*, 21–37.
12. Martin, F.; Burtscher, H.; Steigmeier, P.; Kasper, M. *Field Measurement of Particle Size and Number Concentration with the Diffusion Size Classifier (DiSC)*; SAE Technical Paper; SAE international: Warrendale, PA, USA, 2008.
13. Morawska, L.; Afshari, A.; Bae, G.N.; Buonanno, G.; Chao, C.Y.H.; Hänninen, O.; Hofmann, W.; Isaxon, C.; Jayaratne, E.R.; Pasanen, P.; et al. Indoor aerosols: from personal exposure to risk assessment. *Indoor Air* **2013**, *23*, 462–487. [CrossRef] [PubMed]
14. Bau, S.; Zimmermann, B.; Payet, R.; Witschger, O. A laboratory study of the performance of the handheld diffusion size classifier (DiSCmini) for various aerosols in the 15–400 nm range. *Environ. Sci.* **2015**, *17*, 261–269. [CrossRef] [PubMed]
15. Alphasense Sensors for Air Quality Networks. 2013. Available online: http://www.alphasense.com/index.php/air/ (accessed on 11 November 2016).
16. *CO_2, SprintIR, SprintIR Fast 20–100% CO_2 Sensor*. Available online: http://www.co2meter.com/products/sprintir-100-percent-co2-sensor (accessed on 20 December 2016).
17. AG, Testo. Available online: http://testo-partikel.de/index.php/features/jquery-superfish-menu (accessed on 20 December 2016).
18. *ARCAA Vicon+System*. Available online: https://wiki.qut.edu.au/display/ARCAA/Vicon+system (accessed on 20 December 2016).
19. *9565-VelociCalc*. Available online: http://www.tsi.com/uploadedFiles/Product_Information/Literature/Spec_Sheets/9565-VelociCalc_US-5001361-specsheet.pdf (accessed on 20 December 2016).
20. Hintze, J.L.; Nelson, R.D. Violin plots: A box plot-density trace synergism. *Am. Stat.* **1998**, *52*, 181–184. [CrossRef]

sensors

MDPI

Article

Real-Time Multi-Target Localization from Unmanned Aerial Vehicles

Xuan Wang [1,2,*], Jinghong Liu [1] and Qianfei Zhou [1,2]

[1] Changchun Institute of Optics, Fine Mechanics and Physics, Chinese Academy of Sciences, Changchun 130033, China; liujinghong@ciomp.ac.cn (J.L.); zhouqianfei@ciomp.ac.cn (Q.Z.)
[2] University of Chinese Academy of Sciences, Beijing 100049, China
* Correspondence: wangxuan@ciomp.ac.cn; Tel.: +86-431-8617-6159

Academic Editors: Felipe Gonzalez Toro and Antonios Tsourdos
Received: 3 September 2016; Accepted: 19 December 2016; Published: 25 December 2016

Abstract: In order to improve the reconnaissance efficiency of unmanned aerial vehicle (UAV) electro-optical stabilized imaging systems, a real-time multi-target localization scheme based on an UAV electro-optical stabilized imaging system is proposed. First, a target location model is studied. Then, the geodetic coordinates of multi-targets are calculated using the homogeneous coordinate transformation. On the basis of this, two methods which can improve the accuracy of the multi-target localization are proposed: (1) the real-time zoom lens distortion correction method; (2) a recursive least squares (RLS) filtering method based on UAV dead reckoning. The multi-target localization error model is established using Monte Carlo theory. In an actual flight, the UAV flight altitude is 1140 m. The multi-target localization results are within the range of allowable error. After we use a lens distortion correction method in a single image, the circular error probability (CEP) of the multi-target localization is reduced by 7%, and 50 targets can be located at the same time. The RLS algorithm can adaptively estimate the location data based on multiple images. Compared with multi-target localization based on a single image, CEP of the multi-target localization using RLS is reduced by 25%. The proposed method can be implemented on a small circuit board to operate in real time. This research is expected to significantly benefit small UAVs which need multi-target geo-location functions.

Keywords: multi-target localization; UAV; real time; lens distortion correction; RLS

---

1. Introduction

Real-time multi-target localization plays an essential and significant role in disaster emergency rescue, border security and so on. Over the past two decades, considerable research efforts have been devoted to multi-target localization. UAV electro-optical stabilized imaging systems are equipped with many kinds of sensors, including visible light cameras, infrared thermal imaging systems, laser range finders and angle sensors. Target localization needs to measure the attitude of the UAV, the attitude of the electro-optical stabilized imaging system and the distance between the electro-optical stabilized imaging system and the target.

The target localization methods from UAVs are divided into two categories. One category is target localization using a group of UAVs [1–5]. The other category is target localization using a single UAV [6–10]. This research aims to improve the effectiveness and efficiency of target localization from a single UAV. Particularly, this research proposes a new hybrid target localization scheme which integrates both zoom lens distortion correction and an RLS filtering method. The proposed scheme has many unique features which are designed to geo-locate targets rapidly.

Previous studies on geo-locating targets from a fixed-wing UAV have several limitations. Deming [1] described a probabilistic technique for performing multiple target detection and localization based on data from a swarm of flying optical sensors. Minaeian [2] described a vision-based crowd detection and Geographic Information System (GIS) localization algorithm for a cooperative team of one UAV and a number of unmanned ground vehicle (UGV)s. Morbidi [3] described an active target-tracking strategy to deploy a team of unmanned aerial vehicles along paths that minimize the uncertainty about the position of a moving target. Qu [4] described a multiple UAV cooperative localization method using azimuth angle information shared between the UAVs. Kwon [5] described a robust, improved mobile target localization method which incorporates the Out-Of-Order Sigma Point Kalman Filter (O3SPKF) technique.

In [1–5], target location methods based on data fusion technology have to use a group of UAVs. Target localization using a group of UAVs has some issues, including the high computational complexity of data association, complexity of UAV flight plans, difficulties in efficient data communication between UAVs and high maintenance costs due to the use of multiple UAVs. This paper presents a method for determining the location of objects using a gimbaled EO camera on-board a fixed-wing unmanned aerial vehicle (UAV). We focus on geo-locating targets using a single fixed-wing UAV due to the low maintenance costs. A single fixed-wing UAV (as opposed to rotary wing aircraft) has unique benefits including adaptability to adverse weather, good durability and high fuel efficiency.

In [6], Yan used absolute height above sea level of a UAV to geo-locate targets. In contrast, our system focuses on geo-locating targets in the video stream and does not require absolute height above sea level of the UAV data. In [7], Ha used scale invariant feature transform (SIFT) to extract feature points of the same target in different frames. The author calculated the relative height between the target and the UAV by three dimensional reconstruction. The location accuracy depends on the accuracy of this three dimensional reconstruction. The method thus requires a large amount of computation. In contrast, the accuracy of our system compared to that described in [7] is almost the same, but our system has a great advantage in reduced computational complexity, so our system can geo-locate 50 targets at the same time and improve the efficiency of multi-target localization. This is very important in military reconnaissance and disaster monitoring applications which require good real-time performance.

In [8], Barber introduced a system for vision-based target geo-localization from a fixed-wing micro-air vehicle. In flight tests, the UAV geo-locates the stationary target when the UAV orbits the targets. In [9], the UAV flies in an orbit in order to improve the geo-location accuracy. In [10], the authors assume that the UAV's altitude above the target is known. The target's altitude is obtained from a geo-referenced database made available by the Perspective View Nascent Technology (PVNT) method. In [11], target-location need an accurate geo-referenced terrain database. In [12], all information collected by an aerial camera is accurately geo-located through registration with pre-existing geo-reference imagery. In contrast, our system focuses on geo-locating a specific object in the video stream and does not require any preexisting geo-referenced imagery.

In all the above references, the UAVs are equipped with fixed-focal lenses. The authors do not take into account the effect of zoom lens distortion on multi-target localization. Many electro-optical stabilized imaging system are equipped with zoom lenses. The focal length of a zoom lens is adjusted to track targets at different distances during the flight. The zoom lens distortion varies with changing focal length. Real-time zoom lens distortion is impossible to correct by using calibration methods because the large amount of transformation calculation has to be repeated when the focal length is changed.

The primary contributions of this paper are: (1) the accuracy of multi-target localization has been improved due to the combination of a real-time zoom lens distortion correction method and a RLS filtering method using embedded hardware (a multi-target geo-location and tracking circuit board); (2) UAV geo-locates targets using embedded hardware (the multi-target geo-location and tracking circuit board) in real-time without orbiting the targets; (3) 50 targets can be located at the same time using only one UAV; (4) the UAV can geo-locate targets without any pre-existing geo-referenced imagery, or a terrain database; (5) the circuit board is small, and therefore, can be applied to many kinds of small UAVs; (6) multi-target localization and tracking techniques are combined, therefore, we can geo-locate multiple moving targets in real-time and obtain the target motion parameters such as velocity and trajectory. This is very important for UAVs performing reconnaissance and attack missions.

The rest of paper is organized as follows: Section 2 briefly presents the overall framework of the multi-target localization system. Section 3.1 presents the reference frames and transformations required for the multi-target localization system. Section 3.2 presents our multi-target geo-location model. Section 4 presents the methods to improve the accuracy of multi-target localization. Section 4.1 presents the distortion correction method. Section 4.2 presents the RLS filter method. Section 5 presents the results of multi-target localization for aerial imaged captured from a flight test and evaluates their accuracy. Section 6 presents the conclusions.

2. Overall Framework

The real-time multi-target geo-location algorithm in this paper is programmed and implemented on a multi-target geo-location and tracking circuit board (model: THX-IMAGE-PROC-02, Changchun Institute of Optics, Fine Mechanics and Physics, Chinese Academy of Sciences, Changchun, China, see Figure 1a) with the TMS320DM642 (Texas Instruments Incorporated, Dallas, TX, USA) @ 720 MHz Clock Rate and 32 Bit Instructions/Cycle and 1 GB double data rate synchronous dynamic random access memory (DDR SDRAM). This circuit board also performs the proposed zoom lens distortion correction and the RLS filtering in real-time. The multi-target geo-location and tracking circuit board is mounted on an electro-optical stabilized imaging system (see Figure 1b). This aerial electro-optical stabilized imaging system consists of a visible-light camera, a laser range finder, an inertial measurement unit (IMU), a global positioning system (GPS), and a photoelectric encoder. They are in the same gimbal so that they rotate altogether in the same direction in any axis.

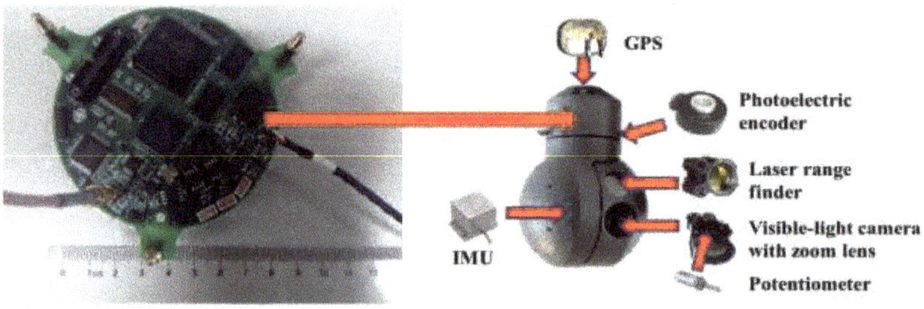

Figure 1. (**a**) Multi-target geo-location and tracking circuit board; (**b**) Electro-optical stabilized imaging system. The arrows in the Figure 1b represent the installation locations of main sensors in an electro-optical stabilized imaging system.

The electro-optical stabilized imaging system is mounted on the UAV (model: Changguang 1, Changchun Institute of Optics, Fine Mechanics and Physics, Chinese Academy of Sciences, Changchun, China) to stabilize the videos and any eliminate video jitters caused by the UAV therefore greatly reducing the impact of external factors.

The UAV system incorporates the electro-optical stabilized imaging system, UAV, data transmission module and ground station, which is shown in Figure 2. In the traditional target geo-location algorithms [1–12], the image and UAV attitude information are transmitted to a ground station. The target geo-location is calculated on a computer in the ground station. However, the date transmission model sends data with divided time mode, so the image and UAV attitude information are respectively transmitted at different times from the UAV to the ground station, so it is not guaranteed that the image and UAV attitude information will be obtained at the same time in ground station. Therefore, the traditional target geo-location algorithm on the computer in the ground station has poor real-time ability and unreliable target geo-location accuracy.

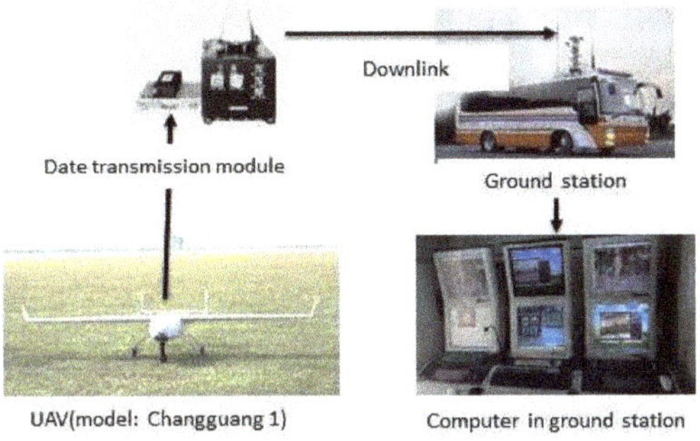

Figure 2. UAV system architecture.

To overcome the shortcomings of traditional target geo-location algorithms such as algorithm complexity, unreliable geo-location accuracy and poor real-time ability, in this paper, the target geo-location algorithm is implemented on a multi-target geo-location and tracking circuit board on the UAV in real-time. Real-time ability is very important for urgent response in applications such as military reconnaissance and disaster monitoring.

The overall framework of the multi-target geo-location method is shown in Figure 3. The detailed workflows of the abovementioned multi-target geo-location method will be introduced as follows: we use UAV to search for the ground targets, which are selected by an operator in the ground station. The coordinates of the multiple targets in the image are transmitted to the UAV through a data transmission model. Then, all the selected targets are tracked automatically by the multi-target geo-location and tracking circuit board using the improved tracking method based on [13]. The electro-optical stabilized imaging system locks the main target in the field of view (FOV) center. Other targets in the FOV are referred to as sub-targets. The electro-optical stabilized imaging system measures the distance between the main target and the UAV using a laser range finder.

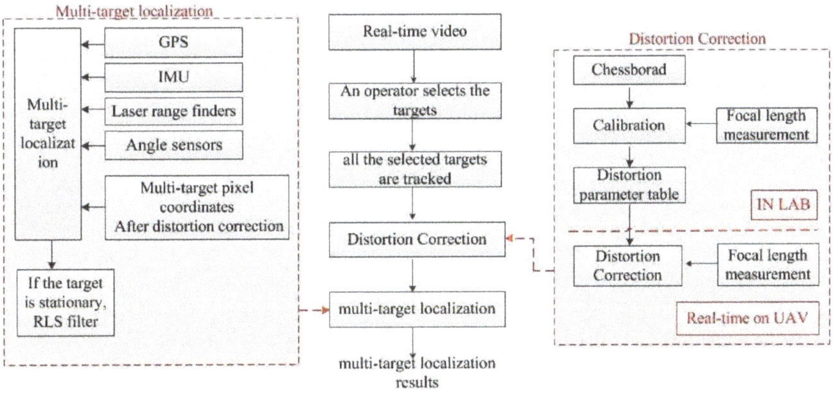

Figure 3. The overall framework of the multi-target geo-location method.

In order to ensure that the image, UAV attitude information, electro-optical stabilized imaging system's azimuth and elevation angle, laser range finder value, and camera focal length are obtained at the same time, the frame synchronization signal of the camera is used as the external trigger signal for data acquisition by the above sensors, so we don't need to implement sensor data interpolation algorithms in the system except for the GPS data. The UAV coordinates interpolation algorithm is shown in Equations (35) and (36).

The multi-target geo-location and tracking circuit board computed the multi-target geo-location after lens distortion correction in real-time. Then, the board used the moving target detection algorithm [14–18] for the tracked targets. If the target which is tracked is stationary, the multi-target geo-location and tracking circuit board uses the RLS filter to improve the target geo-location accuracy. The multi-target geo-location results are superimposed on each frame in the UAV and downlinked to a portable image receiver and the ground station.

This research aims to address the issues of real-time multi-target localization in UAVs by developing a hybrid localization model. In detail, the proposed scheme integrates the following improvements:

(a) The multi-target localization accuracy is improved due to the combination of the zoom lens distortion correction method and the RLS filtering method. A real-time zoom lens distortion correction method is implemented on the circuit board in real time. In this paper, we analyse the effect of lens distortion on target geo-location accuracy. Many electro-optical stabilized imaging systems are equipped with zoom lenses. The focal length of a zoom lens can be adjusted to track targets at different distances during the flight . The zoom lens distortion varies with changing focal length. Real-time distortion correction of a zoomable lens is impossible by using the calibration methods because the tedious calibration process has to be repeated again if the focal length is changed.
(b) The target geo-location algorithm is implemented on a circuit board in real time. The size of the circuit board is very small, therefore, this circuit board can be applied to many kinds of small UAVs. The target geo-location algorithm has the following advantages: low computational complexity and good real-time performance. UAV can geo-locate targets without pre-existing geo-referenced imagery, terrain databases and the relative height between UAV and targets. UAV can geo-locate targets using the embedded hardware in real-time without orbiting the targets.
(c) The multi-target geo-location and tracking circuit board use the moving target detection algorithm [14–18] for the tracked targets. If the target which is tracked is stationary, the multi-target geo-location and tracking circuit board uses the RLS filter to automatically improve the target geo-location accuracy.
(d) The multi-target localization, target detection and tracking techniques are combined. Therefore, we can geo-locate multiple moving targets in real-time and obtain target motion parameters such as velocity and trajectory. This is very important for UAVs performing reconnaissance and attack missions.

The real output rate of the geo-location results is 25 Hz. The reasons are as follows:

(a) The data acquisition frequency of all the sensors is 25 Hz: the visible light camera's frame rate is 25 Hz. The frame synchronization signal of the camera is used as the external trigger signal for all sensors expect GPS (the UAV coordinates interpolation algorithm is shown in Equations (35) and (36)).
(b) Lens distortion correction is implemented in real-time, and the output rate of target location results after the lens distortion correction is 25 Hz.
(c) When it is necessary to locate a new stationary target, the RLS algorithm needs 3–5 s to converge to a stable value (within 5 s, lens distortion correction is implemented in real-time, the output rate is 25 Hz). After 5 s, the geo-location errors of the target have converged to a stable value. We can obtain a more accurate location of this stationary target immediately (it is no longer necessary to run RLS). That output rate is 25 Hz, too.

Our geo-location algorithm can geo-locate at least 50 targets simultaneously. The reasons are as follows:

(a) For a moving target we only use lens distortion correction to improve the target geo-location accuracy. This consumes 0.4 ms on average when calculating the geo-location of a single target and, at the same time, correcting zoom lens distortion (Section 5.4). It consumes 0.4 ms for tracking the multiple targets (Section 3.3). The image frame rate is 25 fps, so the duration of a frame is 40 ms, so our geo-location algorithm can geo-locate at least 50 targets simultaneously.
(b) For a stationary target, only when it is necessary to locate a new stationary target, the RLS algorithm needs 3–5 s to converge to a stable value (within 5 s, lens distortion correction is implemented in real-time, 50 targets can be located simultaneously). After 5 s, the geo-location errors of the target have converged to a stable value. We no longer need to run RLS, so our geo-location algorithm can geo-locate at least 50 targets simultaneously after lens distortion correction and RLS.

Therefore, this algorithm has great advantages in geo-location accuracy and real-time performance. The multiple target location method in this paper can be widely applied in many areas such as UAVs and robots.

3. Real-Time Target Geo-Location and Tracking System

3.1. Coordinate Frames and Transformation

Five coordinate frames (camera frame, body frame, vehicle frame, ECEF frame and geodetic frame) are used in this study. The relative relationships between the frames are shown in Figure 4. All coordinate frames follow a right-hand rule.

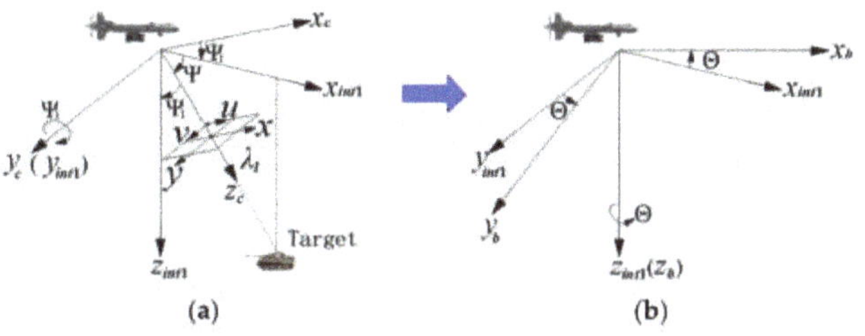

Figure 4. The coordinate frames relation: Azimuth-elevation rotation sequence between camera and body frames. (**a**) camera frame; (**b**) body frame.

3.1.1. Camera Frame

The origin is the camera projection center. The x-axis x_c is parallel to the horizontal column pixels' direction in the CCD sensor (i.e., the u direction in Figure 4). The y-axis y_c is parallel to the vertical row pixels' direction in the CCD sensor (i.e., the v direction in Figure 4). The positive z-axis z_c represents the optical axis of the camera.

3.1.2. Body Frame

The origin is the mass center of the attitude measuring system. The x-axis x_b is the 0° direction of attitude measuring system. The y-axis y_b is the 90° direction of attitude measuring system. The z-axis z_b completes the right handed orthogonal axes set. The azimuth Θ, elevation angle ψ and distance λ_1 output by electro-optical stabilized imaging system are relative to this coordinate frame.

3.1.3. Vehicle Frame v

A north-east-down (NED) coordinate frame. The origin is the mass center of attitude measuring system. The aircraft yaw β, pitch ε and roll angle γ output by the attitude measuring system are relative to this coordinate frame.

3.1.4. ECEF Frame

The origin is Earth's center of mass. The z-axis z_e points to the Conventional Terrestrial Pole (CTP) defined by International Time Bureau (BIH) 1984.0, and the x-axis x_e is directed to the intersection between prime meridian (defined in BIH1984.0) and CTP equator. The axes y_e completes the right handed orthogonal axes set.

3.1.5. WGS-84 Geodetic Frame

The origin and three axes are the same as in the ECEF. Geodetic longitude L, geodetic latitude M and geodetic height H are used here to describe spatial positions, and the aircraft coordinates (L_0, M_0, H_0), output by GPS are relative to this coordinate frame.

The relation between camera frame and body frame is shown in Figure 4. Two steps are required. First, transformation from camera frame to intermediate frame $int1$: rotate 90° (elevation angle ψ) along the y-axis y_c. The next step is transformation from intermediate frame $int1$ to body frame: rotate azimuth angle Θ along the z-axis z_{int1}. In Figure 4a, ψ_1 represents 90°.

The relation between body frame and vehicle frame is shown in Figure 5. Three steps are required. First, transformation from the body frame to the intermediate frame $mid1$: rotate roll angle γ along the x-axis x_b. The next step is transformation from the intermediate frame $mid1$ to the intermediate frame $mid2$: rotate pitch angle ε along the y-axis y_{mid1}. The final step is transformation from the intermediate frame $mid2$ to the vehicle frame: rotate yaw angle β along the z-axis z_{mid2}.

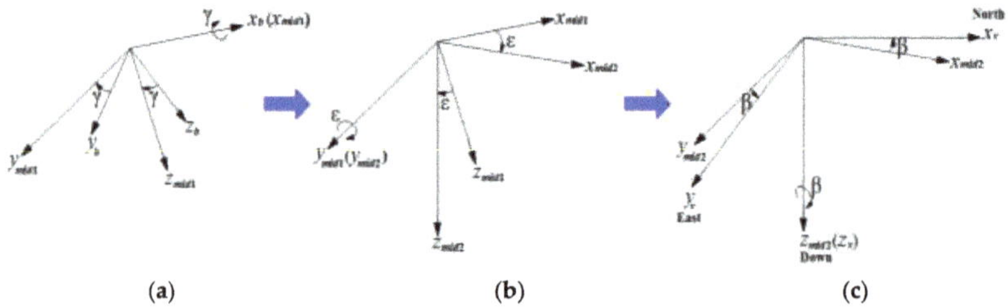

Figure 5. The coordinate frames relation: roll-pitch-yaw rotation sequence between body and vehicle frames. (**a**) body frame; (**b**) intermediate frame; (**c**) vehicle frame.

The relation between vehicle frame and earth centered earth fixed (ECEF) is shown in Figure 6.

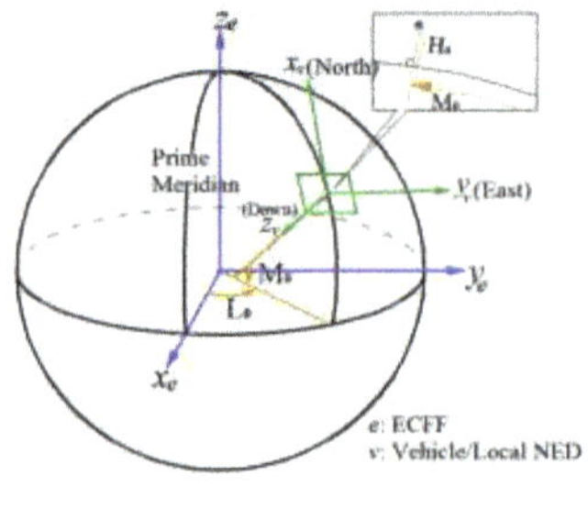

Figure 6. The coordinate frames relation: Vehicle, ECEF and geodetic frames.

3.2. Multi-Target Geo-Location Model

As shown in Figure 4a, the main target is at the camera field of view (FOV) center, whose homogeneous coordinates in the camera frame are $[x_c, y_c, z_c, 1]^{\ T} = [0,\ 0,\ \lambda_1, 1]$, Through the transformation among five coordinate frames ranging from camera frame to WGS-84 geodetic frame, the geographic coordinates of main target in the WGS-84 geodetic frame can be determined, as shown in Figure 7.

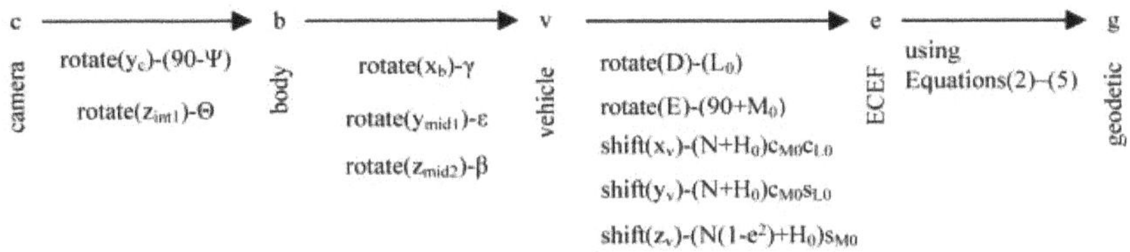

Figure 7. Coordinate transformation process of multi-target geo-location system.

First, we calculate the coordinates of the main target in the ECEF:

$$\begin{bmatrix} x_e \\ y_e \\ z_e \\ 1 \end{bmatrix} = \begin{bmatrix} -c_{L_0}s_{M_0} & -s_{L_0} & -c_{L_0}c_{M_0} & (N+H_0)c_{M_0}c_{L_0} \\ -s_{L_0}s_{M_0} & c_{L_0} & -c_{M_0}s_{L_0} & (N+H_0)c_{M_0}s_{L_0} \\ c_{M_0} & 0 & -s_{M_0} & (N(1-e^2)+H_0)s_{M_0} \\ 0 & 0 & 0 & 1 \end{bmatrix} \times$$
$$\begin{bmatrix} c_\varepsilon c_\beta & -c_\gamma s_\beta + s_\gamma s_\varepsilon c_\beta & s_\gamma s_\beta + c_\gamma s_\varepsilon c_\beta & 0 \\ c_\varepsilon s_\beta & c_\gamma c_\beta + s_\gamma s_\varepsilon s_\beta & -s_\gamma c_\beta + c_\gamma s_\varepsilon s_\beta & 0 \\ -s_\varepsilon & s_\gamma c_\varepsilon & c_\gamma c_\varepsilon & 0 \\ 0 & 0 & 0 & 1 \end{bmatrix} \times \begin{bmatrix} c_\Theta s_\Psi & -s_\Theta & c_\Theta c_\Psi & 0 \\ s_\Psi s_\Theta & c_\Theta & s_\Theta s_\Psi & 0 \\ -c_\psi & 0 & s_\Psi & 0 \\ 0 & 0 & 0 & 1 \end{bmatrix} \times \begin{bmatrix} x_c \\ y_c \\ z_c \\ 1 \end{bmatrix} \tag{1}$$

where $c_* = \cos(*)$, $s_* = \sin(*)$.

Then we derive the geodetic coordinates of main target from earth centered earth fixed-world geodetic system (ECEF-WGS) transformation equations [19]:

$$U = \arctan\frac{az_e}{b\sqrt{x_e^2 + y_e^2}} \tag{2}$$

$$L = \begin{cases} \arctan\frac{y_e}{x_e},\ when\ x_e > 0 \\ \frac{\pi}{2},\ when\ x_e = 0,\ y_e > 0 \\ -\frac{\pi}{2},\ when\ x_e = 0,\ y_e < 0 \\ \pi + \arctan\frac{y_e}{x_e},\ when\ x_e < 0,\ y_e > 0 \\ -\pi + \arctan\frac{y_e}{x_e},\ when\ x_e < 0,\ y_e < 0 \end{cases} \tag{3}$$

$$M = \arctan\frac{z_e + be^2\sin^3 U}{\sqrt{x_e^2 + y_e^2} - ae^2\cos^3 U} \tag{4}$$

$$H = \frac{\sqrt{x_e^2 + y_e^2}}{\cos M} - N \tag{5}$$

In Equations (1)–(5), the semi-major axis of ellipsoid is $a = 6378137.0m$, the semi-minor axis of ellipsoid is $b = 6356752.0m$, the first eccentricity of spheroid $e = \sqrt{a^2 - b^2}/a$, the second eccentricity of spheroid is $e' = \sqrt{a^2 - b^2}/b$, and the radius of spheroid curvature in the prime vertical is $N = a/\sqrt{1 - e^2 sin^2 M}$.

The key of sub-target geo-location is to build a geometrical geo-location model. The coordinates (x_c, y_c, z_c) of sub-targets in the camera frame are solved on the basis of their pixel coordinates, and then their geodesic coordinates are calculated in accordance with the coordinate transformation Equations (1)–(5), Suppose the ground area corresponding to a single image is flat and the relative altitudes between targets and electro-optical stabilized imaging system are the same. Based on the image forming principles for single-plane array charge coupled device (CCD) sensors, a multi-target geo-location model can be established, as shown in Figure 8.

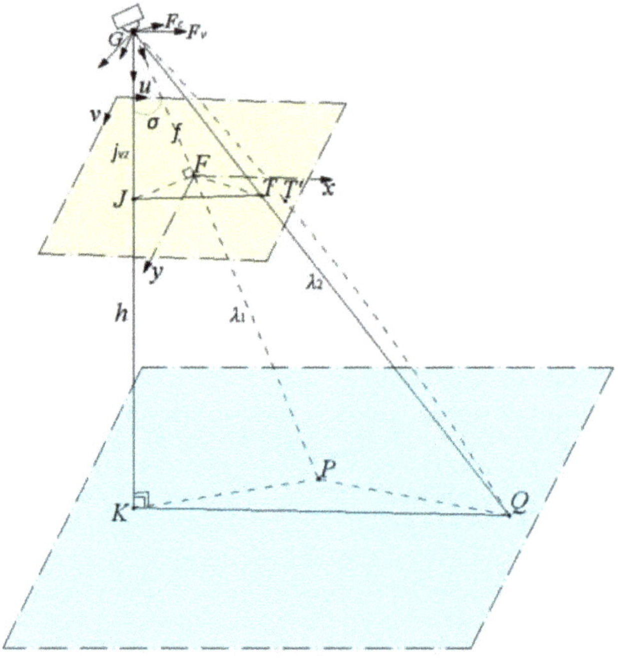

Figure 8. The location of any target in image model.

Suppose that no image distortion exists, the image point T of main target P is at the image center, and the three points, namely projection center G, sub-target Q and its image point T, are on the same line. Then a pin-hole imaging model will be formed and the altitude of a target relative to electro-optical stabilized imaging system will be:

$$h = \lambda_1 \cos\alpha = \lambda_2 \cos\beta \tag{6}$$

where: h is the relative altitude, λ_1 is the distance from electro-optical stabilized imaging system to main target, and λ_2 is the distance from electro-optical stabilized imaging system to a sub-target.

Suppose the line-of-sight (LOS) vectors of the main target P, the sub-target Q and the point K beneath the camera are $\vec{s} = \vec{GF}$, $\vec{t} = \vec{GT}$, $\vec{j} = \vec{GJ}$ respectively, α is the angle between $\vec{s}$ and $\vec{j}$, and β is the angle between $\vec{t}$ and $\vec{j}$, then [20]:

$$\cos\alpha = \frac{\vec{s} \bullet \vec{j}}{\|\vec{s}\| \, \|\vec{j}\|} \tag{7}$$

$$\cos\beta = \frac{\vec{t} \bullet \vec{j}}{\|\vec{t}\| \, \|\vec{j}\|} \tag{8}$$

F_c is the basis vectors for camera frame in a 3-dimensional vector space, the coordinates of LOS vectors $\vec{s}$ and $\vec{t}$ in the camera frame are given by Equations (9) and (10):

$$\vec{s} = F_c^T \begin{bmatrix} 0 \\ 0 \\ f \end{bmatrix} \tag{9}$$

$$\vec{t} = F_c^T \begin{bmatrix} u - u_0 \\ v - v_0 \\ f \end{bmatrix} \tag{10}$$

where f is the camera focal length, the unit is mm. The pixel coordinate of the point F is (u_0, v_0). The pixel coordinates of the point T is (u, v).

F_v is the basis vectors for vehicle frame in a 3-dimensional vector space. In vehicle frame, the LOS vector $\vec{j}$ goes down axis z_v, the coordinates of $\vec{j}$ in vehicle frame is given by Equation (11):

$$\vec{j} = F_v^T j_v = F_v^T \begin{bmatrix} 0 \\ 0 \\ j_{v_z} \end{bmatrix} \tag{11}$$

The coordinates of $\vec{j}$ in the camera frame are solved as:

$$j_c = R_{cv} j_v = R_{cb} R_{bv} j_v = \begin{bmatrix} c_\Theta c_\Psi & c_\Psi s_\Theta & -s_\Psi \\ -s_\Theta & c_\Theta & 0 \\ c_\Theta s_\Psi & s_\Theta s_\Psi & c_\Psi \end{bmatrix} \begin{bmatrix} c_\varepsilon c_\beta & c_\varepsilon s_\beta & -s_\varepsilon \\ c_\beta s_\gamma s_\varepsilon - c_r s_\beta & c_r c_\beta + s_r s_\varepsilon s_\beta & c_\varepsilon s_\gamma \\ s_r s_\beta + s_r s_\beta s_\varepsilon & c_r s_\varepsilon s_\beta - c_\beta s_\gamma & c_r c_\varepsilon \end{bmatrix} j_v \tag{12}$$

where $c_* = \cos(*)$, $s_* = \sin(*)$.

R_{bv} is the rotation matrix transformation from the vehicle frame to the body frame. R_{cb} is the rotation matrix transformation from the body frame to the camera frame. R_{cv} is the rotation matrix transformation from the vehicle frame to the camera frame.

σ is the angle between the z_v axis of the vehicle frame and the z_c axis of the camera frame. According to the geometric relationship in Figure 6, we obtain:

$$\| \vec{j} \| = j_{v_z} \tag{13}$$

$$f = j_{v_z} \cos\sigma \tag{14}$$

Using Euler parameters, or quaternions, we have the definition:

$$\eta = \cos\frac{\sigma}{2} \tag{15}$$

It can also be shown that [21]:

$$\eta = \pm\frac{1}{2}(1 + \mathrm{tr}C_{cv})^{\frac{1}{2}} \tag{16}$$

This may be manipulated into:

$$2\eta^2 - 1 = \frac{\mathrm{tr}C_{cv} - 1}{2} \tag{17}$$

Therefore:

$$j_{v_z} = \frac{f}{\cos\sigma} = \frac{f}{2\cos^2\left(\frac{\sigma}{2}\right) - 1} = \frac{f}{2\eta^2 - 1} \tag{18}$$

and it follows that:

$$j_{v_z} = \frac{2f}{trC_{cv} - 1} \tag{19}$$

By substituting the j_{v_z} value into Equations (11) and (12), the coordinate j_c of $\vec{j}$ in the camera frame can be obtained. Then j_c is substituted into Equations (7) and (8), to obtain $\cos\alpha$ and $\cos\beta$. Finally, according to the known main target distance λ_1 and Equation (6), the relative altitude h and the sub-target distance λ_2 can be determined. Based on the sub-target distance λ_2 and the LOS vector $\vec{t}$ of the sub-target in the camera frame, the coordinates of the sub-target in this frame can be determined:

$$\begin{bmatrix} x_c \\ y_c \\ z_c \end{bmatrix} = \lambda_2 \frac{\vec{t}}{\|\vec{t}\|} \tag{20}$$

Finally, the geodetic longitude L, the geodetic latitude M and geodetic height H of the sub-target can be calculated by substituting x_c, y_c and z_c into Equations (1)–(5).

3.3. Targets Tracking

The operator selects multiple targets in the first image and then these targets are tracked using the tracking algorithm. In recent years, many excellent tracking algorithms were proposed [22–31]. Due to the limited hardware resources in the TMS320DM642 (Texas Instruments Incorporated, Dallas, TX, USA), the tracking algorithm for UAV applications must be simple as well as highly efficient to meet the performance demands of real-time multiple target tracking. We use a simple two stage method to improve the real-time performance of the correlation tracking algorithm described in [13]. The main improvements are as follows:

In the low resolution stage, we calculate the average of four adjacent pixels in the original image to generate a low resolution image, whose resolution is half that of the original image. The low resolution template is generated in the same way. The formula of the normalized cross correlation (NCC) algorithm is as follows:

$$R(u,v) = \frac{\sum_{x=1}^{m}\sum_{y=1}^{m} T(x,y)S(x+u,y+v)}{\sqrt{\sum_{x=1}^{m}\sum_{y=1}^{m} T^2(x,y)}\sqrt{\sum_{x=1}^{m}\sum_{y=1}^{m} S^2(x+u,y+v)}} \tag{21}$$

where the size of low resolution template T is $m \times m$, the size of the low resolution search area S is $n \times n$, (u,v) is the left corner point coordinate in the search area, $0 \le u \le n-m$, $0 \le v \le n-m$. T is moving on the S during the matching operation. When $R(u,v)$ reaches the maximum value $R(u_0,v_0)$, the point (u_0,v_0) is the best matching point in the low resolution search area.

In the original stage, we only need to search a small area in the original image. The size of the template is 2 m × 2 m. The size of the small search area is (2 m + 2) × (2 m + 2). The left corner point coordinate in the search area is $(2u_0 - 1, 2v_0 - 1)$. Then, the best matching point in the original image can be calculated using the NCC algorithm. In our implementation, m is set to 28, n is set to 46.

In [13], the time of template image matching is implemented is 0.62 ms. After improvement, it consumes 0.4 ms on the multi-target localization circuit board. The improved method can meet the real-time requirements of multi-target tracking.

4. Methods to Improve the Accuracy of Multi-Target Localization

4.1. Distortion Correction

The above multi-target geo-location model is established under the assumption that image distortion does not exist. In fact, due to the lens design and manufacturing errors of imaging systems, the image will be distorted [32–35], so the projection rays between the image point and object point can't completely meet the requirement of linear propagation in the total field of view (TFOV) and instead, will bend to some extent. As shown in Figure 6, the image point of the main target moves from the ideal position T to a distorted point position T', and the three points, namely projection center G, image point T' and object point Q, are not in a straight line. Thus it does not conform to the ideal pinhole imaging model. The calculation of target geo-location data based on the ideal pinhole imaging model will lead to a big error, so the lens distortion must be corrected. Distortion correction involves at first deriving the position T of the target in the ideal image from its image point T' in the distorted image according to the distortion model of camera, and then to calculate the geodetic coordinates of the target by using the pixel coordinates of the ideal image point T.

Real-time distortion correction of a zoom lens is impossible by using the calibration methods because the tedious calibration process has to be repeated again if the focal length is changed. In this research, we divide the zoom lens distortion procedures into two steps: lens distortion parameter estimation in the laboratory and real-time zoom lens distortion correction on the UAV.

4.1.1. Lens Distortion Parameter Estimation in the Laboratory

Lens distortion parameter estimation is performed in the laboratory. We use UAV electro-optical stabilized imaging system to take images which contain a chessboard pattern in the laboratory. The lines in the chessboard pattern are straight in the real world, but the images generally contain curved lines caused by the lens distortion. We take lots of images with different focal lengths of a zoom lens. We use the images to construct the distortion parameter table, which is shown in Figure 3.

We extract the chessboard image edges using the Canny edge detector. The thresholds of the Canny edge detector are provided in terms of percentages of the gradient norm of the image.

For a zoom lens, the typical range of distortion coefficient k_1 is given by $[-\frac{1}{D^2}, \frac{1}{D^2}]$, D is the diagonal of the image [36]. In pixel coordinates, the image size is $w \times h$ pixels, the distortion center is (u_0, v_0), the image center is (u_c, v_c), where $u_c = 0.5w$, $v_c = 0.5h$. The range of u_0 is $[u_c - 0.05w, u_c + 0.05w]$. The range of v_0 is $[v_c - 0.05h, v_c + 0.05h]$ [37,38].

We sampled N_2 samples of u_0 in the range of $[u_c - 0.05w,\ u_c + 0.05w]$. We sampled N_3 samples of v_0 in the range of $[v_c - 0.05h, v_c + 0.05h]$. We sample N_1 samples of k_1 in the range of $[-\frac{1}{D^2}, \frac{1}{D^2}]$ in each distortion center (u_0, v_0), so $N_1 \times N_2 \times N_3$ possible distortion parameters are generated from these samples in a certain camera focal length. The distortion parameter $\left(k_1^i, u_0^j, v_0^p\right)$ is shown in Equation (22) to Equation (24):

$$k_1^i = -\frac{1}{D^2} + i \times \delta k_1 \tag{22}$$

$$u_0^j = 0.45w + j \times \delta u_0 \tag{23}$$

$$v_0^p = 0.45h + p \times \delta v_0 \tag{24}$$

where $i = 1, 2, \ldots, N_1$; $j = 1, 2, \ldots, N_2$; $p = 1, 2, \ldots, N_3$; $\delta k_1 = \frac{2}{N_1 D^2}$; $\delta u_0 = \frac{0.1w}{N_2}$; $\delta v_0 = \frac{0.1h}{N_3}$;

For each distortion parameter (k_1^i, u_0^j, v_0^p), the pixel coordinates of the corrected chessboard image's edge points (u_n, v_n) are computed by using Equation (25) to Equation (28):

$$x_d = (u_d - u_0)d_x \tag{25}$$

$$y_d = (v_d - v_0)d_y \tag{26}$$

$$u_n = u_0 + \frac{x_d}{d_x(1 + k_1 x_d^2 + k_1 y_d^2)} \tag{27}$$

$$v_n = v_0 + \frac{y_d}{d_y(1 + k_1 x_d^2 + k_1 y_d^2)} \tag{28}$$

d_x, d_y are pixel size, which units are μm. The distortion center is (u_0, v_0), and the unit is pixels. The pixel coordinates of the distorted image are (u_d, v_d), in units of pixels. The pixel coordinates of the undistorted (corrected) image is (u_n, v_n), and the units are pixels. (x_d, y_d) is the projection coordinates of a distorted point, which units are μm.

The gradient direction $\alpha(u_n, v_n)$ of the corrected chessboard image's edge points (u_n, v_n) are computed using Equation (29) to Equation (31):

$$G_u = \frac{I_{v_n,u_n+1} - I_{v_n,u_n} + I_{v_n+1,u_n+1} - I_{v_n+1,u_n}}{2} \tag{29}$$

$$G_v = \frac{I_{v_n+1,u_n} - I_{v_n,u_n} + I_{v_n+1,u_n+1} - I_{v_n,u_n+1}}{2} \tag{30}$$

$$\alpha(u_n, v_n) = \arctan\left(\frac{G_v}{G_u}\right) \tag{31}$$

I is the chessboard image brightness value, G_u, G_v are the first-order derivatives of the corrected image's edge points brightness.

We compute the Hough transform of the corrected chessboard image. The N strongest peaks in the Hough transform correspond to the most distinct lines. The distance between a line N_q and the origin is $dist(q)$. The orientation of a line N_q is $\beta(q)$, where q =1,2,..., N.

If the angular difference between the edge point orientation $\alpha(u_n, v_n)$ and the line N_q orientation $\beta(q)$ is less than a certain threshold (in our implementation, it is set to 2°. This threshold can meet the distortion correction accuracy requirements. We compute the distance d_q from edge point (u_n, v_n) to the line N_q:

$$d_q = |u_n \cos(\beta(q)) + v_n \sin(\beta(q)) - dist(q)| \tag{32}$$

If d_q is less than a certain threshold. In our implementation, it is set to 2 pixels. This threshold can meet the distortion correction accuracy requirements. The edge point (u_n, v_n) votes for the line N_q, the votes of the edge point (u_n, v_n) is:

$$\text{votes} = \frac{1}{1 + d_q} \tag{33}$$

We compute the sum of all edge points votes. In this focal length, the best distortion parameters (k_1, u_0, v_0) are obtained by maximizing the straightness measure function:

$$max\left\{\sum_{q=1}^{N} votes\left(dist(q), \beta(q), k_1^i, u_0^j, v_0^p\right)\right\} \tag{34}$$

where votes$\left(dist(q), \beta(q), k_1^i, u_0^j, v_0^p\right)$ is the votes of the line N_q in the corrected chessboard image using distortion parameter (k_1^i, u_0^j, v_0^p).

We apply the above algorithm to calibrate the best distortion parameter (k_1, u_0, v_0) with different lens focal lengths. Then, the best zoom lens distortion parameters (k_1, u_0, v_0) in all focal lengths are gained through curve fitting using Matlab Tools. We store the distortion parameter talbe for all focal length in the flash chip on the multi-target geo-location and tracking circuit board.

4.1.2. Real-Time Lens Distortion Correction on the UAV

The zoom lens is connected to the potentiometer through the gears. The relationship between focal length and resistance has been calibrated in the laboratory. We can get the focal length by measuring the resistance value of the potentiometer in real-time. During the flight of the UAV, we use the focal length measuring sensor (model: S10HP-3 3-turn Potentiometer, SAKAE, Nagoya, Japan, resistance error: ±1%) to measure the camera focal length and we find the distortion parameter (k_1^i, u_0^j, v_0^p) in the flash chip on the multi-target localization circuit board. The pixel coordinates (u_n, v_n) of the corrected real-time image are computed using Equation (25) to Equation (28). We use (u_n, v_n) to calculate the geodetic coordinates of the targets.

4.2. RLS Filter

For stationary targets on the ground, the location result in different frames should be the same. Therefore, a popular technique to remove the estimation error is to use a recursive least squares (RLS) filter. The RLS filter minimizes the average squared error of the estimate. The RLS filter uses an algorithm that only requires a scalar division at each step, making the RLS filter suitable for real-time implementation, so we use RLS to reduce the standard deviation and improve the accuracy of multiple stationary target localization.

Suppose the original geo-location data of t images are x_k $(k = 1, 2, \ldots, t)$. The RLS algorithm flowchart is shown in Figure 9. In Figure 9, I_{1X1} is a 1×1 unit matrix. After the RLS filtration of the original data x_k, the obtained data are X_k $(k = 1, 2, \ldots t)$, where X_k can be longitude L, latitude M or geodetic height H.

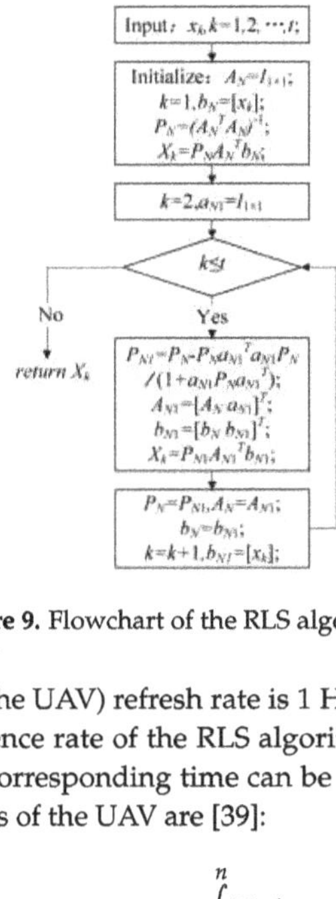

Figure 9. Flowchart of the RLS algorithm.

The GPS data (coordinates of the UAV) refresh rate is 1 Hz, but the video frame rate is usually above 25 Hz. To raise the convergence rate of the RLS algorithm, when the UAV speed is known, the coordinates of the UAV at the corresponding time can be determined through dead reckoning. In the WGS-84 ECEF, the coordinates of the UAV are [39]:

$$x_e = x_{e0} + \int_0^n V_x dt \tag{35}$$

$$y_e = y_{e0} + \int_0^n V_y dt \tag{36}$$

where (x_{e0}, y_{e0}) are the coordinates at the initial time, and V_x is UAV speed in direction X, V_y is UAV speed in direction Y. The influence of the UAV geodetic coordinate position and speed on the reckoned coordinates of the UAV is analyzed as follows:

(1) In the WGS-84 geodetic frame, the higher the latitude of the UAV is, the smaller the projection of 1° longitude onto the horizontal direction. Therefore, in the high latitude area, the measurement accuracy of GPS is high, and the accuracy of the reckoned coordinates is high.
(2) The smaller the UAV speed is, the smaller the distance UAV moves in the same time interval, the higher accuracy of the reckoned coordinates is.

According to Equations (35) and (36), the error resulting from the updating rate of GPS data can be compensated to converge the RLS algorithm rapidly to a stable value. Therefore, we can geo-locate multiple stationary ground targets quickly and accurately.

5. Experiments and Discussion

The targets location data were obtained during a UAV flight in real time. The evaluation is based on UAV videos captured from a Changji highway from 9:40 to 11:10. The resolution of the videos is 1024×768 and the frame rate is 25 frames per second (fps).

5.1. The Zoom Lens Distortion Parameter Estimation Results

To evaluate the proposed lens distortion parameters estimation approach, we use a plane containing a chessboard pattern and a zoom lens camera which are shown in Figure 10. The size of the pattern is 450 mm × 450 mm. We take the images which contained the chessboard pattern for several focal lengths: $f = 5, 8, 10, 15, 20, 25, 30, 40, 50, 60, 70, 80, 90, 100$. Then, we perform the lens distortion parameters estimation approach in Section 4.1 to estimate the lens distortion parameters.

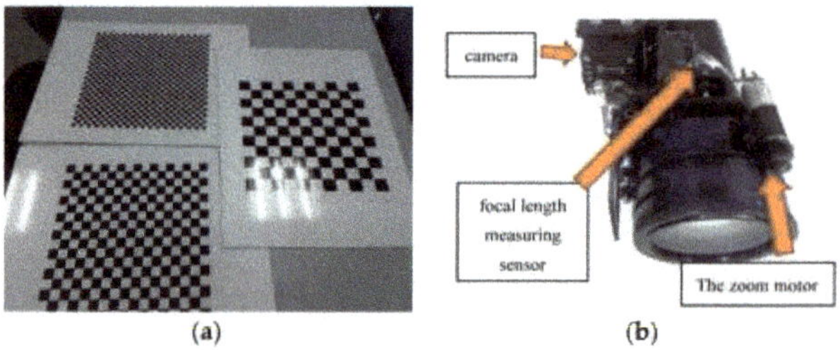

(a) (b)

Figure 10. (**a**) A plane containing a chessboard pattern; (**b**) The zoom lens camera (not yet mount in the electro-optical stabilized imaging system).

The relationships between the distortion coefficient k_1 and the focal length f are shown in Figure 11a. The relationships between the distortion center (u_0, v_0) and the focal length f are shown in Figure 11b.

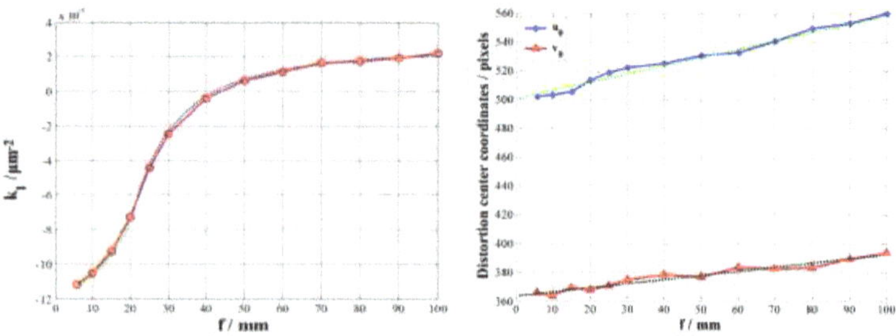

(a) (b)

Figure 11. (**a**) The relationships between distortion coefficient k_1 and the focal length f; (**b**) The relationships between the distortion center (u_0, v_0) and the focal length f.

We fit the curve between the data. If 5.8 mm $\leq f \leq$ 20 mm The relationships between the distortion coefficient k_1 and the focal length f is shown in Equation (37):

$$k_1(f) = \rho \times f^2 + \sigma \times f + \tau \tag{37}$$

where $\rho = 1.369 \times 10^{-10}$, $\sigma = -1 \times 10^{-9}$, $\tau = -1.108 \times 10^{-7}$.

If 20 mm $\leq f \leq$ 100 mm. The relationships between the distortion coefficient k_1 and the focal length f is shown in Equation (38):

$$k_1(f) = \frac{\delta}{(f + \eta)} + \psi \tag{38}$$

where $\delta = -5.245 \times 10^{-5}$, $\eta = 2.768$, $\psi = 2.564 \times 10^{-8}$.

We fit the curve between the data. The relationships between the distortion center (u_0, v_0) and the focal length f is shown in Equations (39) and (40):

$$u_0(f) = \lambda_1 \times f + \mu_1 \tag{39}$$

$$v_0(f) = \lambda_2 \times f + \mu_2 \tag{40}$$

where $\lambda_1 = 0.5722$, $\mu_1 = 500.4904$, $\lambda_2 = 0.2897$, $\mu_2 = 363.5341$.

Based on the above fitting formula, we calculate the zoom lens distortion parameter $\left(k_1^i, u_0^j, v_0^p\right)$ in all focal length to construct the distortion parameter table in the laboratory. We store the distortion parameter table in the flash chip in multi-target localization circuit board.

5.2. Targets Location Experimental Design and Instrument Description

This test is divided into the following four parts:

(1) Monte Carlo simulation analysis of multi-target geo-location error. Through this analysis, the expected error of multi-target geo-location can be determined.
(2) Geo-location test of a single aerial image. We substitute an actual aerial image and its position/attitude data into multi-target geo-location program for target location resolution. We use a high-precision GPS receiver to measure the geo-location data of various ground targets as the nominal values for target geo-location. We compare the calculated values with these nominal values to obtain the multi-target geo-location accuracy of the image. We correct the geo-location error arising from lens distortion, and then compare the calculated geodesic coordinates of each target with its nominal geodesic coordinates to determine the multi-target geo-location accuracy after the distortion correction.
(3) Geo-location test of multi-frame aerial images. For many stationary ground targets, we use the RLS algorithm to adaptively estimate the multi-frame image geo-location data and then by comparing the RLS filtration results with nominal values, we determine the target geo-location accuracy after the RLS filtration.
(4) Real-time geo-location and tracking of multiple ground-based moving targets. We derive the motion trail of each target from the geo-location data and time interval of every image. Then we calculate the speed of each target.

Here, a GPS receiver of the Geo Explorer 3000 series is used for ground measurements. This instrument has 14 channels, including 12 L1 codes and carriers and two satellite-based augmentation systems (SBAS). It is integrated with real-time two-channel SBAS tracking technology, and supports real-time differential correction. It can achieve real-time sub-meter geo-location accuracy—An accuracy of 50 cm is available through Trimble Delta Phase postprocessing.

5.3. Test 1: Monte Carlo Analysis of Multi-Target Geo-Location Accuracy Error

Error analysis is an important step to judge if a geo-location method is good or not. It is very difficult to analyze the target geo-location error through complete differential based on the measurement equation of airborne electro-optical stabilized imaging system, so Monte Carlo analysis is introduced to analyze the multi-target geo-location error. The Monte Carlo method is based on the law of great number and the Bernoulli's theory [40]. On the basis of this method, a model of multi-target geo-location error can be established:

$$[\Delta L\,\Delta M\,\Delta H]^T = F'(X) - F'(X - \Delta X) \tag{41}$$

where: ΔL, ΔM and ΔH are the geo-location errors of each target, and ΔX is geo-location parameter error.

We use five aerial images (32 targets) for multi-target geo-location and distortion correction test. (eight targets in the 1st image, six targets in the 2nd image, five targets in the 3rd image, five targets in the 4th image, and eight targets in the 5th image). We use eight targets in the 1st image in Test 1. The image size is 1024 pixels × 768 pixels, and the pixel size is 5.5 μm × 5.5 μm. The position/attitude data of electro-optical stabilized imaging system collected through GPS, attitude measurement and laser finding at the time of image shoot are shown in Table 1. The root-mean-square error (RMSE) of each parameter is determined in accordance with the maximum nominal error stipulated in the relevant measurement equipment specifications.

Table 1. Localization and attitude data of UAV electro-optical stabilized imaging system.

Designation	Symbol	Unit	Nominal Value	Error
UAV longitude	$L0$	°	112.680649	2×10^{-4}
UAV latitude	$M0$	°	35.125225	1.5×10^{-4}
UAV altitude	$H0$	m	1140	15
UAV pitch angle	ε	°	2.1	0.4
UAV roll angle	γ	°	0.0	0.4
UAV yaw angle	β	°	290.5	1.5
UAV speed	V	m/s	39	0.5
Electro-optical stabilized imaging system's azimuth angle	Θ	°	89.9	0.2
Electro-optical stabilized imaging system's elevation angle	Ψ	°	−112.9	0.2
Laser range finder	λ1	m	965	5
Coordinates of sub-target	(u,v)	pixel	(386,304), (352,379), (511,277), (379,524) (854,463), (756,584), (685,706)	10
Camera's focal length	f	mm	50.0	0.2

Based on the nominal values and errors of above parameters, the sample model for 10,000 random variable arrays can be established in the Matlab software. By using the Equations (1)–(20) and (41), the geodetic coordinate RMSE of every target in this test can be determined through the Monte Carlo method, as shown in Table 2.

5.4. Test 2: Multi-Target Geo-Location Using a Single Aerial Image with Distortion Correction

The data in Table 1 are substituted into Equations (1)–(20) to calculate the geodesic coordinates of each target in a single image, as shown in Table 3.

Table 2. Errors of multi-target expected location.

	Longitude RMSE/(°)	Latitude RMSE/(°)	Altitude RMSE/m
Main target	0.000307°	0.000166°	25.33 m
Sub-target 1	0.000231°	0.000245°	25.33 m
Sub-target 2	0.000340°	0.000199°	26.12 m
Sub-target 3	0.000292°	0.000265°	25.33 m
Sub-target 4	0.000235°	0.000338°	25.33 m
Sub-target 5	0.000311°	0.000345°	24.56 m
Sub-target 6	0.000532°	0.000166°	26.13 m
Sub-target 7	0.000237°	0.000383°	25.34 m

Table 3. Calculated values in the geodesic coordinates of each target.

	Longitude/(°)	Latitude/(°)	Altitude/m
Main target	112.679882	35.125060	171.81
Sub-target 1	112.679058	35.124549	171.82
Sub-target 2	112.680080	35.123993	171.81
Sub-target 3	112.679133	35.125134	171.81
Sub-target 4	112.678723	35.124990	171.82
Sub-target 5	112.680868	35.123948	171.81
Sub-target 6	112.679510	35.123628	171.82
Sub-target 7	112.678275	35.124593	171.82

The calculated geodetic coordinates of each target in Table 3 are compared with its nominal geodetic coordinates measured by GPS receiver on the ground, in order to obtain its geo-location error in a single image, as shown in Table 4. It is found that the geo-location error of each target is within the expected error range through the comparison between the data in Table 4 and those in Table 2. The geodetic height geo-location errors of all the targets are about 18 m—that's basically in line with the assumption in the Section 3.2 that "the ground area corresponding to a single image is flat and the relative altitudes between targets and electro-optical stabilized imaging system are the same".

Table 4. Geo-location error of each target in a single image.

	Longitude Error/°/(°)	Latitude Error/°/(°)	Altitude Error/m
Main target	0.000217	0.000027	17.81
Sub-target 1	−0.000081	0.000180	17.82
Sub-target 2	−0.000261	−0.000111	18.81
Sub-target 3	0.000194	0.000207	17.81
Sub-target 4	0.000090	0.000294	17.82
Sub-target 5	−0.000221	−0.000303	16.81
Sub-target 6	−0.000485	−0.000007	18.82
Sub-target 7	−0.000095	0.000344	17.82

The latitude and longitude geo-location errors of sub-targets are bigger than those of main targets for the following reasons: (1) the error arising from the slope distance difference between a main target and a sub-target. The longer the slope distance of a target to the image border, the bigger the geo-location error [39]; (2) the coordinate transformation error caused by the attitude measurement error and the angle measurement error (measured by the electro-optical stabilized imaging system) during the calculation of altitude and distance of a sub-target relative to the electro-optical stabilized imaging system; (3) the pixel coordinate error of a sub-target found in target detection; and (4) the pixel coordinate error of a sub-target caused by image distortion.

The geo-location errors of a target can be reduced in three ways: by reduction in the flight height H_0 or an increase in the platform elevation angle Ψ (the horizontal forward direction is 0°) to shorten the slope distance, which needs to consider the flight conditions; the selection of a high-precision attitude measuring system and the improvement in angle measurement accuracy of the electro-optical stabilized imaging system, for which one needs to consider the hardware cost; and the distortion correction. Therefore, the influence of distortion correction on multi-target geo-location accuracy will be mainly discussed.

Sub-target pixel coordinate error is mainly caused by image distortion. The correction method is discussed in Sections 4.1 and 5.1. We calculate the corrected pixel coordinates using Equation (25) to Equation (28). Then, we use corrected pixel coordinates to calculate geodetic coordinates of target. Geo-location errors of multi-target after distortion correction are shown in Table 5.

Table 5. Geo-location errors of multi-target after distortion correction.

	Longitude Error/°	Latitude Error/°	Altitude Error/m	Pixel Coordinates/Pixel
Main target	0.000217°	0.000027°	17.81 m	(512.00, 383.99)
Sub-target 1	−0.000110°	0.000163°	17.82 m	(391.06, 306.08)
Sub-target 2	−0.000259°	−0.000067°	18.81 m	(360.11, 378.32)
Sub-target 3	0.000173°	0.000207°	17.81 m	(511.86, 280.58)
Sub-target 4	0.000016°	0.000285°	17.82 m	(386.18, 516.48)
Sub-target 5	−0.000256°	−0.000243°	16.81 m	(838.99, 458.38)
Sub-target 6	−0.000439°	0.000103°	18.82 m	(746.46, 574.63)
Sub-target 7	−0.000117°	0.000306°	17.82 m	(678.45, 691.36)

It can be seen through the comparison between the data in Table 5 and those in Table 4 that, after the distortion correction, the latitude and longitude errors of each target are generally smaller than those before the correction, while the geo-location error of geodetic height remains basically unchanged. For the sub-targets farther from the image center, a more significant reduction in longitude and latitude errors can be obtained. Therefore, lens distortion correction can improve the geo-location accuracy of sub-targets and thus raise the overall accuracy of multi-target geo-location.

Target geo-location accuracy and missile hit accuracy are usually evaluated through the circular error probability (CEP) [41]. CEP is defined as the radius of a circle with the target point as its center and with the hit probability of 50%. In ECEF frame, the geo-location error along X direction is x. The geo-location error along Y direction is y. x and y can be calculated from longitude error and latitude error listed in Tables 4 and 5. Suppose both x and y are subject to the normal distribution, the joint probability density function of (x, y) can be expressed as:

$$f(x,y) = \frac{1}{2\pi\sigma_x\sigma_y\sqrt{1-\rho^2}} \times \exp\left\{-\frac{1}{2(1-\rho^2)}\left[\frac{(x-\mu_x)^2}{\sigma_x^2} - \frac{2\rho\times(x-\mu_x)\times(y-\mu_y)}{\sigma_x\sigma_y} + \frac{(y-\mu_y)^2}{\sigma_y^2}\right]\right\} \tag{42}$$

where μ_x and μ_y are the mean geo-location errors along X and Y directions, respectively; σ_x and σ_y are the standard deviations of the geo-location errors along X and Y directions, respectively; ρ is the correlation coefficient of geo-location errors along X and Y directions, $0 \le |\rho| \le 1$.

Suppose $x = rcos\theta$, $y = rsin\theta$ and $\mathrm{r} = \sqrt{x^2 + y^2}$, the R satisfying the following equation will be CEP:

$$\frac{1}{2\pi\sigma_x\sigma_y\sqrt{1-\rho^2}}\int_0^R\int_0^{2\pi} r\exp\left\{-\frac{1}{2(1-\rho^2)}\left[\frac{(rc_\theta-\mu_x)^2}{\sigma_x^2} - \frac{2\rho(rc_\theta-\mu_x)(rs_\theta-\mu_y)}{\sigma_x\sigma_y} + \frac{(rs_\theta-\mu_y)^2}{\sigma_y^2}\right]\right\}drd\theta = 0.5 \tag{43}$$

where: $c_\theta = \cos\theta$, $s_\theta = \sin\theta$.

If the mean geo-location errors μ_x, μ_y are unknown, they can be substituted by the sample mean geo-location errors $\hat{\mu}_x$, $\hat{\mu}_y$ respectively. If the standard deviations of the geo-location errors σ_x, σ_y are unknown, they can be substituted by the sample standard deviation of the geo-location errors $\hat{\sigma}_x$, $\hat{\sigma}_y$ respectively. If the correlation coefficients of the geo-location errors ρ are unknown, they can be substituted by the sample correlation coefficients of the geo-location errors $\hat{\rho}$. If the total mean geo-location error μ_r is unknown, it can be substituted by the total sample mean geo-location error $\hat{\mu}_r$. If the total standard deviation of the geo-location error σ_r is unknown, it can be substituted by the total sample standard deviation of the geo-location errors $\hat{\sigma}_r$.

Suppose the number of geo-location error samples is n: (x_1, y_1), (x_2, y_2), (x_3, y_3), ...,(x_n, y_n). The sample mean geo-location errors along X and Y directions are:

$$\hat{\mu}_x = \frac{1}{n}\sum_{i=1}^{n} x_i \tag{44}$$

$$\hat{\mu}_y = \frac{1}{n}\sum_{i=1}^{n} y_i \tag{45}$$

The total sample mean geo-location error is:

$$\hat{\mu}_r = \frac{1}{n}\sum_{i=1}^{n} r_i \tag{46}$$

The sample standard deviations of the geo-location errors along X and Y directions are:

$$\hat{\sigma}_x = \sqrt{\frac{1}{n-1}\sum_{i=1}^{n} (x_i - \hat{\mu}_x)^2} \tag{47}$$

$$\hat{\sigma}_y = \sqrt{\frac{1}{n-1}\sum_{i=1}^{n} (y_i - \hat{\mu}_y)^2} \tag{48}$$

The total sample standard deviations of the geo-location errors is:

$$\hat{\sigma}_r = \sqrt{\frac{1}{n-1}\sum_{i=1}^{n} (r_i - \hat{\mu}_r)^2} \tag{49}$$

The sample correlation coefficient of the geo-location errors is:

$$\hat{\rho} = \frac{\sum_{i=1}^{n} [(x_i - \hat{\mu}_x) \times (y_i - \hat{\mu}_y)]}{\sqrt{\sum_{i=1}^{n} (x_i - \hat{\mu}_x) \times \sum_{i=1}^{n} (y_i - \hat{\mu}_y)}} \tag{50}$$

When the number of geo-location error samples is more than 30, the confidence of CEP calculation result can reach 90% [41]. Therefore, through the multi-target geo-location and distortion correction test for five aerial images, 32 samples of target geo-location errors are obtained before and after the distortion correction, respectively, including eight in the 1st image, six in the 2nd image, five in the 3rd image, five in the 4th image, and eight in the 5th image. The normality test and independence test of sample data reveal that, the samples conform to normal distribution but are not independent (the sample correlation coefficient before the distortion correction is $\hat{\rho}_1 = 0.6618$, and the sample correlation coefficient after the distortion correction is $\hat{\rho}_2 = 0.5607$).

Geo-location errors of the eight targets in the 1st image (which are parts of 32 targets) before the correction are shown in Table 4. Geo-location errors of the eight targets in the 1st image (which are parts of 32 targets) after the correction are shown in Table 5.

The multi-target geo-location errors before the distortion correction are processed through the Equations (43)–(50) to calculate the CEP, whereas the mean geo-location errors along the X and Y directions are 17.41 m and 20.34 m, respectively, the standard deviations of the geo-location errors along the X and Y directions are 7.77 m and 10.05 m, respectively, and the total mean geo-location error and total standard deviation are 28.98 m and 6.08 m, respectively. The calculation of Equation (46) through numerical integration finds that, among the 32 samples, 16 are in a circle with a radius of 28.74 m, which means the probability is 50%. The sizes of data samples inside and outside the solid circle in Figure 10 also show that, the CEP1 of multi-target geo-location before the distortion correction is 28.74 m. Then the multi-target geo-location errors after the distortion correction are processed through Equations (43)–(50) to calculate the CEP, where the mean geo-location errors along X and Y directions are 17.04 m and 18.63 m, respectively, the standard deviations of the geo-location errors along X and Y directions are 6.86 m and 8.25 m, respectively, and the total mean geo-location error and total standard deviation are 26.91 m and 5.31 m, respectively. The calculation of Equation (43) through numerical integration finds that, among the 32 samples, 16 are in a circle with a radius of 26.80 m, which means the probability is 50%. The sizes of samples inside and outside the dotted circle in Figure 10 also show that, the CEP2 of multi-target geo-location after the distortion correction is 26.80 m, 7% smaller than that before the distortion correction. Note: We use original geo-location error (32 samples scattered in the four quadrants) to calculate the CEP circle. Because there are 32 samples scattered in the four quadrants. It's too scattered and not conducive to the analysis of the error. Therefore, we take the absolute value of each geo-location error, so all points are in the first quadrant in Figure 12.

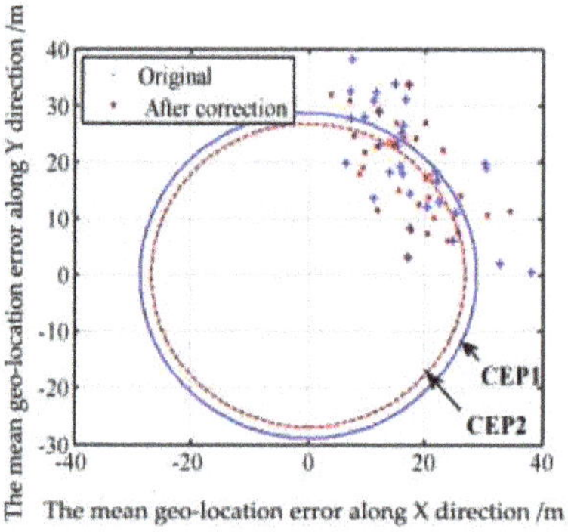

Figure 12. Sample distribution of target location error before and after distortion correction.

In order to compare the performance of our multi-target geo-location algorithm with that of other algorithms, these localization accuracy results have been compared with the accuracies of geo-location methods reported in [7–9], as shown in the Table 6. It can be seen from the Table 6 that, the target geo-location accuracy in this paper is close to that reported in [7]. However, the geo-location accuracy in [7] depends on the distance between the projection centers of two consecutive images (namely baseline length). To obtain higher geo-location accuracy, the baseline length shall be longer, so in [7], the time interval between two consecutive images used for geo-location is quite big. Meanwhile, as the SIFT algorithm is needed to extract feature points from multi-frame images for the purpose of 3D reconstruction, the algorithm in [7] has a heavy calculation load in prejudice of real-time implementation. In [8], the geo-location accuracy is about 20 m before filtering. The real-time geo-location accuracy of the two methods is almost the same. However, our UAV flight altitude is much higher than that in [8]. In [9], geo-location accuracy is about 39.1 m before compensation. Our real-time geo-location accuracy is higher than in [9]. In [9], after the UAV flies many around in an orbit, the geo-location accuracy can be increased to 8.58 m. However, this accuracy cannot be obtained in real-time. The algorithms in [7–9] are implemented on a computer in the ground station. Since the images are transmitted to, and processed on a computer in the ground station, a delay occurs between the data capture moment and the time of completion of processing. In comparison, the geo-location algorithm in this paper is programmed and implemented on multi-target localization circuit board (model: THX-IMAGE-PROC-02, Changchun Institute of Optics, Fine Mechanics and Physics, Chinese

Academy of Sciences, Changchun, China) with a TMS320DM642@ 720-M Hz Clock Rate and 32-Bit Instructions/Cycle and 1 GB DDR. It consumes 0.4 ms on average when calculating the geo-location of a single target and, at the same time, correcting zoom lens distortion. Therefore, this algorithm has great advantages in both geo-location accuracy and real-time performance. The multi-target location method in this paper can be widely applied in many areas such as UAVs and robots.

Table 6. Geo-location accuracy comparison between the proposed algorithm and the algorithms in reference [7].

Algorithm	Error Mean/m	Error Standard/m	Flight Altitude/m
Reference [7]	26.00 [1]	8.00 [1]	over 1000 m [1]
Reference [8]	20.00 [2]	-	100 m–200 m [2]
Reference [9]	39.1 [3]	-	300 m [3]
Proposed (before correction)	28.98	6.08	1140
Proposed (after correction)	26.91	5.31	1140

[1] Data from Reference [7]; [2] Data from Reference [8]; [3] Data from Reference [9].

5.5. Test 3: RLS Filter for Geo-Location Data of Multiple Stationary Ground Targets

The targets in the above tests are all stationary ground targets. After lens distortion correction, we use the 1st aerial images (eight targets) as the initial frame for target tracking. For 150 frames starting from the 1st image, eight targets are tracked, respectively. The coordinates of the UAV are calculated using Equations (25) and (26). The update rate of UAV coordinate is synchronized with the camera frame rate. The geo-location data of each target are adaptively estimated by RLS algorithm respectively. The geo-location results of eight targets before and after RLS filtration are shown in Figure 13 (the dots "•" in different colors in Figure 13 represent original geo-location data of the eight targets, while □, ◁, ▷, ☆, ▽, ○, ◇ and △ represent the geo-location data of main target and sub-targets 1, 2, ..., 7 after RLS filtration). It can be observed from Figure 13 that, after RLS filtration, the dispersion of geo-location data decreases sharply for each target, converging rapidly to a small area adjacent to the true position of the target. Figure 14 shows how the plane geo-location error of sub-target 2 changes with the number of image frames (the corresponding time) in the RLS filtration process. It can be seen from Figure 14 that after filtering the geo-location data of 100~150 images (the corresponding time is 3~5 s), the geo-location errors of the target have converged to a stable value. So, we can obtain a more accurate stationary target location immediately after 150 images (it is no longer necessary to run RLS).

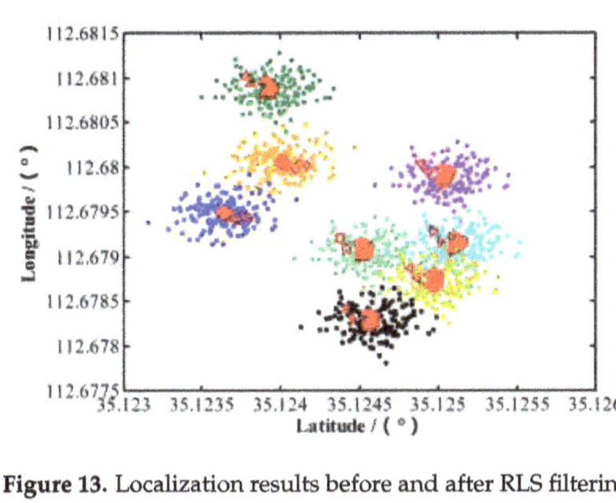

Figure 13. Localization results before and after RLS filtering.

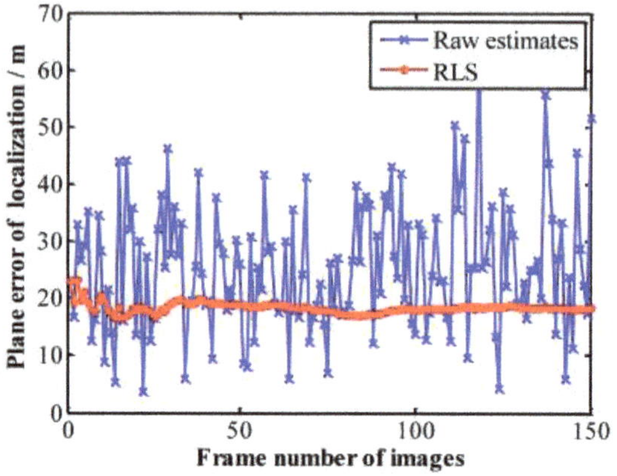

Figure 14. Plane localization errors after RLS filtering.

By comparing the results after the stabilization of RLS filtration with the nominal geodetic coordinates of each target, the geo-location errors of geodetic coordinates of the targets after RLS filtration can be determined, as shown in Table 7.

Table 7. Errors of multi-target location after RLS filtering.

	Longitude Error	Latitude Error	Altitude Error
Main target	0.000167°	0.000016°	15.33 m
Sub-target 1	−0.000005°	0.000131°	15.33 m
Sub-target 2	−0.000193°	−0.000086°	16.08 m
Sub-target 3	0.000150°	0.000151°	15.33 m
Sub-target 4	0.000073°	0.000217°	15.33 m
Sub-target 5	−0.000164°	−0.000230°	14.58 m
Sub-target 6	−0.000361°	−0.000007°	16.08 m
Sub-target 7	−0.000065°	0.000255°	15.33 m

It can be seen from the comparison of results between the data in Table 7 and those in Table 5 that, after RLS filtration, the longitude and latitude errors of each target are much smaller than the multi-target geo-location errors of a single image which is only processed by lens distortion correction. The geo-location error of geodetic height also decreases slightly.

After lens distortion correction, we use the other four aerial images (24 targets) as the initial frame for targets tracking, respectively (eight targets in the 1st image, six targets in the 2nd image, five targets in the 3rd image, five targets in the 4th image, and eight targets in the 5th image).

For 150 frames starting from the 2nd image, six targets are tracked, respectively. The geo-location data of each target are adaptively estimated by the RLS algorithm, respectively. For 150 frames starting from the 3rd image, five targets are tracked, respectively. The geo-location data of each target are adaptively estimated by the RLS algorithm, respectively. For 150 frames starting from the 4th image, five targets are tracked, respectively. The geo-location data of each target are adaptively estimated by the RLS algorithm, respectively. For 150 frames starting from the 5th image, eight targets are tracked, respectively. The geo-location data of each target are adaptively estimated by the RLS algorithm, respectively.

The multi-target geo-location data after RLS filtration are processed through the Equations (43)–(50) to calculate the CEP, where the mean geo-location errors along the X and Y directions are 13.78 m and 14.53 m, respectively, and the standard deviations of the geo-location errors along the X and Y directions are 5.79 m and 7.34 m, respectively. Among the 32 samples obtained through numerical integration of Equation (43), 17 are in a circle with a radius of 21.52 m, which means the probability is 53%. The sizes of samples inside and outside the dotted circle in Figure 13 also show that, the CEP3 of multi-target geo-location after RLS filtration is 21.52 m, 25% smaller than the CEP1 of multi-target geo-location of a single image. Note: We use original geo-location error (32 samples scattered in the four quadrants) to calculate the CEP circle. Because there are 32 samples scattered in the four quadrants. It's too scattered and not conducive to the analysis of the error. Therefore, we take the absolute value of each geo-location error, so all points are in the first quadrant in Figure 15.

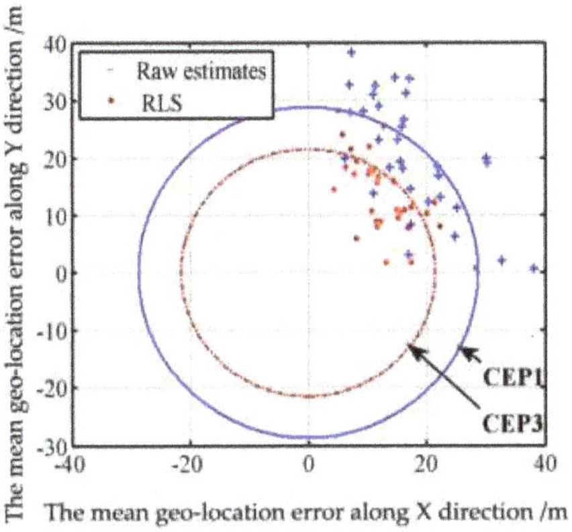

Figure 15. Sample distribution of target location before and after RLS.

5.6. Test 4: Real-Time Geo-Location and Tracking of Multiple Moving Ground Targets

This test localizes and tracks four targets moving on a Changji highway in the video taken by the UAV, as shown in Figure 16 (part of the image). The size of every image is 1024 pixels × 768 pixels, the pixel size is 5.5 μm × 5.5 μm, and the focal length $f = 73.6$ mm. The target in the image center is chosen as main target, and three targets in other positions are chosen as sub-targets. The pixel coordinates of each target in the 1st image and the locations and attitudes data of the electro-optical stabilized imaging system at the corresponding time are shown in Table 8.

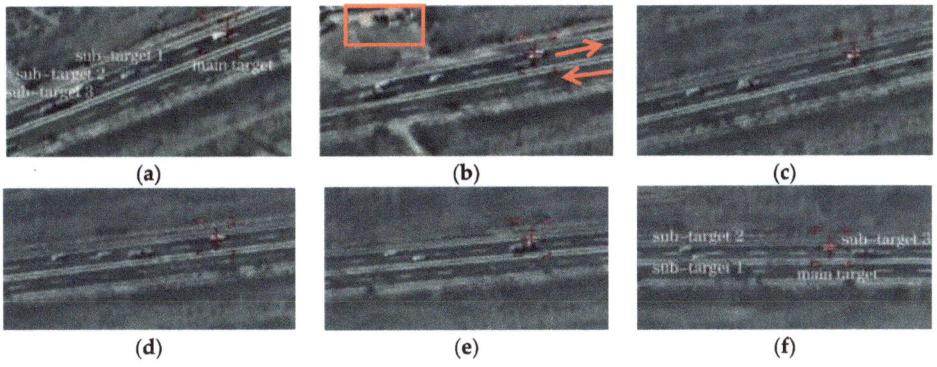

Figure 16. Multiple moving targets in aerial video. (**a**–**f**): frame1, frame 3, frame 4, frame 6, frame 7, frame 9.

Table 8. Locations and attitudes data of electro-optical stabilized imaging system.

Designation	Symbol	Unit	Value
UAV longitude	L_0	°	120.906624
UAV latitude	M_0	°	42.608521
UAV altitude	H_0	m	2505
UAV pitch angle	ε	°	2.0
UAV roll angle	γ	°	−1.8
UAV yaw angle	β	°	350.2
Electro-optical stabilized imaging system's azimuth angle	Θ	°	−117.8
Electro-optical stabilized imaging system's elevation angle	Ψ	°	−46.7
Laser range finder	$\lambda 1$	m	3240
Coordinates of sub-target1	(u,v)	pixel	(453, 342)
Coordinates of sub-target2	(u,v)	pixel	(476, 251)
Coordinates of sub-target3	(u,v)	pixel	(504, 213)
Camera's focal length	f	mm	73.6

Nine chronological images are selected from video images to calculate the geo-location data of each target in every image. The resultant spatial position distribution of all the targets is shown in Figure 17a. With the main target position in the 1st image as the origin of coordinates, the spatial positions of all the targets are projected earthwards to obtain their planar motion trails, as shown in Figure 17b. The time span of those nine frames is 30 s.

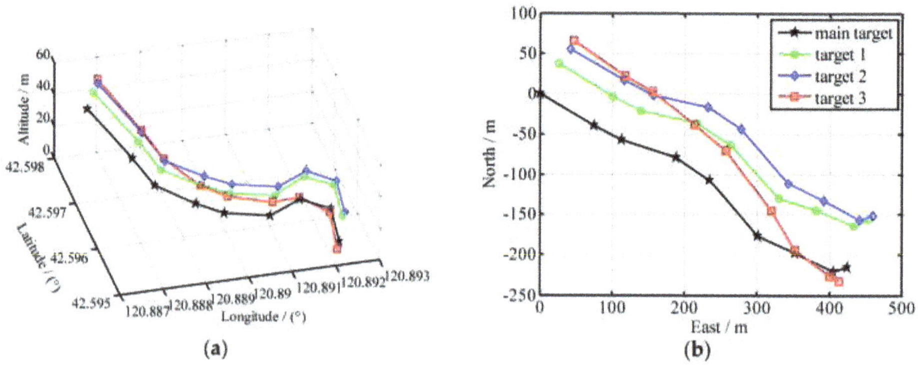

Figure 17. (**a**) The spatial localization of each target; (**b**) The target motion trajectory on the ground.

Since both the UAV yaw and the electro-optical stabilized imaging system azimuth are not 0 in general, aerial images have been rotated and distorted somewhat. The aerial orthographic projection of the abovementioned highway, as shown in Figure 18, is acquired from Google Maps. It can be known from Figure 18 that, the actual direction of that highway is northwest–southeast. In China, cars drive on the right side of the road, so the house in Figure 18 is near the highway which is from northwest to southeast direction. In Figure 16b, we can see this house, so all the cars are on the road travelling in a northwest to southeast direction. This coincides with the localization results of Figure 17.

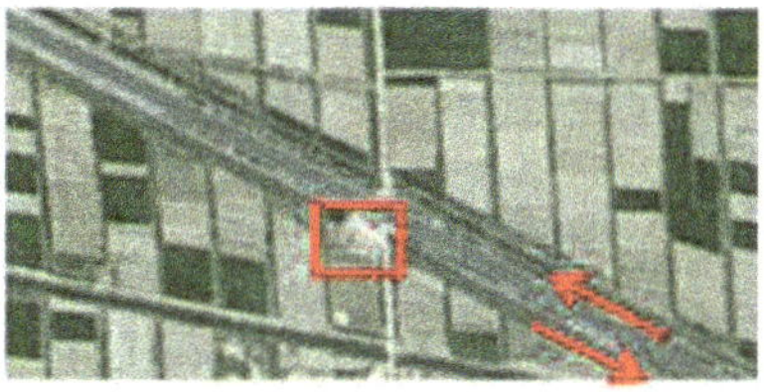

Figure 18. Ortho image of the Changji highway in aerial imagery.

On the basis of Figure 17, the motion of each target has been analyzed as below: the targets in the 1st image, according to their positions from front to back, are sub-target 3, sub-target 2, sub-target 1 and main target in succession; in the 3rd–4th images, the sub-target 2 begins to catch up with and overtake the sub-target 3; in the 4th–6th images, the sub-target 1 begins to catch up with and overtake the sub-target 3; in the 7th–9th images, the main target begins to catch up with and overtake the sub-target 3; at last, all the targets, according to their positions from front to back, are sub-target 2, sub-target 1, main target and sub-target 3 in succession. This coincides completely with the motion law of all the targets in the video image shown in Figure 14, demonstrating that this geo-location algorithm can correctly locate and track multiple moving targets. The speed of each target can also be determined. This test has further verified the correctness of our multi-target geo-location model.

6. Conclusions

In order to improve the reconnaissance efficiency of UAV electro-optical stabilized imaging systems, a multi-target localization system based on a UAV electro-optical stabilized imaging system is proposed. First, a target location model and the way to improve the accuracy of multi-target localization are studied. Then, the geodetic coordinates of multiple targets are calculated using homogeneous coordinate transformation. On the basis of this, two methods which can improve the precision of the target localization are proposed: (1) the lens distortion correction method based on the distortion ratio; (2) the RLS filtering method based on UAV dead reckoning. The localization error model is established using Monte Carlo theory. The analysis of the multiple target location algorithm is carried out. The range of the localization error is obtained. In actual flight, the UAV flight altitude is 1140 m. The multi-target localization results are in the range of allowable error. After we use lens distortion correction method in a single image, CEP of the multi-target localization is reduced by 7%. The RLS algorithm can adaptively estimate the location data based on multi-frame images. Compared with multi-target localization based on the single frame image, CEP of the multi-target localization using RLS is reduced by 25%.

The average time to calculate the location data and distortion correction for a single target is 0.4 ms. The normal video rate is 25 fps, so the proposed localization algorithm can locate the 50 targets at the same time in real time. The proposed method significantly reduced the image data processing time, it is convenient to implement the multi-target localization by using other embedded system [42].

However, when a target is out of field of view and re-enters, the operator has to identify the target again. The future research will aim to address this problem. We will try to apply the tracking-learning-detection (TLD) [43] for automatic target detection when a target re-enters the field of view. Due to the difficulty of constructing TLD on the TMS320DM642 (Texas Instruments Incorporated, Dallas, TX, USA), we will leave all these problems for our future research.

Acknowledgments: We acknowledge the financial support that was given under the Project supported by the National Defense Pre-Research Foundation of China (Grant No. 402040203). We acknowledge Academic Editor for his careful revision of the languages and grammatical structures in this article.

Author Contributions: Xuan Wang, Qianfei Zhou and Jinghong Liu initiated the research and designed the experiments. Xuan Wang and Qianfei Zhou wrote the paper.

Conflicts of Interest: The authors declare no conflict of interest.

References

1. Deming, R.W.; Perlovsky, L.I. Concurrent multi-target localization, data association, and navigation for a swarm of flying sensors. *Inf. Fusion* **2007**, *8*, 316–330. [CrossRef]
2. Minaeian, S.; Liu, J.; Son, Y.-J. Vision-based target detection and localization via a team of cooperative UAV and UGVs. *IEEE Trans. Syst. Man Cybern. Syst.* **2016**, *46*, 1005–1016. [CrossRef]
3. Morbidi, F.; Mariottini, G.L. Active target tracking and cooperative localization for teams of aerial vehicles. *IEEE Trans. Control Syst. Technol.* **2013**, *21*, 1694–1707. [CrossRef]
4. Qu, Y.; Wu, J.; Zhang, Y. Cooperative Localization Based on the Azimuth Angles among Multiple UAVs. In Proceedings of the 2013 International Conference on Unmanned Aircraft Systems (ICUAS), Atlanta, GA, USA, 28–31 May 2013; pp. 818–823.
5. Kwon, H.; Pack, D.J. A robust mobile target localization method for cooperative unmanned aerial vehicles using sensor fusion qualitye. *J. Intell. Robot. Syst. Theory Appl.* **2012**, *65*, 479–493. [CrossRef]
6. Yan, M.; Du, P.; Wang, H.L.; Gao, X.J.; Zhang, Z.; Liu, D. Ground multi-target positioning algorithm for airborne optoelectronic system. *J. Appl. Opt.* **2012**, *33*, 717–720.
7. Han, K.; de Souza, G.N. Multiple Targets Geo-Location Using SIFT and Stereo Vision on Airborne Video Sequences. In Proceedings of the 2009 IEEE RSJ International Conference on Intelligent Robots and Systems, St. Louis, MO, USA, 10–15 October 2009; pp. 5327–5332.
8. Barber, D.B.; Redding, J.; McLain, T.W.; Beard, R.W.; Taylor, C. Vision-based target geo-location using a fixed-wing miniature air vehicle. *J. Intell. Robot. Syst.* **2006**, *47*, 361–382. [CrossRef]
9. Subong, S.; Bhoram, L.; Jihoon, K. Vision-based real-time target localization for single-antenna GPS-guided UAV. *IEEE Trans. Aerosp. Electron. Syst.* **2008**, *44*, 1391–1401.
10. Dobrokhodov, V.N.; Kaminer, I.I.; Jones, K.D.; Ghabcheloo, R. Vision-Based Tracking and Motion Estimation for Moving Targets Using Small UAVS. In Proceedings of the 2006 American Control Conference, Minneapolis, MN, USA, 14–16 June 2006.
11. Yap, K.C. Incorporating Target Mensuration System for Target Motion Estimation Along a Road Using Asynchronous Filter. Master's Thesis, Naval Postgraduate School, Monterey, CA, USA, December 2006.
12. Kumar, R.; Sawhney, H.; Asmuth, J.; Pope, A.; Hsu, S. Registration of Video to Geo-Referenced Imagery. In Proceedings of the Fourteenth International Conference on Pattern Recognition, Queensland, Australia, 20 August 1998; pp. 1393–1400.
13. Huang, D.; Wu, Z. The Application of TMS320c64x DSP Assembly Language in Correlation Tracking Algorithms. In Proceedings of the 2010 3rd International Congress on Image and Signal, Yantai, China, 16–18 October 2010; Volume 4, pp. 1529–1532.
14. Zhang, Y.; Tong, X.; Yang, T.; Ma, W. Multi-model estimation based moving object detection for aerial video. *Sensors* **2015**, *15*, 8214–8231. [CrossRef] [PubMed]
15. Wang, Z.; Xu, J.; Huang, Z.; Zhang, X.; Xia, X.-G.; Long, T.; Bao, Q. Road-aided ground slowly moving target 2D motion estimation for single-channel synthetic aperture radar. *Sensors* **2016**, *16*. [CrossRef] [PubMed]
16. Danescu, R.; Oniga, F.; Turcu, V.; Cristea, O. Long baseline stereovision for automatic detection and ranging of moving objects in the night sky. *Sensors* **2012**, *12*, 12940–12963. [CrossRef] [PubMed]
17. Steen, K.A.; Villa-Henriksen, A.; Therkildsen, O.R.; Green, O. Automatic detection of animals in mowing operations using thermal cameras. *Sensors* **2012**, *12*, 7587–7597. [CrossRef] [PubMed]
18. Rodriguez-Gomez, R.; Fernandez-Sanchez, E.J.; Diaz, J.; Ros, E. FPGA Implementation for Real-Time Background Subtraction Based on Horprasert Model. *Sensors* **2012**, *12*, 585–611. [CrossRef] [PubMed]
19. Bowring, B.R. Transformation from spatial to geographical coordinates. *Surv. Rev.* **1976**, *23*, 323–327. [CrossRef]
20. Li, Y.; Yun, J.; Jun, W. Building of rigorous geometric processing model based on line-of-sight vector of ZY-3 imagery. *Geomat. Inf. Sci. Wuhan Univ.* **2013**, *38*, 1451–1455.
21. Hughes, P.C. *Spacecraft Attitude Dynamics*; Courier Corporation: North Chelmsford, MA, USA, 2004.
22. Fu, C.; Duan, R.; Kircali, D.; Kayacan, E. Onboard robust visual tracking for UAVs using a reliable global-local object model. *Sensors* **2016**, *16*. [CrossRef] [PubMed]

23. Liu, Z.; Wang, Z.; Xu, M. Cubature information SMC-PHD for multi-target tracking. *Sensors* **2016**, *16*. [CrossRef] [PubMed]
24. Perlovsky, L.I.; Deming, R.W. Maximum likelihood joint tracking and association in strong clutter. *Int. J. Adv. Robot. Syst.* **2013**, *10*, 1–9.
25. Tian, W.; Wang, Y. Analytic performance prediction of track-to-track association with biased data in multi-sensor multi-target tracking scenarios. *Sensors* **2013**, *13*, 12244–12265. [CrossRef] [PubMed]
26. Ghirmai, T. Distributed particle filter for target tracking: With reduced sensor communications. *Sensors* **2016**, *16*. [CrossRef] [PubMed]
27. Kyristsis, S.; Antonopoulos, A.; Chanialakis, T.; Stefanakis, E.; Linardos, C.; Tripolitsiotis, A.; Partsinevelos, P. Towards autonomous modular UAV missions: The detection, geo-location and landing paradigm. *Sensors* **2016**, *16*. [CrossRef] [PubMed]
28. Enayet, A.; Razzaque, M.A.; Hassan, M.M.; Almogren, A.; Alamri, A. Moving target tracking through distributed clustering in directional sensor networks. *Sensors* **2014**, *14*, 24381–24407. [CrossRef] [PubMed]
29. Boudriga, N.; Hamdi, M.; Iyengar, S. Coverage assessment and target tracking in 3D domains. *Sensors* **2011**, *11*, 9904–9927. [CrossRef] [PubMed]
30. Dellen, B.; Erdal Aksoy, E.; Wörgötter, F. Segment tracking via a spatiotemporal linking process including feedback stabilization in an n-D lattice model. *Sensors* **2009**, *9*, 9355–9379. [CrossRef] [PubMed]
31. Sun, B.; Jiang, C.; Li, M. Fuzzy neural network-based interacting multiple model for multi-node target tracking algorithm. *Sensors* **2016**, *16*. [CrossRef] [PubMed]
32. Drap, P.; Lefèvre, J. An exact formula for calculating inverse radial lens distortions. *Sensors* **2016**, *16*. [CrossRef] [PubMed]
33. Wackrow, R.; Ferreira, E.; Chandler, J.; Shiono, K. Camera calibration for water-biota research: The projected area of vegetation. *Sensors* **2015**, *15*, 30261–30269. [CrossRef] [PubMed]
34. Sanz-Ablanedo, E.; Rodríguez-Pérez, J.R.; Armesto, J.; Taboada, M.F.Á. Geometric stability and lens decentering in compact digital cameras. *Sensors* **2010**, *10*, 1553–1572. [CrossRef] [PubMed]
35. Bosch, J.; Gracias, N.; Ridao, P.; Ribas, D. Omnidirectional underwater camera design and calibration. *Sensors* **2015**, *15*, 6033–6065. [CrossRef] [PubMed]
36. Miklavcic, S.; Cai, J. Automatic curve selection for lens distortion correction using Hough transform energy. *IEEE Workshop Appl. Comput. Vis.* **2013**, *26*, 455–460.
37. Goljan, M.; Fridrich, J. Estimation of lens distortion correction from single images. *Proc. SPIE* **2014**, *9028*. [CrossRef]
38. Lee, T.Y.; Chang, T.S.; Wei, C.H.; Lai, S.H.; Liu, K.C.; Wu, H.S. Automatic distortion correction of endoscopic images captured with wide-angle zoom lens. *IEEE Trans. Biomed. Eng.* **2013**, *60*, 2603–2613. [PubMed]
39. Liu, J.; Sun, H.; Zhang, B.; Dai, M.; Jia, P.; Shen, H.; Zhang, L. Target self-determination orientation based on aerial photoelectric imaging platform. *Opt. Precis. Eng.* **2007**, *15*, 1305–1309.
40. Sheng, W.; Long, Y.; Zhou, Y. Analysis of target location accuracy in space-based optical-sensor network. *Acta Opt. Sin.* **2011**, *31*, 0228001. [CrossRef]
41. Le, Z.; Wuzhou, L.; Yangfeng, J. Localization accuracy evaluation method based on CEP. *Command Control Simul.* **2013**, *35*, 111–114.
42. Kuo, D.; Gordon, D. Real-time orthorectification by FPGA-based hardware acceleration. *Proc. SPIE* **2010**, *7830*. [CrossRef]
43. Kalal, Z.; Mikolajczyk, K. Tracking-Learning-Detection. *IEEE Trans. Pattern Anal. Mach. Intell.* **2012**, *34*, 1409–1422. [CrossRef] [PubMed]

sensors

MDPI

Article

Coordinated Target Tracking via a Hybrid Optimization Approach

Yin Wang [1,2,*] and Yan Cao [2]

1 State Key Laboratory of Virtual Reality Technology and Systems, Beihang University, Beijing 100191, China
2 College of Astronautics, Nanjing University of Aeronautics and Astronautics, Nanjing 210016, China; caoyan626@126.com
* Correspondence: yinwangee@nuaa.edu.cn; Tel.: +86-138-1399-5163

Academic Editors: Felipe Gonzalez Toro and Antonios Tsourdos
Received: 30 December 2016; Accepted: 23 February 2017; Published: 27 February 2017

Abstract: Recent advances in computer science and electronics have greatly expanded the capabilities of unmanned aerial vehicles (UAV) in both defense and civil applications, such as moving ground object tracking. Due to the uncertainties of the application environments and objects' motion, it is difficult to maintain the tracked object always within the sensor coverage area by using a single UAV. Hence, it is necessary to deploy a group of UAVs to improve the robustness of the tracking. This paper investigates the problem of tracking ground moving objects with a group of UAVs using gimbaled sensors under flight dynamic and collision-free constraints. The optimal cooperative tracking path planning problem is solved using an evolutionary optimization technique based on the framework of chemical reaction optimization (CRO). The efficiency of the proposed method was demonstrated through a series of comparative simulations. The results show that the cooperative tracking paths determined by the newly developed method allows for longer sensor coverage time under flight dynamic restrictions and safety conditions.

Keywords: unmanned aerial vehicles; UAV cooperation; persistent tracking; evolutionary algorithm

1. Introduction

Recently, unmanned aerial vehicles (UAV) have been widely used in both defense and civilian applications. In many of these applications, the UAV is required to continuously track a moving target, such as in the surveillance and tracking of ground objects. This task becomes challenging when the target is moving in a complex and dynamic environment, where the target may be occluded by ground features, such as buildings and mountains. In the case where the tracking cannot be achieved by using a single UAV, multiple UAVs may be deployed to improve the tracking robustness. The methods to coordinate the movements of multiple UAVs can be generally grouped into two categories, namely centralized and distributed algorithms, respectively [1]. In centralized methods, the mission of each UAV is dispatched from a central station or agent based on global information collected from the all of the aircraft. These methods generally tend to produce better solutions since the global information is available. In contrast, the distributed approaches determine the motion of each UAV by using the local information obtained from the UAV itself and its neighboring UAVs, which enjoys the advantage of computational efficiency and robustness when compared with their centralized counterparts. In a moving target tracking problem, besides generating the tracking path for each UAV, the motion of the target should be estimated since such information cannot be obtained from prior knowledge. As the purpose of this paper is to determine the optimal path of each UAV for cooperative tracking tasks, the motion pattern of the target is assumed to be available when it can be observed by the tracking UAVs.

The aim of cooperative tracking is to keep the target under the coverage of the UAV sensors at all time. To address this problem, Yang et al. adopted the Lyapunov Vector Field Control (LVC)

approach to make the unmanned aerial vehicles converge to a circular area on top of the target [2]. However, this method does not take environmental factors into account, such as non-fly zones and the occlusion of sensors' field of view. Besides, in their method prior knowledge about the motion of the target is required, which is not available in most cases. In order to cope with this limitation, Belkhouche et al. [3] proposed to apply active avoidance strategies for obstacles in the environment, and determine feasible fly-zones by calculating the speed and position of the UAV relative to the obstacle. Cheng et al. presented a time-related optimal path planning algorithm to determine the cooperative tracking trajectories for each UAV [4]. Quintero and colleagues adopted a dynamic programming technique to minimize the distance error covariance for solving the cooperative tracking task, however the relative angle between UAVs and the target was ignored [5]. Chen et al., presented a hierarchical method, combining the Lyapunov Guidance Vector Field (LGVF) and Tangent Guidance Vector Field (TGVF) approaches, to solve the path planning problem for coordinated tracking [6]. In this method, the TGVF algorithm is firstly utilized to determine an optimal path from the initial point to a limiting circle, and then UAV would converge to the desired limit circle by optimizing the heading command through the use of the LGVF approach.

As discussed previously, due to the complexity and uncertainty of the ground environment, the target may be out of the sight of the UAV sensors, thus the motion of the target should be estimated. Shaferman et al. considered the sensor field-of-view (FOV) convergence in the urban environment, and determined an optimal path by maximizing the sensor convergence time to the target [7]. In order to improve the tracking performance of unmanned aerial vehicles (UAVs) under unknown conditions. Yu and his colleagues proposed an online path planning algorithm by estimating the location of the target based on its past states using a Bayesian filter based method [8]. Zhang et al., applied the Model Predictive Control (MPC) concept to solve the cooperative path planning problem with unknown target motions [9]. Beard and his co-workers present a decentralized cooperative aerial surveillance method by solving the associated combinational optimization problem, where both flight dynamics and environmental constraints are taken into account [10]. He and Xu proposed a model predictive control-based solution for the cooperative tracking path planning problem by considering urban area occlusions, however, the secure distances between UAVs were not taken into account [11].

Evolutionary algorithms, such as genetic algorithm approaches, also applied to solve the cooperative path planning problem [12–14], achieve good results in many offline path planning applications. Wise et al., compared commonly used methods for cooperative tracking path planning problems and in their work [15], readers can find a comprehensive review of the relevant literature.

This work present a novel cooperative tracking path planning approach for the visual tracking tasks of multiple UAVs, by coupling the receding horizon control principle and chemical reaction optimization framework [16]. The proposed method does not require any prior information about the motion and trajectory of the targets, and is capable of dealing with the occlusion of the sensors' field of view. Followed by this Introduction, the kinematics model of UAV and sensor visible regions model are described in Section 2. Section 3 presents the proposed method in great detail, and we demonstrate the performance of the proposed method through a series of comparative simulations in the following section. Finally, we conclude this work and discuss possible future work directions in Section 5.

2. Problem Description

2.1. Kinematics Model of UAV

Assuming the UAVs are flying at a constant altitude with a fixed speed, the motion of the UAV can be reduced to a 2-dimensional Dubins model. Figure 1 shows the 2D planar view of the Dubins coordinate system, where X_I, Y_I are the Cartesian inertial reference frame. v_a represents the velocity of the aircraft and ψ denotes the heading of the UAV.

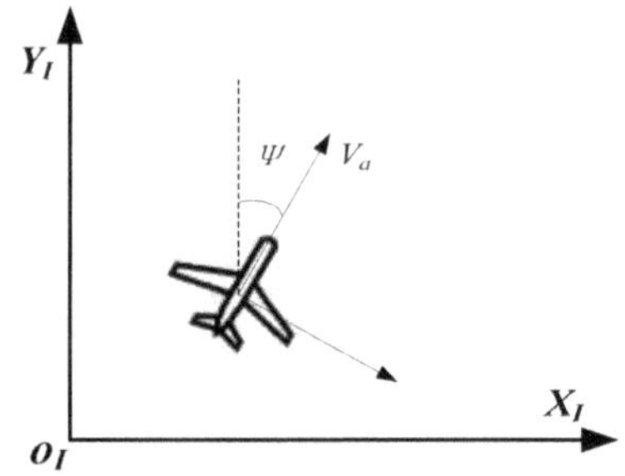

Figure 1. Schematic diagram illustrates the coordinate of the Dubins model.

In this paper, the angle of attack and the angle of side slip are neglected, then the kinematics equations of each aircraft defined in the inertial frame can be represented as:

$$\begin{aligned} \dot{x}_i(t) &= v_a \cos(\psi_i(t)) \\ \dot{y}_i(t) &= v_a \sin(\psi_i(t)) \\ \dot{\psi}_i(t) &= \tfrac{g \tan \phi_i(t)}{v_g}, \quad |\phi_i(t)| \leq \phi_{\max} \end{aligned} \tag{1}$$

where $x_i(t)$ and $y_i(t)$ represent the position of the i-th UAV at time instant t. $\phi_i(t)$ denotes the roll angle of the UAV, which is bounded to the range within $-\phi_{\max}$ and $\phi_{\max}$. Assuming that each UAV is equipped with an autopilot so that it is able to follow the given roll angle, let T_s denote the sampling rate and we apply zero-order hold to the input signal, then the discrete kinematics of the UAV can be found as:

$$\begin{aligned} \dot{\psi}_i(k) &= \tfrac{g \tan \phi_i(k)}{v_g} \\ x_i(k+1) &= \tfrac{v_g}{\dot{\psi}_i(k)} \left[\sin(\dot{\psi}_i(k) T_s + \psi_i(k)) - \sin(\psi_i(k)) \right] + x_i(k) \\ y_i(k+1) &= \tfrac{v_g}{\dot{\psi}_i(k)} \left[\cos(\psi_i(k)) - \cos(\dot{\psi}_i(k) T_s + \psi_i(k)) \right] + y_i(k) \\ \psi_i(k+1) &= \dot{\psi}_i(k) T_s + \psi_i(k) \end{aligned} \tag{2}$$

If the input roll angle is zero, Equation (2) can be rewritten as:

$$\begin{aligned} x_i(k+1) &= v T_s \cos(\dot{\psi}_i(k)) + x_i(k) \\ y_i(k+1) &= v T_s \sin(\dot{\psi}_i(k)) + y_i(k) \\ \psi_i(k+1) &= \psi_i(k) \end{aligned} \tag{3}$$

2.2. Sensor Visible Regions in Complex Environment

The sensors carried by the UAV can be categorized into to two groups, in terms of the installation conditions, namely gimbaled sensors and body fixed sensors. As shown in Figure 2a, gimbaled installed sensors are capable of pointing to a fixed direction regardless of the attitude of the UAV. The sensor converge area is a cone-shaped region under the aircraft. The pointing orientation of body fixed sensors, on the other hand, are usually affected by the state of the UAV, such as the roll and pitch angles, as shown in Figure 2b. To simplify the problem, in this work UAVs are assumed to be equipped with gimbaled sensors. Hence the sensor convergence on the ground would not be affected by the changes of UAV's attitude.

In many applications, the target to be tracked is moving in a complex environment, which may be occluded by terrain features, such as buildings in an urban environment. Figure 3 illustrates a simulated urban environment, where the 3D rectangular structures represent buildings. Due to the occlusion of buildings, not all of the convergence area of the sensor is visible. The actual visible area depends on the height and the field of view of the sensor. As shown in Figure 3, assuming the sensor is located at (sx, sy, sz), the visible regions on the ground can be determined calculated by using the line of slight (LoS) methods. Rather than using conventional geometrical analysis approaches, such as LoS algorithms, to define the visible areas of the sensors, we employ a fast yet robust spatial analysis method developed in [17], which greatly improves the efficiency of this procedure.

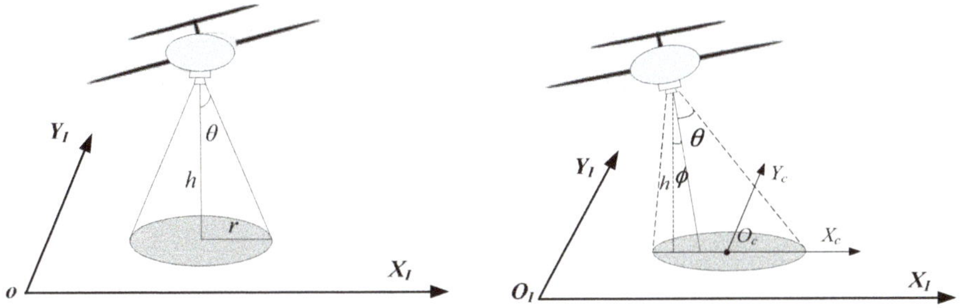

(a) (b)

Figure 2. Illustration of the sensor convergence areas with different types of sensors. (**a**) illustrates the visibility regions of gimbaled sensors when the aircraft is turning; (**b**) depicts the visibility regions of body-fixed sensors when the aircraft is turning.

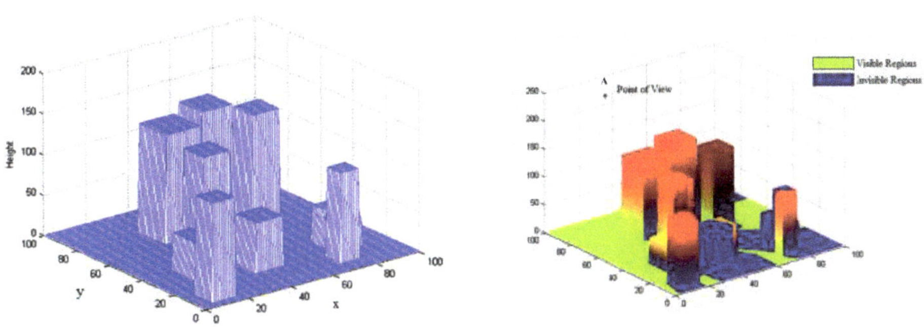

(a) (b)

Figure 3. Illustrates of the sensor visible areas in complex environments. (**a**) The simulated urban environment with buildings (rectangular structures); (**b**) shows the visibility regions of a sensor located at the point **A**.

Figure 3b depicts the visible areas looked from the point **A**, the blue regions are invisible areas due to the occlusion of buildings or terrains. Hence, the detectable regions of the sensor can be defined as:

$$R_{vis}(p) = R_{con}(p) \cap R_{in}(p) \tag{4}$$

where R_{con} is the convergence region of the sensor positioned at p, and R_{in} denotes all of the possible visible areas visible from p. Given the field of view angle of the gimbaled sensor θ, and the height of the sensor h, the convergence region on the ground can be calculated as:

$$R_{con} = \pi r^2 \quad , \quad r = h \tan\theta \tag{5}$$

2.3. The Motion of the Ground Target

The ground object to be tracked is assumed to move within the 2-dimensional plane, and the associated motion can be described by a state vector $x_t = [x_t, y_t, \dot{x}_t, \dot{y}_t, \ddot{x}_t, \ddot{y}_t]$. Let ***F***() represent the state-transition function of the target dynamic, the motion of the target can be described as follows:

$$x_t(k+1) = \boldsymbol{F}[x_t(k)] + \boldsymbol{W}(k) \tag{6}$$

where ***W***() denotes the state noise and is usually assumed to be subject to a zero-mean Gaussian distribution, i.e., p(**W**)~N(0, **Q**), **Q** is the covariance matrix. The possible motion of the target can be predicted by filtering approaches [13].

3. The Proposed Method

In this paper, we present a solution to the problem of tracking by multiple coordinated UAVs based on the concept of the Receding Horizon Control (RHC) method, which minimizes the uncertainties arising from the application environment and the motion of the target. The associated optimal control sequence estimation problem is then solved by using a novel bio-inspired optimization approach, based on the framework of chemical reaction optimization.

3.1. The Receding Horizon Control in Coordinated Tracking

Since the motion of the target is not available as prior knowledge, it requires predicting the trajectory of the target based on previous observation. Receding horizon control (RHC) approach is an advanced model predictive controller allows for determining the optimal solution at current time instant while taking the future states of the target into account. Let $u[k{:}k + L]$ denote the control sequence starts from the time k, and L is a constant indicating the window size of the RHC sequence. $\mathbf{X}[k]$ represents the state of the dynamic system, including the positions of the UAV and the target, respectively. The main steps of RHC method for solving the target tracking problem are as follows:

Step 1: Determine the optimal control sequence based on the state $\mathbf{X}$ $[k]$:

$$u^*[k:k+L] = \text{argmin} J(\boldsymbol{X}[k], u[k:k+L]) \tag{7}$$

where u^* denotes the optimal control sequence minimizing the cost function J. Since the dynamic model of the UAV is obtained from prior knowledge, the trajectory of the UAV can be updated in an iterative fashion based on the input control signal and its current state. Because the motion of the target is unknown, its future positions can only be estimated based on previous observation. The cost function J returns the goodness of the tracking based on the relative position between the target and the UAV, which would be discussed in detail in Section 3.2.

Step 2: Apply the optimal input sequence u^* to the system, and update the states of the system.

Step 3: Repeat Steps (1) and (2) for the next time instant.

3.2. The Cost of the Tracking Path

3.2.1. Dynamic Constraint of the UAV

In practice, the roll angle of the UAV should be within a reasonable scale when changing its orientation, due to safety considerations. Hence, the input roll angle should satisfy the following constraint:

$$\phi \in (-\phi_{\max}, \phi_{\max})$$

where $\phi_{\max}$ is a constant representing the maximal roll angle of the UAV during turning process.

3.2.2. The Target Observation Time

The objective of the coordinate target tracking is to maintain the target located within the detectable region of at least one UAV at any time instant. In addition, it favors the targets located next to the center of the sensor visible area. Let (rx, ry) denote the relative position of the target with respect to the center of the visible area of the sensor:

$$J_s = \sum_{i=1}^{N} \sum_{k=1}^{k+L} sen(\mathbf{X}[k]) \tag{8}$$

$$Sen = \begin{cases} -Ce^{-\frac{rx^2+ry^2}{2\sigma^2}}, & \text{the target is within the visible region of at least one sensor} \\ P, & \text{the target is out of the visible region for all sensors} \end{cases}$$

where the function *Sen*() quantifies whether the target can be observed by the sensors of UAVs based on the relative position between the UAVs and the target, This term reaches its minimum if the target is located within the center of the visible area of all of the sensors, and approaches zero if the target is moving to the boundaries of the sensor visible regions. P is a positive constant used to penalize the case where the target is out of the visible region of all of the sensors.

3.2.3. The Anti-Collision Constraint

In this paper, we assume the UAVs applied to track the target are distributed on the same flight level, thus it is necessary to prevent the UAV from colliding with each other during tracking:

$$J_{AC} = \sum_{i=1}^{N} \sum_{j \in \Omega(i)} erfc[k \cdot (\|\boldsymbol{x}_i - \boldsymbol{x}_j\| - d)] \tag{9}$$

where k is a constant, d indicates the desired distance between UAVs, $\boldsymbol{x}_i$ and $\boldsymbol{x}_j$ represent the positions of i-th and j-th UAVs in the inertia reference frame, respectively. *erfc* denotes the Gaussian error function which is defined as:

$$erfc(x) = \frac{2}{\sqrt{\pi}} \int_x^{\infty} e^{-x^2} dx \tag{10}$$

This term would reach its minimum when the positions of the UAVs are evenly distributed around the target.

3.2.4. The Regulation Term

In order to ensure the planned path is smooth, the change of the input signal is used as the regulator to penalize a jagged path:

$$J_{reg} = \sum_{i=1}^{N} \sum_{j=k+1}^{k+L} (u_i(j) - u_i(j-1)) \tag{11}$$

This term approaches zero when a straight path is generated, and it would be large in magnitude if the resulting path is highly curved. The total cost can be obtained by summing the aforementioned terms together, leading to the following function:

$$J = w_1 J_s + w_2 J_{AC} + w_3 J_{reg} \tag{12}$$

where w_1, w_2 and w_3 are constants controlling the influence of each term on the total energy.

3.3. Chemical Reaction Optimization-Based Coordinated Tracking

The Chemical Reaction Optimization (CRO) algorithm, first reported by Lam and Li [16], is a nature-inspired optimization algorithm based on the principle of chemical reactions. It mimics the process of chemical reactions occuring in a closed container. CRO is a type of population-based algorithm where the basic individuals are molecules. Each molecule is described by some attributes, including the molecular position, i.e., a possible solution of the optimization problem, the potential energy (PE) and the kinetic energy (KE). The molecule structure represents a possible solution of the optimization problem, the potential energy (PE) is associated with the cost of the resulting solution based on the objective function. In general, a small value of the potential energy may indicate a better solution of the optimization problem. Kinetic energy (KE) defines the tolerance of the system for accepting a worse solution than the existing one. In addition to molecules, the container (buffer), where the chemical reaction takes place, is another key component of the CRO algorithm. It is able to supply or absorb the energy resulting during the process of the molecular reactions. Different from many existing population-based optimization algorithms, the total number of molecules may not be the same in each iteration of the optimization process, but the entire amount of energy (the energies of molecules and the energy of the buffer) of the system stays constant throughout the reaction process.

3.3.1. Coding Scheme

As shown in Figure 4, the molecular structure is defined as a sequence of bounded input signals, i.e., the desired roll angle to the autopilot in the time from t_k to t_{k+1} for each UAV. Thus, the dimensions of each molecule would be $N \times L$, where N represents the total number of UAVs employed for coordinated tracking, and L is the number of steps in the predicted ahead current state.

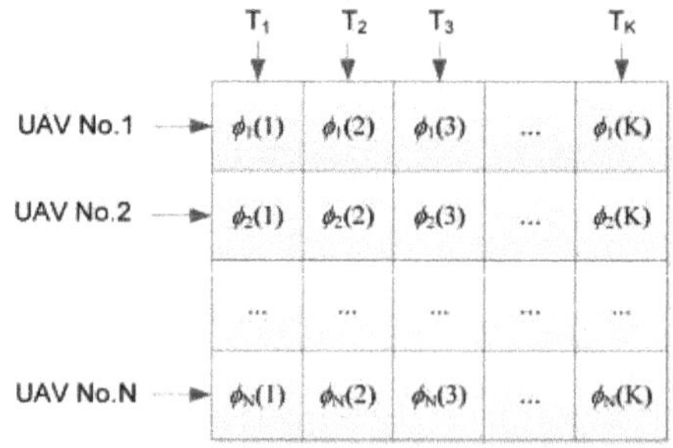

Figure 4. The coding scheme of the CRO algorithm.

3.3.2. On-Wall Ineffective Collision

In this elementary process, the new path is determined by slightly deforming the current path, which can be considered as a local search around the current solution. Let KE_ω and PE_ω represent the kinetic and potential energy of a molecule before it hitting the wall of the container, respectively. The potential energy can be calculated based on Equation (12), and the kinetic energy for each molecule is initialized randomly. Due to the collisions, a certain amount of the energy may transfer to the wall of the container (i.e., the buffer), and thus the potential $PE_{\omega'}$ and kinetic energy $KE_{\omega'}$ of such molecule at the new statue can be computed as:

$$KE_{\omega'} = (PE_\omega + KE_\omega - PE_{\omega'}) \times \alpha, \quad \text{if } (PE_\omega + KE_\omega - PE_{\omega'}) \geq 0 \tag{13}$$

Let *KElossRate* (*KElossRate* ϵ [0, 1]) be a constant defining the percentage of energy that lost before and after the molecule hits the wall, α is a random number evenly distributed from *KElossRate* to 1. Based on the law of conversion of energy, the energy absorbed by the buffer can be expressed as:

$$KE_{\omega'} = (PE_{\omega} + KE_{\omega} - PE_{\omega'}) \times (1 - \alpha) \tag{14}$$

If the kinetic energy of the molecule before wall collision is large enough, the after collision potential energy may larger than PE_{ω}, leading to a worse solution when compared with the existing one. It allows the local search algorithm the ability of 'hill-climbing', which increases the probability of avoiding local optimal solutions.

3.3.3. Decomposition

Unlike the process of the 'on-wall ineffective collision', the molecule breaks into parts after it hits the wall of the container ($\omega \rightarrow \omega_1' + \omega_2'$). For simplicity we only consider two parts. Because more solutions are created through the decomposition process, the sum of KE and PE energies of the original molecule may be less than the sum of potential energies of the resulting molecules:

$$PE_{\omega} + KE_{\omega} < PE_{\omega'_1} + PE_{\omega'_2} \tag{15}$$

In this case, such decomposition is not valid since it disobeys the law of energy conservation. In order to increase the possibility of a correct decomposition process, the buffer may supply a certain amount of the energy to the decomposition process so that the following condition can be satisfied:

$$PE_{\omega} + KE_{\omega} + \delta_1 \times \delta_2 \times buffer \geq PE_{\omega'_1} + PE_{\omega'_2} \tag{16}$$

where δ_1 and δ_2 are two independently random variables uniformly distributed in the range of [0, 1]. The remaining energy is translated as the kinetic energies for two new molecules as:

$$\begin{aligned} PE_{\omega'_1} &= (PE_{\omega} + KE_{\omega} + \delta_1 \times \delta_2 \times buffer - PE_{\omega'_1} - PE_{\omega'_2}) \times \delta_3 \\ PE_{\omega'_2} &= (PE_{\omega} + KE_{\omega} + \delta_1 \times \delta_2 \times buffer - PE_{\omega'_1} - PE_{\omega'_2}) \times (1 - \delta_3) \end{aligned} \tag{17}$$

where δ_3 is another independent random variable range from 0 to 1. The decomposition process provides the molecule the ability of exploring more solution regions with respect to the initial position ω.

3.3.4. Inter-Molecular Ineffective Collision

The reaction process is very similar to its single molecule reaction counterpart, where the collision takes place between molecules. The number of the molecules is assumed to be unchanged before and after the collision, and the molecular energies of the associated reaction process should satisfy the condition:

$$PE_{\omega_1} + KE_{\omega_1} + PE_{\omega_2} + KE_{\omega_2} \geq PE_{\omega'_1} + PE_{\omega'_2} \tag{18}$$

The kinetic energies of the new molecules are:

$$\begin{aligned} KE_{\omega'_1} &= E_{\text{inter}} \times \delta_4 \\ KE_{\omega'_2} &= E_{\text{inter}} \times (1 - \delta_4) \\ E_{\text{inter}} &= PE_{\omega_1} + KE_{\omega_1} + PE_{\omega_2} + KE_{\omega_2} - PE_{\omega'_1} + PE_{\omega'_2} \end{aligned} \tag{19}$$

The inter-molecule collision allows a median range local search, thus increasing the probability of detecting a better solution in a local area.

3.3.5. Synthesis

Synthesis is the opposite reaction process of decomposition, where two (multiple) molecules merge into a single molecule. As the new molecule is produced from multiple molecules, it is more likely to meet the following requirement for such a process:

$$PE_{\omega_1} + KE_{\omega_1} + PE_{\omega_2} + KE_{\omega_2} \geq PE_{\omega'} \quad (20)$$

The remaining energy of the new molecule is:

$$KE_{\omega'} = PE_{\omega_1} + KE_{\omega_1} + PE_{\omega_2} + KE_{\omega_2} - PE_{\omega'} \quad (21)$$

In this reaction process, the new molecule is a combination of multiple old molecules, thus the newly produced molecule may exhibit a large deviation from the original ones in the solution space.

The workflow of the CRO based optimization framework is shown in Figure 5.

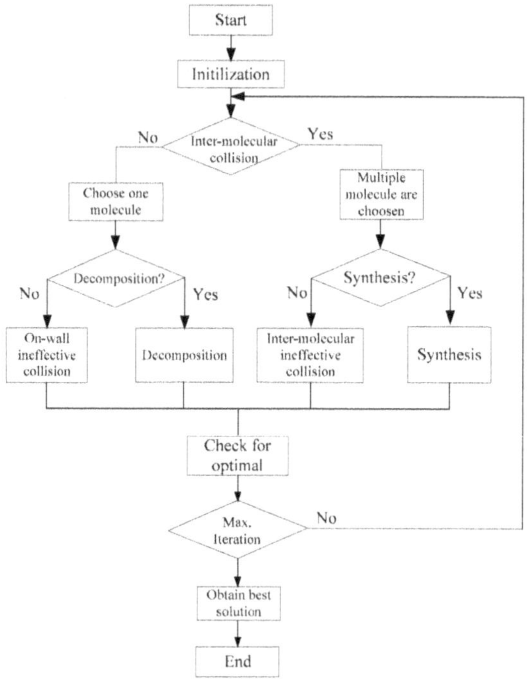

Figure 5. Schematic diagram that illustrates the CRO optimization framework.

4. Simulation Results and Analysis

This section presents comparative simulation results to demonstrate the efficiency and accuracy of the proposed technique in the coordinated UAV tracking problem. The simulations were conducted using 3D synthetic environmental data, which allows for generating a variety of virtual environments with different resolutions and urban features. The simulations were carried out in MATLAB (R2010b, MathWorks, Natick, MA, USA) on a standard specification PC (Dell Precision T3500, DELL, Round Rock, TX, USA, Inter(R) Xeon(R) CPU operating at 2.67 GHz).

The dimensions of the virtual environment are defined within the range of 1.5 km $\times$ 1.5 km, and 30 buildings with different dimensions are randomly generated. Three UAVs are employed to track one moving target, and the motion of the UAVs and target are described using the dynamic model discussed in Section 2. The flight speed v_g for each UAV is set to be 30 m/s, the flight level is 200 m, the maximum roll angle ϕ_{max} is set to 45° and the minimal distance between each UAV is 50 m. The trajectory of the target is randomly generated in the flexible region on the ground, with a constant speed at 15 m/s. The RHC window size L is set as 5 s, and the total simulation time is 100 s, and the sampling time is 1 s. The parameter settings of the CRO algorithm are shown in Table 1.

Table 1. Parameter settings of CRO algorithm.

KELossRate	0.2	InitialKE	200
MoleColl	0.3	PopSize	100
buffer	0	α	500
B	10	c_1	0.15
c_2	2.5		

Figures 6 and 7 illustrate the performance of the proposed method for tracking the moving target in two scenarios. The total visibility time for these two scenarios are 95 s and 94 s (as shown in Figures 6e and 7e, respectively), which approach the total flight time (100 s). This indicates that the proposed method is able to generate an optimal tracking path for each UAV. The costs of the proposed method at each sampling time are depicted in Figures 6f and 7f, respectively. The cost grows rapidly when the target is out of the visibility region for all of the sensors, and for the rest the time the costs are almost constant, implying that the proposed algorithm can determine the optimal path within the sampling time (1 s). As shown in Figures 6d and 7d, the control input for each UAVs, i.e., the roll angle, was changing frequently during the tracking process, since the UAVs are required to maneuver to track the moving target while avoid the occlusions of the buildings. The control inputs determined by the proposed method are generally smooth and changing slowly, which can be tracked by the autopilot of the UAV.

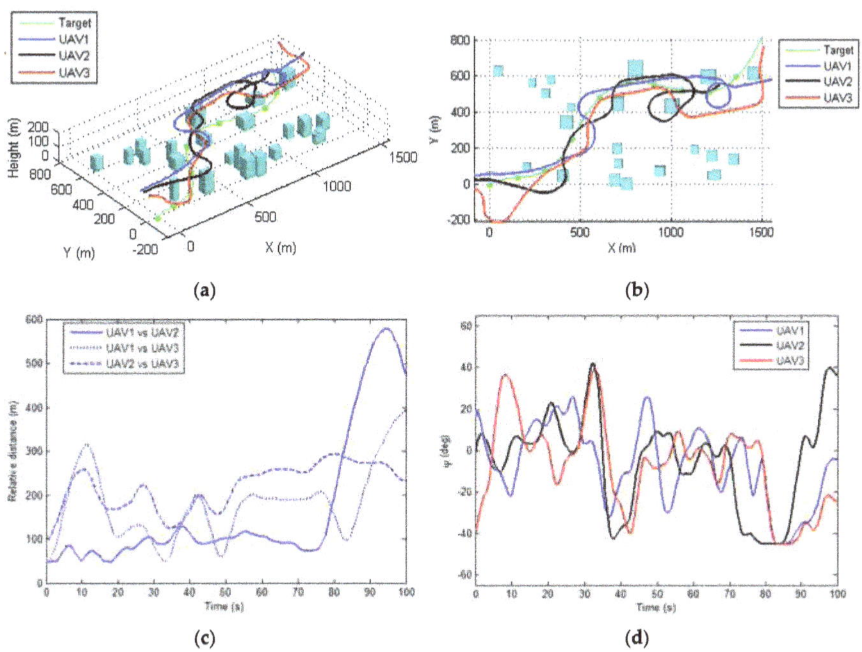

Figure 6. *Cont.*

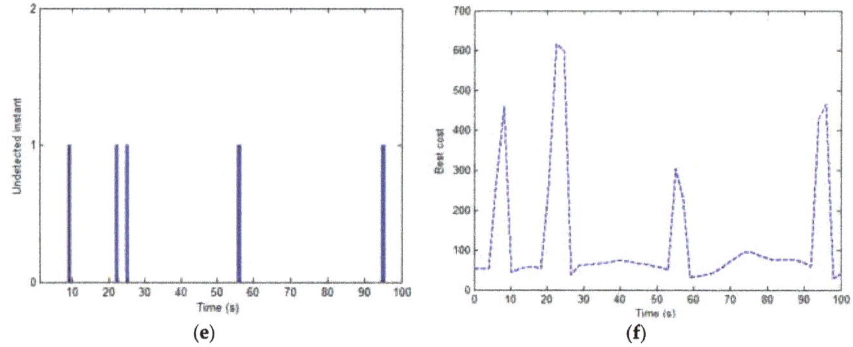

Figure 6. The performance of the proposed method for scenario 1. (**a**) 3D view of the planned path for each UAV; (**b**) Top view of the resulting paths; (**c**) The relative distance among each UAV during tracking; (**d**) The roll-command for each UAV (**e**) The undetected time-instant; (**f**) The total cost of the tracking process.

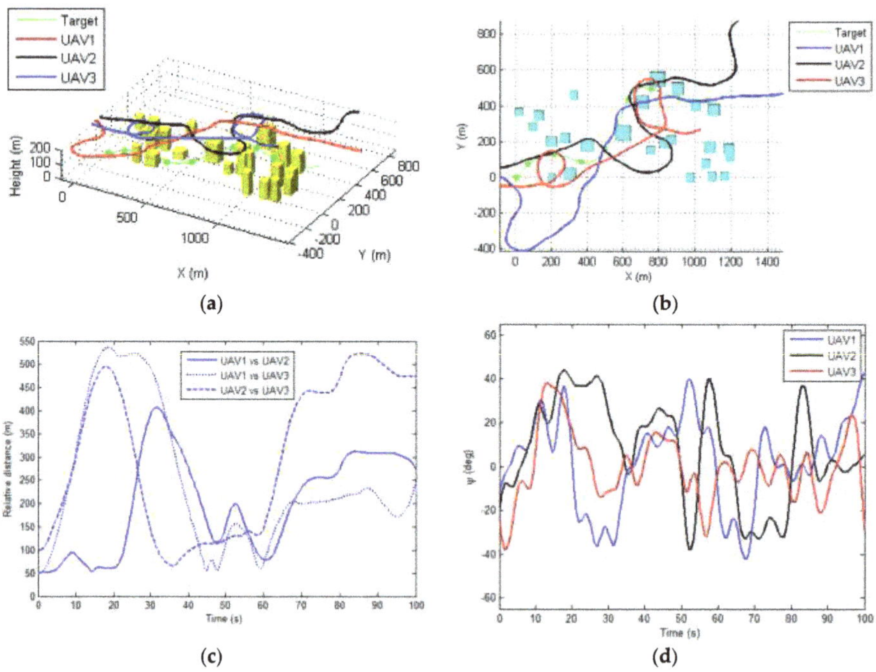

Figure 7. *Cont.*

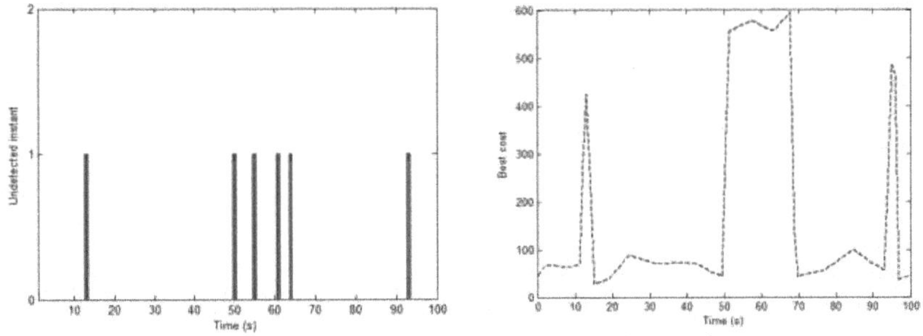

(e) (f)

Figure 7. The performance of the proposed method for scenario 2. (**a**) 3D view of the planned path for each UAV; (**b**) Top view of the resulting paths; (**c**) The relative distance among each UAV during tracking; (**d**) The roll-command for each UAV (**e**) The undetected time-instant; (**f**) The total cost of the tracking process.

Figures 6c and 7c illustrate the relative distances between each UAV during the tracking process for two scenarios, respectively. It can be seen that the generated path by the proposed method always satisfies the safety constraint (i.e., the minimal distance between each UAV). In order to demonstrate the efficiency of the proposed CRO based cooperative tracking path planning algorithm, this section firstly compares the proposed approach with commonly used evolutionary algorithms based optimizer, i.e., the genetic algorithm (GA) [13] and particle swarm optimization (PSO) [14], for determination of the tracking paths.

Figures 8 and 9 illustrate the performances of the proposed method, GA-based method and PSO- based algorithm for the tracking task, respectively. Figure 8a,b and Figure 9a,b illustrate the convergence of the proposed method and the other evolutionary-based optimizers from randomly selected sampling times for two scenarios. It can been seen that the proposed approach outperforms to GA- and PSO-based optimizer in terms of convergence speed and optimality (in terms of the cost value). Figures 8c and 9c depict the visibility time of the ground target of the aforementioned algorithms obtained from 10 repeat simulations. The overall performance of the proposed method is better than GA- and PSO-based planners in terms of the visibility time. This is because the CRO is a variable population-based metaheuristic approach, which allows for generating more solutions near the optimality and thus increasing the convergence speed. Figures 8d and 9d are the average execution times for each decision point (i.e., the sampling point), where the proposed method requires more time than the GA- and PSO-based planners. However, this execution time (around 0.55 s for the proposed method) is acceptable when compared the sampling time (1 s).

Next, the proposed method is compared with a recently developed Lyapunov guidance vector field (LGVF)-based approach for planning the cooperative tracking path [18]. Figures 10 and 11 illustrate the results of using the LGVF-based approach for two scenarios. It can be observed that the proposed method achieves better solution to the cooperative path planning problem in terms of the visibility time. The minimal distances between UAVs are not always satisfactory (i.e., less than the safety distance, i.e., 50 m), as shown in Figures 10c and 11c. The statistics of the proposed method and the LGVF-based approach obtained from 10 repeat simulation tests are shown in Tables 2 and 3, respectively. The mean, maximum and minimum visibility times of the proposed method are better than those of the LGVF-based planner. This is because it is difficult for the LGVF-based optimizer to incorporate nonlinear and non-convex constraints, such as the visibility of sensors and anti-collision constraints in the tracking optimization problem, and thus cannot always reach the optimal solution.

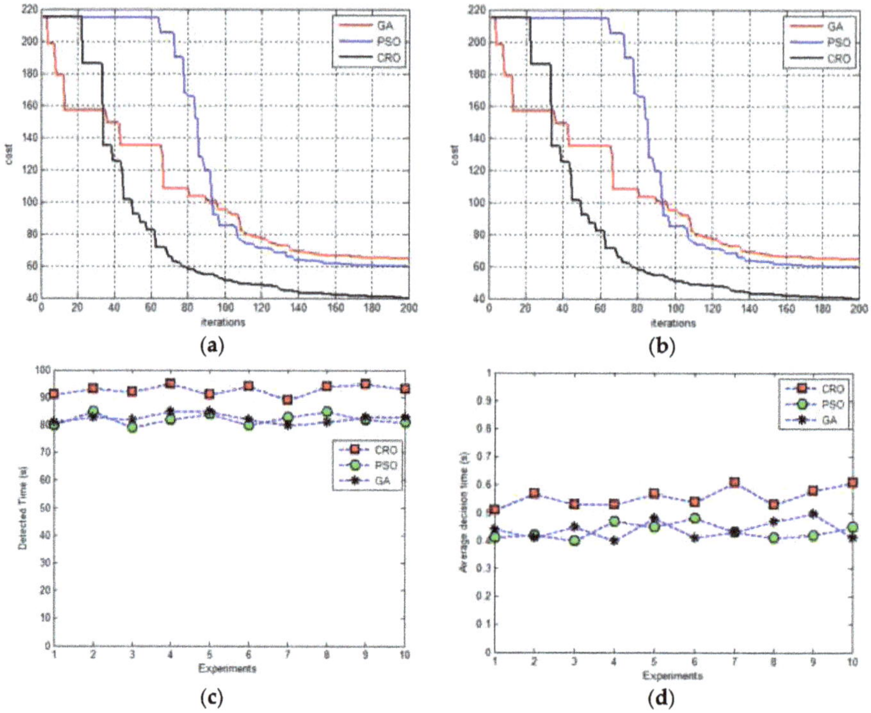

Figure 8. The performances of using different evolutionary optimizers for determination of the tracking path for scenario 1. (**a**) and (**b**) illustrate the convergence of different methods in selected decision time, respectively; (**c**) shows the visibility time of the ground time for the proposed method, PSO and GA based algorithms, respectively; (**d**) is the average execution time of the aforementioned methods.

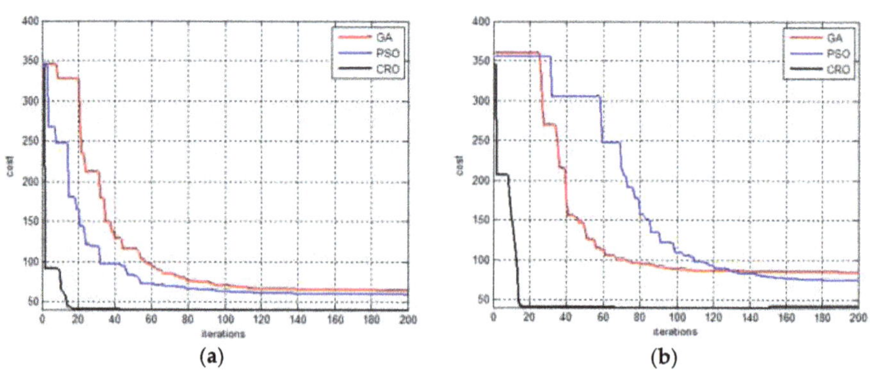

Figure 9. *Cont.*

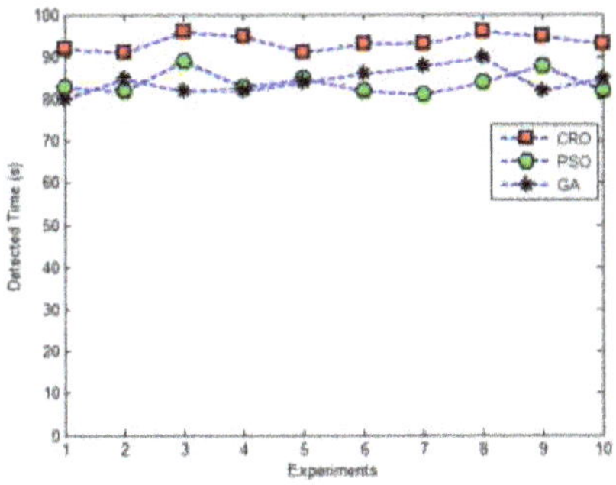

(**c**)

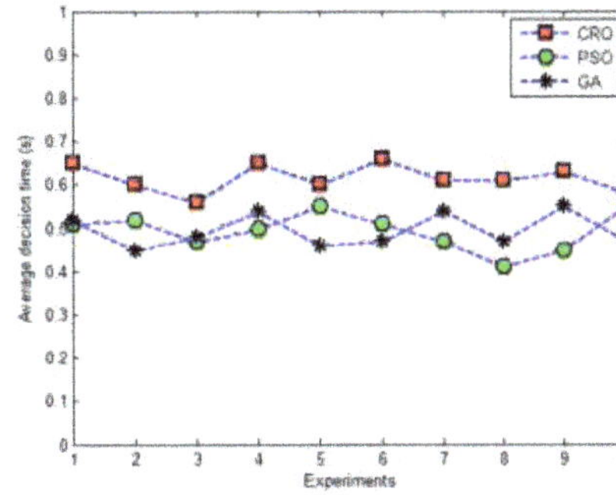

(**d**)

Figure 9. The performances of using different evolutionary optimizers for determination of the tracking path for scenario 2. (**a**) and (**b**) illustrate the convergence of different methods in selected decision time, respectively; (**c**) shows the visibility time of the ground time for the proposed method, PSO and GA based algorithms, respectively; (**d**) is the average execution time of the aforementioned methods.

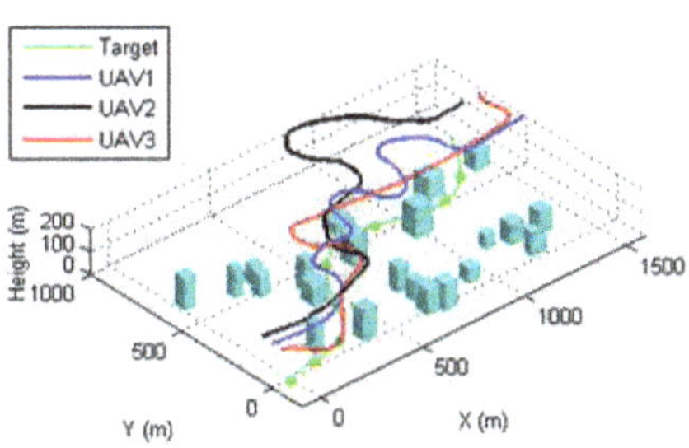

(**a**)

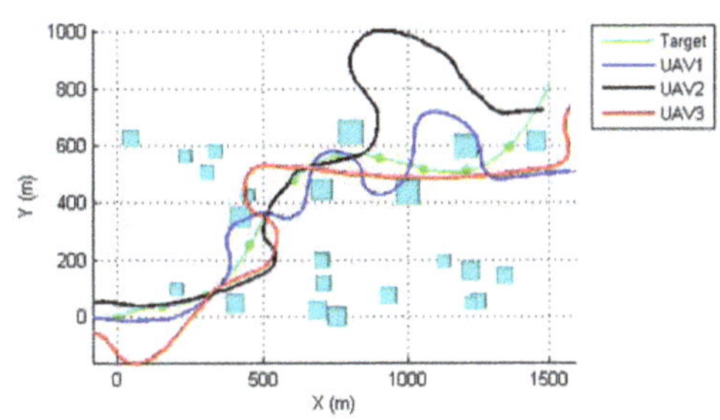

(**b**)

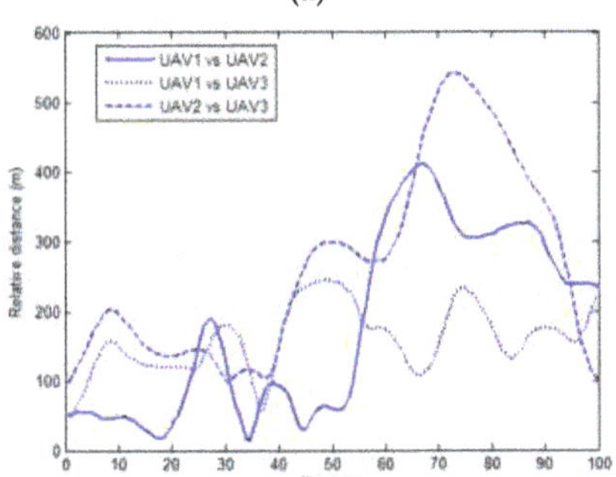

(**c**)

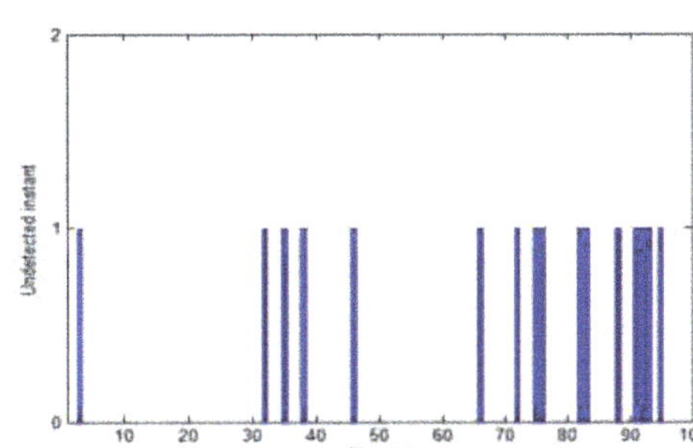

(**d**)

Figure 10. The results obtained by using LGVF based approach for scenario 1. (**a**) 3D view of the planned path for each UAV; (**b**) Top view of the resulting paths; (**c**) The relative distance between each UAV during tracking; (**d**) The undetected time-instant.

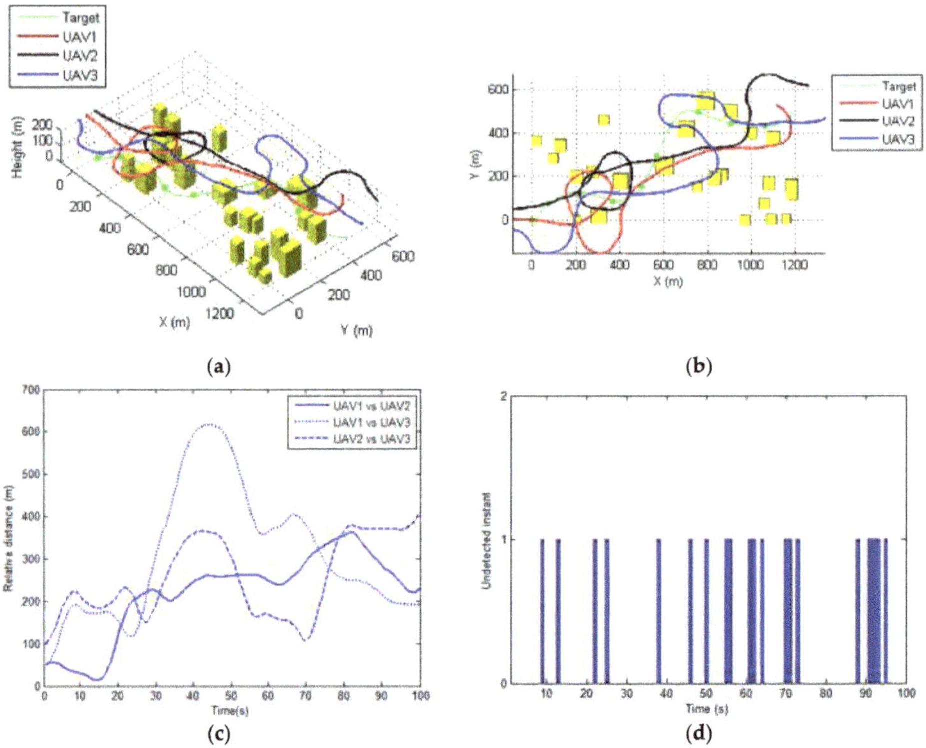

Figure 11. The results obtained by using LGVF based approach for scenario 2. (**a**) 3D view of the planned path for each UAV; (**b**) Top view of the resulting paths; (**c**) The relative distance between each UAV during tracking; (**d**) The undetected time-instant.

Table 2. Statistics results of the tracking performances for scenario 1.

Performance	Proposed Method	LGVF-Based Algorithm
Mean visibility time (s)	95	82
Maximum visibility time (s)	96	79
Minimum visibility time (s)	91	84
Variation of visibility time (s)	4	3
Range of relative distances (m)	[51, 588]	[11, 548]

Table 3. Statistics results of the tracking performances for scenario 2.

Performance	Proposed Method	LGVF-Based Algorithm
Mean visibility time (s)	94	83
Maximum visibility time (s)	97	81
Minimum visibility time (s)	91	89
Variation of visibility time (s)	4	7
Range of relative distances (m)	[54, 608]	[15, 611]

5. Conclusions

Tracking moving targets using UAVs is a challenging task due to the uncertainties arising from the motion of the target and the environmental features. This paper presents a novel method, coupling the concept of model predictive control into the framework of chemical reaction optimization, to solve the coordinated tracking path planning problem in complex urban environments. The sensor visible regions under urban environment, flight dynamics and anti-collision constraints are considered. Particularly, the LoS occlusion caused by dense buildings is discussed, and the FOV for a gimbaled sensor is also analyzed. A series of comparative simulations have shown that the proposed method outperforms other metaheuristic and mathematical methods in terms of tracking ability and safety, while achieving similar computational efficiency. The simulation results imply that the proposed method is capable of improving the performance of target tracking in urban environments, with more visible time, higher security and stronger robustness compared to other methods.

Acknowledgments: This work was funded partially by the Opening Foundation of State Key Laboratory of Virtual Reality Technology and Systems of China under Grant No. BUAA-VR-15KF-11 and National Natural Science Foundation (NSFC) of China under Grant No. 61503185.

Author Contributions: For research articles with several authors, Yin Wang conceived and designed the experiments and wrote this paper. Yin Wang and Yan Cao performed the simulation and analyzed the results.

Conflicts of Interest: The authors declare no conflict of interest.

References

1. Horri, N.M.; Dahia, K. Model Predictive Lateral Control of a UAV Formation Using a Virtual Structure Approach. *Int. J. Unmanned Syst. Eng.* **2014**, 2, 1–11. [CrossRef]
2. Yang, K.; Sukkarieh, S. 3D smooth path planning for a UAV in cluttered natural environments. In Proceedings of the IEEE/RSJ International Conference on Intelligent Robots and Systems, Nice, France, 22–26 September 2008; pp. 794–800.
3. Belkhouche, F. Reactive Path Planning in a Dynamic Environment. *IEEE Trans. Robot.* **2009**, *25*, 902–911. [CrossRef]
4. Cheng, P. Time-optimal UAV trajectory planning for 3D urban structure coverage. In Proceedings of the IEEE/RSJ International Conference on Intelligent Robots and Systems, Nice, France, 22–26 September 2008; pp. 2750–2757.
5. Quintero, P.; Papi, F.; Klein, D.J.; Chisci, L.; Hespanha, J.P. Optimal UAV coordination for target tracking using dynamic programming. In Proceedings of the 49th IEEE Conference on Decision and Control, Atlanta, GA, USA, 15–17 December 2010; pp. 4541–4546.
6. Chen, H.D.; Chang, K.C.; Agate, C.S. UAV path planning with Tangent-plus-Lyapunov vector field guidance and obstacle avoidance. *IEEE Trans. Aerosp. Electron. Syst.* **2013**, *49*, 840–856. [CrossRef]
7. Shaferman, V.; Shima, T. Unmanned aerial vehicles cooperative tracking of moving ground target in urban environments. *J. Guid. Control Dyn.* **2008**, *5*, 1360–1371. [CrossRef]
8. Yu, H.; Beard, R.W.; Argyle, M.; Chamberlain, C. Probabilistic path planning for cooperative target tracking using aerial and ground vehicles. In Proceedings of the 2011 American Control Conference, San Francisco, CA, USA, 29 June–1 July 2011; pp. 4673–4678.
9. Zhang, C.G.; Xi, Y.G. Robot path planning in globally unknown environments based on rolling windows. *Sci. China* **2001**, 2, 131–139.
10. Beard, R.W.; McLain, W.; Nelson, B. Decentralized Cooperative Aerial Surveillance Using Fixed-Wing Miniature UAVs. *Proc. IEEE* **2006**, *7*, 1306–1324. [CrossRef]
11. He, Z.; Xu, J.X. Moving target tracking by UAVs in an urban area. In Proceedings of the IEEE International Conference on Control and Automation, Hangzhou, China, 12–14 June 2013; pp. 1933–1938.
12. Besada-Portas, E.; de la Torre, L.; de la Cruz, J.M.; de Andrés-Toro, B. Evolutionary Trajectory Planner for Multiple UAVs in Realistic Scenarios. *IEEE Trans. Robot.* **2010**, *26*, 619–634. [CrossRef]
13. Wang, L.; Su, F.; Zhu, H.; Shen, L. Active sensing based cooperative target tracking using UAVs in an urban area. In Proceedings of the International conference on Advanced Computer Control, Shenyang, China, 27–29 March 2010; Volume 2, pp. 486–491.
14. Vincent, R.; Mohammed, T.; Gilles, L. Comparison of Parallel Genetic Algorithm and Particle Swarm Optimization for Real-Time UAV Path Planning. *IEEE Trans. Ind. Inform.* **2013**, *9*, 132–141.

15. Wise, R.A.; Rysdyk, R.T. UAV coordination for Autonomous Target Tracking. In Proceedings of the AIAA Guidance, Navigation and Control Conference and Exhibit, Keystone, CO, USA, 21–14 August 2006; pp. 1–22.
16. Lam, A.S.; Li, V.K. Chemical Reaction Optimization: A tutorial. *Mim. Comput.* **2012**, *4*, 3–17. [CrossRef]
17. Gal, O.; Doytsher, Y. Fast and Accurate Visibility Computation in a 3D Urban Environment. In Proceedings of the Fourth International Conference on Advanced Geographic Information Systems, Applications, and Services, Valencia, Spain, 30 January–4 February 2012; pp. 105–110.
18. Yao, P.; Wang, H.; Su, Z. Real-time path planning of unmanned aerial vehicle for target tracking and obstacle avoidance in complex dynamic environment. *Aerosp. Sci. Technol.* **2015**, *47*, 269–279. [CrossRef]

sensors

MDPI

Article

A Hybrid Vehicle Detection Method Based on Viola-Jones and HOG + SVM from UAV Images

Yongzheng Xu [1,2], Guizhen Yu [1,2], Yunpeng Wang [1,2], Xinkai Wu [1,2,*] and Yalong Ma [1,2]

1 Beijing Key Laboratory for Cooperative Vehicle Infrastructure Systems and Safety Control, School of Transportation Science and Engineering, Beihang University, Beijing 100191, China; yongzhengxu@buaa.edu.cn (Y.X.); yugz@buaa.edu.cn (G.Y.); ypwang@buaa.edu.cn (Y.W.); mayalong@buaa.edu.cn (Y.M.)

2 Jiangsu Province Collaborative Innovation Center of Modern Urban Traffic Technologies, SiPaiLou #2, Nanjing 210096, China

* Correspondence: xinkaiwu@buaa.edu.cn; Tel.: +86-010-8231-6713

Academic Editors: Felipe Gonzalez Toro and Antonios Tsourdos
Received: 25 June 2016; Accepted: 15 August 2016; Published: 19 August 2016

Abstract: A new hybrid vehicle detection scheme which integrates the Viola-Jones (V-J) and linear SVM classifier with HOG feature (HOG + SVM) methods is proposed for vehicle detection from low-altitude unmanned aerial vehicle (UAV) images. As both V-J and HOG + SVM are sensitive to on-road vehicles' in-plane rotation, the proposed scheme first adopts a roadway orientation adjustment method, which rotates each UAV image to align the roads with the horizontal direction so the original V-J or HOG + SVM method can be directly applied to achieve fast detection and high accuracy. To address the issue of descending detection speed for V-J and HOG + SVM, the proposed scheme further develops an adaptive switching strategy which sophistically integrates V-J and HOG + SVM methods based on their different descending trends of detection speed to improve detection efficiency. A comprehensive evaluation shows that the switching strategy, combined with the road orientation adjustment method, can significantly improve the efficiency and effectiveness of the vehicle detection from UAV images. The results also show that the proposed vehicle detection method is competitive compared with other existing vehicle detection methods. Furthermore, since the proposed vehicle detection method can be performed on videos captured from moving UAV platforms without the need of image registration or additional road database, it has great potentials of field applications. Future research will be focusing on expanding the current method for detecting other transportation modes such as buses, trucks, motors, bicycles, and pedestrians.

Keywords: vehicle detection; unmanned aerial vehicle; Viola-Jones; HOG; road orientation

1. Introduction

Unmanned aerial vehicles (UAVs) have been widely used in many fields, such as chemical vapour detection [1], nature conservation monitoring [2] and wildlife emergency response [3]. Particularly, UAVs hold great promise for transportation applications, as demonstrated by many transportation studies [4–7]. One important application of UAV technology for transportation is to enhance the traffic and emergency monitoring which has been serving as a backbone of Intelligent Transportation Systems (ITS). Because UAVs are highly portable, UAVs can collect traffic data in the areas where the geographic locations of potential transportation-related problems are only crudely known, or conventional data collection technologies based on point detections cannot be applied to gather the data needed for transportation studies.

For traffic and emergency monitoring, one of the essential but challenging tasks is vehicle detection. Many traditional methods, like background subtraction [8], frame difference [9], and optical flow [10] have been applied for vehicle detections from UAV videos. However, these methods are sensitive to scene complexity and can only detect moving vehicles. Note fail detection of stopped vehicles has limited their applications for vehicle detection under congested traffic conditions. Also, these methods are sensitive to background motions. To improve vehicle detection, many new object detection algorithms have been proposed. Two most famous object detection schemes are the Viola-Jones (V-J) object detection scheme with Ada-boost classifier using Haar-like features [11] and linear support vector machine (SVM) using histogram of oriented gradients (HOG) features (HOG + SVM) [12]. A series of studies have demonstrated that these two methods achieved very promising results on vehicle detection [13–17]. However, when applying V-J and HOG + SVM methods to UAV images, the detection effectiveness and efficiency have been significantly downgraded due to the following two reasons:

(1) Both V-J and HOG + SVM are sensitive to objects' in-plane rotation, therefore they only can detect vehicles when the orientations of vehicles are known and horizontal. Because vehicle orientations in UAV images are usually unknown and even changing, the detection accuracy (i.e., effectiveness) of these two methods has been significantly lowered. As shown in Figure 1a using original V-J method to detect vehicles in an UAV image with non-horizontal roadways, many vehicles cannot be detected. Note some methods have been proposed to address this issue (e.g., Jones & Viola [18], Cao et al. [14], Leitloff et al. [15], Moranduzzo and Melgani [19,20], Liu and Mattyus [21], etc.), but most of these methods are either time-consuming or need extra resources which limit their applications.
(2) The efficiency (i.e., detection speed) of both V-J and HOG + SVM is downgrading with the increase of the detection load (i.e., number of vehicles which need to be detected) in a frame. As shown in Figure 1b through our tests, the detection speeds of both methods are monotonically decreasing with the increase of the number of detected vehicles. But the descending rates of the detection speeds of these two methods demonstrate different characteristics. As shown in Figure 1b, the V-J method overall has a higher descending rate, but it detects much faster than HOG + SVM when the number of detected vehicles is relatively small. By contrast, HOG + SVM has lower detection speed than V-J when the number of detected vehicles is small, but it performs much better when the number of detected vehicles is large. These different characteristics suggest an intuitive idea that the overall efficiency could be improved by switching these two methods based on the number of vehicles which need to be detected, as the black line suggested in Figure 1b.

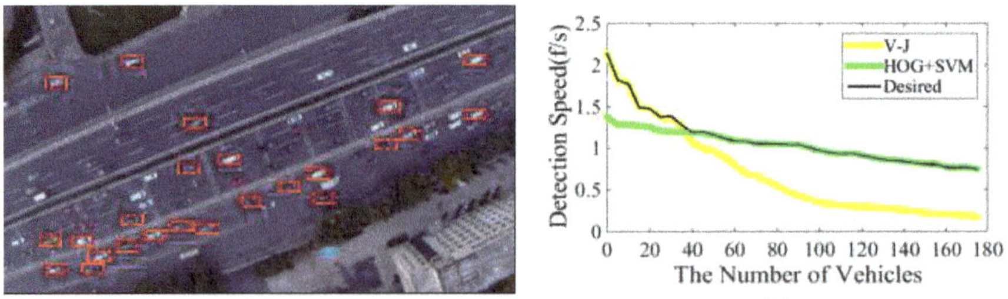

(a) (b)

Figure 1. (**a**) V-J vehicle detection on an UAV image with unknown road orientation (red squares indicate successful detection); (**b**) Detection speeds for V-J and HOG + SVM (the black highlight line is the suggested hybrid method).

This research aims to improve the effectiveness and efficiency of vehicle detection from UAV images by addressing the above mentioned two issues. Particularly, this research proposes a new hybrid vehicle detection scheme which integrates both V-J and HOG + SVM methods. The proposed scheme has two unique features which are designated to solve the above mentioned two issues:

(1) To address the challenge that both V-J and HOG + SVM are sensitive to on-road vehicles' in-plane rotation, we adopt a roadway orientation adjustment method. The basic idea of this method is to first measure the orientation of the road, and then rotates the road according to the detected orientation so the road and on-road vehicles will be horizontal after rotation so the original V-J or HOG + SVM methods can be applied to achieve fast detection and high accuracy. More importantly, different with some existing solutions for the issue of unknown road orientation (e.g., [14,15,18]), the proposed road orientation adjustment method does not need any additional extra resource and only needs to rotate the image one time, so the new method significantly saves computational time and reduces false detection rates.
(2) To address the issue of descending detection speeds for both V-J and HOG + SVM and achieve better efficiency, we integrate V-J and HOG + SVM methods based on their different descending trends of detection speed and propose a hybrid and adaptive switching strategy which sophistically searches for, if not the optimal, at lease improved solution, by switching V-J and HOG + SVM detection methods based on the change of detection speed of these two methods during the detection. This switching strategy, combined with the road orientation adjustment method, significantly improves the efficiency and effectiveness of vehicle detections from UAV images.

The rest of the paper is organized as follows: Section 2 briefly reviews some background information by introducing the basic concepts of V-J and HOG + SVM methods, followed by the methodological details of the proposed hybrid vehicle detection scheme in Section 3. Section 4 presents a comprehensive evaluation of the proposed method using diverse scenarios. Section 5 presents a discussion on some extensions and limitations of the proposed method. Finally, Section 6 concludes this paper with some remarks.

2. Background

A large amount of research has been performed on vehicle detection from UAV images over the years. Many of them applied some traditional methods, such as background subtraction, frame difference, optical flow, etc. For example, Azevedo [8] applied a median-based background subtraction method to fast detect vehicles; Shastry and Schowengerdt [9] applied a frame difference method, combining with the image registration process to detect moving vehicles; and Yalcin [10] proposed a motion-based optical flow method to detect moving vehicles. However, methods like frame difference, background subtraction and optical flow are sensitive to scene complexity therefore have difficulties in detecting slow-moving or stopped vehicles when traffic is congested. Also, some methods, like optical flow method, are sensitive to background motions.

In recent years, object detection algorithms have become popular for vehicle detection from UAV videos. Generally speaking, object detection algorithms are less sensitive to image noise, background motions and scene complexity, therefore are more robust for vehicle detections from UAV videos. For example, Viola and Jones [11] developed the famous V-J object detection scheme; Cao et al. [13] applied the SVM using HOG features for vehicle detection; Leitloff et al. [15] proposed a V-J-based two-stage method to improve detection results; Tuermer et al. [16] used Disparity Maps to limit the search space to road regions and then applied a standard HOG detector to detect vehicles; and Felzenszwalb [22] developed an objects detection framework by applying discriminatively trained deformable part model (DPM). Among them, two most widely applied methods are the V-J and HOG + SVM methods. However, these two methods have two major issues which limit their applications and downgrade their performance as mentioned above. This research aims to improve these two methods. Before we discuss the details of our method, we will briefly overview the background theories of these two methods first.

2.1. Viola-Jones Object Detection Scheme

The V-J scheme is based on multiple cascaded Haar-like classifiers [11]. The basic concept is to use a conjunctive set of weak classifiers to form a strong classifier. The core of this scheme is the Haar-like features, which are essentially drawn from the spatial response of Haar basis functions and derivatives to a given type of feature at a given orientation within the image. In practice, Haar-like features are computed as the sum of differences of the pixel intensities between different rectangular regions at a specific location in a detection window (Figure 2). Rectangular features can be computed very rapidly using an intermediate representation of the image called integral image (also called summed area table [11]). However, these individual Haar-like features are weak discriminative classifiers, which only give the right answer a little more often than a random decision. To construct a "strong" discriminative classifier, many "weak" classifiers are combined as a conjunctive cascade; and Gentle AdaBoost [23], a machine learning meta-algorithm, is applied to train a cascaded classifier over a set of thousands of positive and negative training images.

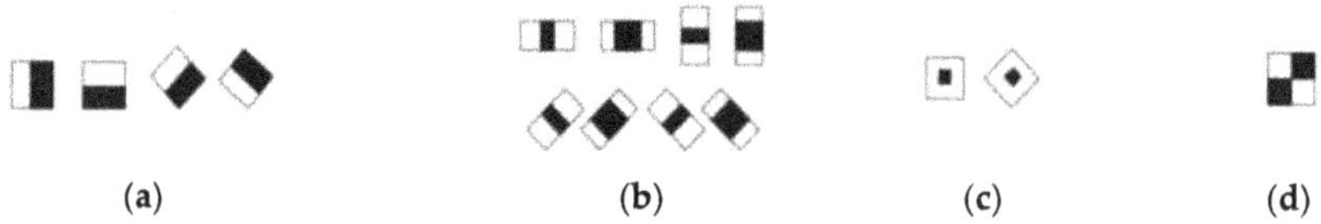

Figure 2. Haar-like features. (**a**) Edge features; (**b**) Line features; (**c**) Center-surround features; (**d**) Special diagonal line features.

The evaluation of the strong classifiers generated by the AdaBoost learning process can be done quickly, but it is not fast enough to process in real-time. For this reason, the strong classifiers are arranged in a cascade in order of complexity. In each cascade, each successive classifier is trained only on those selected samples which pass through the preceding classifiers. If at any stage in the cascade a classifier rejects the sub-window under inspection, no further processing is performed. The cascade therefore has the form of a degenerate tree. This degenerative decision-tree structure can eliminate negative regions as early as possible during detection to focus attention on promising regions of the image. Therefore, this detection strategy dramatically increases the processing speed of the detector, provides an underlying robustness to changes in scale, and maintains achievable real-time performance.

Before applying the V-J method, the critical first step is to build a sample library which provides a training set to include both positive and negative images. As mentioned in [24,25], the number of samples plays a key role in training classifier. In our paper, over 16,800 positive samples (see Figure 3a) were manually collected from 600 UAV images extracted from 100 min of the video data (one image per 10 s). From each UAV image, about 10–100 vehicles under different traffic conditions were manually extracted to form 16,800 positive training samples. These 16,800 samples do not contain duplicated ones and each sample only contains one vehicle. Note that some samples which contain the same vehicle but different backgrounds are treated as different positive samples. All vehicle samples were rotated to align with the horizontal direction. In addition, over 26,000 negative samples (see Figure 3b) which do not contain vehicles were also manually collected.

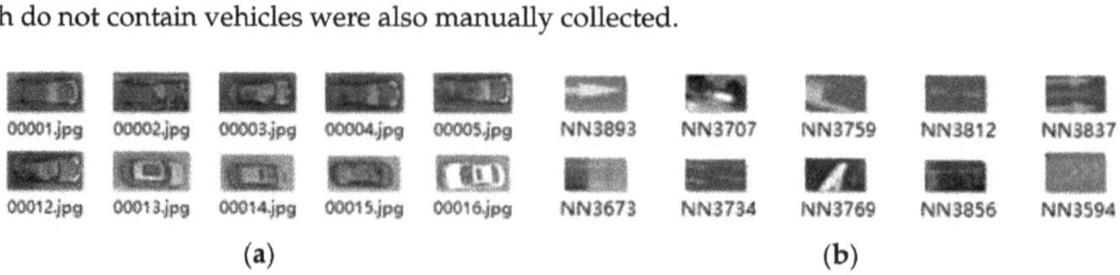

Figure 3. Samples images. (**a**) Positive training samples; (**b**) Negative training samples.

All positive images were further transformed into gray scale images and normalized to a compressed size of 40 × 20 during our tests. Each image was used to calculate a complete set of Haar-like features, and in total, 479,430 Haar-like features were extracted from all images. These 479,430 features were further trained into 469 most significant features by applying the Gentle AdaBoost algorithm [23]. Finally, an 18-stage cascaded classifier was formed based on the 469 significant Haar-like features.

2.2. Linear SVM Classifier with HOG Feature

The histogram of oriented gradients (HOG) is a feature descriptor used in computer vision and image processing for the purpose of object detection. HOG was first described by Dalal and Triggs [12] and achieved good performance for pedestrian detection. The HOG descriptor has many advantages, for example, it is invariant to geometric and photometric transformations, since it operates on local cells. Generally, the extraction of HOG features includes five steps [26] as described as follows:

- Step 1: Gradient computation. Step 1 computes the gradient values and orientations of all pixel units in the image by applying the 1-D centered point discrete derivative mask with the filter kernel $[-1, 0, 1]$ in one or both of the horizontal and vertical directions.
- Step 2: Orientation binning. Step 2 is to create the cell histograms. In this step, the image was divided into cells, and the 1-D histogram H_i is generated by binning local gradients according to the orientation of each cell. Each pixel within the cell will cast a weighted vote for an orientation-based histogram channel based on the values found in the gradient computation.
- Step 3: Descriptor blocks. Step 3 groups cells together into larger, spatially connected blocks F_i.
- Step 4: Block normalization. Step 4 is to normalize blocks in order to account for changes in illumination and contrast. A cell can be involved in several block normalizations for the overlapping block, since each block consists of a group of cells. By concatenating the histograms of all blocks, the feature vector V_{HOG} is obtained. The HOG descriptor is then the concatenated vector of the components of the normalized cell histograms from all the block regions.
- Step 5: SVM classifier. The final step is to feed the descriptors into the SVM classifier. SVM is a binary classifier which looks for an optimal hyperplane as a decision function.

HOG + SVM will use the same sample library (16,800 positive samples and 26,000 negative samples) as for the V-J vehicle detector. To extract the HOG feature, each sample image needs to be first normalized to a compressed size of 40 × 20 pixels; then blocks (8 × 8) with 4-pixel stride step and 9 histogram bins (each bin is corresponding to 20° of orientation) were used to calculate the HOG feature vector. So for an image with 40 × 20 pixels, a total of 36 blocks can be identified (note 36 is calculated by $\left(\frac{(40-8)}{4}+1\right) \times \left(\frac{(20-8)}{4}+1\right) = 36$). The final HOG feature descriptor can be described by:

$$V_{HOG} = [F_1, F_2, \cdots, F_i, \cdots F_{36}] \tag{1}$$

where V_{HOG} is the HOG descriptor, and F_i is the normalized block vector for i-th block. As each block contains four cells, and each cell contains nine bins, $\mathrm{F}_i = \left[\overline{h}_{1,i}, \overline{h}_{2,i}, \ldots, \overline{h}_{j,i}, \ldots, \overline{h}_{36,i}\right]$, where $\overline{h}_{j,i}$ is the j-th normalized value in i-th block. Figure 4 shows the flowchart to obtain HOG descriptor.

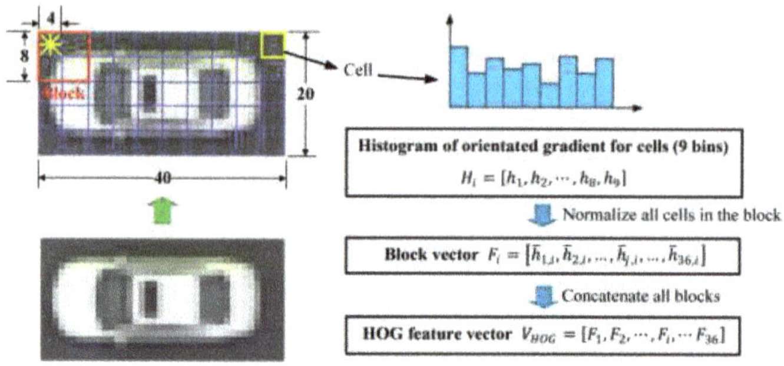

Figure 4. The flowchart to obtain HOG descriptor.

2.3. Limitations

V-J and HOG + SVM have been widely applied in many fields, but one of critical issues of these methods is that both methods are sensitive to object's in-plane rotation. So when applying them to vehicle detection, especially for UAV images, many vehicles with unknown orientations cannot be detected. A simple way to address this issue is to rotate images. For example, Cao et al. [14] rotated each video image nine times (each time 20 degree) in order to cover 180 degrees, but repeating detections of the same image significantly increase the detection time and lead to more false detections. Some researchers tried to train multiple detectors for objects of different angles. For example, Viola and Jones [18] built 12 different detectors for face detection to cover different views. However, training multiple detectors brings heavy workload. Leitloff et al. [15], on the other hand, used road database to get the road orientation and rotated the image according to the road orientation. Because on-road vehicles run in the same direction with the road, the roads and on-road vehicles in rotated images will be aligned with the original vehicle detector; therefore, the original V-J or HOG + SVM can be applied. However, the need of additional geometric information limits its applications. Another issue of V-J and HOG + SVM is the descending detection speed with the increase number of detected vehicles. As shown in Figure 1b, the detection speeds of both methods are monotonically decreasing as the number of detected vehicles increases. However, the descending rates of the detection speeds of these two methods demonstrate different characteristics. Particularly, V-J method detects much faster than HOG + SVM when the number of detected vehicles is relatively small; by contrast, HOG + SVM performs better when the number of detected vehicles is large. The proposed method will take advantage of these features to develop a hybrid vehicle detection scheme which integrates V-J and HOG + SVM to achieve better efficiency and effectiveness.

3. Methodology

3.1. Overall Framework

This research aims to address two critical issues of V-J and HOG + SVM by developing a hybrid vehicle detection scheme. In detail, the proposed scheme integrates the following two improvements:

(1) A roadway orientation adjustment method. The proposed detection scheme adopts a roadway orientation adjustment method to address the roadway rotation issue. The idea is straightforward: first, measure the orientation of the road using the line segment detector (LSD) [27]; second, rotate the road according to the detected orientation so the road and on-road vehicles will be horizontal after rotation; and third, apply the original V-J or HOG + SVM methods to achieve fast detection and high accuracy. A highlight of this approach is that the proposed road orientation adjustment method only needs to rotate the image one time and does not need any additional

extra resource. Therefore, the new method significantly saves computational time and reduces false detection rates.

(2) A detector switching strategy. The proposed scheme further develops an adaptive switching strategy to integrate V-J and HOG + SVM methods based on their different descending trends of detection speed in order to improve the efficiency. Basically, this strategy "intelligently" switches the detection methods between V-J and HOG + SVM to choose the one which has faster detection speed. As shown in Figure 1b, the detection speed essentially is determined by the workload, i.e., the number of vehicles which need to be detected; and V-J and HOG + SVM shows different characteristics of detection speed. So the proposed switching method will detect the detection speeds of both methods periodically and choose the one with faster detection speed.

The overall framework of the proposed vehicle detection method is shown in Figure 5. The details of the above mentioned two major functions will be introduced in the following sections.

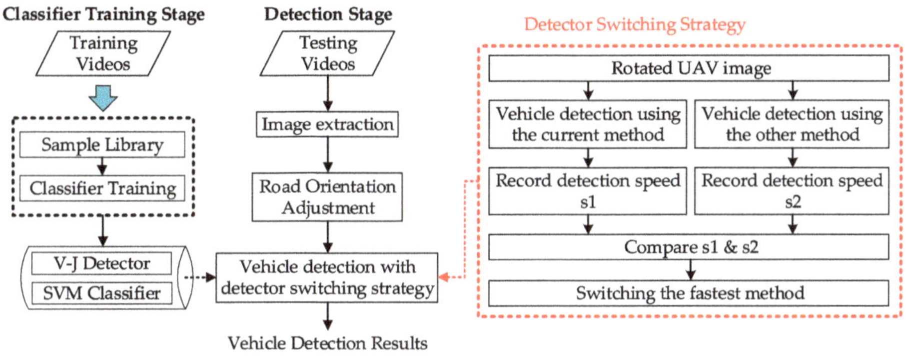

Figure 5. Proposed vehicle detection framework.

3.2. Road Orientation Adjustment Method

As mentioned before, when directly applying V-J or HOG + SVM to detect vehicles from UAV images, the detection rate is significantly low. The reason, as mentioned before, is that the original V-J and HOG + SVM schemes can only detect vehicles with orientations which are aligned with the detectors (i.e., vehicles in training sets). To address this issue, this research proposes a road orientation adjustment method.

Essentially, the proposed road orientation adjustment method is to rotate the image according to the orientation of the road, i.e., the angle between the road and the horizontal of the image. After rotation, the road and on-road vehicles will be aligned with the vehicle detectors. As shown in Figure 6, the general procedure includes:

(1) First, original color images are extracted from aerial videos and then transformed into gray scale images (Figure 7a);
(2) Second, the LSD algorithm [27] is applied to detect straight edge segments (Figure 7b);
(3) Third, the orientation of each detected line segment θ_i is calculated and the relative histogram $H(A)$ of these line orientations is derived (Figure 7c). The angle corresponding to the maximum distribution frequency of relative histogram will be considered as the orientation of the road;
(4) Last, to minimize in-plane rotation jitters, the final rotation angle ω_t for frame t is smoothed by the first-order lag filtering algorithm, which considers the rotation angle ω_{t-1} for the last frame $t-1$. After rotation, the directions of roads and on-road vehicles are aligned with the horizontal direction of the image (Figure 7d). The details of some key techniques will be elaborated in the following.

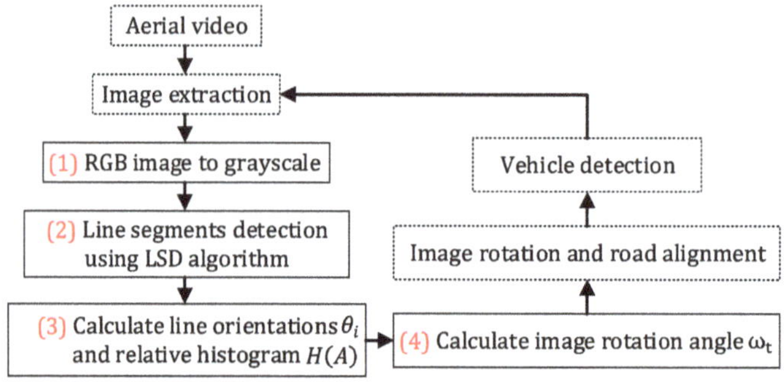

Figure 6. Flowchart of the proposed road orientation adjustment method.

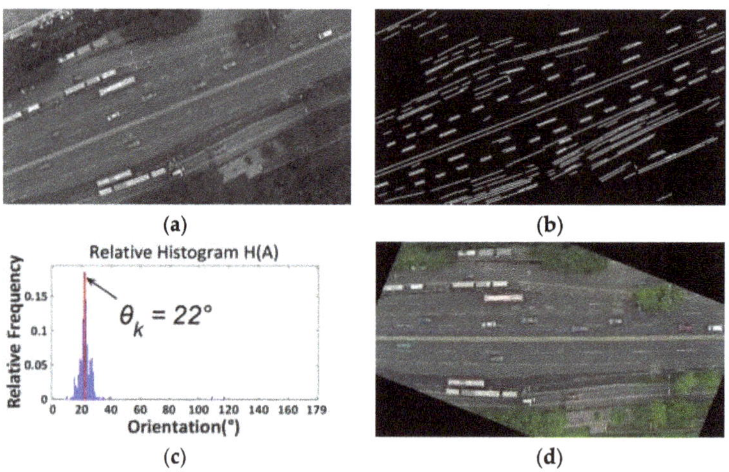

Figure 7. (**a**) Grayscale image; (**b**) Line segments detection; (**c**) Relative histogram (blue bins: distribution frequencies of relative histogram which correspond to different angles; red bin: the maximum distribution frequency of relative histogram which corresponds to the orientation of the road); (**d**) Rotated image.

(1) Straight Line Segments Detection: The proposed method first applies the LSD algorithm [27] to detect straight line segments. LSD is a linear-time line segment detector giving subpixel accurate results. Unlike other line detection algorithms, such as Hough transform [28], which requires tedious parameter tunings and is very likely to be affected by other redundant edges, the LSD algorithm can work on digital image without parameter tuning, therefore it is more robust and efficient. The LSD algorithm is open source [29] and the implementation of this algorithm is available in the Open Source Computer Vision (OpenCV) version 3.1 (Nizhny Novgorod, Russia). As shown in Figure 7b, many line segments are detected. The orientation of each detected line segment can be estimated as described in Equation (2):

$$\varphi_i = \begin{cases} arctan\left(-\frac{r_{i\_2}-r_{i\_1}}{c_{i\_2}-c_{i\_1}}\right), & c_{i\_1} \neq c_{i\_2} \\ 90°, & c_{i\_1} = c_{i\_2} \end{cases} \quad (2)$$

where φ_i is the orientation of detected line i; φ_i is an integer and $\varphi_i \in [0°, 180°)$; (c_{i_1}, r_{i_1}) and (c_{i_2}, r_{i_2}) represent the pixel coordinates of the start and end points (P1 and P2, see Figure 8) of line i in the image coordinate system.

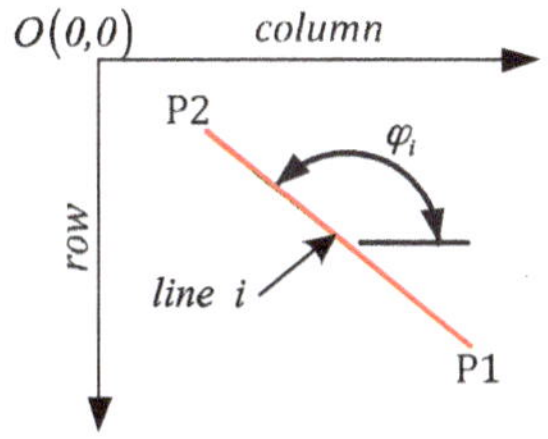

Figure 8. Orientation φ_i of line i.

(2) Road Orientation Estimation by Relative Histogram: As shown in Figure 7b, the road is parallel to the majority of detected line segments. Therefore, to estimate the orientation of the road, essentially, we need to identify the angle of the majority of line segments which have similar orientations. More precisely, we aim to identify an angle range of 1°. A relative frequency histogram is applied to identify the angle. The detailed steps are described as the following:

Step 1: Identify n: the total number of lines;
Step 2: Define 180 class intervals: $\theta_1 = [0°, 1°)$, $\theta_2 = [1°, 2°), \ldots, \theta_i = [(i-1)°, i°), \ldots, \theta_{180} = [179°, 180°)$;
Step 3: Determine the frequency, $h(\theta_i)$, i.e., the number of lines with the angle within the angle interval of class θ_i;
Step 4: Calculate the relative frequency (i.e., proportion) of each class by dividing the class frequency by the total number n in the sample, i.e., $H(\theta_i) = h(\theta_i)/n$;
Step 5: Draw a rectangle for each class with the class interval as the base and the height equal to the relative frequency of the class to form a relative histogram (Figure 7c);
Step 6: Identify θ_k, which is corresponding to the highest rectangle in relative histogram, and $\Theta = k°$ is considered as the orientation of the road.

(3) Image Rotation Angle Estimation by First-Order Lag Filtering: To minimize the impact of the jitters caused by UAVs, the first-order lag filtering algorithm is further applied to calculate the weighted average of the estimated road orientations of current and previous frames. The final image rotation angle for current frame j, is calculated by:

$$\omega_j = (1 - w) \times \omega_{j-1} + w \times \Theta \tag{3}$$

where ω_{j-1} is the image rotation angle for the last frame $j-1$, and w is a predetermined weight.

The final step is to rotate the image by ω_j. After rotation, the road will become horizontal. Figure 9 presents another example for a suburban road.

A highlight of this method is that each UAV image only needs to be rotated once. A visual comparison of vehicle detections using V-J scheme without and with road orientation adjustment is presented in Figure 10. Detected vehicles are marked with red rectangles. As shown in Figure 10a, because the orientation of the road is not horizontal, many vehicles could not be detected by the V-J method. On the contrary, as shown in Figure 10b, after applying the pre-processing step of road orientation adjustment, most vehicles can be detected. Detailed evaluation will be presented in Section 4.

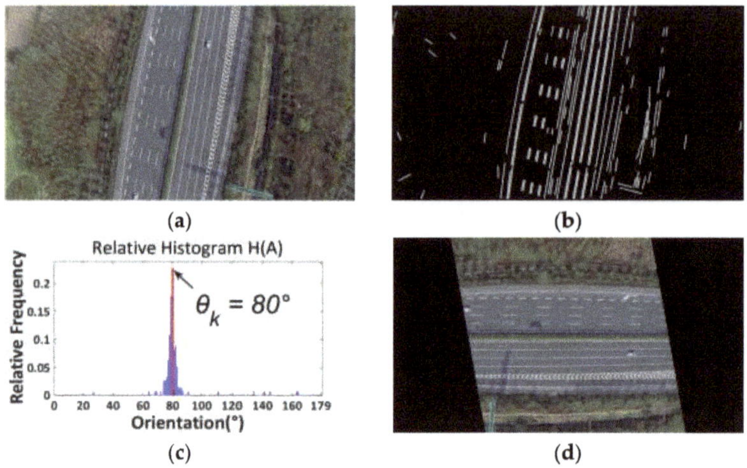

Figure 9. Suburban road. (**a**) Color image; (**b**) Line segments detection using LSD; (**c**) Relative histogram (blue bins: distribution frequencies of relative histogram which correspond to different angles; red bin: the maximum distribution frequency of relative histogram which corresponds to the orientation of the road); (**d**) Rotated image.

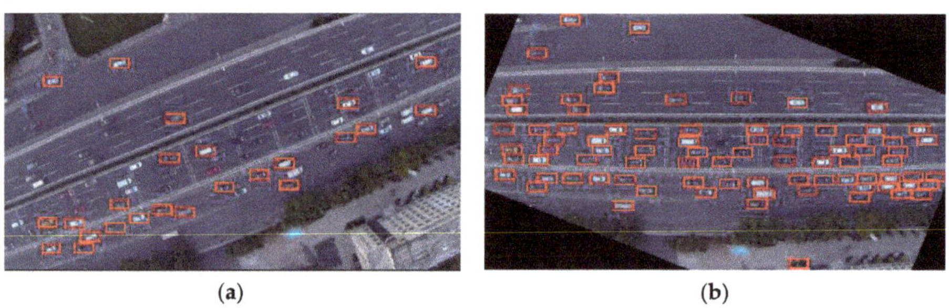

Figure 10. Vehicle detections using the V-J vehicle detector: (**a**) without road orientation adjustment; (**b**) with road orientation adjustment.

3.3. Detector Switching Strategy

The proposed vehicle detection scheme further adopts a vehicle detector switching strategy to improve detection speed. This strategy is based on the speed characteristic lines (see Figure 1b) of the V-J and HOG + SVM methods. The comparison between the detection speeds of both methods (Figure 1b) shows that when the number of vehicles in an UAV image is small, V-J should be applied to achieve faster detection speed, while when the number of vehicles in an UAV image is large, HOG + SVM should be selected to gain better detection speed. Based on this observation, we propose a switching strategy which can "intelligently" choose the faster detection method between V-J and HOG + SVM during the detection. The idea is straightforward. Since we won't be able to know the number of vehicles in the image until we finish detecting, an intuitive idea is to directly compare the detection speeds of both methods and choose the one with faster detection speed. But performing both methods to each frame could be time consuming and is really not necessary. Also, since traffic conditions (i.e., the number of vehicles in the image) are relatively stable within a short period (such as a few seconds) based on the research in [30], it would be much more reasonable and efficient to switch detection methods every several seconds (note there are 24 frames each second). Figure 11 presents the overall flowchart of the proposed switching strategy.

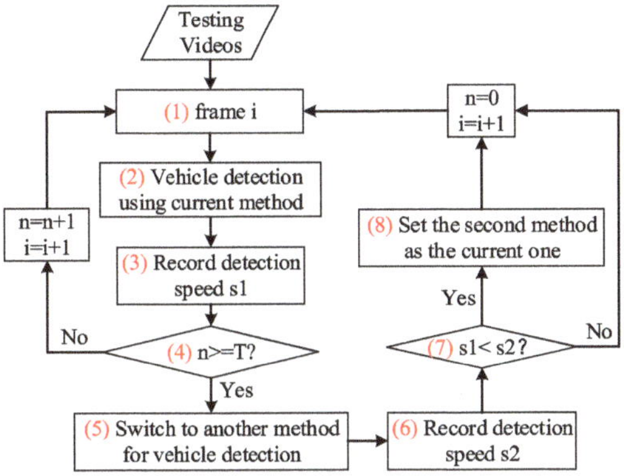

Figure 11. Flowchart of the proposed detector switching strategy.

From the flowchart, we can see that for any frame i, the proposed switching strategy first detects vehicles in the image using the current detection method, i.e., the detection method used to detect vehicles for previous frame $i-1$ (i.e., step (2) in the flowchart). After detection, the detection speed, $s1$, will be recorded (i.e., step (3) in the flowchart). Then the program will check if the accumulative number of frames (n) which apply the current detection method for vehicle detection has reached a predetermined cycle value T (i.e., step (4) in the flowchart). If NO, then the program moves to frame $i+1$ and repeat steps (2)–(4); and if YES, the program will apply the other detection method to detect the vehicles in frame i again (i.e., step (5) in the flowchart) and record the detection speed as $s2$ (i.e., step (6) in the flowchart). If $s1 < s2$ (i.e., step (7) in the flowchart), the program will switch to the new detection method and set it as the "current" method (i.e., step (8) in the flowchart) and apply it to detect vehicles for the following T frames. Otherwise, if $s1 \geq s2$, the program will keep applying the current method to detect vehicles for the following T frames, i.e., repeat previous steps. To be clear, only the image of the T_{th} frame needs to be detected twice, other $T-1$ images from previous $T-1$ frames need to be detected only once. Section 4 presents the testing results. Note during our testing, the value of T is set as 24 (namely, the detection speed comparison is conducted every 1 s). Furthermore, a sensitivity analysis has been conducted to evaluate the impact of different values of T in Section 5.

4. Evaluation

4.1. UAV Data Collection

The UAV system used in this research is equipped with a quadcopter (model: Phantom 2) airborne platform and a Gopro Hero Black Edition 3 aerial camera (see Figure 12). A 3-axis gimbal is mounted on the UAV to stabilize the videos and eliminate video jitters caused by UAV therefore greatly reducing the impact from external factors, such as wind. In addition, an On-Screen Display (OSD), an image transmission module and a video monitor are installed in the system for data transmission and airborne flying status monitoring and control.

The evaluation was based on low-altitude UAV videos captured from five different scenarios with diverse traffic and weather conditions (Table 1). These diverse testing scenes are specifically chosen in order to test the effectiveness of the proposed method. For each scenario, three 15-min videos were recorded, but only 10-min video in the middle were used due to the UAV ascent and descent (Figure 13), so the total video time for each scenario is 30 min. Among them, 20 min of videos were chosen for building the sample library; and the remaining 10 min were used for testing. The resolution of the videos is 1920 × 1080 and the frame rate is 24 frames per second (f/s). Note, all UAV videos were captured with the UAV hovering over a fixed location. Due to UAV motions, the orientations of the roads and on-road vehicles in the images are unknown and changing frequently.

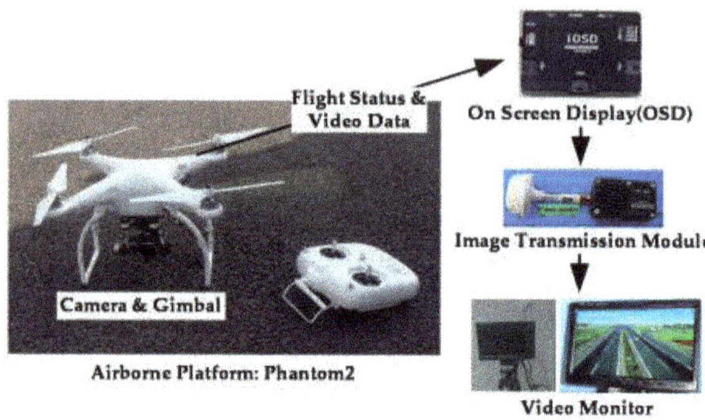

Figure 12. UAV system architecture.

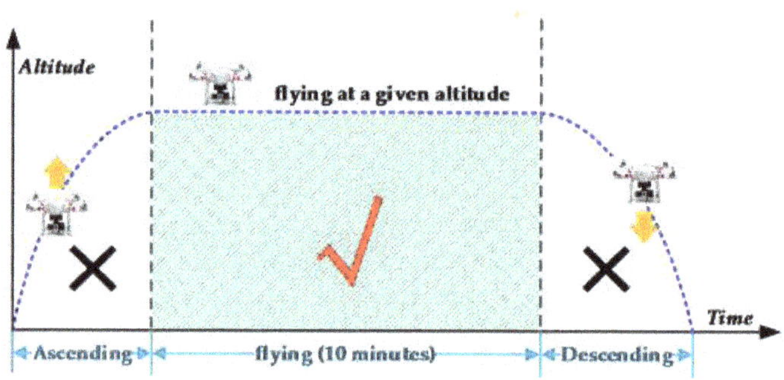

Figure 13. Altitude-Time graph of the quadcopter.

Table 1. Description of UAV Video Datasets.

Scenarios	Traffic Condition; Weather; Location; Time; Flight Altitude
Freeway	Non-congested; cloudy; G6 Jingzang Expressway; 15:40–16:40, 18 November 2014; 150 m.
Urban road 1	Non-congested; foggy; Xueyuan Road; 15:30–16:30, 26 March 2014; 170 m.
Urban road 2	Congested; cloudy; North Fourth Ring Road; 16:20–17:20, 6 November 2014; 150 m.
Urban road 3	Non Congested; cloudy; Xueyuan Road; 15:10–16:10, 12 August 2014; 100 m.
Urban road 4	Congested ; cloudy; North Fourth Ring Road; 16:30–17:30, 5 May 2015; 115 m.

During evaluation, in order to avoid the situation where the same vehicle has been detected multiple times in different frames, we extract detection images each 20 s from the 10-min video for comparison. Because the length of the road segment in an image is about 160 meters, most likely a vehicle will pass the road segment in 20 s. This significantly reduces the possibility of one vehicle being detected multiple times. Note when traffic is congested, it is still possible that some slow-moving vehicles will be detected more than once. All the experiments were conducted using C++ implementation on a laptop computer (model: ThinkPad T440P, Lenovo, Beijing, China) with an Intel i5-4300M @ 2.60 GHz CPU and 8 GB DDR3 memory.

4.2. Performance Evaluation

The performance of vehicle detection is evaluated by the following four typical indicators: *detection speed* (f/s), *Correctness*, *Completeness*, and *Quality*, defined in Equation (4):

$$\begin{aligned} Correctness(Cor.) &= \frac{true\ positives}{true\ positives + false\ positives} \\ Completeness(Com.) &= \frac{true\ positives}{true\ positives + false\ negatives} \\ Quality(Qua.) &= \frac{true\ positives}{true\ positives + false\ positives + false\ negatives} \end{aligned} \quad (4)$$

where *true positives* are the number of "true" detected vehicles; *false positives* are the number of "false" detected objects which are non-vehicle objects; *false negatives* are the number of vehicles missed. In particular, *Quality* is considered as the strictest criterion, which contains both possible detection errors (false positives and false negatives). Note that a successful "detection" is defined as a correct detection of a vehicle in one frame.

4.3. Results and Comparison

Conceptually, by incorporating the road orientation adjustment method, the proposed vehicle detection method will be insensitive to road orientation changes and therefore can achieve high *Completeness* and *Quality*. Furthermore, by combining the vehicle detector switching strategy, the proposed method can achieve fast vehicle detection. To fairly verify these two points, the proposed method is compared with nine other methods:

(1) ViBe, a universal background subtraction algorithm [31];
(2) Frame difference [9] (referred as Frame Diff in Table 2);
(3) Original V-J method (referred as V-J in Table 2) [11];
(4) Rotate each image every 20° from 0° to 180° and detect nine times using the original V-J method (referred as V-J + 9 in Table 2) [14];
(5) Original V-J method combines with the proposed road orientation adjustment method only (referred as V-J + R in Table 2);
(6) Original HOG + SVM (referred as SVM in Table 2) [12];
(7) Rotate each image every 20° from 0° to 180° and detect nine times using the original HOG + SVM method (referred as SVM + 9 in Table 2) [14];
(8) Original HOG + SVM method combines with the proposed road orientation adjustment method only (referred as SVM+R in Table 2);
(9) Apply the proposed vehicle detector switching strategy to integrate V-J and HOG + SVM (without road orientation adjustment) (referred as V-J + SVM + S in Table 2);
(10) Apply the proposed vehicle detector switching strategy and road orientation adjustment method to integrate V-J and HOG + SVM (the proposed method, V-J + SVM + R + S).

As ViBe [31] and frame difference [9] are sensitive to background motions, image registration [9] is performed first to compensate UAV motions. This registration process converts the spatio-temporal video into temporal information, thereby can correct UAV motion and attitude errors. The time for image registration is included in the detection time for these two methods.

Also, as mentioned above, for each scenario, a 10-min video with the resolution of 1920 × 1080 was used for testing. The detection speed for each method was computed as an average of each 10-min video. Note vehicle detection was performed on the entire image (1920 × 1080). After detection, we only extracted images every 20 s in order to avoid that the same vehicle in different frames has been detected multiple times. So for each scenario, totally of 30 detected images were used for computing *Correctness*, *Completeness*, and *Quality*.

Table 2. Vehicle Detection Results.

Scene	Metrics	ViBe	Frame Diff	V-J	V-J + 9	V-J + R	SVM	SVM + 9	SVM + R	V-J + SVM + S	V-J + SVM + R + S
Freeway	Cor. (%)	65.00%	61.11%	85.71%	87.50%	88.57%	91.30%	88.24%	92.27%	89.01%	89.37%
	Com. (%)	86.67%	78.57%	75.00%	92.11%	93.94%	55.26%	85.71%	87.04%	66.76%	94.24%
	Qua. (%)	59.09%	52.38%	66.67%	81.40%	83.78%	52.50%	76.92%	81.13%	61.67%	84.74%
	Speed (f/s)	5.59	9.96	1.16	0.064	0.96	1.17	0.083	1.03	1.17	1.08
Urban road 1	Cor. (%)	49.06%	77.78%	89.47%	73.53%	88.00%	82.05%	80.81%	85.64%	88.86%	87.89%
	Com. (%)	66.67%	48.84%	70.83%	92.59%	91.67%	69.57%	87.43%	90.96%	70.73%	91.46%
	Qua. (%)	39.39%	42.86%	65.38%	69.44%	81.48%	60.38%	72.40%	78.92%	62.89%	81.23%
	Speed (f/s)	6.64	10.40	2.08	0.16	1.57	1.15	0.079	1.03	2.01	1.80
Urban road 2	Cor. (%)	77.87%	75.26%	85.06%	92.16%	95.65%	98.67%	96.34%	98.01%	98.04%	96.39%
	Com. (%)	86.36%	68.87%	89.16%	94.95%	94.62%	77.08%	91.20%	92.96%	77.74%	94.24%
	Qua. (%)	69.34%	56.15%	77.08%	87.85%	90.72%	76.29%	88.14%	91.25%	76.55%	91.02%
	Speed (f/s)	6.17	10.01	0.50	0.049	0.43	1.02	0.065	0.87	1.00	0.94
Urban road 3	Cor. (%)	51.11%	53.49%	64.29%	40.54%	77.78%	66.39%	61.45%	75.94%	64.43%	78.77%
	Com. (%)	79.31%	85.19%	69.23%	88.24%	92.72%	84.21%	89.94%	90.70%	71.01%	92.16%
	Qua. (%)	45.10%	48.94%	50.00%	38.46%	73.30%	59.04%	57.50%	70.46%	51.01%	73.82%
	Speed (f/s)	6.22	10.26	1.37	0.063	0.97	1.16	0.077	1.04	1.36	1.25
Urban road 4	Cor. (%)	67.89%	77.53%	91.59%	86.76%	90.32%	90.99%	86.99%	90.92%	91.05%	90.09%
	Com. (%)	80.43%	51.49%	80.33%	93.65%	88.89%	78.91%	91.73%	93.96%	78.98%	88.64%
	Qua. (%)	58.27%	44.81%	74.81%	81.94%	81.16%	73.19%	80.67%	85.90%	73.29%	80.76%
	Speed (f/s)	5.27	9.64	0.60	0.056	0.46	0.85	0.066	0.75	0.83	0.79
Average	Cor. (%)	62.19%	69.03%	83.22%	76.10%	88.06%	85.88%	82.76%	88.56%	86.28%	88.50%
	Com. (%)	79.89%	66.59%	76.91%	92.31%	92.37%	73.01%	89.21%	91.13%	73.04%	92.15%
	Qua. (%)	54.24%	49.03%	66.79%	71.82%	82.09%	64.28%	75.13%	81.53%	65.08%	82.32%
	Speed (f/s)	5.98	10.05	1.14	0.079	0.88	1.07	0.074	0.942	1.27	1.17

The testing results of ten methods are presented in Table 2. The average metrics listed in the bottom of Table 2 show that our method achieved the best *Quality* (82.32%) with fast speed (1.17 f/s). Some comparisons are presented as follows:

(1) *Vibe & Frame Difference*: These two methods achieved fast detection speed but with low *Quality* (54.24% & 49.03%) which are too low to be accepted for real-world applications. The reason is that some non-vehicle objects (such as tricycles and moving pedestrians) lead to many false positives. Besides, slow-moving or stopped vehicles and some black vehicles which have similar colors with the road surface cannot be detected during detection.
(2) *V-J vs. V-J + 9*: The *Completeness* (76.91%) of V-J is low, this is because many vehicles that are not parallel to horizontal cannot be detected, thus generating many false negatives. The *Completeness* (92.31%) of V-J + 9 is significantly higher than the original V-J. After images were rotated every 20° from 0° to 180° and detected 9 times, vehicles of different orientations can be detected. However, repeating detections of the same image lead to more false positives and greatly increase detection time. The *Correctness* of V-J + 9 (76.10%) is lower than that of the original V-J (83.22%). The *detection speed* of V-J + 9 (0.079 f/s) is also significantly slower than V-J (1.14 f/s).
(3) *SVM vs. SVM + 9*: The comparison of SVM and SVM + 9 also demonstrates similar results as V-J vs. V-J + 9. SVM + 9 achieved higher *Completeness* (89.21%) than SVM (73.01%) but with low *Correctness* and *detection speed*.
(4) *V-J vs. V-J + R*: V-J + R achieves higher *Completeness* (92.37%) than V-J (76.91%); because by incorporating road orientation adjustment, on-road vehicles of unknown orientations will be aligned with the horizontal direction which can be detected by the original V-J detector. V-J + R achieves higher *Quality* (82.09%) than V-J (66.79%), but the detection speed of V-J + R is slower than the original V-J due to two reasons: (1) the road orientation adjustment step will cost some time for road orientation detection and image rotation; and (2) after image rotation and road alignment, many more vehicles in the UAV image need to be detected.
(5) *SVM vs. SVM + R*: The method SVM + R also achieves higher *Quality* (81.53%) than the original SVM method (64.28%). Similar to V-J + R, the detection speed of SVM + R is slower than the original SVM.
(6) *V-J + R vs. V-J + 9*: V-J + R achieves slightly higher *Completeness* (92.37%) than V-J + 9 (92.31%), because those rotated vehicles in V-J + 9 are in fact not exactly aligned with the horizontal direction, therefore may not adapt the original V-J detector well. Also, V-J + R achieves faster detection speed (0.88 f/s) than V-J + 9 (0.079 f/s). The comparisons demonstrate that the proposed road orientation adjustment method can improve both the *Completeness* and *Quality* compared with the original V-J and HOG + SVM methods, but leads to a slightly slower detection speed.
(7) *SVM + R vs. SVM + 9*: Similarly, SVM + R achieved higher *Completeness* (91.13%) than SVM + 9 (89.21%) and higher detection speed.
(8) *V-J vs. SVM vs. V-J + SVM + S*: The proposed switching method (i.e., V-J + SVM + S) achieves faster detection speed (1.27 f/s) than both the original V-J (1.14 f/s) and SVM (1.07 f/s) methods, because the proposed switching strategy can automatically choose the faster method between V-J and HOG + SVM during the detection. Note that, without the road orientation adjustment, the proposed switching method only achieves low *Quality* (65.08%), which is similar to V-J (66.79%) and SVM (64.28%).
(9) *V-J + SVM + R + S*: Our method, which combines the road orientation adjustment method, achieves the best *Quality* (82.32%) than other nine methods. The detection speed of our method is slower than V-J + SVM + S, which is a trade-off between high *Quality* and fast *detection speed*. However, our method is still faster than the original V-J and SVM methods. The detection speed of 1.17 f/s is acceptable for real-time applications.

Overall, the proposed method achieves good vehicle detection performance with fast speed. Particularly, our method is insensitive to on-road vehicles' in-plane rotation. Therefore, the proposed method can be performed on videos captured from moving UAV platforms (for example, UAVs flying along the road) without the need of image registration [9,31] or additional road database [15,16], thus has great potentials in wild field applications.

5. Discussion

5.1. Road Orientation Adjustment Method for Roadways with More Than One Orientation

The proposed road orientation adjustment method can be applied for roadways with more than one orientation. Here we present some testing results for roads with two orientations. As shown in Figures 14 and 15, the proposed road orientation adjustment method was applied to detect road orientations for: (1) an interchange with one freeway crossing over an arterial; and (2) a regular 4-leg intersection. As shown in the figures, two peaks were found in the relative frequency histograms. These two peaks essentially indicate the orientations of roads. For the case of interchange (Figure 14), two different orientations are around 0° and 90°; and for the case of 4-leg intersection (Figure 15), the orientations for the two orientations are around 0° and 92°. Note conceptually the method can be applied to detect many orientations. But when the road orientations are more than two, the peaks in the relative histogram could be difficult to identify.

Besides, it should be mentioned that the proposed road orientation adjustment method can only extract road orientations of straight roads. It will be difficult to extract the orientations of curve roads or very smaller roads. This also leads to some false detections (false negatives and false positives) during our vehicle detection when applying our method to detect vehicles on curve or small roads.

(a) (b) (c)

Figure 14. Roads orientation detection for an interchange: (**a**) Color image; (**b**) Line segments detection using LSD; (**c**) Relative frequency histogram (blue bins: distribution frequencies of relative histogram which correspond to different angles; red bins: two maximum distribution frequencies of relative histogram which correspond to two orientations of an interchange).

(a) (b) (c)

Figure 15. Roads orientation detection for an intersection: (**a**) Color image; (**b**) Line segments detection using LSD; (**c**) Relative frequency histogram (blue bins: distribution frequencies of relative histogram which correspond to different angles; red bins: the first two maximum distribution frequencies of relative histogram which correspond to two orientations of an intersection).

5.2. Straight Line Detection Using Other Algorithms

In this paper, we also compared the adopted LSD method with other line detection algorithms [28,32,33]. Particularly, we compared the LSD method to the Canny edge detector [34] followed by a Hough transform [28]. Note the Hough transform method need to tune parameters for each image manually, because using fixed parameters can lead to many false positives or false negatives. The comparison result is presented in Figure 16. As shown in the figure, the adopted line segment detector (LSD) can achieve much better performance. As seen in Figure 16c, many "redundant" lines were detected by Hough transform; by contrast, LSD achieved very "clean" line detection results (Figure 16d).

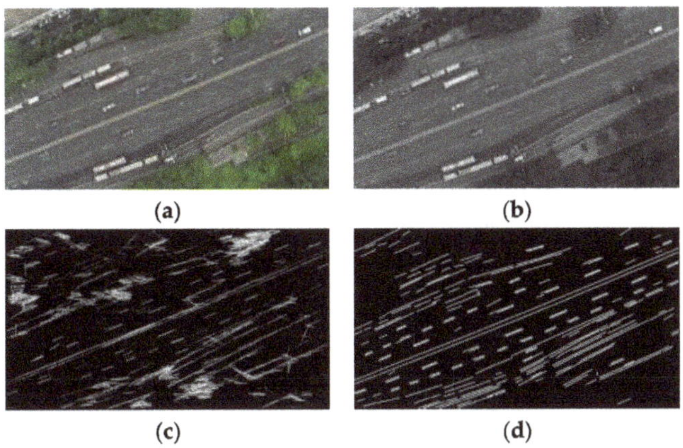

Figure 16. Line segments detection. (**a**) Color image; (**b**) Grayscale image; (**c**) Canny followed by Hough transform; (**d**) line segment detector (LSD).

5.3. Road Orientation Adjustment on Imagery with Low Radiometric Quality

The video images used for our testing have relatively high resolution of 1920 × 1080. But technically, our method can also work with images with low radiometric quality. Particularly, we performed our roadway orientation method on a low radiometric quality satellite image (see Figure 17). It can be seen from the figure that our method performed well on the image with low radiometric quality. Further comprehensive evaluation might be needed for future research.

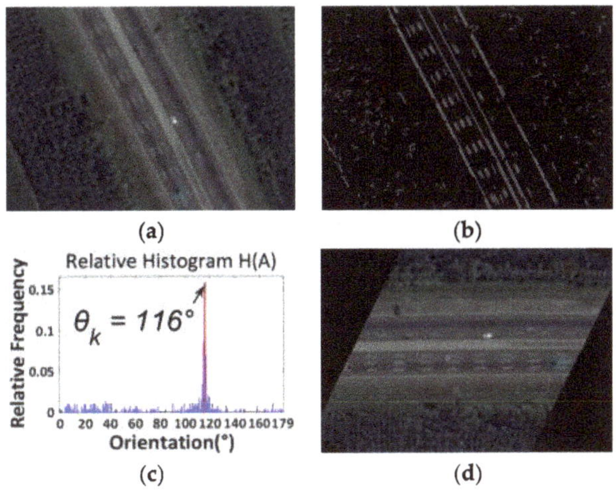

Figure 17. Low radiometric quality satellite image. (**a**) Color image; (**b**) Line segments detection using LSD; (**c**) Relative histogram (blue bins: distribution frequencies of relative histogram which correspond to different angles; red bin: the maximum distribution frequency of relative histogram which corresponds to the orientation of the road); (**d**) Rotated image.

5.4. Detection Using Oriented and Mosaicked Images

To be clear we did not use oriented, mosaicked images for detection in this research because mosaicked images may contain "ghost" vehicles [35], as marked by red arrows in Figure 18. For those "ghost" vehicles, only parts of the vehicle body can be seen. The reason for "ghost" vehicles is that the mosaicked image in Figure 18 was created by two different frames without synchronization (i.e., the two images were captured on different moments). Therefore those moving vehicles passing over the junction of the two images were cut off due to image mosaic. Obviously, those "ghost" vehicles will affect the accuracy of vehicle detections.

Figure 18. Mosaicked image. "Ghost" vehicles (marked by red arrows).

5.5. Vehicle Detection for Turning Vehicles

One drawback of the proposed vehicle detection method is its incapability of detecting turning vehicles. The roadway orientation adjustment can only rotate the image according to the orientation of the road. For turning vehicles, their orientations are changing during turning process and not aligned with the V-J and HOG + SVM detectors. This creates difficulties for vehicle detection. The original V-J and HOG + SVM methods also have this problem.

It is worth mentioning some recently developed state-of-the-art methods [19–21], which can detect vehicles in arbitrary directions. For example, Moranduzzo and Melgani [19] developed an automatic vehicle detection method, which is insensitive to vehicles' in-plane rotation. Conceptually, this method is superior to our method, because this method can be used to detect turning vehicles, while our method cannot, but that method needs to extract asphalted areas (i.e., road regions) first, which could cause some difficulties in detection since in urban traffic environment with complicated road conditions (different road type, roadway surfaces, weather, and illumination), road regions might be difficult to detect without the need of additional resources, like GIS. Moranduzzo and Melgani [20] further improved their method by using an additional road database, but clearly the need of the support of additional road database might limit their applications. Overall, Moranduzzo and Melgani's method achieved a total accuracy of 63.01%. When extracting road regions using GIS-based road maps, the total accuracy can be improved to 76.61%. Although this accuracy is lower than our method (88.50%), a comprehensive comparison using same images will be necessary. So far, due to lack of source codes of this method, such a comparison is difficult to perform. We will leave this work for future research.

Similarly, Liu and Mattyus [21] developed a multiclass vehicle detection method, which can detect vehicles with arbitrary orientations. To solve vehicle orientation problem, they trained a single classifier based on integral channel features (ICF) [36] which is able to detect vehicles orientated in different directions. Then eight binary classifiers are trained, each corresponding to a specific vehicle orientation. This method achieved a recall rate of 79.0% and a precision rate of 94%. Clearly, this method is competitive to our method with 88.5% recall rate and 92.15% precision rate. However, in Liu and Mattyus' method, training the single classifier needs vehicle samples with different directions (i.e., eight different directions in [21]), therefore the sample library is very large. Furthermore, labeling the training samples requires a lot of experience and is a tedious and time-consuming task. In our opinion, for vehicle detections from UAV images captured over arterial roads, as the majority of vehicles run in the same direction with the road, it is reasonable to detect vehicles by a single detector which is sensitive to only one specific direction, as performed in our method. This could significantly reduce the work load on collecting training samples by nearly one order of magnitude, since only samples of one specific direction are needed. The precision rate of our method (92.15%) is a bit lower than that of [21] (94%); this could be the tradeoff between high precision rate and less work load. But we have to point out that Liu and Mattyus' method can detect vehicles with arbitrary orientations, particularly for turning vehicles, while our method cannot. So conceptually Liu and Mattyus' method is superior to our method. A comprehensive comparison using same images will be necessary; but so far, due to the lack of source codes of this method, such comparison is difficult to perform. We will leave this work for future research.

5.6. Sensitivity Analysis of Switching Interval T

During our testing, the switching time interval, T, was arbitrarily set as 24 (i.e., 1 s time interval) based on the assumption that traffic conditions will not change abruptly over a period of time. But it is very possible that different T values might lead to different detection speed and detection accuracy. To comprehensively analyze the influence of different intervals of T on the detection speed and accuracy, experiments with different $T = \{24,120,240,360,480,600,720\}$ (namely, time intervals = 1, 5, 10, 15, 20, 25, 30 s) have been conducted. The testing was applied on the same datasets as shown in Table 1. The testing results of our method with different T intervals are shown in Table 3. It can be seen from Table 3 that, the influences of interval T on detection speed and accuracy are actually small. The reason would be that the time intervals for our videos are short and the traffic conditions do not change much, as confirmed in research done by [30]. Particularly, we present the speed changes of our method under different T intervals in Figure 19. Note the value of T we used for evaluating the performance of our method (see Table 2) is 24 (i.e., the interval is 1 s). For T = 24, the corresponding detection speeds were marked with red cross-shaped "+" shown in Figure 19 for comparison.

Table 3. Vehicle Detection Results with Different Intervals.

Scene	Metrics	T = 24 (1 s)	T = 120 (5 s)	T = 240 (10 s)	T = 360 (15 s)	T = 480 (20 s)	T = 600 (25 s)	T = 720 (30 s)
Freeway	Cor. (%)	89.37%	88.76%	89.60%	88.96%	88.72%	89.66%	90.52%
	Com. (%)	94.24%	94.34%	93.37%	94.16%	93.72%	93.04%	91.57%
	Qua. (%)	84.74%	84.27%	84.24%	84.30%	83.74%	84.03%	83.55%
	Speed (f/s)	1.0843	1.0811	1.0800	1.0847	1.0882	1.0730	1.0763
Urban road 1	Cor. (%)	87.89%	87.21%	88.00%	87.84%	87.17%	87.50%	87.21%
	Com. (%)	91.46%	91.65%	91.67%	91.23%	91.07%	91.42%	91.09%
	Qua. (%)	81.23%	80.79%	81.48%	81.00%	80.30%	80.86%	80.36%
	Speed (f/s)	1.8014	1.8255	1.8285	1.8329	1.8334	1.8337	1.8339
Urban road 2	Cor. (%)	96.39%	96.31%	96.90%	97.22%	96.59%	96.51%	96.36%
	Com. (%)	94.24%	94.13%	93.98%	93.58%	93.10%	92.28%	92.87%
	Qua. (%)	91.02%	90.86%	91.24%	91.15%	90.14%	89.30%	89.72%
	Speed (f/s)	0.9352	0.9406	0.9442	0.9475	0.9434	0.9453	0.9463
Urban road 3	Cor. (%)	78.77%	78.33%	78.75%	78.49%	77.65%	77.72%	77.47%
	Com. (%)	92.16%	91.26%	92.36%	91.53%	91.45%	91.78%	91.56%
	Qua. (%)	73.82%	72.87%	73.94%	73.18%	72.40%	72.66%	72.31%
	Speed (f/s)	1.2505	1.2519	1.2372	1.2608	1.2610	1.2453	1.2454
Urban road 4	Cor. (%)	90.09%	90.41%	89.93%	90.26%	90.30%	90.33%	90.52%
	Com. (%)	88.64%	89.42%	90.19%	91.13%	91.17%	91.80%	92.70%
	Qua. (%)	80.76%	81.68%	81.91%	82.97%	83.04%	83.58%	84.49%
	Speed (f/s)	0.7893	0.7918	0.7946	0.7981	0.7957	0.7977	0.7984
Average	Cor. (%)	88.50%	88.20%	88.64%	88.55%	88.09%	88.34%	88.41%
	Com. (%)	92.15%	92.16%	92.31%	92.32%	92.10%	92.06%	91.96%
	Qua. (%)	82.32%	82.09%	82.56%	82.52%	81.92%	82.08%	82.09%
	Speed (f/s)	1.1722	1.1782	1.1769	1.1848	1.1843	1.1790	1.1801

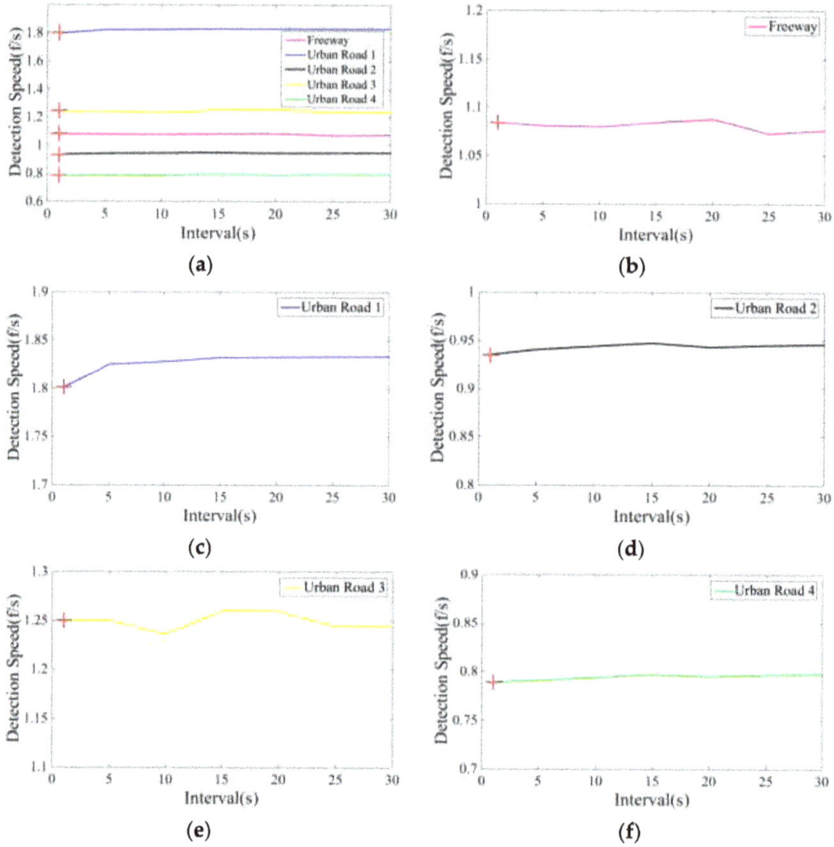

Figure 19. Vehicle detection speed of our method in different test scenes and the red cross shaped "+" presents the detection speed when the value of T is 24 (i.e., the interval is 1 s). (**a**) five scenes; (**b**) freeway; (**c**) urban road 1; (**d**) urban road 2; (**e**) urban road 3; (**f**) urban road 4.

6. Concluding Remarks

In this paper, a new hybrid vehicle detection scheme which integrates V-J and HOG + SVM methods is proposed. As both V-J and HOG + SVM are sensitive to on-road vehicles' in-plane rotation, the proposed scheme first adopts a roadway orientation adjustment method, which rotates each UAV image so that roads and on-road vehicles in images are aligned with the vehicle detector. After rotation, the original V-J or HOG + SVM methods can be applied to achieve higher accuracy. To address the issue of descending detection speed for both V-J and HOG + SVM, the proposed scheme further develops a hybrid and adaptive switching strategy which sophistically integrates V-J and HOG + SVM methods based on their different descending trends of detection speed to achieve better detection efficiency. A comprehensive evaluation shows that the switching strategy, combined with the road orientation adjustment method, can significantly improve the efficiency and effectiveness of the vehicle detection from UAV images. The results also show that the proposed vehicle detection method is competitive compared with other existing vehicle detection methods. Furthermore, it is worth mentioning that the proposed vehicle detection method can be performed on videos captured from moving UAV platforms without the need of image registration or additional road database, therefore it has great potentials of wide field applications.

However, the proposed vehicle detection scheme has difficulties to address turning vehicles. The future research will aim to address this problem. Some recently developed state-of-the-art methods [19–21,37] will be useful references for our future research. Particularly, the hybrid deep convolutional neural networks (DNNs) suggested by Chen et al. [37] would be an interesting direction for our future research. Indeed, it would be very interesting to compare our method with the DNN method, and to seek any possibility of applying the Faster R-CNN [38,39] for multimodal object detection (car, bus, truck, van, pedestrian, etc.) from UAV images. Due to the difficulty of constructing DNN and Faster R-CNN, we will leave all these for our future research.

Supplementary Materials: The supplementary materials are available online at http://www.mdpi.com/1424-8220/16/8/1325/s1.

Acknowledgments: This work is partially supported by the Fundamental Research Funds for the Central Universities and partially by the National Science Foundation of China under Grant #61371076. The authors would also like to thank the insightful and constructive comments from anonymous reviewers.

Author Contributions: Yongzheng Xu and Guizhen Yu designed the overall system and developed the vehicle detection algorithms. In addition, they wrote and revised the paper. Xinkai Wu designed and performed the experiments. Yunpeng Wang analyzed the experiment results. Yalong Ma has made significant contribution on the UAV video data collection.

Conflicts of Interest: The authors declare no conflict of interest.

Abbreviations

UAV	unmanned aerial vehicle
SVM	support vector machine
HOG	histogram of oriented gradient
LSD	line segment detector
DPM	deformable part model
V-J	Viola-Jones object detection scheme
HOG + SVM	linear SVM classifier with HOG feature
Cor.	*Correctness*
Com.	*Completeness*
Qua.	*Quality*

References

1. Rosser, K.; Pavey, K.; Fitzgerald, N.; Fatiaki, A.; Neumann, D.; Carr, D.; Hanlon, B.; Chahl, J. Autonomous Chemical Vapour Detection by Micro UAV. *Remote Sens.* **2015**, *7*, 16865–16882. [CrossRef]
2. Gonzalez, L.F.; Montes, G.A.; Puig, E.; Johnson, S.; Mengersen, K.; Gaston, K.J. Unmanned Aerial Vehicles (UAVs) and Artificial Intelligence Revolutionizing Wildlife Monitoring and Conservation. *Sensors* **2016**, *16*, 97. [CrossRef] [PubMed]
3. Boccardo, P.; Chiabrando, F.; Dutto, F.; Tonolo, F.G.; Lingua, A. UAV Deployment Exercise for Mapping Purposes: Evaluation of Emergency Response Applications. *Sensors* **2015**, *15*, 15717–15737. [CrossRef] [PubMed]
4. Agrawal, A.; Hickman, M. Automated extraction of queue lengths from airborne imagery. In Proceedings of the International IEEE Conference on Intelligent Transportation Systems, Washington, DC, USA, 3–6 October 2004; pp. 297–302.
5. Coifman, B.; Mccord, M.; Mishalani, R.G.; Iswalt, M. Roadway traffic monitoring from an unmanned aerial vehicle. *IEE Proc. Intell. Transp. Syst.* **2006**, *153*, 11–20. [CrossRef]
6. Yu, G.; Zhou, B.; Wang, Y.; Wu, X.; Wang, P. Measuring algorithm for the distance to a preceding vehicle on curve road using on-board monocular camera. *Int. J. Bifurc. Chaos* **2015**, *25*, 1540038. [CrossRef]
7. Angel, A.; Hickman, M.; Mirchandani, P.; Chandnani, D. Methods of analyzing traffic imagery collected from aerial platforms. *IEEE Trans. Intell. Transp. Syst.* **2003**, *4*, 99–107. [CrossRef]
8. Azevedo, C.L.; Cardoso, J.L.; Ben-Akiva, M.; Costeira, J.P.; Marques, M. Automatic Vehicle Trajectory Extraction by Aerial Remote Sensing. *Procedia Soc. Behav. Sci.* **2014**, *111*, 849–858. [CrossRef]

9. Shastry, A.C.; Schowengerdt, R.A. Airborne video registration and traffic-flow parameter estimation. *IEEE Trans. Intell. Transp. Syst.* **2005**, *6*, 391–405. [CrossRef]
10. Yalcin, H.; Hebert, M.; Collins, R.; Black, M.J. A Flow-Based Approach to Vehicle Detection and Background Mosaicking in Airborne Video. In Proceedings of the 2005 IEEE Computer Society Conference on Computer Vision and Pattern Recognition, San Diego, CA, USA, 20–25 June 2005.
11. Viola, P.; Jones, M. Rapid object detection using a boosted cascade of simple features. In Proceedings of the 2001 IEEE Computer Society Conference on Computer Vision and Pattern Recognition, Kauai, HI, USA, 8–14 December 2001; pp. 511–518.
12. Dalal, N.; Triggs, B. Histograms of oriented gradients for human detection. In Proceedings of the 2005 IEEE Computer Society Conference on Computer Vision and Pattern Recognition, San Diego, CA, USA, 20–25 June 2005; Volume 1, pp. 886–893.
13. Cao, X.; Wu, C.; Yan, P.; Li, X. Linear SVM classification using boosting HOG features for vehicle detection in low-altitude airborne videos. In Proceedings of the IEEE International Conference on Image Processing, Brussels, Belgium, 11–14 September 2011; pp. 2421–2424.
14. Cao, X.; Wu, C.; Lan, J.; Yan, P. Vehicle Detection and Motion Analysis in Low-Altitude Airborne Video Under Urban Environment. *IEEE Trans. Circuits Syst. Video Technol.* **2011**, *21*, 1522–1533. [CrossRef]
15. Leitloff, J.; Rosenbaum, D.; Kurz, F.; Meynberg, O.; Reinartz, P. An Operational System for Estimating Road Traffic Information from Aerial Images. *Remote Sens.* **2014**, *6*, 11315–11341. [CrossRef]
16. Tuermer, S.; Kurz, F.; Reinartz, P.; Stilla, U. Airborne Vehicle Detection in Dense Urban Areas Using HoG Features and Disparity Maps. *IEEE J. Sel. Top. Appl. Earth Obs. Remote Sens.* **2013**, *6*, 2327–2337. [CrossRef]
17. Xu, Y.; Yu, G.; Wang, Y.; Wu, X. Vehicle Detection and Tracking from Airborne Images. In Proceedings of the 15th COTA International Conference of Transportation Professionals, Beijing, China, 24–27 July 2015; pp. 641–649.
18. Jones, M.; Viola, P. Fast Multi-view Face Detection. In Proceedings of the 2003 IEEE Computer Society Conference on Computer Vision and Pattern Recognition, Madison, WI, USA, 16–22 June 2003; Volume 11, pp. 276–286.
19. Moranduzzo, T.; Melgani, F. Detecting cars in UAV images with a catalog-based approach. *IEEE Trans. Geosci. Remote Sens.* **2014**, *52*, 6356–6367. [CrossRef]
20. Moranduzzo, T.; Melgani, F. Automatic car counting method for unmanned aerial vehicle images. *IEEE Trans. Geosci. Remote Sens.* **2014**, *52*, 1635–1647. [CrossRef]
21. Liu, K.; Mattyus, G. Fast Multiclass Vehicle Detection on Aerial Images. *IEEE Geosci. Remote Sens. Lett.* **2015**, *12*, 1938–1942.
22. Felzenszwalb, P.F.; Girshick, R.B.; McAllester, D.; Ramanan, D. Object Detection with Discriminatively Trained Part Based Models. *IEEE Trans. Pattern Anal. Mach. Intell.* **2010**, *32*, 1627–1645. [CrossRef] [PubMed]
23. Friedman, J.; Hastie, T.; Tibshirani, R. Additive Logistic Regression: A Statistical View of Boosting. *Ann. Stat.* **2000**, *28*, 374–376. [CrossRef]
24. Cucchiara, R.; Piccardi, M.; Mello, P. Image analysis and rule-based reasoning for a traffic monitoring system. *IEEE Trans. Intell. Transp. Syst.* **2000**, *1*, 119–130. [CrossRef]
25. Tao, H.; Sawhney, H.; Kumar, R. Object tracking with Bayesian estimation of dynamic layer representations. *IEEE Trans. Pattern Anal. Mach. Intell.* **2002**, *24*, 75–89. [CrossRef]
26. Ma, Y.; Wu, X.; Yu, G.; Xu, Y.; Wang, Y. Pedestrian Detection and Tracking from Low-Resolution Unmanned Aerial Vehicle Thermal Imagery. *Sensors* **2016**, *16*, 446. [CrossRef] [PubMed]
27. Grompone, V.G.R.; Jakubowicz, J.; Morel, J.M.; Randall, G. LSD: A Fast Line Segment Detector with a False Detection Control. *IEEE Trans. Pattern Anal. Mach. Intell.* **2010**, *32*, 722–732.
28. Ballard, D.H. Generalizing the Hough Transform to Detect Arbitrary Shapes. *Pattern Recognit.* **1981**, *13*, 111–122. [CrossRef]
29. Gioi, R.G.V.; Jakubowicz, J.; Morel, J.M.; Randall, G. LSD: A Line Segment Detector. *Image Process. Line* **2012**, *2*, 35–55. [CrossRef]
30. Maerivoet, S.; De Moor, B. Traffic Flow Theory. *Physics* **2005**, *1*, 5–7.
31. Olivier, B.; Marc, V.D. ViBe: A universal background subtraction algorithm for video sequences. *IEEE Trans. Image Process.* **2011**, *20*, 1709–1724.
32. Faugeras, O.; Deriche, R.; Mathieu, H.; Ayache, N.J.; Randall, G. The Depth and Motion Analysis Machine. *Int. J. Pattern Recognit. Artif. Intell.* **1992**, *6*, 353–385. [CrossRef]

33. Burns, J.B.; Hanson, A.R.; Riseman, E.M. Extracting Straight Lines. *IEEE Trans. Pattern Anal. Mach. Intell.* **1986**, *8*, 425–455. [CrossRef]
34. Canny, J. A computational approach to edge detection. *IEEE Trans. Pattern Anal. Mach. Intell.* **1998**, *8*, 679–698.
35. Uyttendaele, M.; Eden, A.; Skeliski, R. Eliminating Ghosting and Exposure Artifacts in Image Mosaics. In Proceedings of the 2001 IEEE Computer Society Conference on Computer Vision and Pattern Recognition, Kauai, HI, USA, 8–14 December 2001; pp. 509–516.
36. Dollár, P.; Tu, Z.; Perona, P.; Belongie, S. Integral Channel Features. In Proceedings of the British Machine Vision Conference, London, UK, 7–10 September 2009.
37. Chen, X.; Xiang, S.; Liu, C.L.; Pan, C.H. Vehicle detection in satellite images by hybrid deep convolutional neural networks. *IEEE Geosci. Remote Sens. Lett.* **2014**, *11*, 1797–1801. [CrossRef]
38. Ren, S.; He, K.; Girshick, R.; Sun, J. Faster R-CNN: Towards Real-Time Object Detection with Region Proposal Networks. In Proceedings of the Neural Information Processing Systems Conference, Montreal, QC, Canada, 7–12 December 2015.
39. Faster R-CNN. Available online: https://github.com/rbgirshick/py-faster-rcnn (accessed on 20 June 2016).

sensors

MDPI

Article

Real-Time Robust Tracking for Motion Blur and Fast Motion via Correlation Filters

Lingyun Xu [1,2,3,*], Haibo Luo [1,2], Bin Hui [1,2] and Zheng Chang [1,2]

1 Key Laboratory of Opto-Electronic Information Processing, Chinese Academy of Sciences, Shenyang 110016, China; luohb@sia.cn (H.L.); huibin@sia.cn (B.H.); ChangZheng@sia.cn (C.Z.)
2 Shenyang Institute of Automation, Chinese Academy of Sciences, Shenyang 110016, China
3 University of Chinese Academy of Science, Beijing 100049, China
* Correspondence: xulingyun@sia.cn; Tel./Fax: +86-24-2397-0757

Academic Editors: Felipe Gonzalez Toro and Antonios Tsourdos
Received: 22 June 2016; Accepted: 17 August 2016; Published: 7 September 2016

Abstract: Visual tracking has extensive applications in intelligent monitoring and guidance systems. Among state-of-the-art tracking algorithms, Correlation Filter methods perform favorably in robustness, accuracy and speed. However, it also has shortcomings when dealing with pervasive target scale variation, motion blur and fast motion. In this paper we proposed a new real-time robust scheme based on Kernelized Correlation Filter (KCF) to significantly improve performance on motion blur and fast motion. By fusing KCF and STC trackers, our algorithm also solve the estimation of scale variation in many scenarios. We theoretically analyze the problem for CFs towards motions and utilize the point sharpness function of the target patch to evaluate the motion state of target. Then we set up an efficient scheme to handle the motion and scale variation without much time consuming. Our algorithm preserves the properties of KCF besides the ability to handle special scenarios. In the end extensive experimental results on benchmark of VOT datasets show our algorithm performs advantageously competed with the top-rank trackers.

Keywords: visual tracking; motion blur; fast motion; correlation filter

1. Introduction

Visual object tracking plays an active role in military guidance, robot navigation, medical image processing, virtual augment reality and many other applications. Nevertheless, an efficient tracker faces combined challenges due to varying circumstances. First, the scenarios may suffer from illumination variations, changing camera postures and background clutters. Second, the appearance of target itself may have variations on scale, aspect-ratio, color or deformation in non-rigid object. Moreover, the typical low speed of camera and high speed of target make fast motion a common problem in visual tracking. Meanwhile, in a considerable number of scenarios, critical feature extraction and classical tracking approaches fail affected by motion blur due to the high relative speed [1].

Many research efforts have been focused on building appearance models of targets to estimate the variation of target in the past decades. In a sort of way conventional algorithms extract features from the target patch and evaluate the affine matrix or the consistency to reap the new position or probability distribution of target. Say, NCC [2] is a typical tracking algorithm using images matching which searches the cross-correlation values of the template image and candidate patches to predict the target position with best score. To improve the search strategy of target motion, Akbariet et al. [3] integrate Particle Swarm Optimization and Kalman filter into a framework for tracking. Comanicu D et al. [4] apply mean shift to build a fast search scheme for optimization using grey features. Lucas and Kanade [5] evaluate the Optical Flow between consecutive frames by minimizing the constrained energy function caused by three hypotheses about the consistence of pixels and voxels. Such traditional algorithms almost have no considerations on variations of the target appearance model.

Image sparse representation is hot research recently for it has a merit of representing signal flexibly and can avoid noise effects and appearance changes to some extent when maintaining features of targets. Sparse representation has been applied to tracking task in [6], and later be speeded up in [7]. Although it improves the speed for the task, sparse representation always consumes large computation that may not meet the real-time request in applications. There are also useful algorithms whose starting point is the Bayesian networks frame. In [8], conditional random filed is exploited to evaluate the matching score of target. However, these methods also suffer from the larger variations of target or scenarios, including applications with motion blur and fast motion.

Another view of tracking is to take the task as a binary classification between target foreground and the background. Kalal [9] takes the tracking task as a tracking, detection and learning(TLD) problem, which makes full use of the target and motion information. This mechanism has been proved to be effective and flexible. If the location information of object in previous frame is individually reaped to estimate the current target, the appearance model may accumulate errors and once the object disappears, the target will fail to be tracked permanently. The updating classifier detector by new samples coming in adjusts the appearance model in a more reliable way. Also inspired by this idea, (Struck) [10] introduces a clear output space and proposes a structure output prediction based adaptive visual tracking framework. The TLD tracker and Struck tracker outperform most of the traditional trackers in public test especially when the appearance model of a target changes heavily or gets lost, but they also have difficulties when handling the fast or blur motions of a target and always assume large computation.

Among the state-of-the-art methods, recently the algorithms based on correlation filters shows its advantages both in accuracy and robustness. Through introducing frequency domain transformation, Henriques et al. [11] consider the detection as a binary kernel ridge regression problem. As a tracking-by-detection tracker, the processing speed of Kernelized Correlation Filter (KCF) fantastically is fantastically dozens of times that of TLD or Struck. The multi-channel features and the approach to integrate them together are applied in KCF to build an insensitive stronger classifier for illustration variation and appearance model variation. In this way, the tracker is easier to understand and can be supported by richer powerful features e.g., CNN features and textural features, rather than just using the raw greyscale pixels. It is more efficient than the model with just single feature and has the potential to be widely used for real applications. However, there still remain many challenges, say, the scale variation, majority occlusion motion blur and fast motion.

In this paper, we present a new scheme to deal with fast motion and the motion blur caused by fast motion without deblurring the image. Our algorithm achieves much higher accuracy and robustness competing with the top-rank trackers. Our main contributions can be concluded as follows: (1) analyze the property of frequency-domain feature for blurred image; (2) utilize sharpness point function to evaluate the motion of target; (3) build an active and search scheme on kernelized correlation filter for fast and robust tracking of scenarios with fast and/or blur motion. Our method has achieved excellent performance competing with the current rank-top trackers on extensive experiments; in addition, our method reaches a high speed even in worst scenarios.

2. Related Works

In this part, the related works is discussed for three problems: (1) correlation filter based tracking; (2) blur motion and fast motion handling; (3) Scale variation handling.

2.1. Correlation Filter Based Tracking

Heads from signal processing, correlation filter (CF) recently gives great rise to interests in image detection and classification. In signal processing, the cross correlation of signals assesses the similarity between them and reaches the extreme value when the two functions for signals are the same. This property can be utilized to evaluate the similarity of two image patches. Adopting the principle of achieving the maximum cross correlation response between its model and candidate patch, the Minimum Output Sum of Squared Error (MOSSE) filter is proposed by David S. Bolme et al. [12] to reap the appearance model of target. Instead of computing the inverse matrix as many algorithms do, MOSSE only requires divide operation in frequency domain, which makes its robustness and speed both compelling. Hamed et al. [13] ameliorate the way of selecting training patches to dramatically reduce both the boundary effects and expensive sampling computations for MOSSE.

Despite the advances of MOSSE, it has prime problems in two aspects: (1) the way of sampling consumes much computation; (2) as a classifier, it could be stronger and more flexible. Introducing the kernel trick for classifier improvement and the cycle shift technique for fast sampling, Henriques et al. [14] take the training model as a ridge regression question and then give the corresponding quick solution. To get better interpretability and higher adaptability for diverse scenarios, Henriques et al. also introduce multi-channel HOG features and further create a scheme to integrate selected multiple features into the tracking framework. The KCF tracker has made great improvement both in accuracy and processing speed, meanwhile it has been paid so much attention that recently corresponding researchers devote to fixing it in different aspects. For better feature detection, literature [15] discusses how various convolution neural network features affect the accuracy of CFs, and draws a conclusion that the feature in the first level wins. In [16], a part-based model strategy is applied to cope with deformation of target. A scale adaptive Kernelized Correlation Filter (SAMF) [17] relieves the problem of scale estimation for target by applying a Laplacian Pyramid with a pool of 7 different scales. Some other peer trackers combine long-term tracking with KCF. For instance, Long-term Correlation Tracking [18] implements an active scheme to utilize random ferns for re-detection and Muster [19] build a biologically inspired system for cooperation of CF short term tracking and long term tracking. However, there has not been an effective scheme to handle with fast motion and motion blur under this tracking frame.

2.2. *Blur Motion and Fast Motion Handling*

Fast motion and motion blur are two prominent problems in tracking system, which have also been pervasively encountered in real applications, e.g., moving objects in large scenes, shaking camera. The problems for blur motion and fast motion include: (1) fast motion between camera and target makes estimating motion parameters more difficult even in consecutive frames; (2) motion blur brings about slightly or greatly negative effects on the performance of the traditional feature based detectors. An spontaneous idea to settle these problems is to deblur the blurred candidate patch and then apply original tracking methods. Although this could be theoretically successful, the current algorithms for deblurring image consume large amount of time which makes real-time tracking improbable. Blur-driven Tracker (BLUT) [20] builds a set of blurred images with different parameters as models for sparse representation. Since the blur motion model could be formulated as the convolution of a Point Spread Function (PSF) and the original image, Dai and Wu [21] propose a method to estimate the PSF to track the blurred target. In [22], a matching score function caused by the cost function between blurred image and model is introduced to search the target region.

Our algorithm avoids estimating the parameters of blur model and adaptively handles the problem with slightly or greatly blurred or fast motion. Meanwhile, we have not a higher computational complexity of our method for visual sequences with from barely to greatly fast or blurred motion.

2.3. *Scale Variation Handling*

Since the original KCF algorithm has no scheme for scale variation estimation, in the recent two years certain improved KCF algorithms focus their efforts on this problem. As discussed above, SAMF exploits Laplacian Pyramid of multi levels on the candidate patches and updates the model by resizing the detected target to a certain scale. Multi-kernel Correlation Filter [23] advices to apply the

frequency peak sidelobe ratio (PSR) of target patch for estimating its scale. Spatio-temporal Context tracker (STC) [24] observes that the variation of scale affects the score of the estimated target position in Bayesian confidence map. For fast and accurate scale estimation, our method improves the scale variation of STC via considering the variation of confidence the motion model brings about.

3. Approach

We propose a new method to combine KCF and STC for scale estimation, and practice our new scheme for motion blur and fast motion testing. In the first place, the KCF algorithm is re-formulated and fixed with scale estimation; secondly the foundation of our algorithm is theoretically analyzed. At the end, the detailed steps of the algorithm for fast and blur motions are given.

3.1. Re-Formulate Kernelized Correlation Filter Tracking with Scale Handling

The tracking frame of Kernelized Correlation Filter, like other correlation filter algorithms, is inspired by MOSSE tracker. Now the scheme of KCF tracker is specified. At first, let's consider the ridge regression problem: given the training data set and their labels $\{(\mathbf{x}_i, y_i)\}_{i=1\sim n}$, function $f(\mathbf{z}) = \mathbf{w}^T\mathbf{z}$ could be found to satisfies the equation below,

$$\underset{\mathrm{w}}{\operatorname{argmin}} \sum_i (f(\mathbf{x}_i) - y_i)^2 + \lambda \|\mathbf{w}\|^2 \tag{1}$$

where λ is the penalty coefficient of regularization item, preventing the over-fitting phenomenon. Our aim is to train a model $\mathbf{w}$ to best satisfy the training samples based on the Linear Least Squares criteria. Here is the explicit solution for Equation (1) in complex fields:

$$\mathbf{w} = (X^H X + \lambda \boldsymbol{I})^{-1} X^H y \tag{2}$$

where $X^H = (X^*)^T$ is the complex conjugate matrix of X. Introducing "kernel trick" to support a richer model, the model can be re-formulated as

$$\underset{\alpha}{\operatorname{argmin}} \|\mathbf{K}\alpha - \mathbf{y}\|_2^2 + \alpha \mathbf{K} \alpha \tag{3}$$

where $\mathbf{K}$ is the kernel matrix with dot-product in Hilbert space as its elements $\langle \phi(x_i), \phi(x_j) \rangle = k(x_i, x_j)$. According to the relevant theory, the solution to the classical kernelized ridge regression can be given by:

$$\alpha = (\boldsymbol{K} + \lambda \boldsymbol{I})^{-1} \mathbf{y} \tag{4}$$

Since this solution involves the matrix inverse operation, the computational complexity could increase fast. Luckily, Henriques et al. have proved in [11] that if $\boldsymbol{K}$ is a cyclic matrix and some certain conditions are met by $\boldsymbol{K}$, then the solution in frequency domain can be simplified as:

$$\hat{\alpha} = \frac{\hat{y}}{\hat{K}^{xx} + \lambda} \tag{5}$$

Here $\boldsymbol{K}^{xx}$ is the first row vector of the cyclic matrix $\boldsymbol{K}$. The hat symbol ˆ represents the Fourier transformation of a vector. For convenience, they also give three most typical kernel functions qualified for the theory, namely, Polynomial Kernel, Gaussian Kernel and Linear Kernel. In specially, when selecting a Linear Kernel, the problem is reduced to original regression problem without relative kernel trick correspondingly. Once securing or initializing the coming target patch x', the kernel matrix in frequency domain is calculated:

$$K^{xx\prime} = \left(F^{-1}(\hat{x} \odot \hat{x}'^*) + a\right)^b \tag{6}$$

$$K^{xx\prime} = \exp(-\frac{1}{\delta^2}(\|x\|^2 + \|x'\|^2 - 2F^{-1}(\hat{x} \odot \hat{x}'^*))) \tag{7}$$

Here Equation (6) is for Polynomial Kernel and Equation (7) is for Gaussian Kernel. $\odot$ denotes element-wise multiplication and $*$ means the complex conjugate of a vector. The vector x represents the appearance model and in the first coming frame, x is initialized to x'.

The response for position $\mathbf{z}$ is then estimated by parameters $\boldsymbol{\alpha}$ and $\mathbf{x}$:

$$\hat{f}(z) = \hat{K}^{xz} \odot \hat{\alpha} \tag{8}$$

To catch the variations of target appearance model, the KCF tracker has its scheme to update its template with fixed learning rate η (set as 0.2 in our experiment):

$$\begin{cases} \mathcal{F}(\alpha)^t & = (1-\eta)\mathcal{F}(\alpha)^{t-1} + \eta\mathcal{F}(\alpha) \\ \hat{x}^t & = (1-\eta)\hat{x}^{t-1} + \eta\hat{x} \end{cases} \tag{9}$$

Since KCF has no scheme to handle with scale variation, the target scale could be estimated by referring to the STC tracker. Unlike other discrete scale evaluation methods e.g., SAMF, the STC tracker has a simple but effective scheme to flexibly calculate the scale variation of target. Its principle starts from a statistical point that if other conditions remain unchanged, the score of Bayesian confidence map goes down descend as the scale gets bigger. In detail, the target scale at frame t is given by:

$$s_t = (\frac{p((x_0)_t)}{p((x_0)_{t-1})})^{\frac{1}{2}} \tag{10}$$

where x_0 is the evaluated position of target and $p(x_0)_t$ is the confidence in frame t, namely the score tracker acquires for position x_0. For more reliable and stable estimation of target scale, the updating scheme for s contains an inertia item calculated by mean of $\{s_i\}_{i=1:t}$:

$$\begin{cases} \bar{s} & = (1/n)\sum_i s'_i \\ s_{t+1} & = (1-\eta)s_t + \eta\bar{s} \end{cases} \tag{11}$$

where η is again a fixed learning rate. Above all is the basic idea and theoretical derivations for the KCF tracker with a scale evaluation scheme used in the STC tracker. The two methods are fused and summarized in Algorithm 1.

Algorithm 1. The KCF tracker with target scale estimation

Inputs: Template x, target position x_0 initialized by the given ground truth in the first frame;
Initial scale $s = 1$;
Corresponding image sequence for tracking;
Outputs: Estimated target position x_0' and estimated target scale s' for each frame;
Updated x' and model coefficient $\boldsymbol{\alpha}'$ at each time;

```
Selection: Select kernel function and features type;
Initialize the model coefficient α with Equation (5);
for every coming frame do
Get the candidate image patch by {x, s} and resize the patch to the same size as x via bilinear interpolation;
Calculate the kernel function K^{xx'} with Equation (6) or (7) according to the selected kernel;
Calculate the response f(z) with Equation (8);
Acquire the position of the maximum response x0' and new size s' with Equations (10) and (11);
Update the template by Equations (5) and (9);
end for
Return {x', α', s'}.
```

3.2. Analysis of Motion Blur and Fast Motion in Frequency Domain

Point Spread Function. The degradation model and its influence of motion blur on our tracker are analyzed in this section. If there is relative motion between the photographic apparatus and the object in exposal moment, then the image captured from apparatus always gets blurred. In visual tracking, motion blur caused by fast relative motion between the target and background often makes trouble for building the appearance model of target. We should have a big concern with it. Generally, PSF is employed to approximate the motion model for linear and shift invariant system. Assume the original image is $f(x, y)$, the blurred image could be given by:

$$g(x,y) = f(x,y) \otimes h(x,y) + n(x,y) \tag{12}$$

where $h(x, y)$ is the PSF and $n(x, y)$ is the additive noise. $\otimes$ denotes function convolution. A motion vector could be disassembled into resultant movement concise of motions in vertical and horizontal directions. It's noticeable that the motion between two consecutive frames is so slight compare to other motions that the motion can be regarded as uniform motion. For convenience, let's consider the variation in horizontal direction first. The PSF for horizontal uniform linear motion can be written as:

$$h(x,y) = \begin{cases} 1/L & ,|x| \le L, y = 0 \\ 0 & ,else \end{cases} \tag{13}$$

To get information about the blurred image in frequency domain, the function $h(x, y)$ in Equation (12) is replaced by Equation (13) and its frequency format is derived via Fourier transform:

$$\begin{aligned} & G(u,v) = F(u,v) \odot H(u,v) \\ & = F(u,v) \cdot \iint h(x,y)\exp(-j2\pi(ux+vy))dxdy \\ & = F(u,v) \cdot \int_0^L \tfrac{1}{L}\exp(-j2\pi ux)dx \\ & = F(u,v) \cdot \tfrac{\sin(\pi uL)}{\pi uL}\exp(-j\pi uL) \end{aligned} \tag{14}$$

Observing Equation (14), the spectrum for $g(x, y)$ is concise of vertical parallel stripes. In a more general sense, the motion vector can be decomposed to two vectors in horizontal and vertical directions respectively. The principle for uniform motion in vertical direction can be deduced in the similar way. Actually, it can summarize that the direction of the target motion is orthogonal to the direction of parallel stripes in frequency spectrum. Some examples are listed in Figure 1. The motion between two neighboring frames could be estimated as uniform motion for the mistiming is limited. In our algorithm, the PSF of blur motion is not been estimated. Instead, the direction θ for the motion is calculated and utilized to determine the distance between target positions of two frames to avoid interference objects nearby. This information is abundant for target estimation in our tracker.

Radon Transformation. If the black stripes in spectrum are considered as lines, the position and direction of the dark lines could be evaluated via radon transformation. Once the directions of these dark lines are calculated, the motion direction is obtained by rotating 90°. By calculating the line integral in specified direction, Radon transformation figure out projection of a matrix in the corresponding direction. The radon transformation along angle α is defined as:

$$I_\alpha(x') = \int f(x'\cos\alpha - y'\sin\alpha, x'\sin\alpha + y'\cos\alpha)dy' \tag{15}$$

As we could see, the response to $I_\alpha(x')$ hits the maximum when there is a long straight line in the project direction. In order to get the movement direction, radon transformation of angle 0°~180° is made for the spectrum $G(u, v)$ and acquire the vector combined with every maximum value of each angle. Then these maximum value are united together to get a curve with direction varying. The maximum value of this curve is respected to the motion direction.

Figure 1. Figures from left to right are: original image with no motion, 20°, 40°, 60° movement in the horizontal direction. The figures blew are their corresponding spectrum.

The Motion Model in Frequency Domain. Although motion blur is introduced by relative fast motion, the target may have a high speed for movement without motion blur. In this part, the motion model is discussed for these scenarios to find the elusive target position. CF trackers always employ correlation functions in frequency domain to evaluate the similarity of two image patch. When two functions are equal to each other, namely the two image patches are the same, the value of cross-correlation function hits the maximum value. In signal processing, the cross-correlation function of two signals is written by:

$$R_{xy}(l) = \lim_{T \to \infty} \frac{1}{T} \int_0^T x(t)y(t+l)dt \tag{16}$$

This function is always utilized to test the delay for signal. It can be seen that convolution between PSF of uniform motion and image will not change the maximum response position of the cross-correlation function. Moreover, if a signal is spread from point A to point B, R_{xy} secures the peak value in the position of delay epoch. Determining the delay time can be utilized to measure the rate of movement of the object in engineering application, which also suggests us to estimate the movement of the target. The estimation formulations will be specify in the next section.

The Image Blur Metric. The question remained is how to reduce computation consuming. In this part the choice criteria is presented for image blur metric in our tracker and the chosen one. One compelling advantage of the KCF tracker is the high speed, hence we manage to improve the tracker without much computing. Since the computation is mainly on larger range searching, how to reduce the unnecessary searching should be accounted. If we could have an evaluation of the target's motion, then we could lock the searching range when it's less important. Here we introduce clarity evaluation function of motion blurred image to figure out the intensity of the target movement. Familiar metrics for image clarity are variance metric, autocorrelation-based metric, derivative-based metrics, frequency threshold metric, etc. What we need is one without reference or complex computing, perhaps not too sensitive to introduce a complex scheme for our tracker. In experiment we select JNB [25] metric as our clarity metric function. This metric is based on Sobel operation and can be employed to handle different contents in one image without much computational complexity that satisfies our requirement.

3.3. The Tracker

In the former section the theoretical basis of the proposed algorithm is discussed, in this part we will specify our tracker. It's important to note that all the parameters in our implementation are fixed and need no manual setting. There is a brief illustration for our tracker in Figure 2.

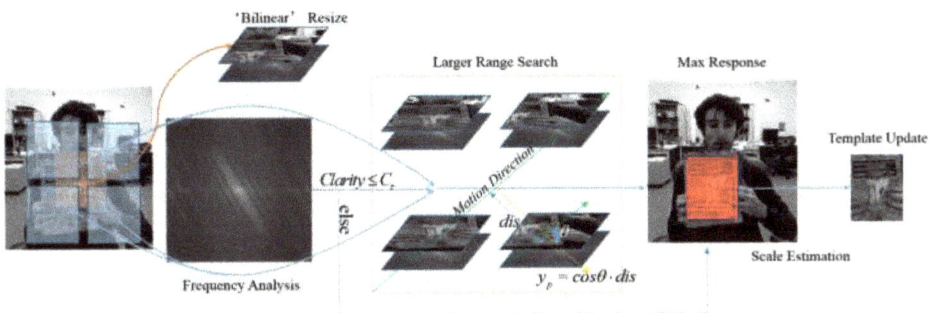

Figure 2. A brief illustration for our method.

Denoting by s and x_0 respectively the scale and position of target in the current frame, s and x_0 from the target information is initialized at the first frame. Just as a slight clarification, all our experimental datasets can be find in Visual Tracking Benchmark [26], the information about target at the first frame is given as common protocol. At frame $t + 1$, the candidate image patch x_s of scale s is searched nearby. Then this patch is resized to the same size as the template and its clarity value c is calculated via JNB metric. The clarity value is inversely proportional to the blur extent the target gets. For further computation, the FFT for x_s is analyzed.

Meanwhile, in frequency domain, Radon Transformation is made for spectrum of x_s and figure out the estimated movement direction. Denoting the clarity values in the last frames by $\{c_i\}_{i=1\sim t}$, the procedure of selecting threshold c_τ is inspired by Ostu threshold method [27], and the value in process is initialized by a fixed value. Specifically, the calculation of c_τ is given by:

$$\begin{aligned} c_l &= \sum_{c_i<c_\tau} c_i/card(\{c_i\} < c_\tau) \\ c_h &= \sum_{c_i>c_\tau} c_i/card(\{c_i\} > c_\tau) \\ c_\tau &= \text{argmax}\{c_l * c_h * (c_l - c_h)^2\} \end{aligned} \tag{17}$$

where c_l and c_h respectively denote the mean value of low and high clarity values. When the clarity value exceeds the adaptive threshold c_τ, calculated by Equation (17), it indicates that the target tends to be blur free or slightly blurred. Hence, the maximum response is calculated for template in the original range and output the target position for next coming frame. Otherwise, the search range is enlarged according to the blur extent by a shifting L above, below, to the right and to the left in the current frame. Here we have:

$$L = \lambda r(c_t + 1)/(c + 1) \tag{18}$$

where λ is a positive coefficient and we set $\lambda = 2.25$ in experiment.

Notice that simply enlarging the searching range may introduce another problem for this tracker: it may take the similar but wrong target nearby as its aim. To solve this problem, consider the cityblock distance between the point p in subwindows and x_0, and the calculated movement direction θ. All the responses of four subwindows are weighted by:

$$\begin{aligned} dis(p, x_0) &= \sum_{i=1}^{2} \left|p_i - x_{0_i}\right| \\ w_p &= \mu dis(p, x_0) \cdot \cos\theta \end{aligned} \tag{19}$$

where μ is a positive parameter that controls the weight of relatively far patches and is determined in experiment by target size. Here we approximate the similarity between real movement direction and estimated movement direction via the cosine function of θ. Then the position of maximum response is outputted as the current target position. At last, we estimate the scale variation for target and update the template by Equations (10) and (11). The procedure of our algorithm is shown in Algorithm 2.

Algorithm 2. The Blur KCF tracker

```
Inputs:   Template x, target position x0 initialized by the given ground truth in the first frame;
          Initial scale s = 1;
          Corresponding image sequence for tracking;
Outputs:  Estimated target position x0' and estimated target scale s' for each frame;Updated x' and model
          coefficient α' at each time;
Selection: Select kernel function and features type;
Initialize the model coefficient α with Equation (5);
repeat;
Calculate the kernel function Kxx' with Equation (6) or (7) according to the selected kernel;
Calculate the response f(z) with Equation (8) and acquire the maximum response Rmax;
Compute the Clarity value c of the candidate image patch via JNB metric;
Find the threshold cT with Equation (17); //Larger range search;
If c ≤ cT then;
Make Radon Transformation for patch to figure out the target movement with Equation (15);
Enlarge the search range with a shift from x0 with Equation (18) of four directions;
for every new candidate image patch;
Calculate the kernel function Kxx';
Calculate the response f'(z) and acquire the maximum response R';
Weighted the response with Equation (19);
If R' > Rmax then Rmax = R';
end;
end if;
Acquire the position of the maximum response x0' and new size s' with Equations (10) and (11);
Update the template by Equations (5) and (9);
Until End of video sequences;
Return {x', α', s'}.
```

4. Experiments and Results

4.1. *Quantitative Evaluation and Speed Analysis*

In this section our algorithm is mainly evaluated on a series of challenging sequences. We find out almost all the sequences with motion blur or/and fast motion in VOT benchmark and run our tracker on them, yet we did not find all these sequences' running results for every state-of-art tracker. In order to compare our tracker with the top trackers, ten sequences from these sequences were selected with full results of all chosen trackers. Besides motion blur and fast motion, these video sequences are also suffered from other difficulties including occlusion, scale variation, illumination variation, in-plane/out-plane rotation, background clutter, etc. The trackers are run on these challenging sequences for general ability and special scenarios handling testing. These sequences are listed in Algorithm 1. As the protocol from [28], there are metrics from different perspective, namely One-Pass Evaluation (OPE), Temporal Robustness Evaluation (TRE) and Spatial Robustness Evaluation (SRE) are considered in our experiment. SRE evaluates the trackers by different initial annotation (slightly shifted from each other); TRE divides the sequences into 20 pieces and evaluates the trackers on different length of sequences; meanwhile, OPE evaluates trackers with other treatment. The overall performance is presented in Figure 3.

In order to evaluate our tracker in average performance, we compared our tracker with state-of-art trackers and list the top ten trackers in each metric. These 28 state-of-art trackers are from the code library of VOT, say, ASLA, BSBT, CPF, KCF, CXT, L1apg, Struck, MIL, SBT, TLD, ORIA, etc. In specify, we compare the center location error for each sequence with two other excellent trackers and the original KCF tracker. CXT, Struck outperform other 25 trackers over the dataset. The details are given in Table 1.

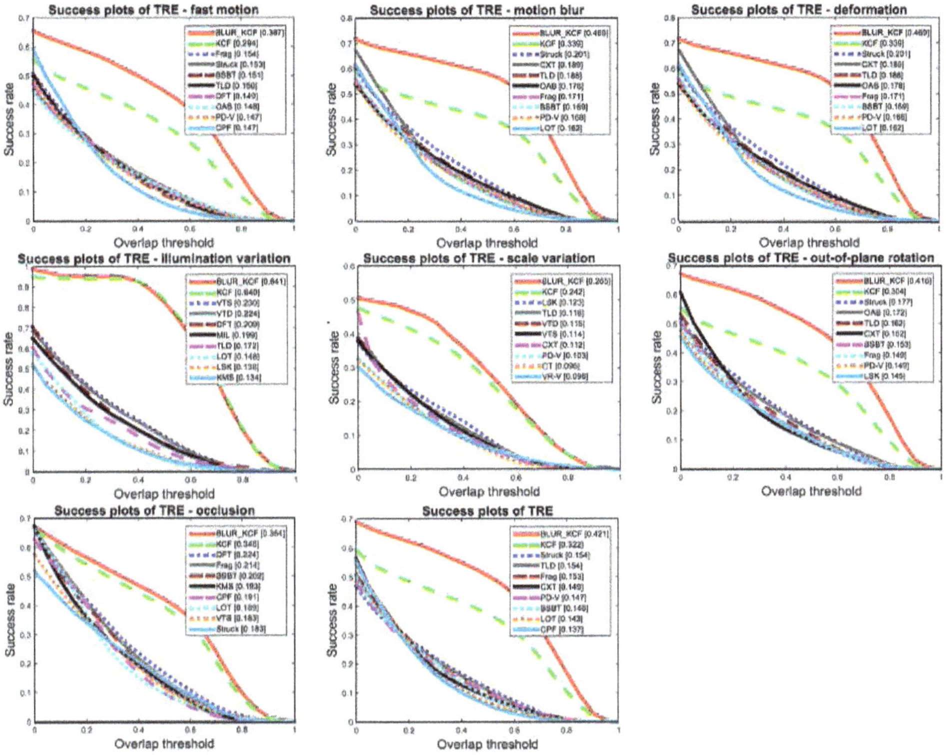

Figure 3. Overall performance in public datasets. These figures show the performance of trackers handing with fast motion, motion blur, deformation, illumination variation, scale variation, out-of-plane rotation and occlusion. The last figure shows the TRE success rate of these trackers under different threshold.

Table 1. Average center location error compared with other top trackers.

Title	Blur Body	Blur Car2	Blur Face	Blur Owl	Clif Bar	Deer	Fleetface	Freeman1	Freeman4	Shaking	Speed
CXT	25.94	26.8	19.29	57.33	33.08	19.99	57.3	20.41	67.46	157.39	9 fps
Struck	12.86	19.36	21.65	12.86	20.08	12.51	43.39	24.7	59.14	65.14	15 fps
KCF	64.12	6.81	8.36	92.2	36.7	21.16	26.37	94.88	27.11	112.5	**360 fps**
Ours	**11.95**	**5.82**	**8.01**	**8.88**	**6.04**	**9.46**	**26.37**	**8.06**	**4.5**	**17.5**	186 fps

These figures show the success rate plot comparison among trackers. In Figure 4, we plot the precision rate comparison between our algorithm and KCF tracker. Precision plot shows the percentage of frames whose location error measured in Euclidean distance is within the corresponding threshold in the x-axis; whereas the success plot shows the percentage of frames whose accuracy measured with $accuracy = \frac{|A_t^P \cap A_t^G|}{|A_t^P \cup A_t^G|}$ is within the overlap threshold. Additionally, our method has been tested with Matlab2014a on a PC, Intel corei5, 3.49 GHZ and 4 GB. Other improved KCF trackers like SAMF and part-based KCF take 5 times computation or more than KCF. Our tracker performs advantageously compared with these top trackers for sequences containing blur and fast motion that consumes not much computing.

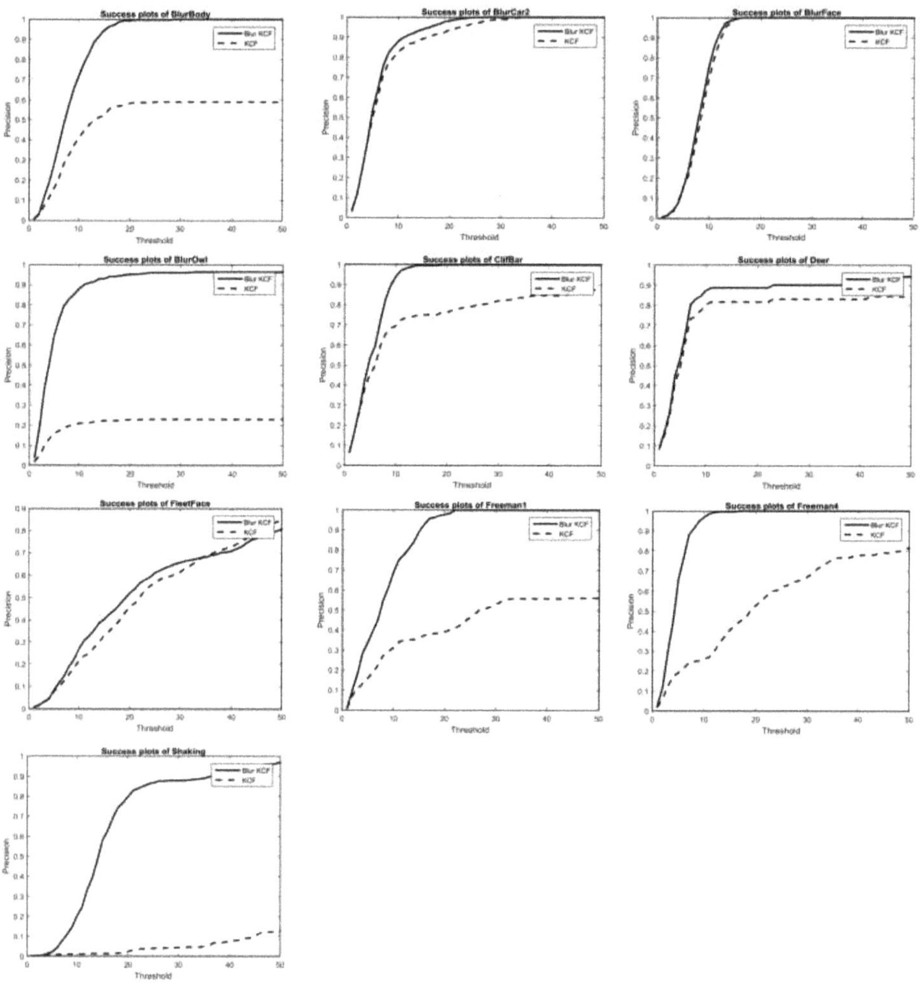

Figure 4. Precision plot in comparison with KCF.

4.2. *Qualitative Evaluation*

To evaluate the practical performance, the results of CXT, Struck, KCF trackers and our algorithm are also printed for qualitative evaluation. Take the BlurBody sequence for example, the camera is always shaking in this sequence as respect to the target boy. When there is a big shaking between the camera and target, the CXT and KCF trackers lost the position of target, Struck and our algorithm handle well this problem. This is because KCF utilize *dense sampling* that only search the candidate patch at the target position of last frame, leading to an error for search when the target moves fast.

As shown in Figure 5 below, the Struck tracker works well in the beginning frames of Freeman4 sequence but fails later by mistaking another similar person nearby as target. Our tracker has not influenced by this scenario since a scheme has been created to avoid the distraction from similar interfering substances by giving the relatively far candidate less weight. In BlurOwl sequence, almost every frame is subjected to varying degree of motion blur and many of them also involve fast motion. The KCF tracker lost target again when the motion is fast as stated and the CXT tracker also get lost when the candidate image patch gets blurred heavily for lacking of a scheme to handle blurred appearance model.

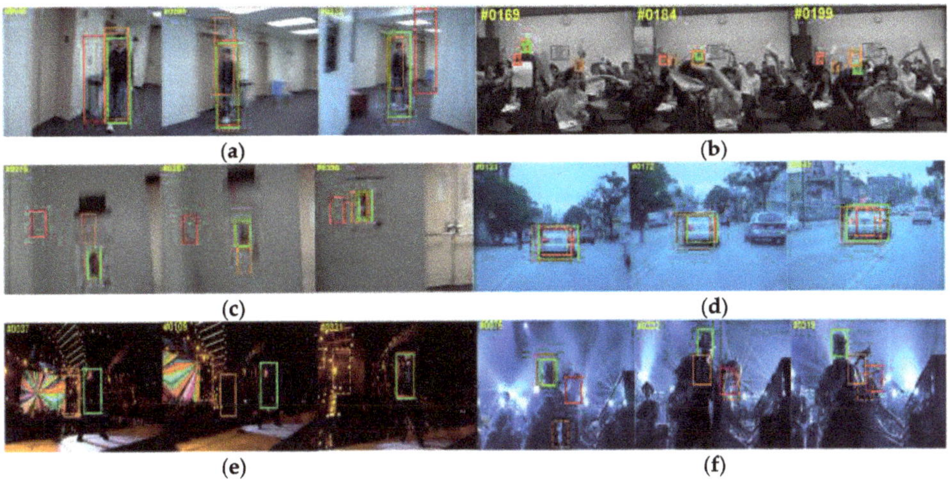

Figure 5. (a) BlurBody; (b) Freeman4; (c) BlurOwl; (d) BlurCar2; (e) Singer2; (f) Shaking. The results marked in orange, dashed orange, red and green are respectively from CXT, Struck, KCF and Ours.

In the sequence of BlurCar2, Singer2, Shaking, besides motion blur and fast motion, the other main challenges for tracking are: scale variation, illustration variation, background clutter and illustration variation, respectively. From the figures in 'BlurCar2', we can see that benefits from the scale estimation strategy, the scale variation with the car was handled well when Struck and KCF get a shift for target. The results in sequence Singer2 show that our method keeps the virtue of KCF for background clutter. In sequence Singer2 and Shaking, accurate direction estimation in frequency domain helps the tracker to avoid the slight shift from target in illustration, the proposed tracker finish the tracking task well till the end when other trackers get lost.

5. Conclusions

In this paper we represent a new algorithm handling the motion blur and fast motion in visual tracking. We analyze the KCF tracker when it suffers from blur or fast motion, and improve the tracker without much computation. We consider the motion model between two frames to avoid inference of other similar objects. In this way our tracker keeps the advantages of original KCF tracker. Since the KCF tracker has no scheme to handle scale variation, we also fix the scale estimation problem by combining KCF and STC trackers together. At last, we compare our tracker with other 28 outstanding trackers in public tests. The results show our tracker performs advantageously compared with them.

Author Contributions: Lingyun Xu has done this research as part of her Ph.D. dissertation under the supervision of Haibo Luo. The main idea of this paper is proposed by Lingyun Xu, and the experiment design and result analysis were done by Lingyun Xu and Haibo Luo. Bin Hui and Zheng Chang edited and approved the final document. All authors were involved in discussions over the past year that shaped the paper in its final version.

Conflicts of Interest: The authors declare no conflict of interest.

References

1. Menegatti, A.E.; Bennewitz, M.; Burgard, W.; Pagello, E. A visual odometry framework robust to motion blur. In Proceedings of the IEEE International Conference on Robotics and Automation (ICRA '09), Kobe, Japan, 12–17 May 2009; pp. 2250–2257.
2. Luo, J.; Konofagou, E. A fast normalized cross-correlation calculation method for motion estimation. *IEEE Trans. Ultrason. Ferroelectr. Freq. Control* **2010**, *57*, 1347–1357. [PubMed]
3. Akbari, R.; Jazi, M.D.; Palhang, M. A Hybrid Method for Robust Multiple Objects Tracking in Cluttered Background. In Proceedings of the 2nd International Conference on Information & Communication Technologies (ICTTA '06), Damascus, Syria, 24–28 April 2006; Volume 1, pp. 1562–1567.
4. Comaniciu, D.; Meer, P. Mean shift: A robust approach toward feature space analysis. *IEEE Trans Pattern Anal. Mach. Intell.* **2002**, *24*, 603–619. [CrossRef]
5. Lucas, B.D.; Kanade, T. An iterative image registration technique with an application to stereo vision. In Proceedings of the 7th international joint conference on Artificial intelligence (IJCAI '81), Vancouver, BC, Canada, 24–28 August 1981; pp. 674–679.
6. Mei, X.; Ling, H. Robust Visual Tracking using L1 Minimization. In Proceedings of the IEEE 12th International Conference on Computer Vision, Kyoto, Japan, 29 September–2 October 2009; pp. 1436–1443.
7. Bao, C.; Wu, Y.; Ling, H.; Ji, H. Real time robust L1 tracker using accelerated proximal gradient approach. In Proceedings of the IEEE Conference on Computer Vision and Pattern Recognition (CVPR), Providence, RI, USA, 16–21 June 2012; pp. 1830–1837.
8. Yang, B.; Nevatia, R. An online learned CRF model for multi-target tracking. In Proceedings of the IEEE Conference on Computer Vision and Pattern Recognition (CVPR), Providence, RI, USA, 16–21 June 2012; pp. 2034–2041.
9. Kalal, Z.; Matas, J.; Mikolajczyk, K. P-N learning: Bootstrapping binary classifiers by structural constraints. In Proceedings of the IEEE Conference on Computer Vision and Pattern Recognition (CVPR), San Francisco, CA, USA, 13–18 June 2010; pp. 49–56.
10. Hare, S.; Saffari, A.; Torr, P.H.S. Struck: Structured output tracking with kernels. In Proceedings of the International Conference on Computer Vision, Barcelona, Spain, 6–13 November 2011; pp. 263–270.
11. Henriques, J.F.; Caseiro, R.; Martins, P.; Batista, J. High-speed tracking with kernelized correlation filters. *IEEE Trans. Pattern Anal. Mach. Intell.* **2015**, *37*, 583–596. [CrossRef] [PubMed]
12. Bolme, D.; Beveridge, J.R.; Draper, B.A.; Lui, Y.M. Visual object tracking using adaptive correlation filters. In Proceedings of the IEEE Computer Society Conference on Computer Vision and Pattern Recognition, San Francisco, CA, USA, 13–18 December 2010; pp. 2544–2550.
13. Galoogahi, H.K.; Sim, T.; Lucey, S. Correlation filters with limited boundaries. In Proceedings of the IEEE Computer Society Conference on Computer Vision and Pattern Recognition, Boston, MA, USA, 7–12 June 2015; pp. 4630–4638.
14. Henriques, J.F.; Caseiro, R.; Martins, P.; Batista, J. Exploiting the circulant structure of tracking-by-detection with kernels. In *Lecture Notes in Computer Science*; Including Subseries Lecture Notes in Artificial Intelligence and Lecture Notes in Bioinformatics; Springer: Berlin, Germany, 2012; Volume 7575, Part 4, pp. 702–715.
15. Danelljan, M.; Gustav, H.; Khan, F.S.; Felsberg, M. Convolutional Features for Correlation Filter Based Visual Tracking. In Proceedings of the IEEE International Conference on Computer Vision Workshop (ICCVW), Santiago, Chile, 7–13 December 2015; pp. 58–66.
16. Liu, T.; Wnag, G.; Yang, Q. Real-time part-based visual tracking via adaptive correlation filters. In Proceedings of the IEEE Conference on Computer Vision and Pattern Recognition, Boston, MA, USA, 7–12 June 2015; pp. 4902–4912.
17. Li, Y.; Zhu, J. A scale adaptive kernel correlation filter tracker with feature integration. In *Lecture Notes in Computer Science*; Including Subseries Lecture Notes in Artificial Intelligence and Lecture Notes in Bioinformatics; Springer: Berlin, Germany, 2015; Volume 8926, pp. 254–265.
18. Ma, C.; Yang, X.; Zhang, C.; Yang, M. Long-term Correlation Tracking. In Proceedings of the IEEE Conference on Computer Vision and Pattern Recognition, Boston, MA, USA, 7–12 June 2015; pp. 5388–5396.
19. Hong, Z.; Chen, Z.; Wang, C.; Mei, X.; Prokhorov, D.; Tao, D. Multi-Store Tracker (MUSTer): A cognitive psychology inspired approach to object tracking. In Proceedings of the IEEE Computer Society Conference on Computer Vision and Pattern Recognition, Boston, MA, USA, 7–12 June 2015; pp. 749–758.

20. Wu, Y.; Ling, H.; Yu, J.; Li, F.; Mei, X.; Cheng, E. Blurred target tracking by blur-driven tracker. In Proceedings of the International Conference on Computer Vision, Barcelona, Spain, 6–13 November 2011; pp. 1100–1107.
21. Dai, S.; Wu, Y. Motion from blur. In Proceedings of the IEEE Conference on Computer Vision and Pattern Recognition (CVPR), Anchorage, AK, USA, 23–28 June 2008; pp. 1–8.
22. Jin, H.; Favaro, P.; Cipolla, R. Visual Tracking in the Presence of Motion Blur. In Proceedings of the IEEE Conference on Computer Vision and Pattern Recognition (CVPR), Boston, MA, USA, 20–25 June 2005; Volume 2, pp. 18–25.
23. Tang, M.; Feng, J. Multi-Kernel Correlation Filter for Visual Tracking. In Proceedings of the IEEE International Conference on Computer Vision (ICCV), Santiago, Chile, 7–13 December 2015; pp. 3038–3046.
24. Zhang, K.; Zhang, L.; Liu, Q.; Zhang, D.; Yang, M.H. Fast visual tracking via dense spatio-temporal context learning. In *Lecture Notes in Computer Science*; Including Subseries Lecture Notes in Artificial Intelligence and Lecture Notes in Bioinformatics; Springer: Berlin, Germany, 2014; Volume 8693, Part 5, pp. 127–141.
25. Ferzli, R.; Karam, L.J. A no-reference objective image sharpness metric based on the notion of Just Noticeable Blur (JNB). *IEEE Trans. Image Process.* **2009**, *18*, 717–728. [CrossRef] [PubMed]
26. Visual Tracking Benchmark. Available online: http://cvlab.hanyang.ac.kr/tracker_benchmark/datasets.html (accessed on 28 August 2016).
27. Otsu, N. A threshold selection method from gray-level histograms. *IEEE Trans. Syst. Man Cybern.* **1979**, *9*, 62–66.
28. Wu, Y.; Lim, J.; Yang, M.H. Online object tracking: A benchmark. In Proceedings of the IEEE Computer Society Conference on Computer Vision and Pattern Recognition, Portland, OR, USA, 23–28 June 2013; pp. 2411–2418.

sensors

MDPI

Article

A Hierarchical Building Segmentation in Digital Surface Models for 3D Reconstruction

Yiming Yan [1,*], Fengjiao Gao [2,*], Shupei Deng [3] and Nan Su [3]

1 Institute of Information Technology, Harbin Engineering University, Harbin 150001, China
2 Institute of Automation of Heilongjiang Academy of Sciences, Harbin 150001, China
3 Department of Information Engineering, Harbin Institute of Technology, Harbin 150001, China; dengshupei2016@126.com (S.D.); sunan_hit@hotmail.com (N.S.)
* Correspondence: yanyiming@hrbeu.edu.cn (Y.Y.); ai.gaofengjiao@gmail.com (F.G.); Tel.: +86-139-3651-3116 (Y.Y.)

Academic Editors: Felipe Gonzalez Toro and Antonios Tsourdos
Received: 17 October 2016; Accepted: 13 January 2017; Published: 24 January 2017

Abstract: In this study, a hierarchical method for segmenting buildings in a digital surface model (DSM), which is used in a novel framework for 3D reconstruction, is proposed. Most 3D reconstructions of buildings are model-based. However, the limitations of these methods are overreliance on completeness of the offline-constructed models of buildings, and the completeness is not easily guaranteed since in modern cities buildings can be of a variety of types. Therefore, a model-free framework using high precision DSM and texture-images buildings was introduced. There are two key problems with this framework. The first one is how to accurately extract the buildings from the DSM. Most segmentation methods are limited by either the terrain factors or the difficult choice of parameter-settings. A level-set method are employed to roughly find the building regions in the DSM, and then a recently proposed 'occlusions of random textures model' are used to enhance the local segmentation of the buildings. The second problem is how to generate the facades of buildings. Synergizing with the corresponding texture-images, we propose a roof-contour guided interpolation of building facades. The 3D reconstruction results achieved by airborne-like images and satellites are compared. Experiments show that the segmentation method has good performance, and 3D reconstruction is easily performed by our framework, and better visualization results can be obtained by airborne-like images, which can be further replaced by UAV images.

Keywords: building segmentation; digital surface model; remote sensing; 3D reconstruction

1. Introduction

For making 3D city maps and visualization [1–3], 3D reconstruction of buildings using remote sensing (RS) images and digital surface models (DSMs), is always a complicated but necessary work. To better generate digital building models, 3D reconstruction of buildings using DSMs has been widely investigated. The main task is to reconstruct all the outer surfaces of a building, which are composed by the roof and all facades.

With the development of airborne light detection and ranging (LiDAR), high accuracy DSM can be obtained after pre-processing the original LiDAR data. 3D building reconstructions using DSMs are generally based on models [4–6]. A number of types of 3D models of buildings are firstly constructed offline, then a matched one for each segmented buildings in the DSM is used for the 3D reconstruction. A limitation of these methods is the over-reliance on the completeness of the offline-constructed models of the buildings, which cannot be easily guaranteed since in modern cities there are a variety of types of buildings. Moreover, 3D modeling and model selection are labor-intensive work, in which complex interactive manual operations are always needed. Furthermore, the facades of buildings are not properly processed in most of these methods for further texture-mapping and visualization.

By using high accuracy DSM and texture-images of the roofs and facades of the buildings, 3D reconstruction can be performed without creating offline models [7–9], but most of these kinds of model-free methods only focus on reconstruction of the roofs of buildings, or do not process the texture-images. In our idea, a DSM and the texture-images of the outer-surfaces of buildings can be associated for a better reconstruction. Firstly the shapes of the roofs of buildings need to be extracted and analyzed.

DSMs can mostly be processed as a digital image. Various image segmentation methods are employed [10–12] for DSM processing, such as classification, edge detection, and some other filter-based methods. Active contour-based methods are also used. For instants, a level-set is suitable for efficiently searching the contours of buildings in DSMs, since buildings have larger values than other objects on the ground. However, roof details cannot be detected if the initialization is not properly done. Moreover, when a high tree is next to a building, good performance on the edges of the adjacent building generally cannot be obtained by level-set approaches. Jia et al. [13] proposed a level-set-based method for building modeling, but it is only applicable to isolated buildings in relatively flat areas. The processing of details on the edges needs to be improved. Iovan et al. [14] proposed a support vector machine (SVM)-based tree extraction method in DSMs with prior knowledge. This could be expanded for further building segmentation since only trees and buildings are higher. Similarly, a supervised classification method was introduced for change-detection of buildings [15]. Buildings are segmented by height, spectra, and shape information from a DSM, assisted by orthophotos of the same area. However, prior knowledge for supervised classification is not easily obtained, and it always determines the segmentation performance. Liu et al. [16] employed an urban DSM segmentation by extended locally excitatory globally inhibitory oscillator networks (LEGION), which is an unsupervised classification-based method without prior knowledge, but it fails to do small features and boundaries of complex buildings. McCann and his team summarized the limitations of various generic image segmentation methods and engineered systems. They concluded that most existing segmentation methods are not successful on histopathology images, since these methods vary in how the segmentation regions are parameterized and whether edge or region information is used [17]. In consideration of that, McCann modeled images as occlusions of random texture for segmentation (ORTSEG), and proposed a new unsupervised mathematical framework method based on local histograms. The method handles the difficult class of edgeless images and has excellent applicability to histology image segmentation and it could also be expanded for building segmentation in DSM, and thus improve the performance on edges, but the limitations of this method are that certain numbers of parameters need to be properly set for different conditions, and any overall terrain undulation could have a negative impact on the class judgments in certain local regions near the building.

Considering both efficient detection and more accurate segmentation on the edges, we introduce a hierarchical building segmentation method. Rough positioning of buildings is performed by level-set in a global DSM first, and then all regions of buildings are processed separately by the ORTSEG method to decrease negative terrain factors, and improve the segmentation accuracy.

Once the roof of a building is extracted from the DSM, only the 3D structure of the roof is obtained, but all facades of the building need to be generated. Since roof contours might be of different height, the part of the facade which is close to the roof is uneven. As a result, certain criteria must be defined referring to the range of each facade. A roof-contour and texture-image guided interpolation (RTGI) method is introduced for facade generation and texture-mapping of buildings by synergizing with texture-images of each outer-surface of the building, which are collected from RS optical images of different views of the building obtained by satellite or unmanned aerial vehicles (UAVs). UAVs are currently performing significant roles for city remote sensing missions, and higher resolution texture-images could be collected by UAVs. After pasting these texture-images onto the corresponding facade generated by RTGI, the visualization results could be greatly improved. Furthermore, UAVs equipped with LIDAR can collect high precision DSMs efficiently in the near future, and better use of our method will be made. The main contributions of this study are:

(1) A novel 3D reconstruction framework without offline models is proposed, and the introduced RTGI could effectively help generate facades of a building. Our framework could reconstruct 3D buildings with limited RS resources, and reduce complicated manfully modeling work.
(2) Based on the level-set and ORTSEG techniques, we propose a hierarchical segmentation method for extracting buildings from DSMs, named LS-ORTSEG. It can greatly reduce the negative factors that result from the uncertain terrain and the limitations of the different segmentation algorithms.

This paper is organized as following: the model free framework for 3D reconstruction is introduced in Section 2. The hierarchical segmentation method LS-ORTSEG is discussed in Section 3, then Section 4 presents our analysis and experimental results. Section 5 concludes this work, and discusses the advantages, limitations and future work of this method.

2. Framework of 3D Reconstruction

As shown in Figure 1, the first step is to segment the roof of buildings from DSM. Details are discussed in Section 3.

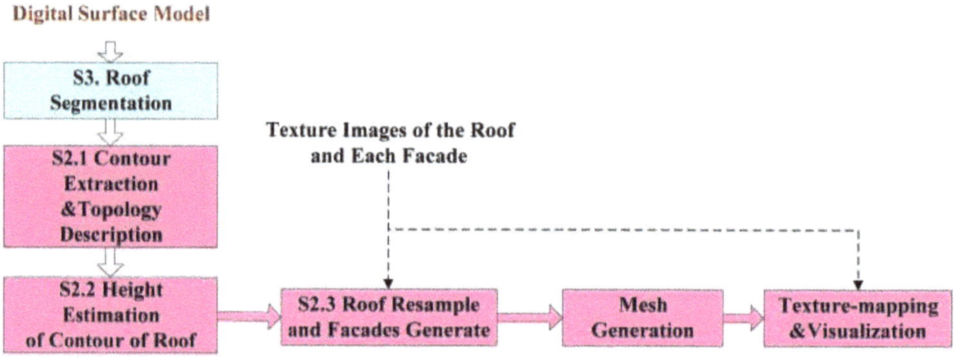

Figure 1. Flowchart of 3D building reconstruction and visualization.

Then, the contour of the roofs are extracted and analyzed, and the numbers of facades are found and then coded. After that, the actual heights of all sampled points of the contour are estimated by comparing the elevations between inner points and outer points. Further, the RTGI method is introduced for generating all the outer-surfaces of the building, guided by the coded number of facades and sizes of different texture-images, and then the 3D points-sets of each outer-surface are obtained. We then gather the 3D points-sets of all the outer-surfaces of the buildings, and generate the mesh. Finally, the texture images are mapped onto the corresponding triangle-area of each outer-surface, and we visualize the mapping result.

2.1. Extraction and Coding of Roof Edge-Lines

Since the edge-line of the roof is the intersection of each facade and the roof, it is important to generate each facade. In DSM data, like in orthographic images, buildings are represented by quantified grids (like pixels) in the horizontal direction, and only the elevation of the blue part of a facade can be obtained, as shown in Figure 2a. When the roof of a building is extracted from the DSM, Su's [18] and Qiu's [19] work are taken for optimizing the extraction results and further determine the topology of the building. Then buildings with irregular top shapes can be divided into several blocks, as shown in Figure 3. The main steps are as follows: (1) line extraction methods like LSD [20], can be used to find the line features of the building; (2) after removing some erroneous lines caused by angle and length-based restrictions, the remaining lines can form a group of parallelogram grids, which divide the irregular shape of roof into different regions; (3) a specific decision criterion is used to judge each region as building or non-building class, considering both the segmented result and the restriction of the parallelogram grids. All the blocks can be coded in a customized order, and then the edge-lines of the roof of each block or a simple shaped building can be directly coded in a clockwise or counterclockwise, as shown in Figure 2b.

(a) (b)

Figure 2. Examples of four types of buildings. (**a**) The four buildings; (**b**) Edge-lines of the roofs.

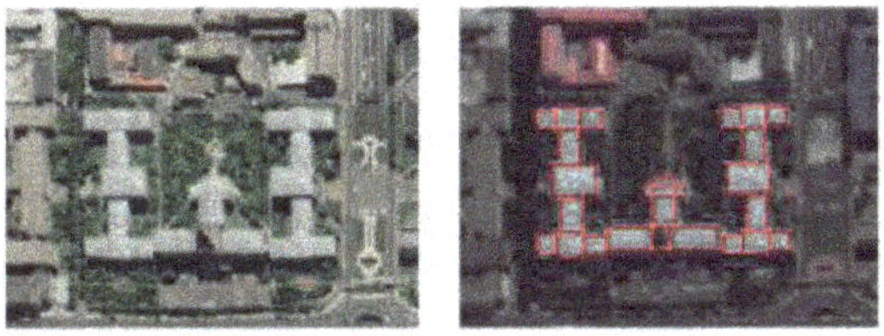

(a) (b)

Figure 3. Blocks of an irregular top building. (**a**) An irregular top building; (**b**) Divided blocks.

2.2. Height Estimation of Edge-Lines

After extracting and coding the roof edge-lines, the heights of the edge-lines need to be estimated for further facade generation. In DSM, only the elevation of buildings is directly obtained, but the heights of buildings need to be estimated. Here we estimate them by comparing the elevation of the roof-contour and foot-contour. Mathematical morphology methods are used for extracting the roof-contour *RC* by an erosion operation after a dilation operation, and the foot-contour *FC* by a dilation operation after an erosion operation. If n and m are respectively the numbers of points of the *RC* and *FC*, the height of each point of the *RC* is estimated by Equation (1):

$$H(i) = RC(i) - \frac{1}{3}\sum_{m=1}^{3} FC_{c3}(m) \quad i = 1, 2, \ldots, n \tag{1}$$

As shown in Figure 4, *RC*(i) is elevation of the i-th point of the *RC*, and $FC_{c3}(m)$ is the elevation of the m-th point of the closest three points around the i-th point of the *RC*.

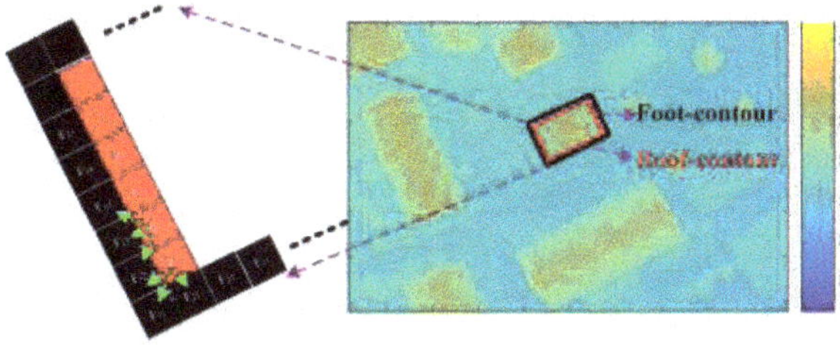

Figure 4. Height estimation.

2.3. Roof-Contour and Texture-Image-Guided Interpolation

Since each edge-line point has the same horizontal coordinate as its corresponding part on the facade below, only the 'Z-coordinate' needs to be interpolated. Assuming the size of a texture image is $M \times N$, then the facade can be interpolated as shown in Figure 5. Firstly, we resample the edge-line to fit the image with the nearest neighbor criterion, and the new number of points of the i-th edge-line P' will be N. Then we interpolate each row n' ($n' = 1, 2, \ldots, P'$) of the side from the foot to the top by Equation (2):

$$S_i^{n'}(p) = \frac{H_i(n')}{M} \cdot p \quad p = 1, 2, \ldots, M \tag{2}$$

Figure 5. Generated building facades with interpolated points.

After finishing interpolating, each facade can be generated by a 3D point-set as seen in Figure 5, and then 3D meshes are generated by the combined point-sets including all outer-surfaces for further detailed texture-mapping. The texture-mapping error $\widetilde{E}$ could be evaluated by Equation (3), where we select N_T testing points, and compare the mean error of the 3D coordinates between the testing points $(p_x^t(i), p_y^t(i), p_z^t(i))$ and the corresponding measured points $(p_x^m(i), p_y^m(i), p_z^m(i))$:

$$\widetilde{E} = \frac{1}{N_T} \sum_{i=1}^{N_T} \sqrt{(p_x^t(i) - p_x^m(i))^2 + \left(p_y^t(i) - p_y^m(i)\right)^2 + (p_z^t(i) - p_z^m(i))^2} \tag{3}$$

3. A Hierarchical Building Segmentation Based on the LS-ORTSEG Method

As illustrated above, building segmentation is significant to our 3D reconstruction framework. Since DSMs used for better 3D reconstruction have quite high resolution, serrated building boundaries can be happen in the DSM, and also branches and leaves of trees around the building might cause bad boundaries. As a result, the edges and details of the roof need to be improved when segmenting. We consider the advantages of the level-set and ORTSEG method, and hierarchically segment the DSM with following steps, shown in Figure 6.

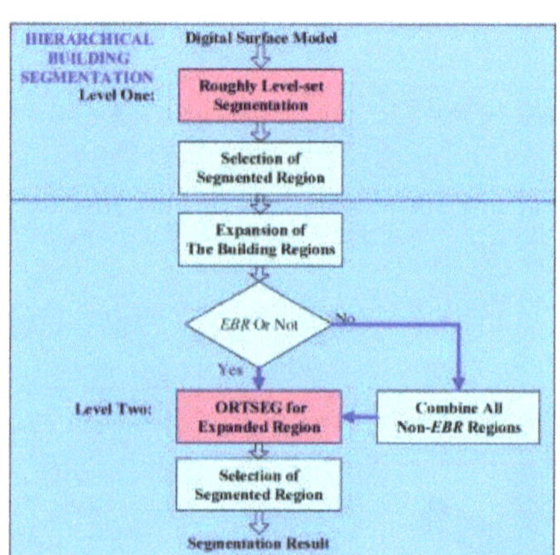

Figure 6. Hierarchical LS-ORTSEG building segmentation.

(1) A classic Chan-Vese model-based level-set is introduced for level-one segmentation of the DSM, using a circle mask in the whole DSM. Then the N_{ls} regions are roughly segmented as buildings.

(2) Restrict building regions (*BR*) adopting some certain geometric properties. Here, we set a threshold TH_{Area} for the area of the segmented region as Equation (4):

$$\underset{i=1,2,..,N_{ls}}{Area}\{i\} = NUM_{pixel}\{i\}\cdot res_{DSM} < TH_{Area} \quad (4)$$

where NUM_{pixel} is the number of pixels in the region and res_{DSM} is the DSM resolution. Then N_{ssr} regions are identified as buildings:

(3) Extract the minimum circumscribed rectangle (*MCR*) of each building region. As shown in Figure 7, the N_{EBR} expanded building regions (*EBR*) are obtained by Equation (5):

$$EBR_i = MCR_i \cup EX_i(\Delta x, \Delta y) \quad (5)$$

where the i-th *EBR* is the union of i-th *MCR* and *EX* in the original DSM region and Δx, Δy are the expansion distance for the horizontal and vertical directions. Note that when a building is near the boundary of the input DSM, the corresponding side of Δx or Δy must be decreased to fit the boundary, and the other side retains the original value. After obtaining all the *EBR*s, we combine all non-*EBR*s as background, by setting the values of all the *EBR*s in the original DSM to a large enough constant, and the segmented results of these positions will be replaced by that of *EBR*s.

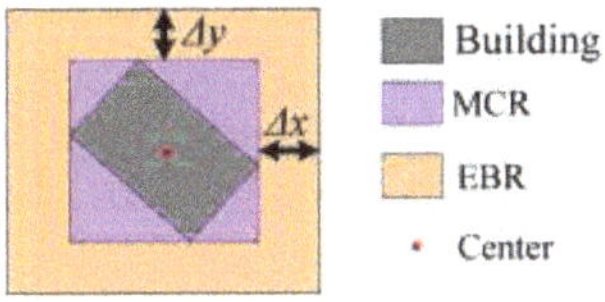

Figure 7. Region expansion for level-two segmentation.

(4) Then, level-two segmentation is performed on both all *EBR*s and the background by the ORTSEG method. The labeling function γ of an image, which defines the class of each pixel with different numbers, splits the image into N_{ort} regions. That models an image as occlusions of random textures $I = O_\gamma\{I_n\}_{n=0}^{N_{ort}-1}$ (Figure 8).

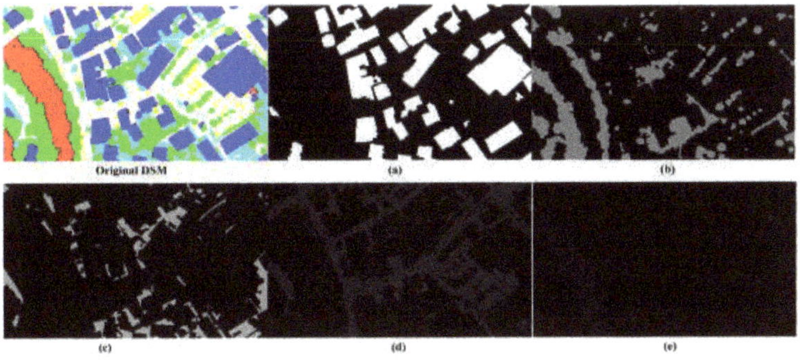

Figure 8. Modeling a DSM as an occlusion of textures. **(a–e)** is one of the five random textures.

McCanna [17] proved that a local histogram could be approximated by a convex combination of the value histograms of those regions. As a result, segmentation is handled as an optimization, which is to search the labeling function γ, for the best approximation of the local histograms $L_\omega H$ by the statistical function $h_{\{n\}}(r)$ and probability density function $\hat{p}_h^{Xn}$ of the N_{ort} regions:

$$\underset{\gamma}{\text{argmin}} \| L_\omega H - \sum_{n=0}^{N_{ORT}-1} [\omega * h_{\{n\}}(\gamma)] \hat{p}_h^{Xn} \| \tag{6}$$

Similarly, DSM can be segmented using ORTSEG. However, since the labeling function γ of ORTSEG is to fit global statistical properties, if ORTSEG were used on the whole DSM, the potential overall undulating terrain might lead to poor performance in the local regions of different buildings. For example, buildings on the mountainside and on the foot could be identified as different labels. As a result, we intend to improve the local accuracy by segmenting all the *EBRs* and the background instead. The main flow of LS-ORTSEG based *EBRs* segmentation is as shown in Figure 9.

```
      For i = 1: N_EBR
(i)   Input EBR(i);
(ii)  Parameters Settings;
(iii) Quantize EBR(i);
      Iteration start
      For iter = 1:max_iteration
(i)   Labeling function γ update, by local histogram
      approximate optimization;
      iter = iter + 1;
      end
      end
```

Figure 9. Main flow of LS-ORTSEG based *EBRs* segmentation.

4. Experimental Section

4.1. Data Statement

Three groups of datasets are collected for our experiments. As shown in Table 1, Dataset 1 was captured by an aircraft over Vaihingen in Germany at a height of 500 m, which can be reached by UAVs. The dataset is collected for segmentation with ground truths, as shown in Figure 10. The Vaihingen dataset was provided by the German Society for Photogrammetry, Remote Sensing and Geoinformation (DGPF) [21]. The DSMs of Datasets 2 and 3 were obtained from the OpenTopography website [22], and the texture-images of the two datasets are from satellites and Google-Earth, for testing the 3D reconstruction performance.

Table 1. Statement of datasets.

SN of Datasets	1	2	3
Position	Vaihingen, Germany	San Diego, CA, USA	San Diego, CA, USA
DSM collection	Airborne	Airborne	Airborne
Texture-images collection	Airborne	Satellite	Satellite
Ground Truths of DSM	YES	NO	NO
Test-points of facades?	NO	By Google Earth	By Google Earth
Resolution of DSM	0.09 m	1.0 m	1.0 m
Resolution of Texture-images	-	1.0 m/0.3 m	1.0 m/0.3 m

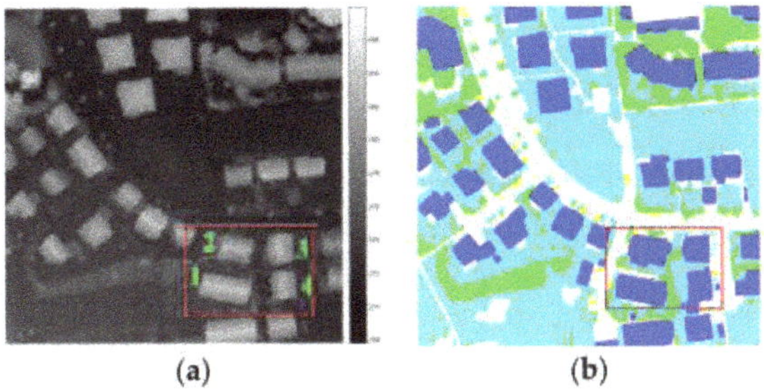

Figure 10. Dataset 1: (a) original DSM shown as image, and 1, 2, 3 and 4 is the serial number of the four buildings; (b) ground truths.

4.2. Building Segmentation Using Hierarchical Method

After investigating the terrain of this scene using the actual map and the DSM data, four adjacent buildings were selected to verify the reliability and robustness of our method. The reasons were as follows: (1) they are adjoined, and the surroundings are similar; (2) their terrain factors are similar but have certain distintions, especially between Buildings 1 and 2, and also between Buildings 3 and 4; (3) they have several kinds of shapes: Buildings 3 and 4 are similar but different from the others; (4) they have different orientations, and DSMs have different degrees of error in different directions.

Level-set, ORTSEG, and LS-ORTSEG are used for segmenting these regions. In ORTSEG, besides the quantize and deconvolution methods that need to be selected, many other parameters need to be set. 'Hist.wSize' (sets the size of the histogram filter) and 'NumTextures' (sets the number of textures to segment to) are the key parameters that affect the segmentation results. 'Hist.wSize' is generally set according to resolution and experience. 'NumTextures' is set to 5 for ORTSEG, since the number of classes is 5 in the ground truth of Dataset 1. Generally, better results should be obtained if we set 'NumTextures' close to the number of classes in the ground truths. However, in LS-ORTSEG, 'NumTextures' are strictly set to 2 since most buildings in these *EBR*s are simple shapedones. Other uniform settings of ORTSEG and LS-ORTSEG are as follows: (1) 'quantizeMethods' are set to '*k*-means'; (2) 'quantize.numColors' = 8; (3) factor.numReps = 25 (it decides how many times to repeat the factorization).

The results obtained with these general settings are shown in Figure 11. Similar performance is obtained for Buildings 1 and 3 by the three methods, but LS-ORTSEG gets better results on Buildings 2 and 4.

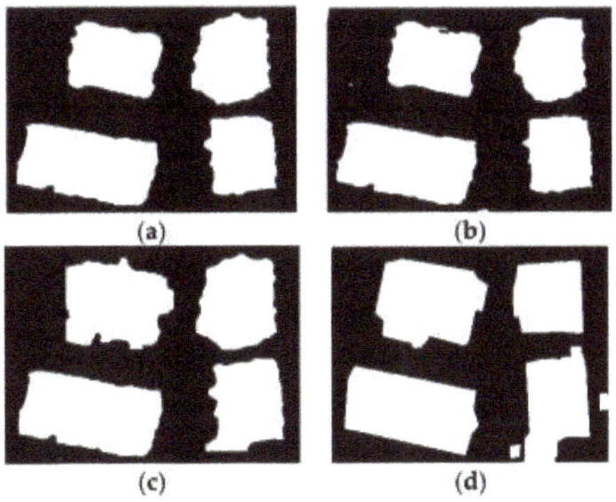

Figure 11. Segmentation results of four buildings by different methods. (a) Level-Set; (b) ORTSEG; (c) LS-ORTSEG; (d) Ground Truths. The 'Hist.wSize' is set to 3 for both ORTSEG and LS-ORTSEG.

ORTSEG and LS-ORTSEG were further compared when using different parameter settings. Accuracy of segmentation is evaluated by Equation (7):

$$\text{Accuracy} = (\text{TP} + \text{TN})/(\text{TP} + \text{FN} + \text{FP} + \text{TN}) \quad (7)$$

where TP is the number of points of the building in the ground truth matrix. FP is the number of non-building points in the ground truth matrix. TN is the number of correctly segmented points of the building according to the ground truth matrix. FN is the number of erroneous segmented points of the building according to the ground truth matrix. Firstly, 'NumTextures' is fixed to 5 for ORTSEG and LS-ORTSEG, and we draw the accuracy of the two methods when 'Hist.wSize' are set from 3 to 9, as shown in Figure 12, which illustrates that LS-ORTSEG gives better accuracy than ORTSEG for most cases.

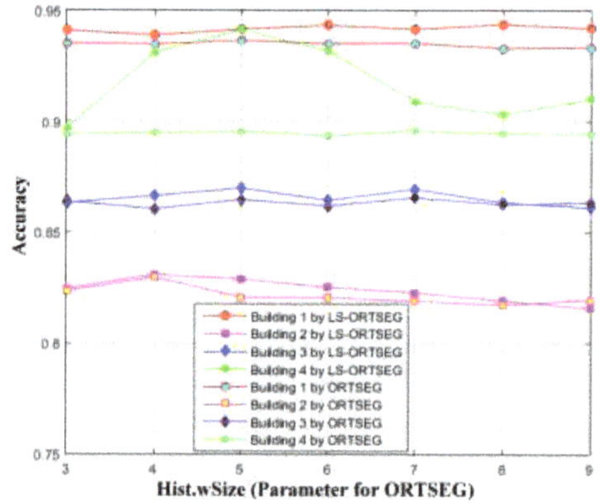

Figure 12. Accuracy using different window sizes. The 'NumTextures' are set to 5 for ORTSEG, and 'Hist.wSize' are set from 3 to 9 for both ORTSEG and LS-ORTSEG.

The accuracy analysis when 'Hist.wSize' is fixed to 5 and 'NumTextures' are set from 2 to 8, is shown in Figure 13. It indicates that 'NumTextures' should be set close to the number of classes of the ground truths for better segmentation. Otherwise, when 'NumTextures' is set too small or too large, the negative impact of the terrain will dominantly destroy the segmentation, and poor accuracy is obtained.

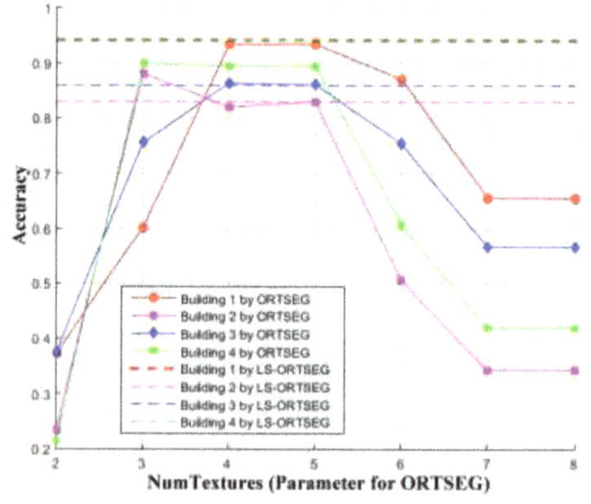

Figure 13. Accuracy using different texture numbers. The 'Hist.wSize' are set to 5, and 'NumTextures' are set from 2 to 8 for both ORTSEG and LS-ORTSEG.

The best accuracy of different methods is listed in Table 2. It illustrates that LS-ORTSEG achieved better segmentation than other methods. Moreover, Datasets 2 and 3 are processed, as shown in Figures 14 and 15. It could be seen that level-set is not as sensitive to the terrain, but it missed some of the buildings. ORTSEG obtained a good result while LS-ORTSEG can further segment the small prominent structures on the roofs, which is crucial to analyze the roof contour.

Table 2. Accuracy of different methods.

Method	Accuracy			
	Building 1	**Building 2**	**Building 3**	**Building 4**
Level Set	92.26%	82.44%	84.57%	90.27%
ORTSEG	93.64%	82.96%	86.57%	89.58%
LS-ORTSEG	94.37%	83.08%	86.99%	94.14%

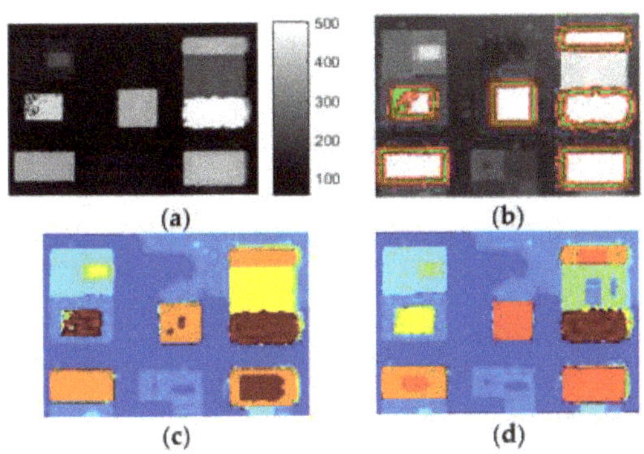

Figure 14. Original DSM and segmentation results of Dataset 2. (**a**) DSM (**b**) Level-set; (**c**) ORTSEG; (**d**) LS-ORTSEG. 'Hist.wSize' = 3, 'NumTextures' = 7 for both ORTSEG and LS-ORTSEG.

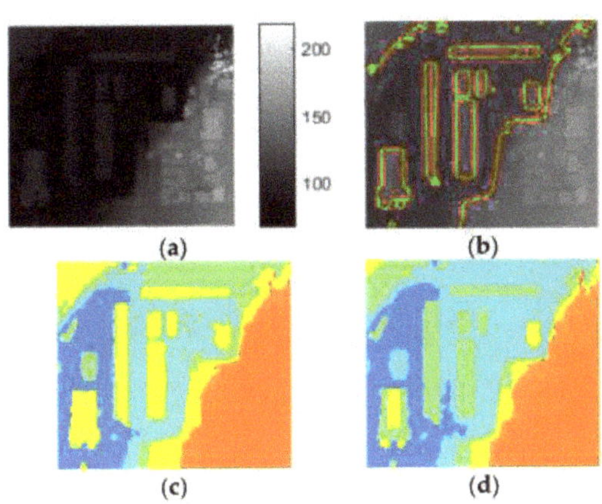

Figure 15. Original DSM and segmentation results of Dataset 3. (**a**) DSM (**b**) Level-set; (**c**) ORTSEG; (**d**) LS-ORTSEG. 'Hist.wSize' = 3, 'NumTextures' = 7 for both ORTSEG and LS-ORTSEG.

4.3. 3D Reconstruction of Buildings

As shown in Figure 16, we selected two buildings from Datasets 2 and 3, respectively, to show the subsequent 3D reconstruction. After extracting the building roofs from the DSM, the facades are generated by RTGI.

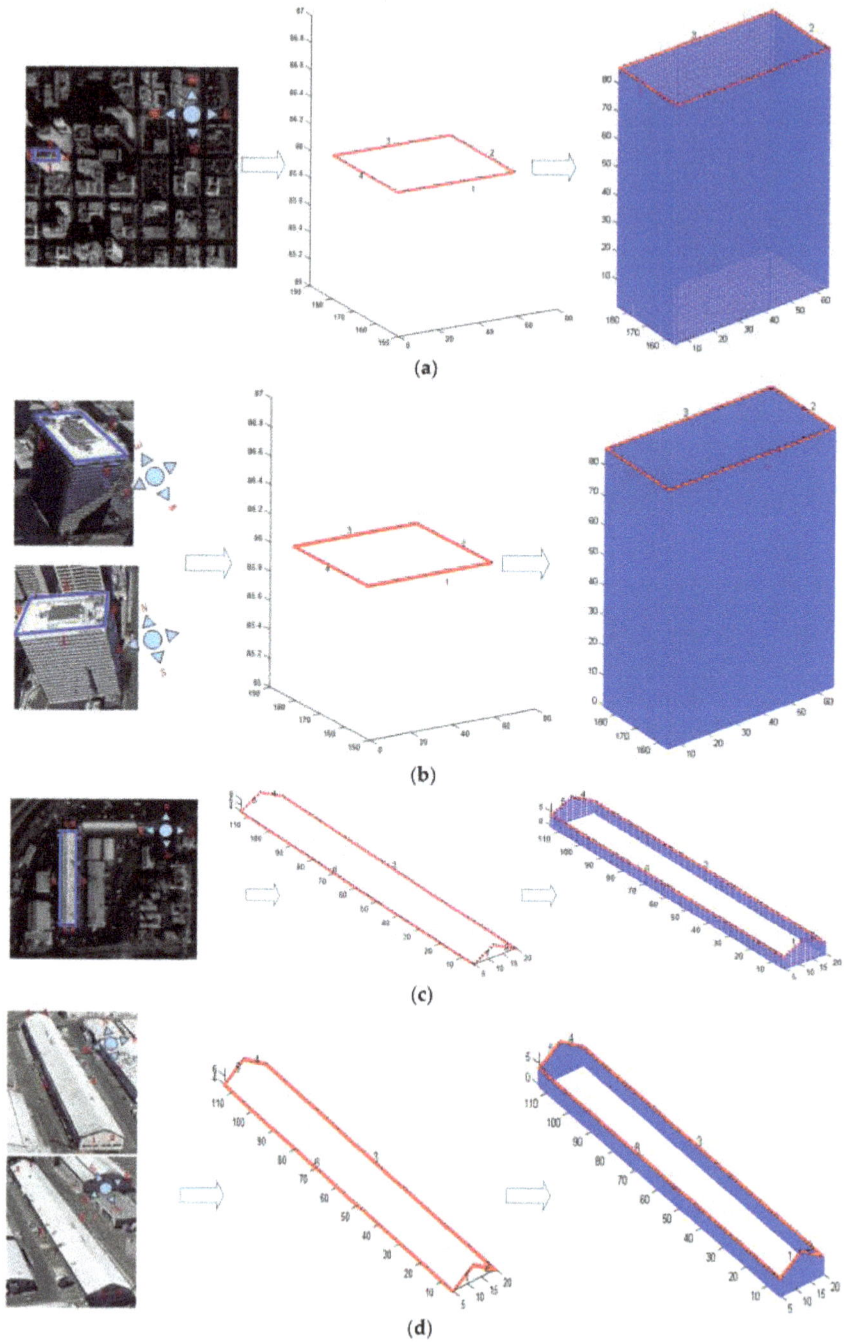

Figure 16. Facade generation of buildings in different cases. (**a**) Facade generation of Building A by satellite texture-images; (**b**) facade generation of Building A by airborne-like texture-images; (**c**) facade generation of Building B by satellite texture-images; (**d**) façade generation of Building B by airborne-like texture-images.

Different texture cases were compared in the experiments. For one case, facade images of buildings captured by IKONOS are cut as texture-images. The resolutions are 1 m/pixel. For another case, facade images of the buildings are taken from Google-Earth to simulate airborne-like images, and the resolutions were resampled to 0.3 m/pixel. The texture-images can be replaced by other UAV images if collected. The heights of the facades of Building A and Building B are 86 m and 4 m, respectively. The interpolations were strictly fit to the texture-images. 3D-points were properly generated for the two buildings.

3D-points of all sides of the buildings were collected to generate the triangular mesh. Then the buildings were reconstructed by the triangular mesh and all texture-images (black texture for invisible facades in different cases), as shown in Figure 17a–d. They illustrate that RTGI could process texture-images using different resolutions adaptively, and higher reconstructed quality were obtained from better texture-images. To verify the accuracy of our method, two groups of testing points are selected: 15 points for Building A, 10 points for Building B, as shown in Figure 18a,b. In the maximum display scale of the buildings in Google-Earth, the texture of the two buildings show a perfect fit and were clearly shown in the screen, so we manually selected and measured the position of the points as a standard, referring to the supervised geometric information of Google-Earth. Then we calculated the error at different directions between the testing points and the selected standard points by Equation (3). The results were acceptable, as listed in Table 3.

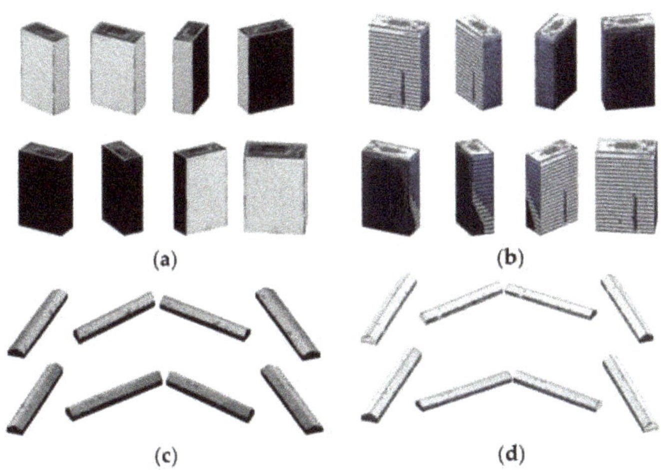

Figure 17. 3D reconstruction of buildings in Dataset 2 and Dataset 3. (**a**) 3D reconstruction of Building A in multiple views by satellite texture-images; (**b**) 3D reconstruction of Building A in multiple views by airborne-like texture-images; (**c**) 3D reconstruction of Building B in multiple views by satellite texture-images; (**d**) 3D reconstruction of Building B in multiple views by airborne-like texture-images.

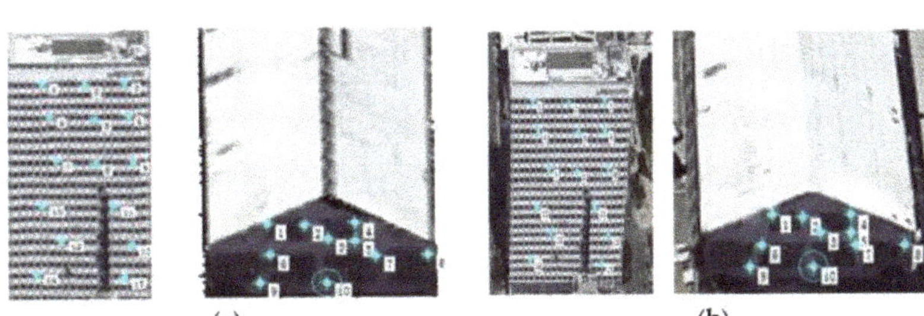

Figure 18. Selected points for accuracy verification. (**a**) Testing points of the two buildings; (**b**) standard points selected from Google-Earth.

Table 3. Reconstruction errors of testing points.

Serial Number of the Testing Points	Building A	Building B
1	0.1	0.4
2	0.9	0.1
3	0.4	0.6
4	0.8	0.0
5	1.6	0.3
6	0.8	0.4
7	0.3	0.4
8	1.1	0.5
9	0.2	0.3
10	0.8	0.2
11	1.1	-
12	0.2	-
13	0.7	-
14	0.6	-
15	1.1	-
Mean Error	0.7	0.3

5. Discussion and Conclusions

In this paper, a hierarchical segmentation method named LS-ORTSEG was proposed to enhance the performance of a model-free framework for building 3D reconstructions with RS resources. In this framework, 3D information of buildings are segmented from DSM, and then together with texture-images, facades of the buildings are generated by the RTGI method, and texture-images are further pasted onto the corresponding outer-surfaces to finish the 3D reconstruction. LS-ORTSEG was used to decrease the negative factors resulting from the uncertain terrain.

Three groups of datasets were used, which were collected from different areas of the world. Some specific buildings were selected to test the performance of our proposed methods. In the experiments, the advantages of our LS-ORTSEG in different cases compared to traditional methods were verified by Datasets 1, 2 and 3. The optimization of parameter settings of LS-ORTSEG could be a future objective. Moreover, the 3D reconstruction performance showed the effectiveness of our framework, and an acceptable 3D reconstruction accuracy was obtained according to the measured coordinates of the testing points. Texture-images from aerial images gave better results than satellite images. Besides, 3D reconstruction results, especially for irregularly shaped roofs, can be found in other work of our team [19,23]. Furthermore, UAV images, which can take texture-images with more ideal angles and spatial resolution, are more easily collected now, making our proposed method having scalability and interesting practical prospects.

Acknowledgments: This work is supported by the Young Scientists Fund of the National Natural Science Foundation of China (Grant No. 61601135) and the Fundamental Research Funds for the Central Universities (Grant No. GK2080260139).

Author Contributions: Yiming Yan and Fengjiao Gao conceived and designed the experiments; Shupei Deng and Nan Su performed the experiments; Moreover, Yiming Yan analyzed the data, contributed analysis tools and wrote the paper.

Conflicts of Interest: The authors declare no conflict of interest. The founding sponsors had no role in the design of the study; in the collection, analyses, or interpretation of data; in the writing of the manuscript, and in the decision to publish the results.

References

1. Jin, L.; Zhu, L.; Gong, H.; Pan, Y.; Du, L. Design and development of wetland evaluation system based on remote sensing and 3D visualization. In Proceedings of the IEEE International Conference on Computer Application and System Modeling, Taiyuan, China, 22–24 October 2010; Volume 2, pp. 170–173.
2. Zhang, L.; Han, C.; Zhang, L.; Zhang, X.; Li, J. Web-based visualization of large 3D urban building models. *Int. J. Digit. Earth* **2014**, *7*, 53–67. [CrossRef]

3. Saito, K.; Spence, R. Mapping urban building stocks for vulnerability assessment—Preliminary results. *Int. J. Digit. Earth* **2011**, *4*, 117–130. [CrossRef]
4. Lafarge, F.; Descombes, X.; Zerubia, J.; Pierrot-Deseilligny, M. Structural approach for building reconstruction from a single DSM. *IEEE Trans. Pattern Anal. Mach. Intell.* **2010**, *32*, 135–147. [CrossRef] [PubMed]
5. Cheng, L.; Gong, J.; Li, M.; Liu, Y. 3D building model reconstruction from multi-view aerial imagery and LIDAR data. *Photogramm. Eng. Remote Sens.* **2011**, *77*, 125–139. [CrossRef]
6. Brenner, C. Building reconstruction from images and laser scanning. *Int. J. Appl. Earth Observ. Geoinf.* **2005**, *6*, 187–198. [CrossRef]
7. Chen, Y.; Cheng, L.; Li, M.; Wang, J.; Tong, L.; Yang, K. Multiscale Grid Method for Detection and Reconstruction of Building Roofs from Airborne LiDAR Data. *IEEE J. Sel. Top. Appl. Earth Observ. Remote Sens.* **2014**, *7*, 4081–4094. [CrossRef]
8. Zeng, C.; Zhao, T.; Wang, J. A Multicriteria Evaluation Method for 3-D Building Reconstruction. *IEEE Geosci. Remote Sens. Lett.* **2014**, *11*, 1619–1623. [CrossRef]
9. Sportouche, H.; Tupin, F.; Denise, L. Extraction and three-dimensional reconstruction of isolated buildings in urban scenes from high-resolution optical and SAR spaceborne images. *IEEE Trans. Geosci. Remote Sens.* **2011**, *49*, 3932–3946. [CrossRef]
10. Tian, P.; Sui, L. Building contours extraction from light detect and ranging data. In Proceedings of the 2011 Symposium on Photonics and Optoelectronics (SOPO), Wuhan, China, 16–18 May 2011; pp. 1–3.
11. Gamba, P.; Houshmand, B. Digital surface models and building extraction: A comparison of IFSAR and LIDAR data. *IEEE Trans. Geosci. Remote Sens.* **2000**, *38*, 1959–1968. [CrossRef]
12. Katartzis, A.; Sahli, H. A stochastic framework for the identification of building rooftops using a single remote sensing image. *IEEE Trans. Geosci. Remote Sens.* **2008**, *46*, 259–271. [CrossRef]
13. Jia, B.; Zhang, Y.; Chen, Y.; Wu, Z. A novel level set framework for LOD2 building modeling. In Proceedings of the 2012 19th IEEE International Conference on Image Processing (ICIP), Orlando, FL, USA, 30 September–3 October 2012; pp. 1781–1784.
14. Iovan, C.; Boldo, D.; Cord, M. Detection, characterization, and modeling vegetation in urban areas from high-resolution aerial imagery. *Sel. Top. Appl. Earth Observ. Remote Sens.* **2008**, *1*, 206–213. [CrossRef]
15. Qin, R.; Huang, X.; Gruen, A.; Schmitt, G. Object-Based 3-D Building Change Detection on Multitemporal Stereo Images. *Sel. Top. Appl. Earth Observ. Remote Sens.* **2015**, *8*, 2125–2137. [CrossRef]
16. Liu, C.; Shi, B.; Yang, X.; Li, N.; Wu, H. Automatic buildings extraction from LiDAR data in urban area by neural oscillator network of visual cortex. *IEEE J. Sel. Top. Appl. Earth Observ. Remote Sens.* **2013**, *6*, 2008–2019. [CrossRef]
17. McCann, M.T.; Mixon, D.G.; Fickus, M.C.; Castro, C.A.; Ozolek, J.A.; Kovacevic, J. Images as occlusions of textures: A framework for segmentation. *IEEE Trans. Image Process.* **2014**, *23*, 2033–2046. [CrossRef] [PubMed]
18. Su, N.; Zhang, Y.; Tian, S.; Yan, Y.; Miao, X. Shadow Detection and Removal for Occluded Object Information Recovery in Urban High-Resolution Panchromatic Satellite Images. *IEEE J. Sel. Top. Appl. Earth Observ. Remote Sens.* **2016**, *9*, 1–15. [CrossRef]
19. Qiu, M.; Zhang, Y. Feature guided multi-window area-based matching method for urban remote sensing stereo pairs. In Proceedings of the 2015 IEEE International Geoscience and Remote Sensing Symposium (IGARSS), Milan, Italy, 26–31 July 2015; pp. 4510–4513.
20. Von Gioi, R.G.; Jakubowicz, J.; Morel, J.M.; Randall, G. LSD: A fast line segment detector with a false detection control. *IEEE Trans. Pattern Anal. Mach. Intell.* **2010**, *32*, 722–732. [CrossRef] [PubMed]
21. Cramer, M. The DGPF test on digital aerial camera evaluation—Overview and test design. *Photogramm. Fernerkund. Geoinf.* **2010**, *2*, 73–82. [CrossRef] [PubMed]
22. Krishnan, S.; Crosby, C.; Nandigam, V. OpenTopography: A services oriented architecture for community access to LIDAR topography. In Proceedings of the 2nd International Conference on Computing for Geospatial Research & Applications, Washington, DC, USA, 23–25 May 2011.
23. Yan, Y.; Chen, X.; Gao, F.; Zhang, Y.; Shen, Y.; Su, N.; Tian, S. A roof-contour guided multi-side interpolation method for building texture-mapping using remote sensing resource. In Proceedings of the IEEE International Geoscience and Remote Sensing Symposium, Milan, Italy, 26–31 July 2015; pp. 2998–3001.

MDPI

St. Alban-Anlage 66

4052 Basel, Switzerland

Tel. +41 61 683 77 34

Fax +41 61 302 89 18

http://www.mdpi.com

Sensors Editorial Office

E-mail: sensors@mdpi.com

http://www.mdpi.com/journal/sensors

www.ingramcontent.com/pod-product-compliance
Lightning Source LLC
Chambersburg PA
CBHW041750220926
51847CB00014B/161
9783038428558